D0622211

About the editor . . .

Currently an Associate Professor of Civil Engineering at Howard University, Washington, D.C., **Gajanan M. Sabnis** has extensive experience in structural design. During his career he has worked as a structural researcher for American Cement Corporation, and as a structural designer for Bechtel Power Corporation. Dr. Sabnis is a Fellow of the American Concrete Institute (ACI) and the American Society of Civil Engineers (ASCE), and he is actively involved in their committees that relate to deflections of structures. Author of numerous published works, Dr. Sabnis edited a special publication — *Deflections of Concrete Structures,* SP-43 — put out by the American Concrete Institute in 1974.

Handbook of COMPOSITE CONSTRUCTION ENGINEERING

Edited by
Gajanan M. Sabnis, Ph.D., P.E.
Associate Professor of Civil Engineering
Howard University
Washington, D.C.

VAN NOSTRAND REINHOLD COMPANY
NEW YORK CINCINNATI ATLANTA DALLAS SAN FRANCISCO
LONDON TORONTO MELBOURNE

Van Nostrand Reinhold Company Regional Offices:
New York Cincinnati Atlanta Dallas San Francisco

Van Nostrand Reinhold Company International Offices:
London Toronto Melbourne

Library of Congress Catalog Card Number: 78-18354
ISBN: 0-442-27735-0

Manufactured in the United States of America

Published by Van Nostrand Reinhold Company
135 West 50th Street, New York, N.Y. 10020

Published simultaneously in Canada by Van Nostrand Reinhold Ltd.

15 14 13 12 11 10 9 8 7 6 5 4 3 2 1

Library of Congress Cataloging in Publication Data

Main entry under title:

Handbook of composite construction engineering.

Includes index.
1. Composite construction. I. Sabnis, Gajanan M.
TA664.H36 624'.18 78-18354
ISBN 0-442-27735-0

To my wife, Sharda,
and to our son,
Rahul

Contributors

Dan E. Branson, Ph.D., P.E., Professor of Civil Engineering, University of Iowa, City, Iowa

Donald R. Buettner, Ph.D., P.E., President, Computerized Structure Design, Inc., Milwaukee, Wisconsin

Arthur L. Elliott, P.E., Bridge Engineer, Sacramento, California

James M. Fisher, Ph.D., P.E., Principal, Computerized Structural Design, Inc., Milwaukee, Wisconsin

Richard W. Furlong, Ph.D., P.E., Professor of Civil Engineering, University of Texas, Austin, Texas

James R. Goodman, Ph.D., P.E., Professor of Civil Engineering, Colorado State University, Fort Collins, Colorado

Richard M. Gutkowski, Ph.D., P.E., Associate Professor of Civil Engineering, Colorado State University, Fort Collins, Colorado

Srinivasa H. Iyengar, P.E., Partner, Skidmore, Owings and Merrill, Chicago, Illinois

Gajanan M. Sabnis, Ph.D., P.E., Associate Professor of Civil Engineering, Howard University, Washington, D.C.

Charles G. Salmon, Ph.D., P.E., Professor of Civil Engineering, University of Wisconsin, Madison, Wisconsin

J. K. Sridhar Rao, Ph.D., Associate Professor of Civil Engineering, California State University, Long Beach, California

Foreword

Composite construction or, using a more general term, mixed steel-concrete construction, is an essential way of improving the economy and functionality of many buildings, bridges, and other structures. The advantages of this type of construction are more widely and effectively utilized in some other countries, such as Japan, Britain, and Germany, than in the United States. In part, this is caused by the ambivalent and uncoordinated approach to it by code-writing bodies. Thus, in the building field, flexural composite members are covered in the American Institute of Steel Construction's Specification for the Design, Fabrication and Erection of Structural Steel for Buildings, while columns, concentric and eccentric, are included in the American Concrete Institute's Building Code Requirements for Reinforced Concrete. This lack of consistency results in a situation in which certain composite columns with small eccentricity, when designed by the ACI Code, show smaller design load capacities than the identical steel shapes alone, without concrete, when designed by the AISC Specification. Likewise, because of this uncoordinated approach, the vital area of connections in composite construction is not covered in any American building design specification or code.

Another consequence of this indefinite status of composite or mixed construction is that, with very few exceptions, it is not covered in engineering college curricula. In consequence, the young graduating engineer is hardly aware of the existence, let alone the functional and economic advantages, of this type of construction. Connected with this general situation is the relative paucity of literature and of research activity in this area.

Prior to the publication of this volume, the only fairly comprehensive treatment of this field was the recent state-of-the-art report, *Composite or Mixed Steel-Concrete Construction for Buildings*, prepared by S. H. Iyengar for the Structural Specifications Liaison Committee and published in 1977 by ASCE. It is an excellent but relatively small volume of some 250 pages. It covers only building applications and does not deal with bridges or with composite wood and timber construction.

All this illustrates that the present volume, written by well-known experts, fills a real and important need. It represents the first comprehensive and inclusive treatment of this topic in American construction literature. It is hoped that it will encourage American engineers to take broader advantage of the possibilities inherent in composite construction. It should also encourage more intensive research and a more unified treatment in design codes and specifications.

George Winter
Honorary Member, ASCE and ACI
Professor Emeritus, Cornell University
Ithaca, New York

Preface

Composite construction has played a very important role in structural engineering. Its uses have varied from small structures to large buildings and bridges. The basic principle underlying composite construction is that certain materials may be used more effectively in certain types of stressed conditions; thus, the combination of a material strong in compression with one strong in tension makes a very economical union for its use in structures. With basic materials such as concrete, steel, masonry materials, wood, and timber, a number of successful combinations can be used; however, until now, such information has not been readily available to design or practicing engineers.

The *Handbook of Composite Construction Engineering* assembles in one volume the different types of composite construction. The contributors were chosen for their eminence and experience in the specific types of composite construction. Design problems are presented along with considerable background and state-of-the-art information on the subject. As reference material, these contributions should help design engineers considerably in their professional work.

Basic principles and an in-depth overview of composite construction engineering are presented in Chapter 1, giving the reader an excellent idea of the field. Readers can further explore any of the topics covered in the remaining chapters by using the extensive bibliography at the end of Chapter 1. Chapter 2 deals with steel-concrete composite construction, which is generally shortened to "composite construction" by most engineers. Chapter 3 discusses the application of light-gauge steel, one of the recent concepts being used in practice; the codes of practice will probably include some of this material in the very near future.

Chapters 4 and 5 take the reader into other types of composite construction: reinforced and prestressed concrete. These chapters cover principles and examples of such construction with a large list of references. Chapter 6 presents yet another concept in composite construction, which results in considerable economy in structural columns. Chapters 7 and 8 deal with applications of composite construction in buildings and bridges. Chapter 9 is concerned with another important feature in composite construction using different structural elements in structure—namely, walls and frames. To date, this topic has received very little attention from the profession for practical application.

Chapter 10 deals with composite construction in wood and timber, which has recently been the subject of extensive research and development. Practical design applications in this field are cited as well as results of recent research.

At the end of each chapter, an extensive list of references is given so that more information may be sought, especially for related topics that could not be covered in this volume.

Keeping up with the trend of using SI units of measurements, we have used both the FPS and SI units throughout the *Handbook*. In the examples, equivalents of only the main numbers are given to avoid confusion yet to make them useful in both units. The use of both units should extend the use of this book, not only in this country, but also in many parts of the world.

Gajanan M. Sabnis
Silver Spring, Maryland

Contents

Handbook of COMPOSITE CONSTRUCTION ENGINEERING

1

Fundamentals and Overview of Composite Action in Structures

GAJANAN M. SABNIS, Ph.D., P.E.
Associate Professor of Civil Engineering
Howard University
Washington, D.C.

J. K. SRIDHAR RAO, Ph.D.
Associate Professor of Civil Engineering
California State University
Long Beach, California

1.1 INTRODUCTION

Composite action in structures may be considered as the "interaction of different structural elements and may be developed using either different or similar structural materials." It is often referred to as "composite construction" and includes steel-concrete beams; columns, whether fully encased, partially encased, or held together by means of suitable connectors; concrete-over-concrete beams; wood-concrete beams; layered wood systems; sandwich-beam construction, such as gypsum facing bonded with polyurethane foam core or honeycombed treated craft paper. Another example of composite construction is the interaction of infilled masonry shear wall with structural frame and slabs, or between structural and nonstructural components such as architectural panels. The interaction of different structural components, including beams, frames, shear walls, columns, slabs, and panels of a structure made of similar materials, is sometimes referred to as *composite*. In such cases, however, the structure may be called a monolithic or "stressed skin" construction. Typical examples of the last category are an orthotropic steel girder-deck system or layered wood system. The preceding examples are only illustrative of, and do not include, the full range of possibilities that can be developed by using combinations of materials, forms of structural elements, and types of connections. Other possibilities will be discussed later in this chapter.

The most common composite construction in buildings and bridges is the composite steel-concrete wherein a steel beam and a reinforced concrete slab (cast-*in-situ* or precast) are so interconnected with shear connectors that they act together as a unit.

The steel *beam* may be fully encased in concrete, partially encased, or placed below the slab. In some cases, the beam may have a concrete haunch above it. If concrete encasement is monolithic and of at least a specified minimum thickness, its natural bond with the steel beam will provide some composite action and additional stiffness as well as strength. To insure composite action, shear connectors such as studs, steel bars, or rolled shapes can be welded to the top flange of the steel beam and embedded in the concrete slab.

The method of construction may also vary. To obtain the full economic potential, consideration must be given to methods of erection. The use of temporary supports during construction, termed as *propped composite construction*, has some advantages. The advantage of prestressing may be accomplished by jacking, by cables through steel beams, or by "preflexing" by precambering beams in pairs, preflexing and clamping the ends, casting concrete slabs, and removing the clamps. By precompressing the steel, more tensile capacity is reserved for service loads. In continuous beams, a compressive stress in negative moment region is helpful for the concrete slab (see Fig. 1.15). To achieve optimum results, a judicious choice of materials, design method, structural system, and construction method must be made.

There can be several types of steel beams: rolled beams with or without cover plates, built-up welded plate girders, hybrid girders with webs and flanges of different yield strengths, open-web joist, a three-dimensional open-web joist, hollow box girders, etc. Often, an unsymmetrical section, such as a rolled steel joist with a cover plate on the bottom flange only, may provide optimum economy.

A reinforced concrete *slab* is either precast or cast-in-place and may be cast on light-gauge steel decking that serves as the form. This may be either slab with ribs parallel to the steel beam or at right angles to the steel frame. When the slab spans in the same direction as the beam, it is not essential to weld the connectors through the light-gauge formed deck. The decking should be firmly attached to the steel beam by stud welding or other connectors to enable the steel beam to act as a reinforcement for the slab and to prevent shear failure on the plane of contact between the decking and the top of the steel girders. In the case of deck reinforced slabs with ribs at right angles to the steel beam, the shear connector spacing is determined in part by the rib spacing of the slab system.

Both normal-weight and lightweight concretes have been used in deck reinforced concrete slabs. The slab design is independent of the composite beam action and is carried out in the same manner as the design of noncomposite floors.

The necessary connection between the slab and the beam is provided by either the bond between concrete and steel for full encasement or by mechanical shear connectors. Adequate bond is necessary to transfer longitudinal shear from the slab to beam so that the two will act as a unit. In addition to longitudinal shear transfer, bond devices must insure that steel and concrete cannot separate vertically. Vertical separation will occur as both the steel and concrete seek their own flexural radii of curvature. Whenever there are no shear connectors and the surfaces between steel beam and concrete slab are smooth, slip occurs so that the beam and slab act separately. However, the load sharing by the two component elements requires compatibility between the radii of curvature at their vertical separation, and causes two neutral axes, instead of the one desired in the composite case. Some longitudinal slip between steel and concrete can be tolerated even in composite construction, permitting the use of partial interaction theories.

1.1.1 Types of Composite Construction

Some composite systems formed by similar and dissimilar construction materials are presented here and discussed later on in detail.

Various systems of *similar materials* are:

1. *Monolithic structure* in cast-in-place structural concrete members (foundations, columns, beams, frames, shear walls, slabs, panels).
2. *Composite structure* between precast structural concrete elements and cast-in-place concrete elements. An example is precast concrete beam with a cast-in-place reinforced concrete slab.
3. *Orthotropic structures*, in which steel beams interact with steel deck (e.g., in a bridge), may be considered composite construction. This also happens when a simple roof truss structure consists of a metal deck, rafters, purlins, steel truss, and wind bracing—e.g., World Trade Center Building in New York City; IBM Building in Pittsburgh.

The principle of composite action in the stressed-skin "continuum" type of structure is to consider combining the different elements of the structure into one complex but integrated continuum in order to resist loads and external environmental forces. Instead of assigning each structural member a single, isolated, and specific task, the continuum attempts to use composite action by uniting all the structural members by means of proper connections, thus creating a single structural member that acts as a multidirectional load-carrying element in the total structure—e.g., orthotropic steel bridge plate-deck in which the metal sheet is stressed in various ways. The concept can be used for a simple roof structure consisting of metal deck, purlins, steel truss, and wind bracing. Brandel (1967) suggests reshaping the material of the truss chord into a metal sheet to fulfill simultaneously the functions of deck, purlins, partial truss, and bracings. Buckling considerations will require corrugation of such a metal sheet.

Examples of structural members and systems of dissimilar materials are:

1. Composite steel—concrete construction
2. Composite wood—concrete construction
3. Infilled masonry—structural frame interaction
4. Composite sandwich and laminated structures.
5. Miscellaneous.

A report by an ASCE (1974) subcommittee has discussed many of these types. Composite steel-concrete construction also includes: encased columns and frames; concrete-filled steel tubes; concrete slabs with permanent formwork; ferrocements with steel plates at top and bottom, with or without shear connectors (the steel plates may be checkered plate with protrusions to provide a slip-resistant surface); and composite form-reinforced slab.

Another very important area of composite construction is the interaction between masonry walls and structural frame-shear walls. The seismic response of tall buildings, especially with openings in masonry and reinforcements for increasing ductile behavior and shear connections of masonry and frames, is of considerable significance in developing economic design procedures of such composite systems.

In composite structures of foamed and other hollow-core sandwich construction, facing materials provide structural resistance, fireproofing, and environmental effects. Furthermore, composites with fiberglass reinforcement in plastics, fiber-reinforced concrete, or polymerized concrete, together with the proper interaction with steel beams, may be an area of research and application in the future. Dietz (1971), Idorn and Fordos (1974), and Jones (1972) provide some development in this area. Sandwich panels with foam-in-place cores and light-gauge cold-formed metal faces are becoming more and more popular as building enclosures, due to their structural efficiency, thermal and sound insulation qualities (thereby conserving energy), mass productivity, transportability, durability, and reusability. Such sandwich panels, which are used as "prefabricated structures" in commercial and industrial buildings, were used in buildings associated with the Alaska pipeline project. Kuenzi (1960), Allen (1969), Harstock (1969), and Harstock and Chong (1976) have discussed various aspects of design of foam-filled sandwich panels. Sabnis and Aroni (1971) discuss the application of sandwich construction in housing.

Rao (1971) has discussed various forms of composite structural systems in buildings and bridges; the analysis is supplemented with an extensive annotated bibliography. Shen et al. (1976) and Lee et al. (1975) discuss the behavior and design applications of composite construction using earth reinforced with steel strips. Another significant area is the interaction of corrugated pipes and soil. Composite construction using prefabricated reinforced brick masonry panels with cast-*in-situ* reinforced concrete for low-cost housing is described by Rao (1975) and Verma (1974). An interesting application is in the field of fire protection of steel structures, where vermiculite-gypsum plaster is used on expanded metal lathing, or sprayed asbestos or vermiculite is used. Precast or prestressed floor units spanning between steel beams, with hollow tiles and *in situ* concrete with or without structural topping, have also been used. The system using hollow ceramic brick blocks with concrete and prestressing steel wires and other composite systems are discussed for applications in developing countries by Rao (1975). It is essential to recognize here the wide variety of structural applications and composite interaction among various materials. Later discussions of actual design procedures will also demonstrate the prosperous future of this important structural aspect. A number of examples of miscellaneous composite constructions are discussed in Section 1.12.

1.1.2 Historical Development

The history of composite construction is intimately linked with that of reinforced concrete and reinforced brick masonry, which are other familiar forms of composite structures. Knowles (1973) described the early development of composite construction in buildings as fireproof floors, jack arch construction, and the fireproofing of steel joists by embedding them in concrete. Scott (1925) and Caughey and Scott (1929) described the early tests on encased beams and steel beams–concrete slabs conducted in the United Kingdom in 1914. Gillespie et al. (1923) reported a series of tests of I-beams encased in concrete, which were conducted under the direction of Dominion Bridge Company in Canada.

The early research on composite construction for bridges in the United States was conducted at the University of Illinois, Urbana; a summary is reported by Siess (1949). Viest (1960) described the historical development related to AASHTO code specifications and German specifications using prestressed slabs on steel beams. Later research, done mainly at Lehigh University, has been reported by Fisher (1970) and Slutter (1974) and is detailed in Chapter 2. Johnson (1970) described work in the United Kingdom and included an extensive list of references.

1.2 COMPOSITE VS. NONCOMPOSITE ACTION

Composite construction became generally accepted by engineers for bridges during the 1950s and for buildings during the 1960s. During the past decade, research on continuous composite beams and connections has led to even greater economy through the use of composite design in continuous structures due to increased strength through continuity and also the increased stiffness due to composite action. Local buckling of the compression flange is delayed, and greater resistance to lateral buckling is possible by connecting steel beam to a concrete slab with shear connectors.

Research as reported by Furlong (1968), Gardener and Jacobson (1967), and Stevens (1965) has shown that concrete-filled tubes and cased rolled shapes also possess higher shear strengths than do reinforced concrete columns of comparable size. Dobruszkes et al. (1969) and Roderick (1972) report that the connections between columns and composite beams can be made very ductile.

Okamoto (1973) summarizes Japanese earthquake research in structural connections. Experience in Japan has shown that encasing steel sections in reinforced concrete is particularly beneficial for earthquake-resistant design. More research studies are needed on the suitability of other types of composite construction for earthquake-resistant design.

Davies (1975), Knowles (1973), and Johnson (1975) have given examples of savings in composite construction between steel and concrete for buildings and bridges. Johnson (1975) concludes from a study of current European practice that composite construction is particularly competitive in medium- or long-span structures where a concrete slab is needed for other reasons, where fire protection of steelwork is not required, and where there are economic advantages resulting from rapid construction. Composite construction is especially advantageous in cases where construction time may be reduced. For example, composite slabs using cold-formed corrugated sheeting as the formwork have been used in the United States to a large extent in tall buildings, as demonstrated by Fisher (1970). Johnson and Cross (1973) have reported on the economy achieved due to the use of precast concrete permanent formwork and full-thickness precast floor slabs with cast-*in-situ* concrete, especially in multistory parking facilities. Soor (1973), taking into account the composition action of beams with masonry, was concerned with the economy in composite design using shear-connected precast pretensioned slabs and cast-*in-situ* concrete in a large building project in India. In a project study by Constructional Steel Research and Development Organization (CONSTRADO), London, in 1973, the economy due to earlier construction time was reported in the case of a 29-story office building using a slip-formed concrete core surrounded by a steel frame designed to act composite with floor slabs, which were made of precast concrete planks with an *in situ* concrete topping. The construction time of 27 months using composite construction was 8 months less than that for alternative construction systems.

The level of fire protection also influences the economical choice to be made among structural steel, reinforced concrete, and composite construction. In bridges and multistory garages with spans of 30 to 35 ft (9 to 10 m), where the vulnerability of steel to fire is not a problem, steel beams acting composite with concrete slabs are more economical. Encasing steel columns in concrete is economical since the casing provides a substantial gain in strength. On the other hand, encasement of steel beams in concrete contributes little to the strength of the composite beam, and at the same time requires additional reinforcement to control cracks and to hold beams in place during fire. In such steel beams, therefore, lightweight coating materials are relied upon for fire protection.

In frame construction, composite beams, columns, composite construction, or their combination may be used. Composite columns help reduce the effective slenderness of a steel column, thus increasing its buckling load. Furthermore, the concrete encasement also carries its share of the load according to the actual behavior of composite columns under load. Sometimes, in an otherwise reinforced concrete frame building, in order to maintain a constant size of columns over the entire height, composite columns could be used in the lower part of the structure. A composite column can be constructed without the use of formwork by filling a steel tube with reinforced concrete. This is especially advantageous when fire protection for the steel is not a critical factor.

Many comparisons between composite and other types of construction have been made in the last 30 years. The benefits of composite construction may vary among structures, location, and relative costs of materials and labor in a particular country. Some comparisons are presented here, which indicate the economics of such construction.

Davies (1975) compares steel beam sizes for a typical, simply supported span of 27 ft (8 m) to those for an office floor beam carrying identical dead and live loads for unshored construction. There are savings brought about by the reduced structural floor depth and reduced story height; savings associated with a reduced building volume for heating, cooling, and reduction in fireproof casing; and savings in overall dead loads of the building, which may reduce foundation costs especially in difficult soil conditions.

Johnson et al. (1966) report design studies of composite frames for buildings designed both elastically and plastically according to British specifications as follows:

Comparative Weight and Cost for a Three-Bay Six-Story Building Frame

Type of Frame	*Weight (%)*	*Height (%)*
Elastic noncomposite	100	100
Plastic noncomposite	95	102
Elastic composite	86	91
Plastic composite	66	90

Wilenko (1969) shows that the relative weight of composite, shored construction is 73% and that of a composite, prestressed steel beam is 55% of that required for noncomposite construction in a building frame. He further claims that the prestressing method adds little to the construction costs.

Iwamota (1962) reports on the savings in both weight and costs in the case of a continuous composite bridge construction in Japan, for which the design should also be carried out for eliminating tension in the concrete deck under negative bending. His findings are summarized in the following table. (See next page)

Studies were carried out by Siess (1949) for simple bridge spans from 30 to 90 ft (9 to 27 m), with beam spacings of 5 to 7 ft (1.5 to 2.2 m). The following table presents these comparisons

Comparison of Three-Span Continuous Bridge for Spans of 105 ft (32 m)

Type of Beam	*Weight (%)*	*Cost (%)*
Noncomposite	100	100
Composite	87	92

in terms of steel weight, which includes an allowance for shear connectors:

Relative Weights of Simply Supported Bridge Span

Type of Beam	*Relative Weight (%)*
Noncomposite rolled beam	100
Composite symmetric rolled beam	
Without flange plates	
Unshored	92
Shored	77
With flange plates on bottom flange	
Unshored	76
Shored	64
Composite, using unsymmetrical rolled section, unshored	82
Composite, using double T-section, unshored	82
Composite, using welded section	
Unshored	69
Shored	40-60

It can be seen from the table that a significant weight saving is possible by shored composite construction in which the dead load of steel and concrete is supported by temporary shores until the concrete is cured. In contrast, in unshored construction, the dead load of steel beam and cast-in-place concrete is taken by the steel beam alone. However, in some cases, shoring may be difficult, especially in bridges or floors of buildings where finishing operations are hindered. Several other methods of achieving economic savings in composite construction involve some form of prestressing and are presented later in this chapter (see Section 1.12).

The structural advantages of composite vs. a noncomposite construction may thus be summarized as follows:

1. Depth of steel beam is reduced to support a given load.
2. An increase in the capacity is obtained over that of a noncomposite beam, on a static ultimate load basis (fatigue effects may reduce this increase).
3. For a given load, a reduction in dead loads and construction depth reduces in turn the story heights, foundation costs, paneling of exteriors, and heating, ventilating, and air-conditioning spaces, thus reducing the overall costs of buildings. In bridges, embankment costs would be reduced. The amount of reduction, however, will vary from case to case.

On the other hand, there are several disadvantages. The design methods are more complex for composite than for noncomposite construction. In addition, if recent research related to the effective width of composite beams, load-slip relations, cracking due to temperature differential, creep and shrinkage effects, and complex interaction effects from low loads to failure is to be incorporated in limit-states design, the increase in design time may become an inhibiting factor. Over the past 30 years, the very large amount of theoretical and experimental research, design applications, and construction work carried out has shown the efficiency and economy of composite construction.

1.3 IMPORTANCE OF SHEAR TRANSFER IN COMPOSITE ACTION—ROLE AND TYPES OF SHEAR CONNECTORS BETWEEN STRUCTURAL ELEMENTS

Composite action between steel and concrete or between structural elements of the same material implies interaction between them and a transfer of shear at the connection. In reinforced concrete, this is taken care of by the natural bond between concrete and the deformed reinforcing bars. In the case of fully encased steel joists or beams, there is a large embedded area for a shear transfer. In the common type of composite beam, there is some shear transfer by bond and friction at the interface between steel beam and concrete slab. It cannot be depended upon if there is a single overload or pulsating load that will destroy such bond and cause a separation of the slab from the beam. Hence, shear connectors are needed to give reliable composite action with two objectives:

1. to transfer shear between the steel and the concrete, thus limiting the slip at the interface so that the slab beam system acts as a unit to resist longitudinal bending (with one neutral axis for the composite section)
2. to prevent an uplift between the steel beam and concrete slab, i.e., to prevent separation of the steel and concrete at right angles to the interface.

In the case of reinforced concrete monolithic beam-slab construction, the longitudinal shear is taken up by the concrete and steel stirrups in the web. However, in a steel-concrete composite beam, there is a distinct possibility of longitudinal shear failure in the concrete slab portion. The relatively wide concrete slab is required to receive shear force from the steel beam along a narrow interface. This may cause unacceptably high shear stresses on planes in concrete close to the beam. This is especially critical in haunched concrete slabs above the beam, since the haunch being narrower than the slab itself may be a source of weakness. Considerations of longitudinal shear in concrete place limits on haunch dimensions, as shown in Fig. 1.1. It is important to provide at least some transverse reinforcement at the lower face of the slab since it is in this area that stress concentrations occur at the shear connectors. Johnson (1970) made a survey of a large number of load tests on composite beams up to and including failure and recommended some criteria for design of steel for longitudinal shear. The design basis is further discussed in Section 1.6 and Chapter 2.

The importance of shear connection in concrete slab–steel composite beam may be understood by comparing it to a beam with no shear connection. Johnson (1975) discussed the behavior of flitched beams with no shear connectors with full slip and shear connectors with no slip (i.e., full interaction), as shown in Fig. 1.2. Two beams are made of elastic material,

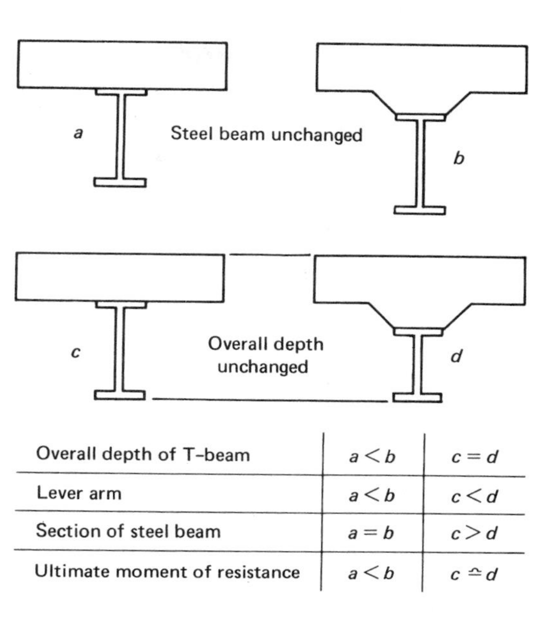

Overall depth of T-beam	$a<b$	$c=d$
Lever arm	$a<b$	$c<d$
Section of steel beam	$a=b$	$c>d$
Ultimate moment of resistance	$a<b$	$c \triangleq d$

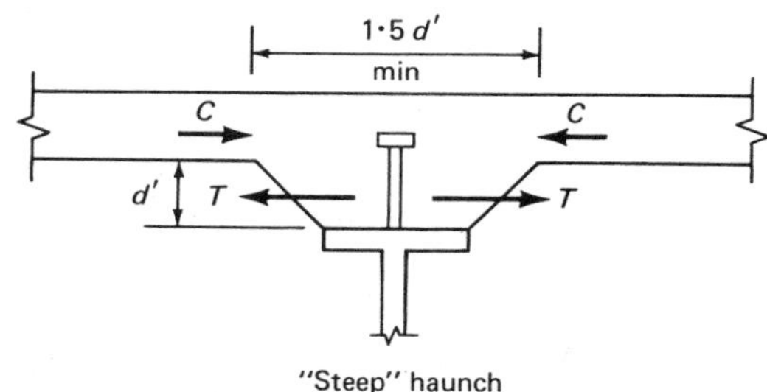

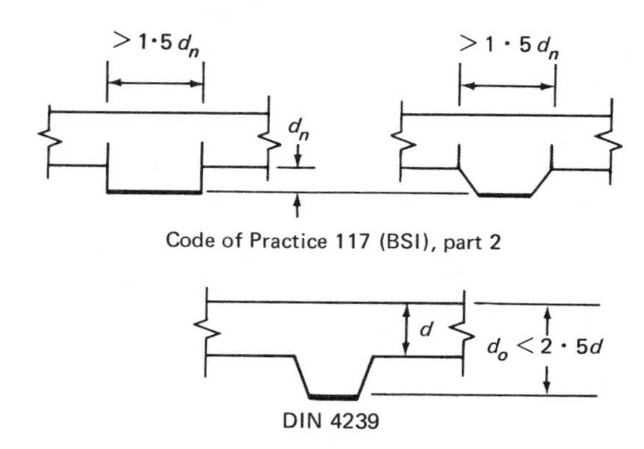

Fig. 1.1. Haunch dimensions—Longitudinal shear considerations.

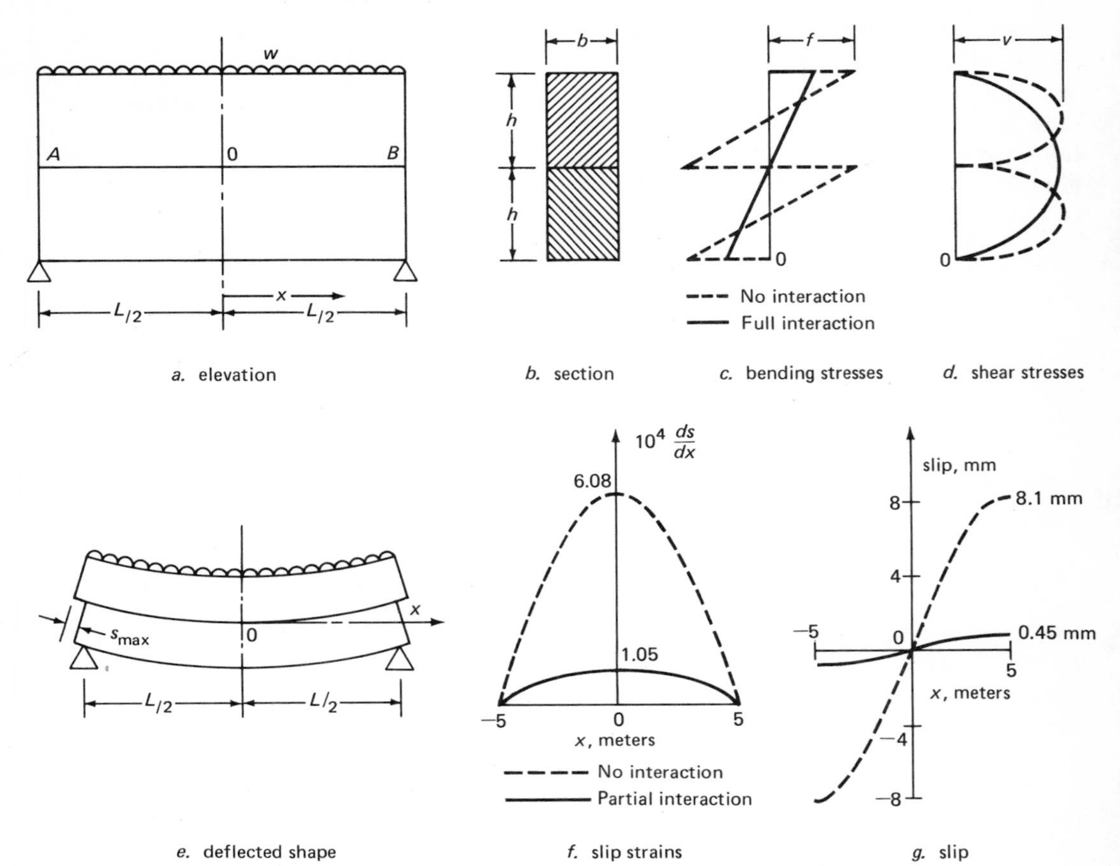

Fig. 1.2. Slip and partial interaction for two rectangular beams (Johnson 1975).

with Young's modulus E, of a cross section bh and a simply supported span L, and subjected to uniformly distributed total load w. It may be shown that the maximum bending stress is reduced by 50%, providing shear connection, while the maximum shear stress is unchanged. Also, the midspan deflection with fully effective shear connectors (no slip at the interface AOB) is 25% of that in the case with no shear connectors. Thus, the provision of shear connectors increases both the strength and stiffness of a beam of given size, which leads to a reduction in the beam size for a given loading and economy as discussed in Section 1.2. In this case, the interface AOB coinciding with the neutral longitudinal shear stress at the interface is equal to the maximum vertical stress. This suggests that the shear connection needs to be as strong in shear as the weaker of the two materials joined. In general, the neutral axis of the composite beam need not coincide with the interface.

1.4 TYPES OF SHEAR CONNECTORS

Various types of shear connectors have been used to resist longitudinal shear and uplift. They are rigid, flexible, bond-type, high-strength friction-grip bolts, and employ epoxy gluing between the two components, etc. Some of these are shown in Fig. 1.3. In a broad sense, connectors may be divided into two categories: (1) rigid, and (2) flexible. It must be pointed out that slip must occur before they are utilized; therefore, the terms are relative. The rigid type is the barlike heavy connector. The flexible ones are the stud-and-channel type of connector. The rigid-bar or channel connector (Fig. 1.3a, b) is limited to shear transfer in one direction only, while the welded-stud connector (Fig. 1.3c) can resist and transfer shear in any direction perpendicular to the shank, making it the more useful connector. The welded-stud connector is particularly suitable for stringers in bridges.

Of the many types of shear connectors shown here, only channels, studs, and spirals are widely used in the United States. *AISC Manual* (1969) lists allowable loads for both the channel and stud connectors (Table 2.4), and AASHTO (1974) refers to the specifications on shear connectors as used in bridges. ACI Code permits the use of closed loop spiral, which is quite useful in resisting the uplift.

Chapters 2 and 8 deal with the detailed calculations of shear connectors as used in steel-concrete composite construction, while Chapter 10 is concerned with shear connectors in wood and timber composite construction.

1.5 COMPOSITE ACTION BETWEEN STRUCTURAL ELEMENTS IN BUILDINGS

Composite action between various structural elements in any given structure always exists merely because they are continuous (monolithic or connected by shear connectors for continuity). Depending on the size of the building, certain simplifications may be made to approximate their interaction and to design isolated structural components in a conservative manner. However, with the trend to use taller and taller buildings, the forces become larger and larger, and will require large-size members unless proper cognizance of the interaction is taken. The use of higher strength materials and such composite action become important factors in making the entire system work economically. Tall buildings require additional considerations such as slenderness, flexibility, and sensitivity to differential effects; thus, height is not the only criterion for the "tallness."

Steel and concrete are the major materials used in composite systems. Although they have several dissimilar physical characteristics, it has been possible to use them together beneficially

Property	*Steel*	*Concrete*
Compressive strength/cost	SF[a]	F
Tensile strength/cost	F	NF
Strength/weight	F	NF
Rigidity/weight	F	SF
Strength/rigidity	F	SF
Damping	NF	F
Creep	F	NF
Ductility	F	NF
Fatigue strength	F	SF
Fireproofing capacity	NF	F
Manufacturing tolerances	F	SF
Factory production	F	SF
In-place construction	F	SF
Flexibility for shapes	NF	F

[a]F = fair, SF = somewhat fair, and NF = not fair.

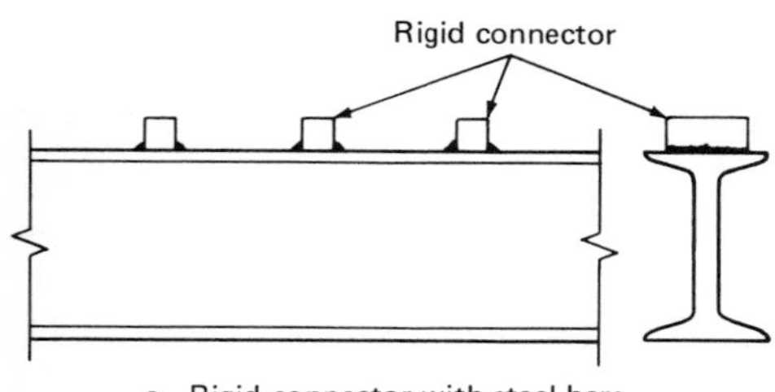

a. Rigid connector with steel bars

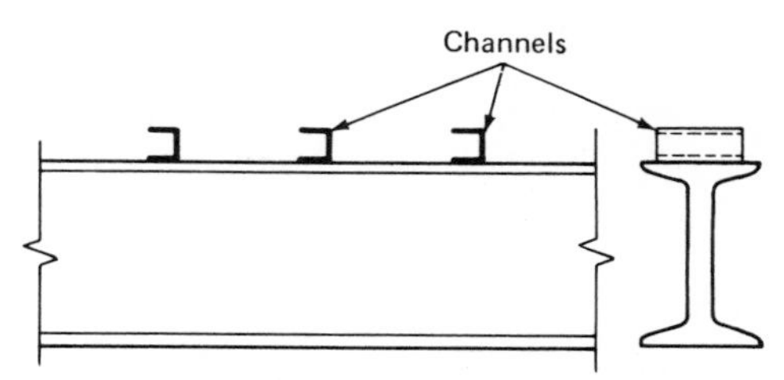

b. Flexible connector with channel

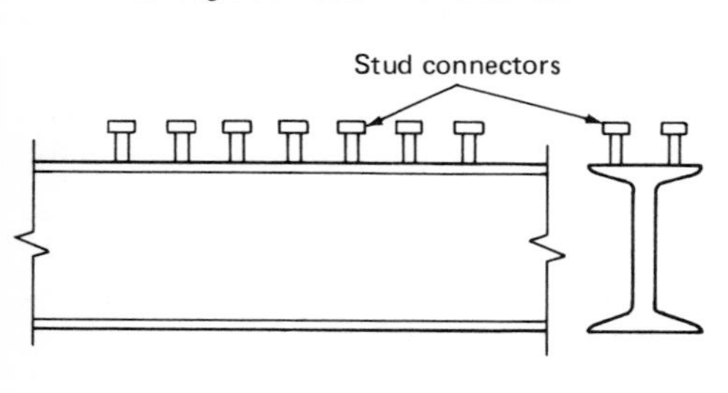

c. Flexible connector with studs

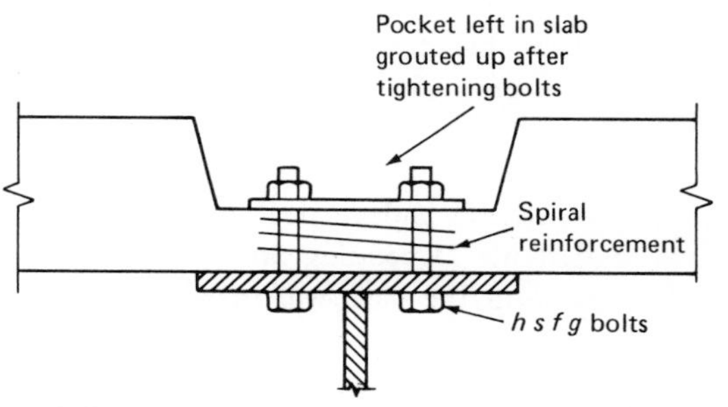

d. High-strength friction-grip bolt connector (*h s f g*)

Fig. 1.3. Types of shear connectors.

in different manners. Hisatoku and Nishikawa (1972) classify the benefits of steel and concrete based on various properties as shown in the table on p. 9.

A number of systems have been developed in the last few decades that successfully combine steel and concrete. The following systems demonstrate their composite action in buildings on a broad basis:

1. *Frame Structure.* Frame structure can be either rigid or flexible depending on the relative rigidity of beams with respect to columns. In the case of a rigid system, both these elements are connected to each other rigidly to transmit lateral loads to the foundation. In addition, slabs provide transverse stiffness from floor to floor to the entire system.
2. *Shear Wall Structure.* Within the preceding frame structure, a system of shear walls may be provided in the direction of lateral forces to help transmit them to the foundation. In addition, the shear walls surround certain services, such as elevators or stairs, and serve a doubly useful purpose. A suitable combination of shear wall and frame system can serve a very useful load-transmitting system for both vertical and horizontal forces. However, in the design against earthquake effects, the need for adequate ductility must be investigated, as shown by Fintel (1974).
3. *Staggered Shear Wall System.* Shear walls in this system are staggered from floor to floor. The walls support floors both above and below, and, by virtue of staggering, they can help create the large open areas needed in commercial buildings.
4. *Tubular System.* The tubular system combines characteristics of the preceding three systems. It acts as a rigid but perforated tube, which, by means of its monolithic action and rigidity, transmits both the transverse and vertical forces to the foundation. Depending on the height of the building, the shear wall (as a core) may be combined with the external tube structure. This is also known as a "frame-tube" structure.

Components that interact and cause a composite action in buildings are (1) column and beam (frame system); (2) slab, which may either be flat or part of the concrete joist floor to give a transverse rigidity; and (3) the shear walls (or the masonry walls) that give considerable in-plane rigidity to the frame. Although details of the actual composite action between various elements will be discussed in Chapters 7 and 9, a brief discussion follows of some of the problems and methods of analysis.

Bare framework of beams and columns with partial restraints at floors (or even none) can be analyzed and designed by proportioning components to their part of resistance to the applied loads. Simplified calculations using hand calculators, the preceding analysis may be performed for heights as great as 10 to 15 stories. The available computer programs, such as STRESS, NASTRAN, STRUDL, or FRAME, can also be used conveniently if the facilities allow. More accurate analysis using these programs and the composite action with slabs (floors) can be performed successfully. Although masonry infill walls can be used as shear walls without great difficulty,* the general tendency is to neglect them. Present codes in most countries do not provide for the additional beneficial effect of such interaction, and the structure in turn has a higher safety factor.

In earthquake regions, the column-beam connections (particularly in the case of a reinforced concrete frame) should be properly designed to provide adequate ductility to the structure in order to achieve a unified resistance to the applied loads. Recent tests at the University of California, Berkeley, have demonstrated that realistic models cannot at present be formulated to predict the nonlinear behavior of tall buildings especially in reinforced concrete structures (e.g., Krawinkler et al. 1975; Mahin and Bertero 1976). Based on this research, it is at least possible to understand the different failure mechanisms in such frames.

Floor slabs interconnecting the frames transversely help to considerably redistribute stresses, especially in places of discontinuity or in the case of asymmetry in the frames. If the analysis is to be done on a plane-frame basis, part of the slab is considered an equivalent beam to

*See Chapter 9.

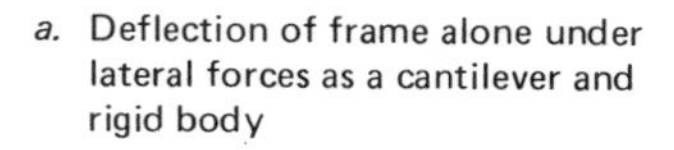

a. Deflection of frame alone under lateral forces as a cantilever and rigid body

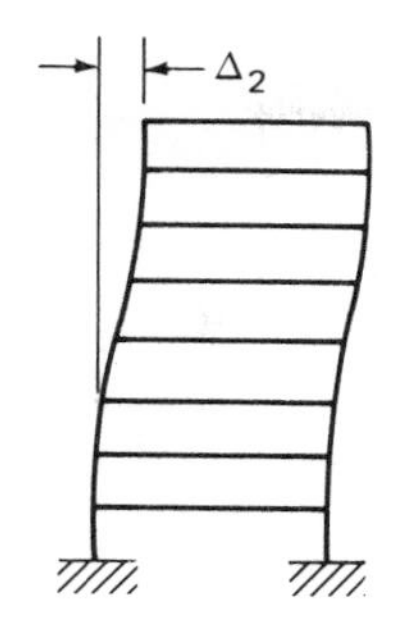

b. Deflection of the same frame using additional shear wall stiffness to resist lateral forces

Fig. 1.4. Behavior of plane frame with shear walls.

resist lateral forces. The effective width of the slab determines the stiffness of the one-dimensional beam elements used in plane-frame analysis. Effectiveness of such interaction can be seen from Fig. 1.4 in which are shown the frame deflections with and without slab to resist the lateral forces.

A beam-column-slab type of frame structure along with the shear wall can be economically used up to a height of about 40 stories. The general approach to the design of such frames has been to disregard it, due to its considerably larger flexibility than that of the shear wall, and to design the shear wall to resist only the horizontal forces. The frame in this case is the main load-bearing member in the vertical direction. However, due to the composite action, the frame is also subjected to forces higher than the applied lateral forces unless proper care is taken. The main advantage of the frame–shear wall interaction is that it increases the rigidity for lateral load resistance and, with the 33% increase allowed by codes for resistance to wind or similar lateral loads, helps reduce the frame moments and in turn the reinforcements or sections to give an economical solution. In general, proper proportioning means that the shear walls would contribute over 50% of the overall stiffness of the structure. Another significant increase in the lateral stiffness in a frame structure can be achieved by combining large girders with shear walls either at an intermediate level or at the top of the building. These girders generally have depths of up to a full floor height, and the enclosed space can be conveniently used for providing utilities, such as water tanks, thus serving both a structural and nonstructural purpose. Frames with various arrangements of the interacting components and their beneficial influence are shown in Figs. 1.4 through 1.6. Several examples and detailed treatment of the topic are given by Fintel (1974).

Although only a two-dimensional analysis has been presented, the three-dimensional nature of the interaction cannot be disregarded. This may be due to either a nonsymmetrical struc-

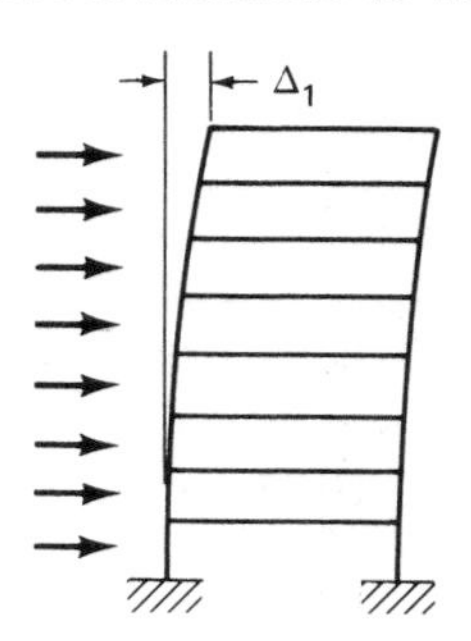

a. Deflection under lateral forces neglecting slab stiffness in plane-frame analysis

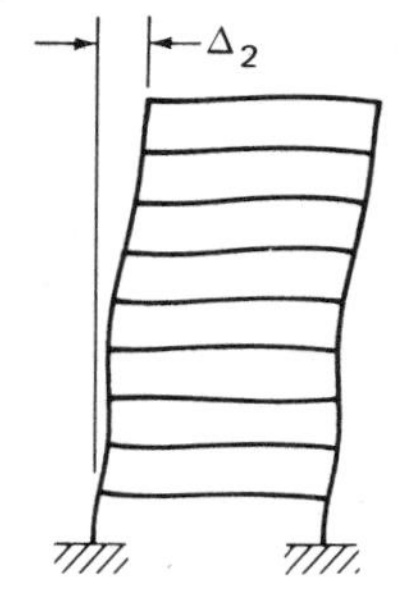

b. Deflection under lateral forces, using slab as effective with frame in plane-frame analysis

Fig. 1.5. Behavior of plane frame (schematic) ($\Delta_1 >> \Delta_2$ with slabs).

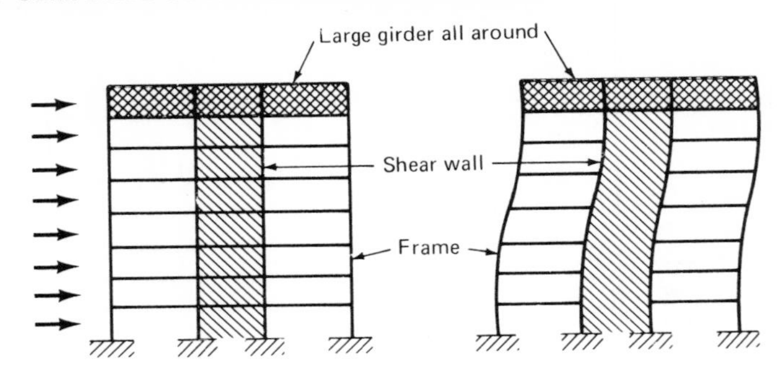

Fig. 1.6. Use of large girder in a frame structure. (*Based on Fintel 1974*).

tural plan or even to a symmetric structure with nonplanar frames. Torsional effects due to such nonsymmetry, must be considered requiring a three-dimensional analysis of the structure. Undoubtedly, the resulting problem is a complex one and requires the help of large-capacity computers. The programs mentioned earlier, such as STRESS and FRAME, can be used, but a large-size matrix is required for the entire structure. Overall stability analyses of such structures have been presented by Danay and Gluck (1975) and Nair (1975).

Beyond 30 to 40 stories, neither of the preceding systems has been found to be economical. Tubular structure, which uses external facade or frame in three dimensions, is used. Development of such a structural tube has made it possible to design and construct buildings taller than 100 stories. In addition to the external tube, a shear wall may also be provided inside as a core connected to the outer tube by slabs at floor levels, which act as diaphragms. In the preliminary design approach, the analysis can be carried out—even by hand computation—using an equivalent beam stiffness and a deflection similar to a cantilever. The local forces in beams and columns are then calculated by distributing the overall shear and bending moment to the discrete frame system. A typical tubular frame and an approximate stress distribution due to wind load are shown in Figs. 1.7 and 1.8. Khan and Amin (1973) and Khan (1974) were responsible for the development and use of a tube-frame method, and they deal with it extensively in several of their publications. Coull and Bose (1975) have proposed a simplified method for analysis of frame-tube structures.

1.6 MATERIALS USED IN COMPOSITE CONSTRUCTION

In this section, materials used primarily in composite construction are listed in order to serve as a background for fundamental mechanics of

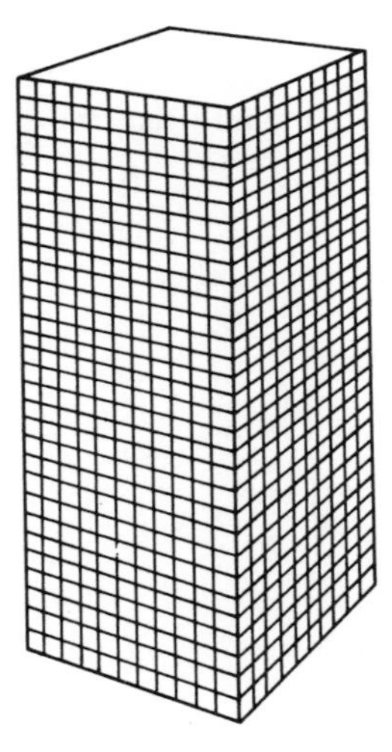

Fig. 1.7. Tubular frame. (*After Khan 1974.*)

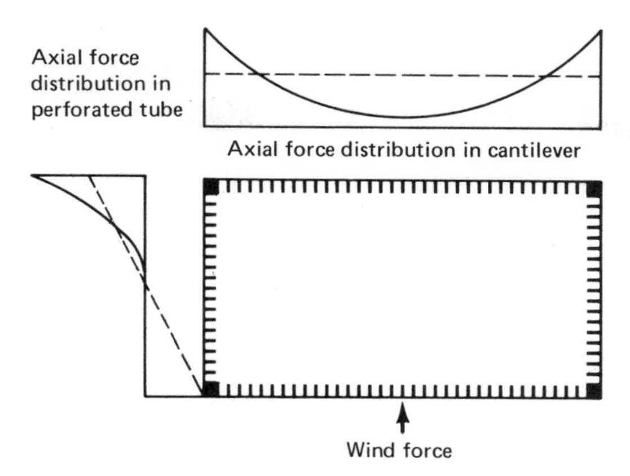

Fig. 1.8. Stress distribution in tubular frame. (*After Fintel 1974.*)

composite systems. Although reinforced concrete, reinforced masonry, and prestressed concrete are composite materials, they do not strictly classify as composite construction, as the steel reinforcement is not structurally self-supporting. In steel-concrete composite construction, on the other hand, beams are able to support their own weight and that of the concrete slab under construction, which also simplifies the construction process. The range of material and the specific associated elements in composite construction discussed earlier are listed in the following:

1. Composite *Steel-Concrete* Construction
   a. *Steel* beams, including rolled sections, castellated beams, truss or open web joist, box beam, plate girders with or without top flange, and hybrid girders.
   b. *Concrete* slabs of normal-weight or light-weight concrete, including haunched slabs, prefabricated planks or slabs, form-reinforced slabs with ribs parallel or perpendicular to the steel beam.
   c. *Steel* shear connectors, including studs, angle, channel, or coiled connectors; high-strength friction bolts; epoxy resins; grooved flanges and checkered plates; etc.
   d. *Fireproofing* with concrete, which includes solid, hollow, and profile protection.
   e. *Steel-concrete* composite columns, including wall columns, encased steel columns, concrete-filled steel tubes, light-gauge cold-formed metal composite and walls, and composite columns with hollow metal tube surrounded by concrete for built-in drainage and structural support.
   f. *Steel-concrete* composite frames, both in buildings and bridge structures; they include:
      (1) Braced or unbraced frames.
      (2) Tubular composite and plate-cladded buildings, steel frames with slitted reinforced concrete shear walls, concrete-encased steel frames.
2. Composite *wood-concrete* construction, in which shear transfer is achieved by adhesives or mechanical connectors. Materials in this case include:
   a. *Wood, timber, and plywood* elements with shear connectors.
   b. *Layered wood* (glulam) in beams, columns, and frames.
   c. *Composite wood* systems as in wood joist floors, walls, and roofs.
   d. Composite of wood with other materials, as in the case of:
      (1) Reinforced and prestressed wood components.
      (2) Wood-mortar and wood-concrete components for walls, floors.
      (3) Open-web composite steel–wood floor slab systems.
      (4) Plywood-lumber composite structural components.
3. *Infilled Masonry-Frame Composite Structures.* They consist of frames (either steel

or reinforced concrete) and walls of brick or concrete block masonry and/or concrete panels. Walls may be reinforced or unreinforced and may have openings. The cladding may also be used as a formwork for the exterior frames, thereby stiffening the structure, although added strength is not allowed in codes at the present time. Various types are:

a. *Steel plate* cladding: welded to frame and to each other using a heavy-gauge corrugated metal.

b. *Precast concrete* cladding as a formwork to the reinforced concrete frame.

c. *Precast concrete* panels attached mechanically to frames.

d. *Masonry infill* walls for in-plane shear rigidity with proper connections between walls and frames.

4. Composite *Sandwich and Laminate* Structures

a. Sandwich panels with *foam-in-place* cores and *metal* facings made of either flat or cold-formed corrugated U- or V-shaped forms.

b. Sandwich panels with *honeycombed* cores of craft paper or cardboard, with facings made from gypsum wall board or plywood.

c. Foamed *polyurethane* or polystyrene cores with bamboo, jute, or other natural fiber as reinforcement and mortar topping.

5. Miscellaneous composite construction

a. *Precast-prestressed concrete* with cast-*in-situ* concrete slab.

b. *Reinforced earth* with galvanized *metal strips*, or other reinforcements.

c. Composite *soil-steel* structures such as pipes and culverts.

d. Prestressed *precast concrete* elements with infilled *ceramic blocks.*

e. Shells made in reinforced *brick masonry* with edge beams of reinforced/prestressed concrete.

1.7 FUNDAMENTAL MECHANICS OF COMPOSITE STRUCTURAL ELEMENTS

The behavior and design of composite structural elements for various loading and environmental conditions may be considered as follows:

1. Strength: flexural, axial, vertical, shear buckling of flanges
2. Serviceability: short-term and long-term deflection, crack control, longitudinal slip, vibrations, fatigue effects
3. Failure modes and ductility: margins of safety at different "limit" states.

In addition, the effects of quasi-static and dynamic loading, temperature, and shrinkage variations in material properties due to quality control and time effects (e.g., creep) should be considered.

The importance of shear connectors in steel-concrete composite construction was discussed earlier to differentiate between the two extreme cases of complete and no interaction. In reality, there is a longitudinal shear-slip relationship, which calls for analysis of partial interaction. This is important in understanding the behavior of composite structural elements; however, it may not be necessary to use it in practice to design the shear connectors. In the present section, only an overview of the important structural behavioral considerations are given. Details of the fundamental mechanics, environmental effects, and design considerations for a number of composite structural elements and systems are given in later chapters.

1.7.1 Shear Connection and Load-Slip Characteristics in Composite Design

If complete interaction is assumed in the elastic range of behavior, the horizontal shear q per unit length of beam at the interface is given by elastic formula

$$q = \frac{VA\overline{y}}{I} \tag{1.1}$$

where

V = vertical shear force

A = transformed area of the concrete above the interface

$\overline{y}$ = distance between the center of area of A and the elastic neutral axis

I = moment of inertia of transformed composite section about the elastic neutral axis.

Equation 1.1 indicates that, for an elastic beam subject to uniformly distributed loads, the spacing of shear connectors should be closer near the supports, since shear varies linearly from end of span of beam to center. However, in the present design practice, the spacing is usually uniform based on the ultimate behavior when the flexural strength of the composite section is reached. Under these conditions, the horizontal slip is relatively large, and therefore shear connectors should be flexible enough to permit a redistribution of the total force in concrete and to load all the connectors equally. The total number of shear connectors (n) between the sections of maximum and zero bending moments must carry the ultimate compression force F_c in the concrete stress block. If b is the width of the compression flange, d the thickness of the slab, and P_c the design value of one shear connector, then assuming neutral axis to be within the flange, N is given by

$$N = \frac{F_c}{P} = 0.85 \frac{f'_c \cdot ba}{P_c} \qquad (1.2)$$

where

a = depth of stress block ($\leqslant d$ = depth)
f'_c = specific cylinder strength of concrete.

Equation 1.2 may also be written in terms of steel forces.

1.7.2 Load-Slip Characteristics

This is an important property of shear connection; it varies widely with the type of shear connector employed. As mentioned, the longitudinal slip between the steel beam and the concrete slab at the interface is ignored in design—i.e., the connection is treated as rigid. Yam and Chapman (1968) proposed an exponential relation for the load-slip characteristics in the form

$$P = a(1 - e^{-ps}) \qquad (1.3)$$

where

P = load
S = slip
a and p = experimentally determined constants for shear connector in a particular type of concrete.

1.7.3 Effect of Slip on Stresses and Deflection

Usually, the interaction between steel and concrete is incomplete due to slip caused by flexibility of shear connectors and the compressibility of concrete. It produces a discontinuity in the strain distribution at the interface. Newmark et al. (1951) developed a partial interaction theory in the elastic range assuming a linear load-slip relation. They showed that fairly large variations in the value of the shear connector modulus K (defined as the ratio of load to slip) affected the deformations only slightly; however, such variations affected the deflections considerably and to a large degree the elastic strains and stresses (e.g. 37% *increase* in the value of K resulted in 1% decrease in bottom flange strain and stress and 2% decrease in the maximum deflection). Also, 5% *decrease* in K resulted in 4% increase in strain and 16% increase in deflection. This theory has been extended by Dai et al. (1970) to take into account the nonlinear load-slip characteristics of shear connectors and also the inelastic behavior of steel and concrete. The partial interaction theory is useful in determining the effects of differential temperature strains and shrinkage in composite beams. Johnson (1975) has shown that the shear connectors reduce end slip considerably and that the effects of slip on deflection of composite beam are lesser than the theoretical calculations, due to higher connector modulus and the presence of bond in actual practice.

1.7.4 Effective Width of T-Beams and Shear Lag Effects

The subject of effective width of wide-beam flanges was investigated for the elastic case of simply supported beams, first by Von Karman (1923) and later on by many others for various loading and boundary conditions. Although early interest was in the applications in reinforced concrete structures, structures of hulls of ships and sheet-stiffener combinations for aerospace structures with essentially elastic behavior, recent interest has been directed to finding the effective width at ultimate loads, and based on deflection calculations from the "limit state of stress."

In the case of box-girder composite bridges, Van Dalen and Narasimhan (1976) analyzed shallow-wide flange composite girders by using a more rigorous folded plate theory rather than the usual simple beam theory with an assumption of reduced width of flange to account for shear-lag effects of deflections and stress due to longitudinal shear at the interface of beam-slab.

A similar problem occurs in aerospace structures. For example, we cite a box beam formed by two channels to which thin sheets are attached by welding or riveting along the edges. If the whole beam is built at one end and loaded by forces applied at the channels at the other end, the distribution of tensile and compressive stresses across the width at the top and bottom, respectively, will not be uniform as per the simple beam theory, but would decrease from the end to the center of the width. Such nonuniformity of stresses at any longitudinal section is due to the sheet being tensile, and compressive stresses in the sheet result from the shear stresses at the edges connected to it by the channels. This is known as "shear lag" since it involves shear deformation in the sheets. In the case of a beam with wide flanges, the web causes the bending stress at the top flange to vary from a maximum at top of web to a minimum at the end because of shear lag. In a ribbed plate or a series of interconnected wide-flange T-beams, plane sections do not remain plane after bending because of the shear lag distortion, as shown in Fig. 1.9. The figure shows the nonuniform longitudinal stress distribution across the section width of the top of the flange. Generally, the stress distribution varies from section to section along the span and is not only a function of the relative dimensions, stiffnesses of the composite structural system, boundary conditions, structural behavior (elastic, nonlinear elastic, creep, shrinkage, temperature effects, etc.), but also depends on the nature and distribution of applied loads.

Because of their complexity, the theoretical solutions that are available are too cumbersome to be used in design. A simplified concept of effective width is generally used in lieu of the complex shear lag problem. However, as shown by Van Dalen and Narasimhan (1976), this simplification of composite beams with shallow-wide flange cold-formed girders of thin section may be grossly erroneous in many cases. The effective width used in design is defined as that width of slab that, when acted on by the actual maximum stress, would have the same static equilibrium effect as the existing variable stress.

Traditionally, the effective width is based on stress distribution along the span and width of

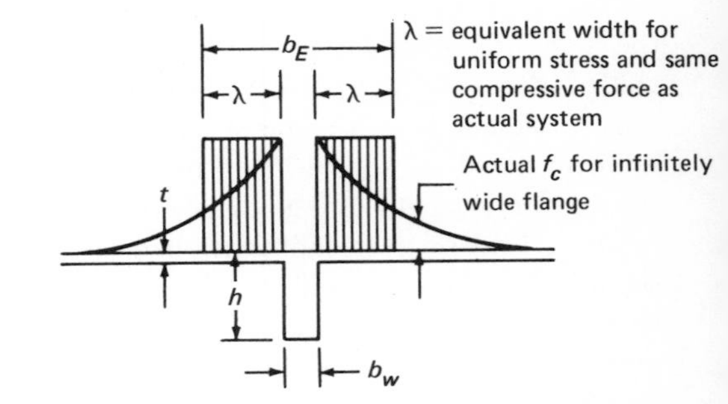

Elastic stress distribution in a monolithic reinforced concrete composite beam-slab

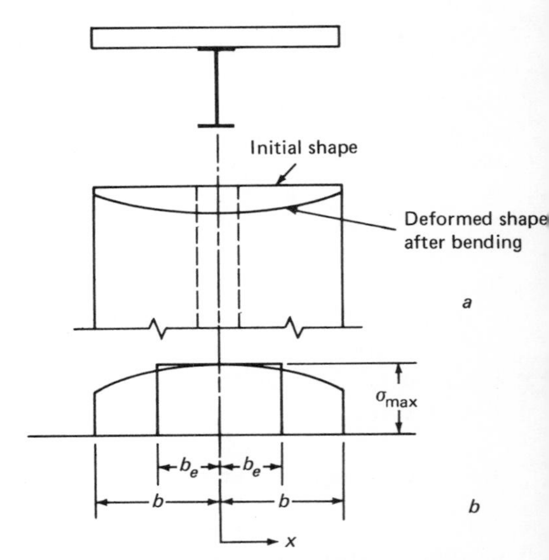

Fig. 1.9. Shear lag in wide-flange beam.

the top flange. Recently, the attention is given to the effect of deflection on the effective width with an application in limit states design, as seen from the work of Adekola (1974), Fraser (1971), and Saillard (1971). Fraser (1971) has shown that the actual deflection of a composite system is overestimated if the effective depth is determined taking into account the in-plane membrane stresses. Adekola (1974) showed that the effective width based on deflection considerations is more rational than that obtained from the maximum equivalent stress concept. Fan and Heins (1976) described an analytical method to predict the elastic and elastoplastic response of composite girder-slab bridges and to determine the effective composite beam width at ultimate load. Design equa-

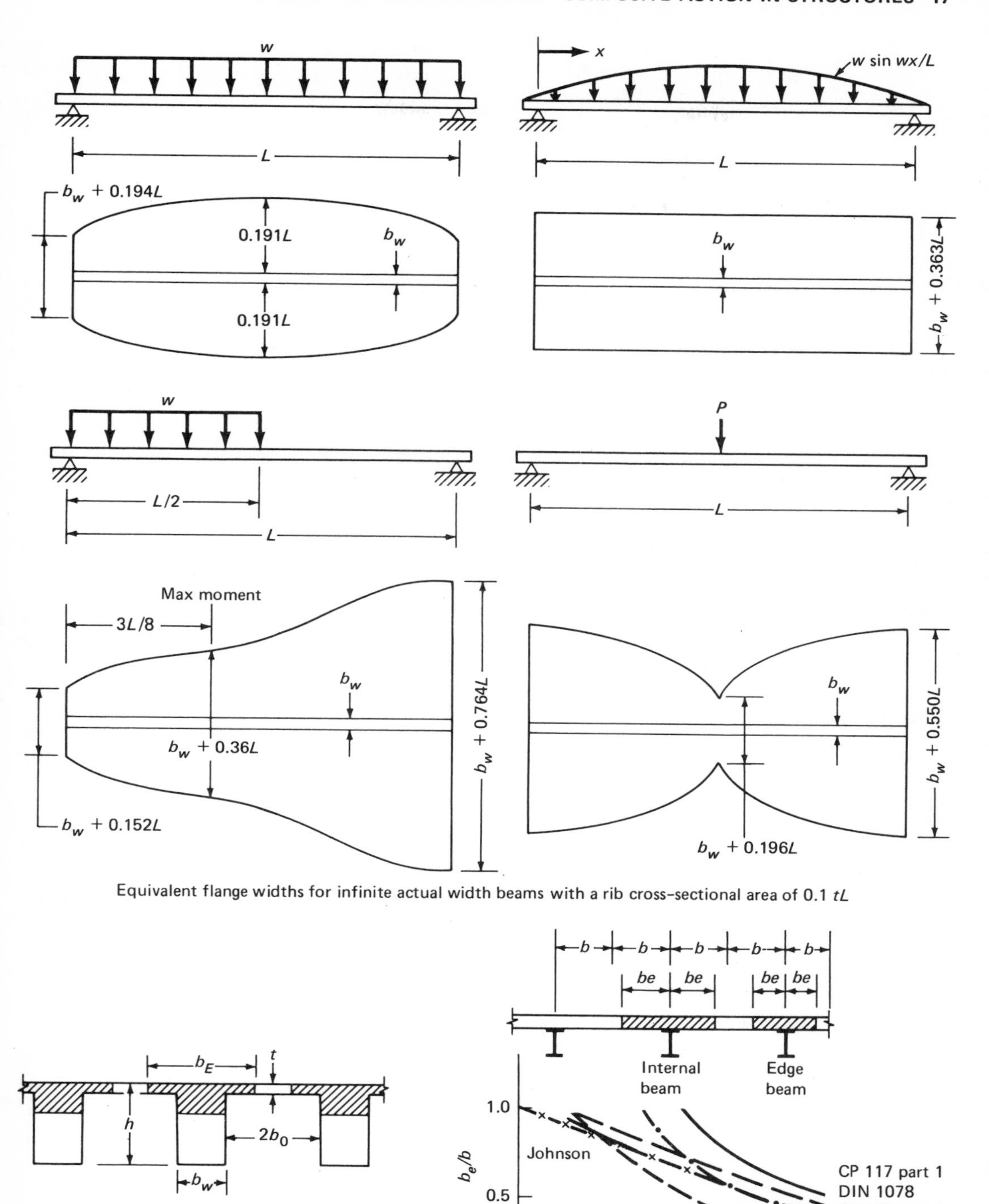

Fig. 1.10. Comparison of effective width formulas.

tions are given to predict the effective widths of composite slab girder bridges for AASHTO ultimate live loads. Based on the preceding analysis, they concluded that AASHTO effective widths, as applied to load factor design, overestimate the widths for interior girders and underestimate the widths for the exterior girders.

According to Chapman and Teraskiewicz (1968), the effective width based on stress distribution obtained experimentally is greater than that based on theoretical elastic analysis. Chapman and Teraskiewicz showed that due to the complexity of the problem and the actual behavior of various types of composite beams with monolithic beam-slabs construction under complex loading and time and environmental conditions, simplified formulas for effective width are needed. The preceding theoretical solutions do not take into account inelastic behavior and transverse cracking, creep, and shrinkage effects or statistical variations in loads and material properties. The European Concrete Committee (CEB) has significant recommendations for determining the effective width. (See, for example, FIP-CACA 1970 and Saillard 1971). A comparison of effective width formulas used in practice by various codes is shown in Fig. 1.10.

1.7.5 Failure Modes and Limit-States Design

There is a growing shift in the design of structure from classical principles using elastic analysis to the more logical philosophy of limit states based on failure modes of strength and serviceability under various conditions of loading. The term *failure* means any relevant state out of a number of possible limit states in a structure. In composite construction, the situation is further complicated by the multiple nature of the system and different coefficients of variability of the material properties for the various components: namely, concrete slab spanning transversely between steel beams; shear connectors; steel beams acting compositely with the slab in the longitudinal direction; and connections between beams and columns. Types of failure include:

1. failure by *flexure* in the composite beam by the formation of sufficient hinges to result in a plastic collapse
2. failure by *flexure* of the slab in the transverse direction by formation of sufficient "yield" lines (for a plastic collapse in the transverse direction)
3. failure of *shear* connection due to reduction or complete removal of composite action caused by excessive slip between slab and beam or complete failure of shear connection leading to beam collapse
4. *shear-bond* failure (for example, in composite steel deck slabs)
5. failure by *longitudinal splitting* of the slab
6. *local shear* failure in the slab in regions of high stress around shear connectors
7. *excessive cracking* and service-load deflections including longitudinal cracking in the slab
8. failure of *composite connections* between beams and columns.

Although a "perfectly designed" composite beam would fail in all the limit states simultaneously, even under the most strictly controlled laboratory conditions, this would be unlikely to happen (because of variations in material properties, structural behavior, and degree of accuracy in design methods). In practice, composite beams are designed for flexure by increasing the factors of safety for other failure modes and ensuring that they do not precede the flexural collapse. In the plastic design of statically indeterminate structures, such as frames or continuous beams, both negative and positive plastic hinges are required to form a complete collapse mechanism. The sequence of hinge formation is important. Since the rotation capacity of positive (sagging) hinges may be limited by excessive strain in concrete, the negative (hogging) hinges (which do not have this restriction) need to be formed first. Ansourian and Roderick (1976) pointed out the need for proper detailing of these areas in composite structures based on tests on connections between the encased beams and the external columns. Although probabilistic methods have been developed for the design of reinforced concrete structures (see Vorlicek and Tichy 1969), and to a lesser extent for steel and other metallic structures (ASCE 1971), development of these methods for composite structures to account for variability in loads, material properties, and fabrication is needed. At present, all calculations for safety in com-

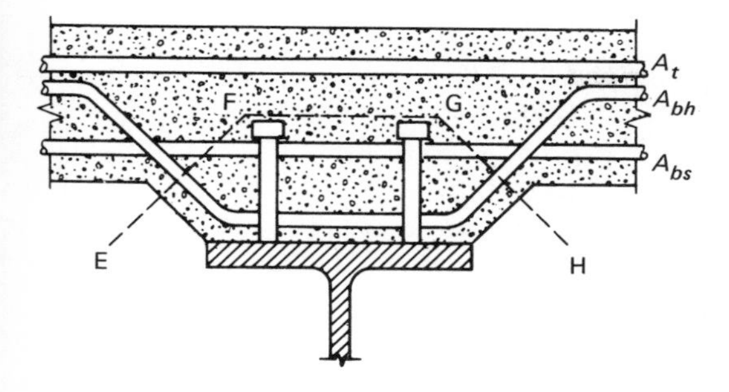

EFGH = shear plane through the haunch
A_t = top reinforcement of the slab
A_{bs} = bottom reinforcement of the slab
A_{bh} = reinforcement in the launch at the bottom

Fig. 1.11. Longitudinal shear in haunched beams. (*Based on Johnson 1975.*)

posite construction are based on deterministic approaches.

Since flexural mode is the basis of design, the role of longitudinal shear, longitudinal cracking in the slab, and partial interaction must be well understood so that we may effectively account for them in design. Figure 1.11 shows the different steel reinforcements that take care of longitudinal shear in haunched beams, especially when the haunches are deep. The behavior of composite beams when there is a failure of shear connectors (either partial or complete slip) is shown in Fig. 1.12 in terms of flexural strains for complete and partial interaction. In the case of partial interaction, the relative movement between the slab and beam caused by deformation of the shear connection can be rigorously analyzed if the time-dependent stress-strain properties of steel, concrete, and shear connectors are considered. The shear connectors must be well designed and fabricated to reduce the effect of slip on failure; otherwise, the failure may be similar to that of noncomposite slab-beam at lower safety levels.

Schuster (1976) and Porter and Ekberg (1976) discuss shear-bond failure in composite steel deck slabs, *vis-à-vis* flexure of an underreinforced and an overreinforced section and the design provisions to guard against this mode. It is characterized by the formation of a diagonal tension crack in concrete at or near one of the load points, followed by a loss of bond between the steel deck and the concrete, and it results in slip between the two, which is observed near the end of the span. It causes a loss of composite action. In a well-designed composite steel deck floor slab, there is some mechanical means of providing positive interlocking between the deck and the concrete. This is usually achieved by one of the following:

1. embossments, indentations, or both
2. transverse wires attached to the deck corrugations (by spot welding)
3. holes placed in the corrugations
4. deck profile (cellular or noncellular deck profile) and steel surface bonding
5. shear connectors.

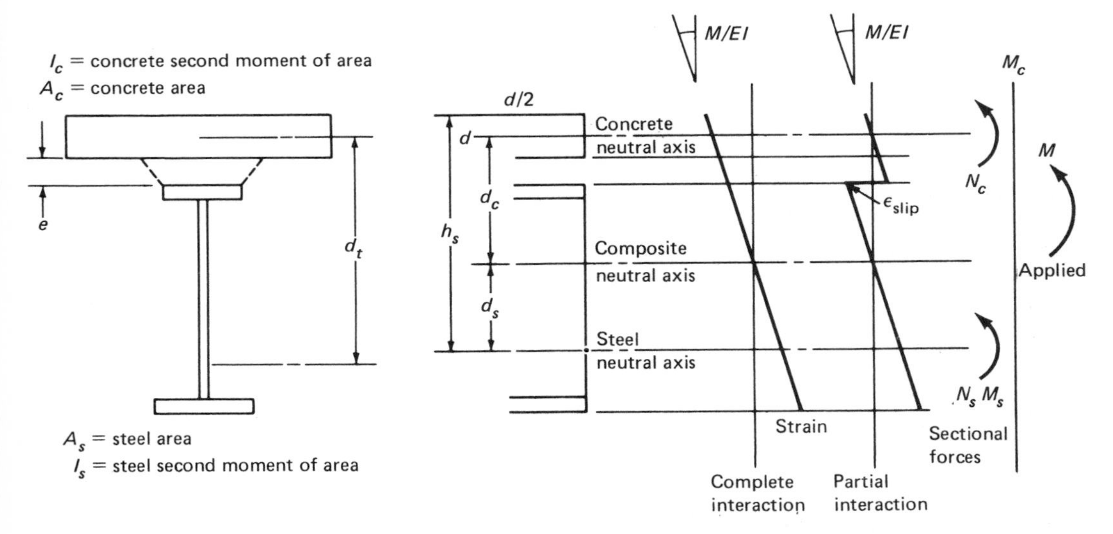

Fig. 1.12. Strain diagrams for complete and partial interaction (Knowles 1973).

In these, the cellular deck profile has closed cells formed by an added sheet of steel connected to the bottom corrugations of the deck.

Porter et al. (1976), in a report on the shear-bond analysis of steel-deck-reinforced slabs, indicated that the shear-bond capacity increases with (1) an increase in effective depth of the composite strength, (2) a decrease in shear span, (3) an increase in compressive concrete strength, (4) a decrease in spacing of the shear transferring devices, and (5) width of the specimen. They also found that end slip is exhibited in some cases prior to ultimate load, the magnitude of which is generally less than 0.06 in. (1.5 mm) even at ultimate.

1.8 DEFLECTIONS, CREEP, SHRINKAGE, AND OTHER EFFECTS

The deflection of a composite steel-concrete beam may be considered to be of three parts: (1) short-time dead load, (2) long-time creep and shrinkage, and (3) short-time live load. It is calculated using composite properties (transformed section) of steel and concrete and using a modular ratio, $n = E_s/E_c$. The modular ratio for deflection due to long-time creep and shrinkage (case 1), however, is taken to be higher than that for live load (case 2). In British specifications (CP 117), the recommended value of n in case 1 and case 2 is 30 and 15, respectively, whereas in American practice it is recommended that a ratio of 3 (instead of 2—i.e., 30/15) be used to increase the value of n in case 1. (For example, see Slutter 1974.) The higher modular ratio results in a smaller moment of inertia, and hence a larger calculated deflection, but only by a relatively small amount.

In computing deflections, it is necessary to take into account creep, shrinkage, thermal effects, and the method of construction. When no temporary supports (props or shores) are used during casting and curing of the concrete slab, the dead loads are resisted by the steel beam alone. If the temporary supports are used during construction, then the dead loads are resisted by the composite section. If the dead loads are sustained, they cause creep, which is taken care of by increasing the conventional value of the modular ratio n. This long-time effective modular ratio is taken as 2 and 3 for building and bridge design, respectively. Dead-load deflection may be one reason for excessive thickening of the slab in beams built without shores and excessive dishing of the slab in beams with shores. Proper cambering must therefore be incorporated in design.

Proper design of composite floor systems is required to avoid the undesirable effects of vibrations. Steady-state vibrations are eliminated by insulating and damping the source. To alleviate transient vibrations, the following maximum depth-to-span ratios have been recommended by Viest (1969):

1:24 for ordinary building applications
1:20 for buildings where vibrations and shock are present
1:25 for bridge.

In view of the damping characteristics of walls, partitions, and so on, depth is taken as the overall depth including the steel beam and the concrete slab. On the other hand, in open areas particularly susceptible to annoying vibrations, only the depth of the steel beam is considered in such limits.

Figure 1.13 shows the effects of differential strain (due to creep, shrinkage, and temperature) in a steel-concrete composite beam. Fig. 1.14 shows the stress changes due to creep. The stresses in concrete decrease whereas those in

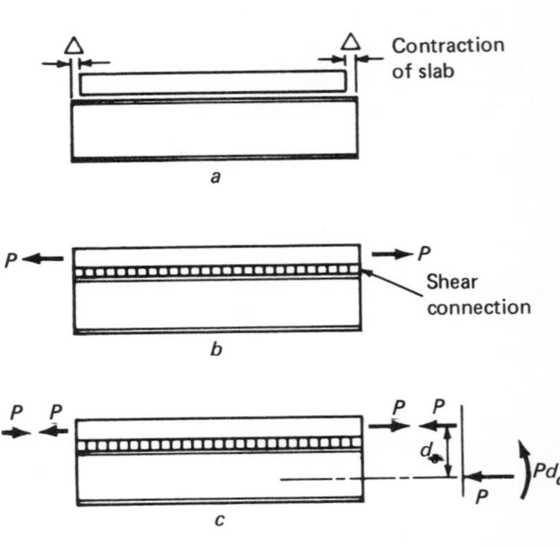

Fig. 1.13. Differential deformation in composite beam: *a*, slab disconnected; *b*, compatibility restored; *c*, equilibrium restored.

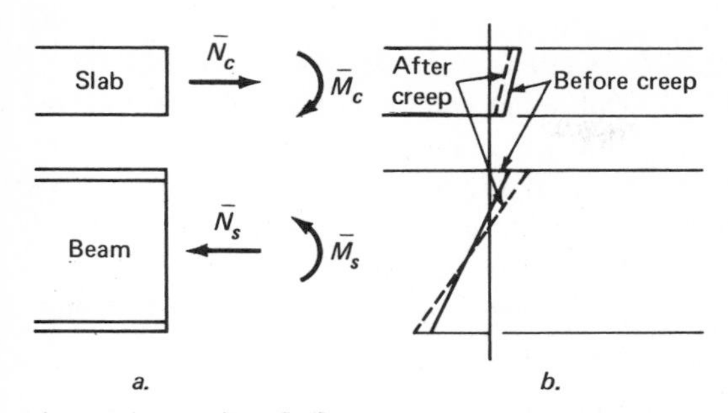

Fig. 1.14. Stress changes caused by creep: *a*, section force changes caused by creep and, *b*, stresses before and after creep has occurred (Knowles 1973).

the steel increase due to creep effects. Branson (1977) discussed a number of methods of accounting for creep, shrinkage, and temperature. The temperature differential caused by steel beams being masked from the direct solar heat on concrete (by difference in the rate of conduction) is significant in bridges and similar structures. In addition to load effects, time-dependent effects occur in indeterminate composite beams. Busemann (1950), Hasse (1969), Mainz and Wolff (1972), and Trost (1968) discuss methods of taking care of creep effects of composite beams. Roll (1971) has discussed a method for computing effects of differential shrinkage and creep.

In composite columns, shrinkage may cause tensile cracking when the percentage of reinforcement is high and when the dead loads are negligible or small, i.e., during construction. However, in concrete-filled steel pipes or in columns with hollow tubular reinforcements in concrete, when the percentage of reinforcement is low, it is possible that the initial stress in steel is within allowable limits; however, with sustained load effects, the steel may yield in compression or even buckle (creep-buckling). These considerations thus point out the need for limiting the maximum and minimum percentage of steel in composite columns.

The substantial increase in deflections due to creep and shrinkage for composite beams with lightweight concrete has been reported by McGarraugh and Baldwin (1971) and Janss (1972). Based on sustained tests, they recommend that, for design purposes, additional time-dependent deflections may be taken as equal to instantaneous deflection when lightweight concretes are used in composite beams. In addition to static short-term and long-term effects, fatigue effects need to be considered, especially in the design of continuous composite bridges. The deflection due to shear will be slightly higher for a composite beam than for a compatible noncomposite steel beam. However, flexural deflection is likely to be substantially greater than the shear deformation, and the latter may be generally ignored.

The slip-forming method of construction of bridges eliminates a large percentage of shrinkage effects as compared to cast-*in-situ* decks. The effects of temperature in long-span composite bridge structures may be significant. Menzies (1968) reported that the effect of temperature changes on the Moat Street Flyover for a period of 2 years was much greater than the effects of creep and shrinkage. Ciolina (1971) studied the effects of creep, shrinkage, and temperature on two-span continuous composite beams and obtained a coefficient of shrinkage of 3.1×10^{-4} compared to the value of 4×10^{-4} recommended for design of composite floor beams by CTICM (1965). Zuk (1965) and Berwanger (1970) report studies of thermal and shrinkage effects in composite bridges for both insulated and uninsulated conditions.

1.8.1 Earthquake-Resistant Design

The earthquake response of steel-concrete composite structures is a subject of much interest and is reported by the ASCE Subcommittee in the State-of-the-Art Survey of Composite Steel-Concrete Construction (1974). The survey concludes that local buckling of the compression flange is delayed by connecting a steel section to a concrete slab with shear connectors, and resistance to lateral buckling is greatly increased. Furthermore, concrete-filled tubes and encased rolled shapes possess much higher shear strengths than reinforced concrete columns of the same size, and the connections between columns and composite beams can be made very ductile. The ASCE report concludes that research on the response of composite beams, columns, and frames to cyclic loading offers much promise to practical applications of composite construction in earthquake-affected regions.

In Japan, unlike in the United States, an earthquake-resistant design of composite structures was developed by using only the built-up steel sections and riveting latticed angles together; concrete portion was reinforced as in conventional reinforced concrete construction. This is known as "steel-reinforced concrete construction." Hot-rolled shapes and welded I- or H-sections have gradually replaced battened steel sections. Naka et al. (1972) and Wakabayashi et al. (1972) discuss superior earthquake-resistant properties, which became widely recognized in Japan after the Kanto Earthquake in 1923. In 1972, the Architectural Institute of Japan standards were revised to make allowable column shear strength based on shear strength of the steel column alone, even though stirrups must also be provided. However, it specified that the allowable strengths of the cross section may be calculated by superposition of the allowable strengths of the steel and reinforced concrete portions.

Systematic studies of the elastoplastic behavior and hysteresis characteristics of structural-steel-reinforced concrete structures were carried out at Tokyo University in 1977. (Based on these studies, the procedures for dynamic structural analysis of composite systems will be modified in the future editions of Architectural Institute of Japan Standards.) In brief, the studies have shown clearly that encasing steel sections in reinforced concrete is particularly beneficial for earthquake-resistant design. However, practically no studies are available on the suitability of other types of composite structural systems for earthquake-resistant design.

1.9 CONSTRUCTION AND ERECTION METHODS

The advantages of composite construction, both for economy and performance, require careful attention to proper construction methods. The methods will depend on type of job, location, and facilities available. Some of these are:

1. Rapid construction to suit an early schedule or rapid occupancy of the structure.
2. Suitability of construction technique independent of adverse weather conditions
3. Minimum interference with existing facilities due to minimum formwork required
4. Optimum combination of prefabricating and *in situ* construction technique
5. Phase loading to cause stressing of structure at different times
6. Possible savings in foundation due to simplification and more compact excavation operations.

These advantages will be discussed in light of the different construction and erection methods used in composite construction of buildings and bridges. In buildings, composite action results in a more rigid, stiffer structure, and the encasement of columns can help against fire protection. On the other hand, high live-to-dead load ratios in the case of a bridge structure may require temporary shoring of the steel structure to support only the dead load of concrete and later to sustain the full dead and live load by composite action.

Different construction methods are followed in different countries and are discussed by Roret, Taylor, Smith, and others in *Proceedings of the ASCE-IABSE (1972) Conference on Planning and Design of Tall Structures.* The countries include France (Roret 1972), England (Taylor 1972 and Smith 1972), Australia (Simon 1972), Japan (Hisatoku and Nishikawa 1972), and the U.S.S.R. (Vasiliev et al. 1972).

Knowles (1973) discusses the advantages of using prestressed steel, especially in buildings where there are limitations on depth. In such construction, temporary props are set up from columns to cause prestressing of the steel beams above. These are removed after the concrete hardens, which reduces the loading on the steel members alone, thus reducing their depths and resulting economy. Figures 1.15a–d based on Knowles' work illustrate this construction technique.

Dziewolski (1972) describes a prestressed composite structure that is achieved by combining the two consecutive floors as shown in Fig. 1.16. In his method, Dziewolski gives a camber to the floor beam during erection, before the concrete is cast; this is done by means of tie bars acting as inclined jacks, which are attached at the upper part of the beam. Figure 1.16 illustrates a floor beam of 15-m (50-ft)

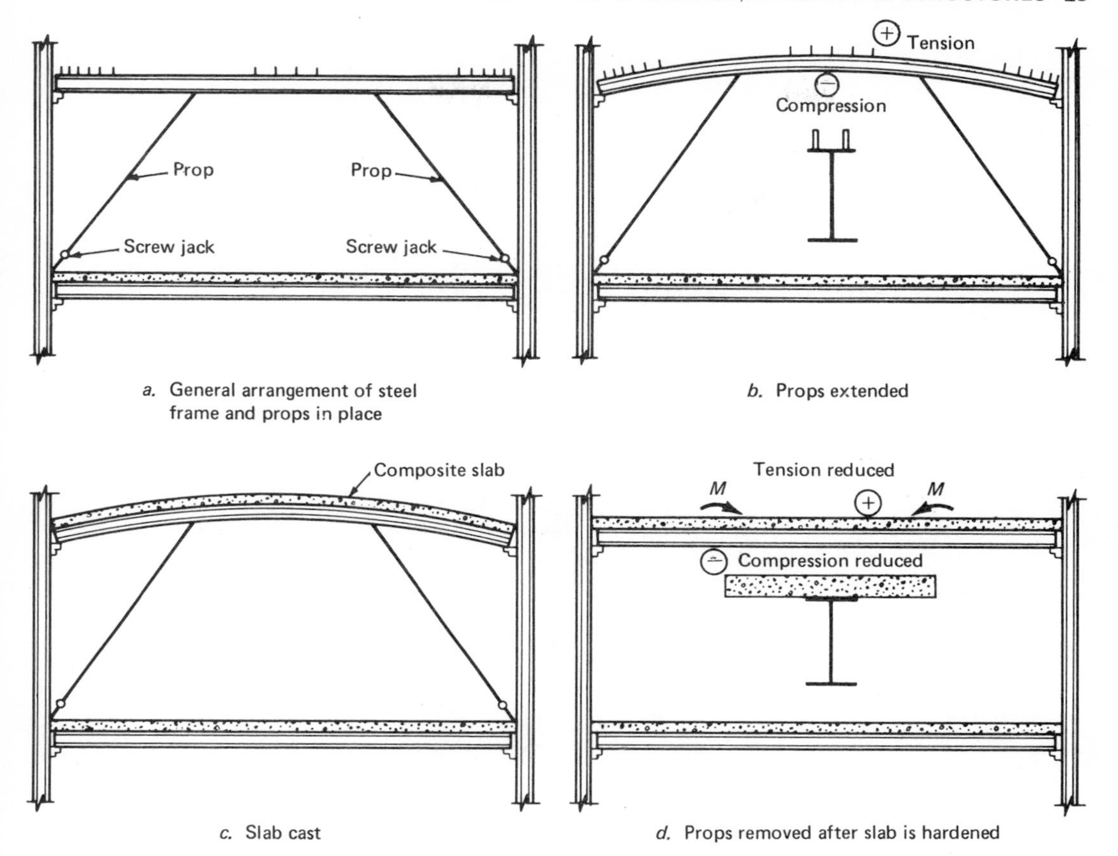

Fig. 1.15. Details of prestressed composite construction in buildings. (*Adapted from Knowles 1973.*)

span in the Havas Conseil building in Neuilly, France. The following table compares the maximum stresses for different loading conditions with those in the regular method of composite construction.

It may be noted that the method can also be

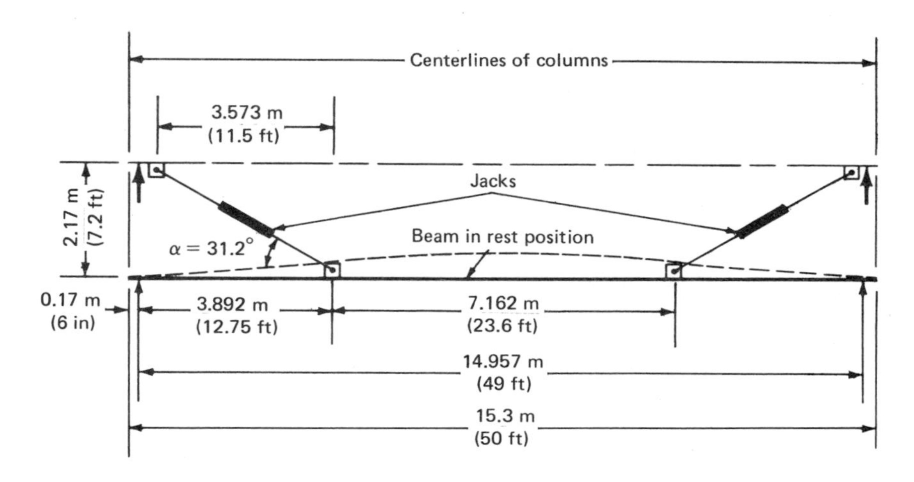

Fig. 1.16. Prestressed composite structure. (*After Dziewolski 1972.*)

	Stress[a] *in* kg/mm²			
	Prestressed Composite Beam		*Composite Beam*	
Loading	*Upper Chord*	*Lower Chord*	*Upper Chord*	*Lower Chord*
Dead weight, steel	-2.89	+1.41	-2.89	+1.41
Prestressing	+31.50	-17.50	-	-
Dead weight, slab	+6.86	-2.07	-20.30	+9.90
Slackening of prestressing	-10.95	+17.66	-	-
Fixed additional load	-8.40	+16.20	-8.40	+16.20
Live load	-3.44	+6.63	-3.44	+6.63
Concrete shrinkage	-8.30	+1.20	-8.30	+1.20
Total	-2.48	+23.53	-43.33	+35.34

[a]Plus sign (+) indicates tension; minus sign (–) indicates compression.

used in bridge construction wherein traffic and the river itself often make it impossible to position shores (or props) beneath the beams.

One major difference between the building and the bridge is the type of loading, which clearly reflects on the construction methods used in bridges. Shored construction is seldom used due to the high live-to-dead load ratios compared to those in buildings. The structural steel portion (noncomposite) is used to carry the initial load of wet concrete on a reduced span by shores. The shores are removed when concrete is hardened; the composite section now has many desirable properties for bearing

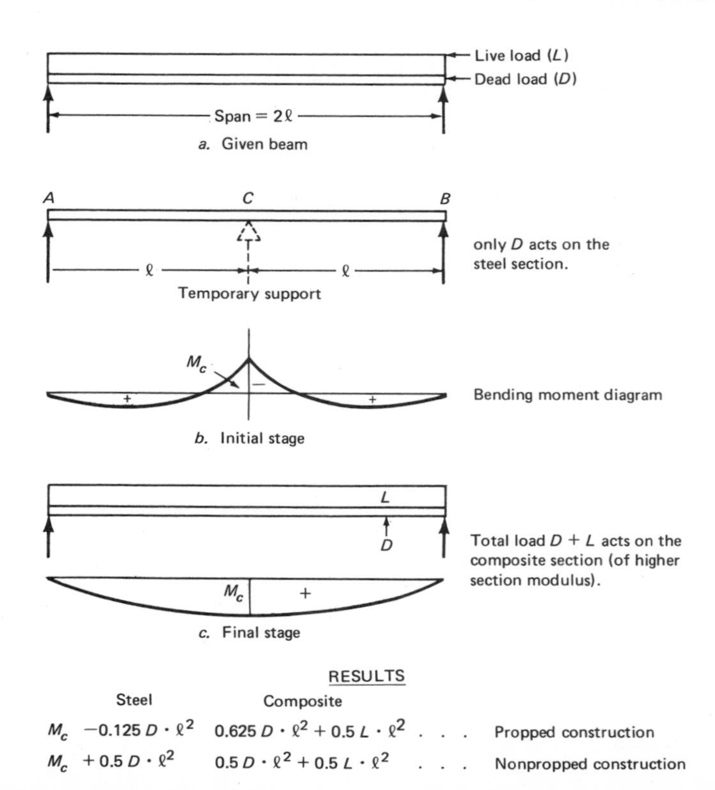

Fig. 1.17. Effect of temporary supports on the moments at the central section of a be-

both the dead and live load on the full span. Such construction will undoubtedly result in an overall economy. A schematic shown in Fig. 1.17 indicates reduction in the bending moments in a beam and the savings thereby.

The economy of shored composite construction was discussed earlier in Section 1.2. Such a construction method also affects design methods and the various stages of loading, which the designer must check properly. Slutter (1974) indicates this aspect in his work. The stages to be considered are:

1. shored dead-load stresses (top and bottom of steel section)
2. unshored dead-plus-live-load stresses (top of slab and bottom of steel section).

It should be noted that the value of the modular ratio n used in calculating the transformed sections in such a construction is taken as three times its elastic value to account for the net effect of time-dependent properties of concrete, such as creep and shrinkage during the sustained dead load. As a result, the effectiveness of concrete slab reduces and that of steel increases due to increase in dead load.

Another method of "prestressed" composite construction is carried out by phased concreting as suggested by Knowles (1973). This is achieved by a careful choice of sequence of concrete placement in a certain portion of the beam—i.e., in the portion where additional stiffness is useful to carry the load. It can be shown that by placing concrete in the middle third of the span, up to 40% of the slab dead weight will be transferred from steel beam to the composite section.

Several other techniques used on a limited basis are discussed in Section 1.12.

1.10 APPLICATIONS IN BUILDINGS AND BRIDGES

From the discussion so far, it is clear that composite construction can have limitless possibilities and many applications. Although principles of composite construction do not vary in terms of application, the construction techniques and the applied loads influence the use of composite construction. Different examples are presented in this section to cover a wide range of applications both in buildings and bridges.

1.10.1 Applications in Buildings

Fintel (1974) gives an extensive list of applications; a condensed form of his presentation is given here to illustrate the various types of composite construction.

Selected List of Framed Buildings and Their Types

Type	*Building Name and Location*	*Year*	*Figure*
1 Shear wall frame	Keio Plaza Hotel, Tokyo, Japan	-	-
2 Shear wall frame	National Life Building Nashville, TN	1970	1.18a
3 Bundled tube	Sears Towers, Chicago, IL	1975	1.18b
4 Tube-in-tube	CBS Building New York, NY	1964	-
5 Coupled shear wall	Americana Hotel, New York, NY	1962	1.19a
6 Composite (concrete tube steel interior)	One Shell Square, New Orleans, LA	1971	1.19b
7 Exterior frame and core	LaSalle Plaza, Chicago, IL	1972	-
8 Framed tube	Xerox Building Rochester, NY	-	1.20

Fig. 1.18a. National Life Building, Nashville, Tennessee. *(Courtesy Portland Cement Association, Skokie, Illinois.)*

In (1) and (2), *frames* are used to take up vertical loads, whereas *shear walls* or braces are designed for horizontal loads (Fig. 1.18a). As mentioned earlier, the concept of shear wall was modified in the later system of tube-in-tube (Fig. 1.18b). Some of the problems met are discussed in detail by Khan (1974). Coupled shear wall system is one in which openings in the shear walls are present. These openings occur generally in vertical rows throughout the height of the building, such as in the Americana Hotel in Fig. 1.19a. Methods of analysis for such systems are dealt with by Fintel (1974). One Shell Square (Fig. 1.19b) is very similar to the tube-in-tube concept, in which the advantages mentioned for both concrete and steel are used.

It must also be mentioned that the preceding examples are just a few notable tall buildings in which systems used can be identified. Many more can be found in structures used as residential (apartment) buildings, office buildings (Fig. 1.20), hotels, schools, multistory garages, and combinations thereof.

Fig. 1.18b. Sears Tower, Chicago, Illinois.

1.10.2 Applications in Bridges

Applications of bridges are found in many references and are treated in detail in Chapter 8. Some applications of the various types of bridges are presented here to complete this overview. These are:

1. Steel composite with concrete
2. Continuous steel-plate girder composite with concrete (Fig. 1.21)
3. Continuous welded plate girder composite with concrete (Fig. 1.22)

Fig. 1.19a. Americana Hotel, New York City. *(Courtesy Portland Cement Association, Skokie, Illinois.)*

4. Poured-in curved steel composite with concrete slab
5. Precast prestressed concrete composite with poured-in concrete (Fig. 1.23)
6. Precast reinforced composite with poured-in slab
7. Concrete box girder composite with concrete deck, all poured in place (Fig. 1.24)
8. Curved composite concrete.

All of the preceding applications are quite slim (shallow) due to their composite action, and these bridges give an excellent aesthetic and streamlined appearance and merge well with the surrounding nature. The first four apply to the composite bridges between steel and concrete, while the rest are in the various types of concretes and their structural forms.

Fig. 1.19b. One Shell Square, New Orleans, Louisiana. *(Courtesy Portland Cement Association, Skokie, Illinois.)*

Fig. 1.20. Xerox Building, Rochester, New York. *(Courtesy Portland Cement Association, Skokie, Illinois.)*

Fig. 1.21. Bridge with a continuous steel-plate girder with composite concrete deck. (*Courtesy California Department of Highways.*)

Fig. 1.22. Bridge with welded steel-plate girders with composite concrete deck. (*Courtesy California Department of Highways.*)

Fig. 1.23. Bridge with precast prestressed concrete girders with cast-in-place composite concrete deck. (*Courtesy California Department of Highways.*)

Fig. 1.24. Bridge with concrete box-girder poured-in-place with compositely acting concrete deck. (*Courtesy California Department of Highways.*)

1.11 COMPOSITE CONSTRUCTION IN WOOD AND TIMBER*

In wood and timber structural systems, recognition and treatment of composite behavior differ in many respects from those given to common steel-concrete systems. Whereas composite action is readily accepted and intentionally incorporated in steel and concrete applications, the state-of-the-art in wood and timber design has not advanced to this point.

Composite action in wood and timber structural systems depends in large measure on the type of connection existing between the various components. With few exceptions, these connections have never been designed specifically for the purpose of achieving composite action. Consequently, the composite action in these systems is less than complete, and code provisions, to date, have virtually ignored its presence. In great part, this was because proper methods of analysis for predicting the degree of composite action simply did not exist. However, in recent years, research studies have demonstrated that many wood and timber systems do exhibit significant composite behavior. Theoretical and experimental studies have resulted in viable, verified mathematical models, which permit proper assessment of this interaction.

*Section 1.11 was contributed by Drs. James R. Goodman and Richard M. Gutkowski, to whom the authors are greatly indebted.

Mechanical fasteners used in wood systems constitute the critical component affecting composite behavior. In practice, most connections fall between two extremes: rigid connections (e.g., some glues) for which complete interaction results, and no connections for which the interaction is negligible. Rigid connections prevent relative movement of the structural components, while in the absence of connectors such motion is restrained only by insignificant frictional forces. For layered wood or wood in combination with other materials, as used primarily in beam or beam-type systems, the degree of interlayer movement caused by connector deformation is significant. Thus, the type of connection is of critical importance, and recognition of partial or incomplete composite action is essential to the analysis of composite wood systems.

For wood and timber systems, performance cannot be properly assessed without consideration of slip motion. Since this motion is negligible in many other composite systems, such as most steel-concrete construction, a comprehensive theoretical treatment is provided in Chapter 10. Fundamental concepts of "incomplete composite action" and abbreviated derivations for newly developed analytical methods are also presented in that chapter, as are design procedures for wood composite systems, including detailed examples for specific applications.

Table 1.1 lists several of the common types of wood structural systems that are treated and

Table 1.1. Composite wood systems.

Type of System	Typical Type of Connector
Layered wood beams	Nails, rigid and nonrigid glues, bolts, other mechanical fasteners
Composite wood columns	Nails, bolts and other mechanical fasteners
Wood joist systems	Nails, nonrigid glues, other mechanical fasteners
Wood-stud wall systems	Nails, nonrigid glues and other mechanical fasteners
Reinforced wood components	Mechanical fasteners, glues
Prestressed wood components	Prestressing cables or steel rod
Wood-concrete components	Shear connectors, epoxy
Open-web composite wood-steel systems	Bolts
Plywood-lumber components	Rigid adhesives, mechanical connectors
Composite glulam bridge systems	Bolts, dowels

summarizes the wood systems for which the benefits of composite action may be considered.

1.12 OTHER TYPES OF COMPOSITE CONSTRUCTION

There are several types of composite construction in which interaction between two construction materials takes place and in which such systems are used as alternate solutions to the specific problems. These different systems have been used for special purposes, often claiming even greater economy in steel weight than the more conventional composite constructions. Many novel systems were proposed and used in the 1930s during the infancy of composite construction.

Various systems of composite construction discussed in this section include:

1. Open-web joist, castellated beams and inverted T-beams, and inverted T-beams plate girder without top flange used with concrete decking
2. Prestressed and preflex steel composite beams
3. Concrete-encased steel beams
4. Composite lattice structures
5. Sandwich construction
6. Hybrid steel beams used with concrete
7. Composite concrete steel (plates or cold-formed) slabs.

Each of these is described herein with an example, advantages, and disadvantages. No attempt will be made to present any theoretical background, which falls back to the basic interaction of components discussed earlier; references cited may be followed for more detailed treatment. It should be noted, however, that these applications have been used only on a limited basis in a particular situation. Additional tests carried out—even on a scale-model basis—will enhance their use and the confidence of the user. McDevitt and Viest (1972) give a survey of many of these in their work.

1. *Open-web beams or girders* lead to more economy since some material from web is taken out; however, the cost of fabrication is generally higher. The compromise to achieve economy is essentially among fabrication, savings in material, and gain in the headroom, since the open portions of web can be conveniently used to support the piping or other utility service lines. Furthermore, more flexibility in design may be achieved due to the independent selection of webs and flanges. It must, however, be recognized that the most important aspect in the design will be governed by the shear connectors between the tensile and comprehensive elements of the system.

The early work was done by Galambos et al. (1970). Lembeck (1965) conducted tests on specimens with webs of joist extended above into compression flange to form shear connectors. Wang and Kaley (1967) tested open-web beams with concrete slab keyed into a top chord to form a dovetail-shaped trough. Later on, the actual shear connectors were tested with the open-web beams by Tide and Galambos (1970). Due to the advantages mentioned previously, this type of construction has been used in multistory buildings.

The beams or joists in the preceding discussion may be replaced by *girders* or heavier type of bridge *trusses* to achieve longer spans. Often, girders take the form of a reinforced or castellated beam or an inverted T-section. In either form, they are designed essentially as deep sections under the action of combined forces. The use of castellated beams has been demonstrated by Cassell et al. (1966), Giriyappa (1966), and Larnach and Park (1964). They have been applied both in buildings (e.g., 21-story office building referred to by Giriyappa) and in a multispan continuous bridge over the Mongaturanga River in New Zealand.

The use of *inverted T-beams* are discussed in the work of Toprac (1965), Toprac and Eyre (1967), and McDermott (1965, 1976). Generally, the top few inches of the beam are embedded in concrete along with some shear connectors. A notable example of this construction is a two-span continuous bridge in Kansas in which a considerable cost savings was demonstrated.

2. *Prestressed steel beam* in composite construction essentially works on the principle of inducing an initial prestress condition, which will later counteract that due to the service load. Prestressing can be achieved in three ways: prestressing components with high-strength tendons, preflex beams, and hybrid beams prestressed internally. The most comprehensive report was published, with an extensive list of references, by ASCE-AASHTO (1968).

Prestressing of steel beams is done with high-strength tendons or cables in two ways: (1) by placing them below the center of gravity of the beam and attaching them to the beam at its end, which results in constant prestress, or (2) by draping the tendons along the length of the beam. The similarity between these methods and those in the field of prestressed concrete is evident. Numerous combinations of the two methods have been used, with applications ranging from roofs and crane girders to bridges and other structures.

Knowles (1973) cites an interesting example of *prestressed tension* flange with concrete encasing the tensioning cables, thus protecting them from any environmental damage. Figure 1.25 indicates this schematically. The enclosure with steel jackets also minimizes the unde-

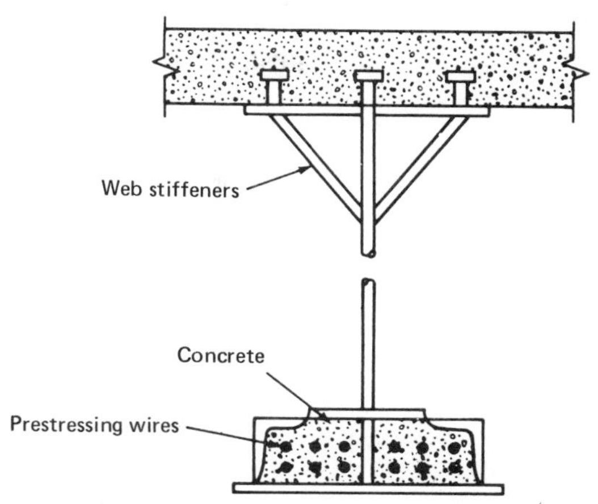

Fig. 1.25. Prestressed tension flange. (Adapted from Knowles 1973.)

sirable effects of creep and shrinkage in concrete. Knowles points out that the advantage of prestressed flange lies in the much greater stress range available in steel, i.e., from f to f, a total of 2.0 f compared to 1.0 f for the non-stressed steel, where f is the allowable stress in compression or tension in the material.

A variation of this is used in *preflex beam*, where the bottom flange of a steel section is encased in concrete subjected to permanent compressive stress. Steel I-section is cambered so that is achieves as much as the stresses of opposite nature as the design stresses. Shear connectors are also attached to this member, and the tension flange is embedded in concrete. Once the concrete hardens to the desired strength, the predeflection is released, which compresses concrete, resulting in a favorable stress condition in the earlier case. This technique makes it possible to use high-strength materials, thus enabling savings in weight and feasibility of long-span structures. Baes and Lipski (1951) developed this technique in Belgium; it was also used later in a modified form by Nicholas (1968).

Different methods of prestressing beam, as discussed by ASCE-AASHTO Committee (1968), are shown in Figs. 1.26 through 1.31. The following description is taken from the ASCE-AASHTO report:

> The system shown in Figure [1.26] has been used on many old-time bridges. It consists of a trussed beam with a steel, or wrought iron rod and turnbuckle. The rod is fastened

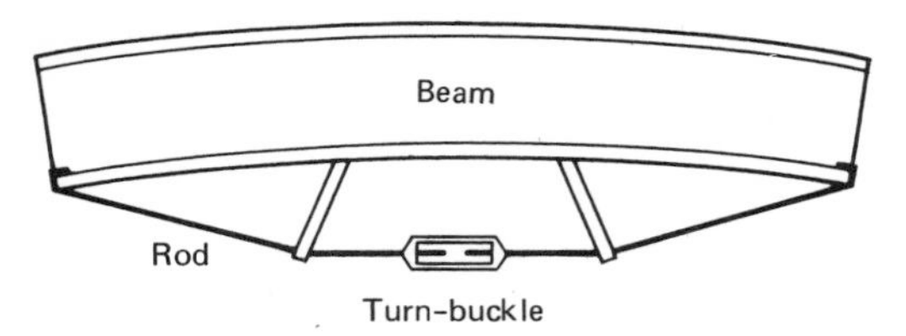

Fig. 1.26. Beam with tensioned rod. (*Courtesy ASCE 1968.*)

to each end of the beam, and is exerting a negative moment primarily by means of the inclined struts located near the third points. Figure [1.27] is a "streamlined" version of the previous example, and is considerably more efficient in that cables are contained within the outside dimensions of the beam. Figure [1.28] shows a sectional, and part-elevation view of a wide-flange beam which has been prestressed by two straight high-strength rods which are reacting against the end bearing block. This design was developed by the Iowa Highway Commission, recognizing that draping of prestressed tendons for steel members is not normally advantageous in simple beams.

Stressing components of hybrid beams is a second method of prestressing steel members. Two possible means of fabricating such beams have been used. One method, similar in principle to that used in pretensioned prestressed concrete by Emanuel and Hulsbos (1962) is shown in Figure [1.29]. Here a direct tensile force is applied to a high-strength plate, and while tension is maintained, the plate is welded to an unstressed T-section of structural steel. Upon completion of welding, the force in the plate is released, and the prestressing is completed. Figure [1.30] shows a second method, in which a structural steel I-beam is deflected in loose contact with one or more

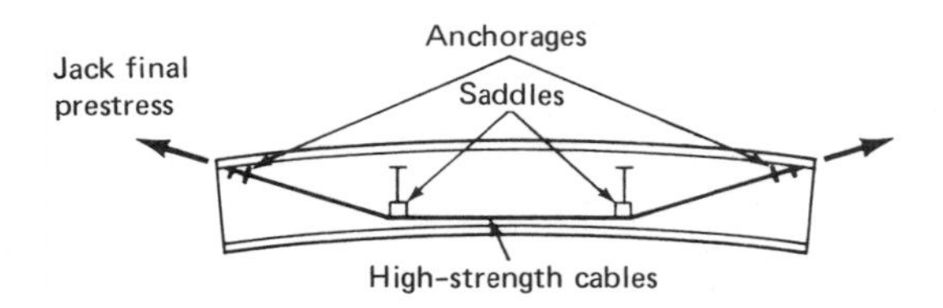

Fig. 1.27. Prestressed beam with draped cable. (*Courtesy ASCE 1968.*)

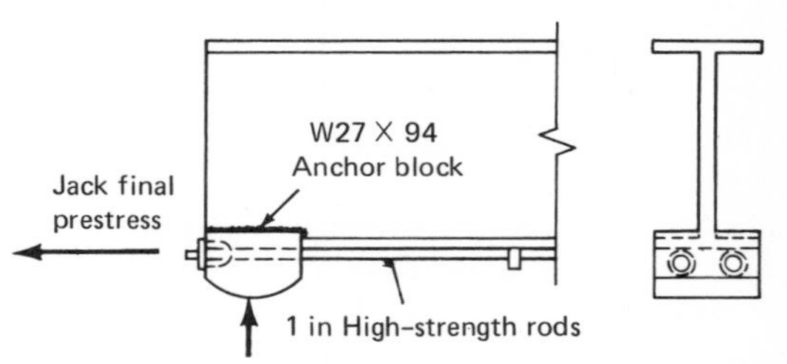

Fig. 1.28. Details of prestressed beam with straight rods. (*Courtesy ASCE 1968.*)

high-strength cover plates. While the system is maintained in a deflected position, the cover plates are welded to the flanges of the beam. Subsequent release of the jack load results in the desired prestress.

Figure [1.31] illustrates a method of prestressing, whereby a steel beam is first deflected, after which a concrete slab is cast against the beam. In the top view the jacking forces are applied, in a downward direction, to a steel beam which has been cambered upward. In the bottom view, a concrete slab is cast in composite fashion with the lower portion of the beam, and following curing, the jacking forces are removed. This action induces compressive forces in the concrete. This method has been patented, and the rights in the United States are presently held by the Preflex Corporation of America.

In the case of *prestressed steel hybrid* beams, the tension is applied to a high-strength cover plate to induce a desired prestress into the remaining portion of a beam made of medium-grade structural steel. The prestress is applied either directly or by welding the high-strength plate after stressing it or by deflecting the remaining beam, both causing the beam to prestress after release of external load. The main advantage of such prestressing is that it permits more efficient use of hybrid sections within the code limitation and specifications for elastic design of homogeneous members.

Fig. 1.29. Prestressing by direct tensioning of high-strength plate. (*Courtesy ASCE 1968.*)

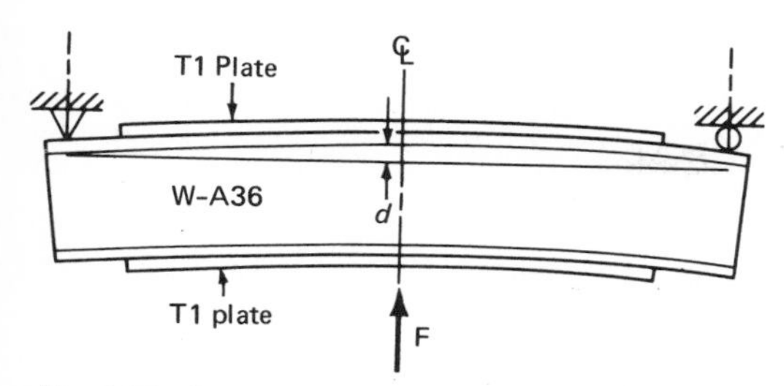

Fig. 1.30. Prestressing by deflecting a beam and attaching cover plates. *(Courtesy ASCE 1968.)*

Concrete-encased steel beams have been used for fire protection and for architectural reasons and to a lesser degree for other advantages. The structural advantage can be obtained only if proper and reliable bond is obtained between the two. Fire protection may also be obtained by other means. A better structural stability of webs can be accomplished by encasement, in which again the bond plays an important role. This is why considerable work was done in investigating bond and to achieve desired results. Viest (1960) has presented a review of the early development of the topic. AISC Specifications (1969) and other codes allowed the design of encased beams based on elastic analysis with some restrictions to obtain bond by mechanical shear connectors and provisions of reinforcement in the encasement. Recent work on encased beams is due to Hawkins (1970), Proctor (1967, 1969), Varghese et al. (1968), and by Naghshineh and Bannister (1967). Their work involved studying the many characteristics, including developing ultimate strength equations, use of lightweight concrete, and lateral torsional stability of webs.

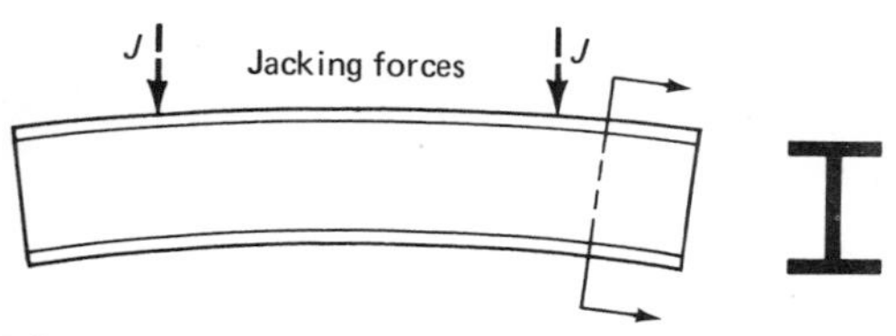

a. Step 1: Jacking forces applied to beam furnished mill with predetermined camber

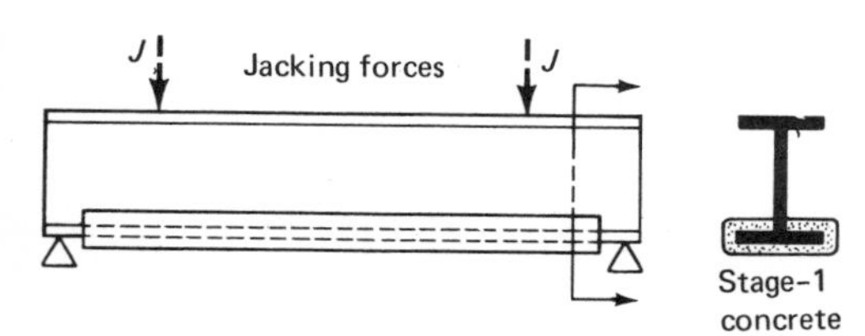

b. Step 2: Stage-1 concrete is placed while jacking forces are maintained

Fig. 1.31. Preflex techniques. *(Courtesy ASCE 1968.)*

Latticed structures can be combined with concrete or similar floor to form a composite and very economical solution. By definition, latticed structure is a system in the form of a network of elements (as opposed to a continuous surface). The elements may be rolled, extruded, or fabricated sections. Another important characteristic is that the system can be divided into plane frameworks much more easily than the continuous system, leading to a more economical solution. The latticed structures even in noncomposite form can achieve economy. The disadvantage of large deflections is reduced by the composite action between them and stiffening concrete (or some other continuous roofing material). An extensive bibliography has been published by ASCE Committee on Latticed Structures (1970).

Sandwich construction is a laminated construction comprised of a combination of series of different materials, which contribute to the total stiffness of the entire system and protect against fire and water penetration. The proper distribution of materials, in line with the developed stresses, make sandwich-constructed structures very lightweight; thus, sandwich construction is important in airborne structures. More development of theories, analysis, and various materials took place in the aerospace industry in which strength-to-weight criteria were most important. An early use of cedar plywood as facings and balsa wood core was attempted in 1937 in the DeHavilland Albatross airplane. Later, the honeycomb cores, both small and large, were developed because of their ease of transportation and fabrication. The key to the development of sandwich construction was the "adhesive," which would act as "shear connectors" and make "core" and "facings" work together as a composite. Vinyl phenolic adhesive was developed around 1944, giving sandwich construction a big boost, although fireproofing and water penetration posed problems, just as they did with epoxy resins later on. Later on, a number of test methods also were standardized by ASTM (1975). At the present time, many combinations of materials can be fabricated if the end product can be justified by practical and economic needs.

In the building industry, sandwich construction was developed later than in the aerospace industry. Literature regarding this late development has appeared in the IAHS Symposia held in Montreal (1971) and Atlanta (1976). The use of craft-paper honeycomb "core" combined suitably with "gypsum" or "stucco" concrete facing seems to be a feasible solution as demonstrated by Sabnis and Aroni (1971), who pointed out the use of these buildings in regions of low rains. The contribution of Sabnis and Aroni and several others in the symposia indicate that the structural sandwich panel could be a construction medium of the future for combining structural and environmental (thermal, sound, water, and fire) protection.

Hybrid beam is defined as a fabricated beam that has a stronger grade of steel in its flanges than in its web. This arrangement results in an economical beam, but the economy is somewhat sacrificed due to the additional cost of fabrication. The section of beam may also remain constant, even with the change in moment and shear in the span, if the higher strength steels are used for the more severely stressed portion of the beam. The use of either composite construction or hybrid beam by itself, or a combination of both, will undoubtedly result in a better alternative. A major summary of work on this topic is found in the report of ASCE-AASHTO Committee on flexural members (1968). Later on, composite action between concrete slab and hybrid beam was investigated by several researchers. According to Milek (1970), there is a slight reduction in flange strength of up to 7% for plate girders of average proportions when A-514 flanges (100 ksi) and A-36 (ksi) were used. Schilling (1968) deals with design equations for reduced allowable stress in flanges. Carskaddan's work (1968) on maximum slenderness ratios of unstiffened webs, work by Lew and Toprac (1968) on static and fatigue tests at the University of Texas, and Carskaddan's work (1969) on stiffened webs of hybrid girders have contributed considerably to the knowledge of the topic. They have also resulted in the adoption of some design specifications for highway bridges (AASHTO 1973) and for buildings (AISC 1970).

Composite concrete steel-plated slabs have been used in specialized applications–namely, storage tanks, pressure vessels, blast-resistant hatch covers, etc. The function of steel plate is twofold: first, to act as a linear to prevent any leakage due to cracking in concrete and, second, to act as a composite with slab to increase its effective strength. (ASCE 1970) reported work at the University of New Mexico on "dish"-type composite plates. Steel formwork (i.e., plate and the surrounding edge plate with shear connectors) served as a dish confining the concrete and increased rigidity as well as the load-carrying capacity. In their study, Casillas et al. (1967) tested a composite concrete slab and steel-plate configuration with shear connectors at the bottom. Their work showed a high degree of composite action even after bond failure between plate and slab. Plate membrane strength, if possible, should be used, which results in greater savings, due to a two-way action. The Tarcanville Suspension Bridge, an overpass over the railroad tracks near Paris, is one of the notable applications. An interesting case is that of a cold-formed deck-concrete floor using this technique, which saves formwork costs and increases the speed of construction.

1.13 CODE PROVISIONS FOR COMPOSITE CONSTRUCTION

Throughout this book, the main theme is to demonstrate economy in design structures using composite action between the two materials. Related to the theme, some features of the applicable codes will be pointed out here. The major codes used are:

1. AISC (structural steel in buildings)
2. ACI (reinforced and prestressed concrete in buildings)
3. AASHTO (highway bridges).

Many other countries, Great Britain and Germany in particular, also have their own code provisions for composite construction; however, specific reference herein will be made only to those just mentioned.

Although investigations of composite steel-concrete structures were carried out as early as the 1920s, the real impact on construction using such interaction did not occur until the mid-1940s, the main reason being the code provisions. The first formal code provisions for composite construction in the United States came about with the adoption of the AASHO Specifications in 1944. Even with its limited provisions, this type of construction became more and more accepted in the highway bridges. At about the same time, i.e., near the end of World War II, Germany developed its code of practice while achieving the most economical design using both steel and concrete for the reconstruction of bridges and buildings destroyed in the war. It is interesting to note that the AASHO code was much simpler in formulation and use than the German code; the complexity of the German code was a result of the inclusion of all possible manners to design the overall structure as economically as possible due to the shortage of materials in the post-war years.

Inclusion of composite design for buildings occurred in the early 1950s when AISC Specifications were issued. Due to the earlier developments in highway bridges, composite construction in buildings was initially considered only for heavy industrial loads. However, starting with the 1961 editions of AISC Specifications, more provisions were made using different loading conditions. In the case of concrete composite construction, the development took place following more and more use of precast and prestressed concretes. In combination with the cast-in-place concrete slab, they showed considerable advantages and economy, and the provisions were introduced in the ACI Code for the first time in 1963 following recommendations of the joint ACI-ASCE Committee 333 (1960). Later editions of ACI Codes consisted of more advantageous use of such construction with the development of lightweight concrete. It has become particularly important where long spans are required, whether in buildings or bridges, which cannot be achieved economically with noncomposite structure.

AASHTO (1974) provides general instructions for designing composite structures for bridges that involve different construction materials: Section 1.3.5 refers to wood-concrete, 1.6.14 to concrete-concrete, 1.7.96 through 1.7.101 to steel-concrete girder, and 1.7.102 through 1.7.109 to box-girder composite structures. In the section on wood-concrete composite, the distribution of loads among the components is given, along with elastic constants related to different modulii of elasticity, which are used in the design and which essentially follow the well-defined elastic methods. AASHTO Specifications clearly stress the "integration" of two concretes in a concrete-concrete composite, by means of full shear transfer using a properly designed reinforcement (i.e., vertical ties) for this purpose. Due to the fatigue type of loading, stresses for shear capacity are kept lower than in the ACI Code, which primarily relates to statically loaded building structures. The AASHTO Specifications also point out that differential shrinkage between two types of concretes should be properly considered to minimize cracking and also the deflection profile.

In the case of a composite structure between steel and concrete, AASHTO recommends strength based on fatigue tests; however, it limits the ultimate strength value of the shear transfer by the lesser of the capacities of either steel or concrete based on the test results. The shear connectors are provided uniformly, based on the research by Slutter and Driscoll (1965), which indicated no significant difference in the ultimate capacity of beam under uniform loading and different spacing of these connectors. Later tests by Lew (1970) also showed similar results. Lew pointed out that uniform spacing of shear connectors showed a definitely lower strength in the case of transfer girders with very small shear span ratio and recommend that the spacing may follow approximately the shear distribution in the span.

AASHTO also deals with the design of multi-box as well as continuous girders, which have been used often in recent years. It also specifies the restriction of 6 ft (1.8 m) or 60% of spacing of the girders, the remaining design criteria being the same as the composite beam system discussed previously. Vincent (1969) has discussed the background of load-factor design criteria, and PCA (1974) discusses bridge design applications of AASHTO 1973 Specifications incorporating load-factor design.

AISC (1969) discusses the design of steel-concrete composite structure in buildings. In addition to giving the design details of this construction, it also gives useful design quantities related to section properties with standard thickness slabs of 4 to 5 in. (100 to 125 mm) and suitable steel sections in the form of tables, which are helpful in the quick preliminary design of a composite section. Different values and conditions help the inexperienced engineer in making a reasonable guess and in reducing the number of design iterations before the final design of section. Type of construction—shored or unshored—should also be taken into account while using the aforementioned tables in AISC specifications. The ACI (1977) Code provisions are similar to the AASHTO Specifications discussed earlier. It allows the use of contact surfaces for shear transfer and specifies values of permissible shear stress in different surface conditions. The ACI Code, which is essentially the same as the AISC Specifications, recommends the use of *all* loads acting on the beam regardless of the type of construction in computing total horizontal shear. The ACI Code further provides the design information to calculate deflections and to maintain cracking control, similar to the rest of the reinforced concrete construction. Branson (1977), Wang and Salmon (1973), and many others outline detailed procedure for calculating both short- and long-term deflections using creep and shrinkage characteristics of composite structure.

1.14 FUTURE OF COMPOSITE CONSTRUCTION

The success of composite construction basically lies in the fact that two or more materials are used in the best possible manner to result in an economical structure. Various types of construction in concrete, steel, and timber in this chapter point out that, in addition to the principle of composite action, other effects, such as shrinkage, creep, and temperature, should also be considered in the use of composite construction.

Composite construction was generally accepted by the engineering profession in the 1950s in bridges in the United States and in many European countries; later on, it extended into the building industry. Considerable research has been carried out in many parts of the world and will be carried out for years to come. The many innovative systems described in Section 1.12 need more attention by engineers to build structures using them and to report more development in the literature to further enhance their use. Many systems were only reported here as "miscellaneous" because of their limited use in the past.

In many bridges, the beams are designed and constructed as simply supported; since the design is simpler, design rules are available through codes and other design aids. As in any other type of construction, continuity certainly will result in economy; the continued research in the field of composite construction will be helpful in obtaining increased strength through continuity and the increased stiffness due to composite action.

There is also an open field for composite structures in the earthquake-prone areas. Experience shows that many desirable properties are possessed by composite structures for earthquake resistance, and only recently interest has been shown in research related to this topic in California and some laboratories in Japan.

Very little information has been available on the fire resistance of composite structures. Proper precautions to prevent differential temperature effects and future research in this field will also be very desirable to enhance the use of composite construction.

The composite structure between precast concrete (and even prestressed concrete in certain applications) and steel sections will also have a greater future due to the potential economy that can be achieved by these components in terms of savings in time, labor, and money. Several aspects of such construction are shown in later chapters, and made available to the design and construction professions.

In recent years, the increased use of curve girders has also drawn engineers' attention to composite construction due to the added stiffness characteristics against torsion. Fatigue behavior, however, needs to be looked into further to ensure proper behavior of such girders; the behavior of shear connectors in fatigue loading needs further research in order to increase the use of curved girders.

The remainder of this book treats in detail the various types of composite construction briefly overviewed in this chapter, along with design examples.

REFERENCES

AASHTO (1973), *Standard Specifications for Highway Bridges*, Eleventh Edition, American Association of State Highway and Transportation Officials.

AASHTO (1974), *Interim Specifications 1974*, American Association of State Highway and Transportation Officials.

ACI 318 (1977), *Building Code Requirements for Reinforced Concrete*, American Concrete Institute, Detroit, Michigan.

ACI-ASCE 333 (1960), Tentative Recommendations for Design of Composite Beams and Girders for Buildings, *Proceedings of ASCE*, ST12, December.

Adekola, A. O. (1974), On the Influence Curves for Effective Widths in Non-Prismatic Composite Beams, *Proceedings, Institution of Civil Engineers*, Vol. 57, Part 2, London, England, March, pp. 51–66.

AISC (1969), *Specifications for Design, Fabrication and Erection of Structural Steel for Buildings*, American Institute of Steel Construction, New York.

Allen, D. H. (1969), *Analysis and Design of Structural Sandwich Panels*, Pergamon Press, New York.

Ansourian, O. and Roderick, J. W. (1976), Composite Connections to External Columns, *Journal of Structural Division*, Proc. ASCE, August, pp. 1609–1625.

ASCE (1970), Survey of Current Structural Research, *ASCE Manual* No. 51.

ASCE (1973), *Probabilistic Methods for Steel and Metal Structures*, ASCE Specialty Conference held in St. Louis, Missouri.

ASCE (1974), Report of the Committee on Composite in Steel-Concrete Construction, *Proceedings of ASCE*, ST5, May, pp. 1085–1139.

ASCE-IABSE (1972), *Current Research on Tall Buildings*, Report No. 10, Proceedings of Planning and Design of Tall Buildings Conference, Lehigh University, June.

ASCE-AASHO (1968), Development and Use of Prestressed Steel Flexural Members, Committee Report, *Proceedings of ASCE*, ST9, September, pp. 2033–2060.

ASCE (1970), *Bibliography on Latticed Structures*, reported by Subcommittee on Latticed Structures, *Proceedings of Specialty Conference on Steel Structures* at Rolla, Missouri, June, pp. 39–46.

ASTM (1975), *Determination of Strengths of Structural Sandwich Constructions*, Volume 25 (C364, C480, C393, C272).

Atlanta (1976), *Proceedings of IAHS Symposium on Low Cost Housing Problems*, Atlanta, Georgia, June.

Baes, L. and Lipski, A. (1951), *La Poutre Preflex*, Acta Technica Belgica, Revue 'C,' No. 4.

Berwanger, C. (1970), Thermal Stresses in Composite Bridges, *Proceedings of American Society of Civil Engineers' Specialty Conference*, University of Missouri, pp. 27–35.

Brandel, H. (1967), New Concepts in Structural Design, in *Structural Systems* by H. Engel, Frederick A. Prager, Inc., New York, pp. 1–14.

Branson, D. E. (1977), *Deformation of Concrete Structures*, McGraw-Hill International Book Co., Center for Advanced Publishing, Dusseldorf and New York.

Busemann, R. (1950), Calculation of Creep Effects in Loaded Composite Beams by Two Creep Fibres (in German), *Bauingenieur*, 25, II.

Cassel, A. C., Chapman, J. C., and Sparkes, S. R. (1966), Observed Behavior of a Building of Composite Steel and Concrete Construction, *Proceedings, The Institution of Civil Engineers*, Vol. 33, April, pp. 637–658.

Caughey, R. A. and Scott, W. B. (1929), A Practical Method for the Design of I-Beams Haunched in Concrete, *Structural Engineer*, August, p. 275.

Chapman, J. C. and Teraskiewicz, J. S. (1968), Research on Composite Construction at Imperial College, *Proc., Conference on Steel Bridges*, British Constructional Steelwork Association, London.

Carskaddan, P. S. (1968), Shear Buckling of Unstiffened Hybrid Beams, *Proceedings of ASCE*, Structural Division, October.

Carskaddan, P. S. (1969), Bending of Deep Girders with A514 Steel Flanges, *Proceedings of ASCE*, Structural Division, October.

Constrado (1973), The Berkeley Hambro Bishopsgate Tower, Project Study 2, Constructional Steel Research and Development Organization, London, Nov.

Coull, A. and Bose, B. (1975), Simplified Analysis of Frame-Tube Structures, *Proceedings of ASCE*, ST11, November, pp. 2223–2240.

Dai, P. K., Thiruvengadam, T. R., and Siess, C. P. (1970), Inelastic Analysis of Composite Beams, *Proceedings of American Society of Civil Engineers' Specialty Conference*, University of Missouri, pp. 9–20.

Danay, A., Gellert, M., and Gluck, J. (1975), *Continuum Method of Overall Stability of Tall Asymmetric Buildings, Proceedings of ASCE*, ST12, December, p. 2505.

Davies, C. (1975), *Steel-Concrete Composite Beams for Buildings*, Halsted Press Book, John Wiley & Sons, New York.

Dietz, A. G. H. (1971), *Engineering Composite Laminates*, M.I.T. Press, Cambridge, Massachusetts.

Dobruszkes, A., Janss, J., and Massonnet, C. (1969),

Experimental Researches on Steel Concrete Frame Connections, 1st Part: Frame Connections Comprising Concrete Columns and a Composite Beam. 2nd Part: Frame Connections Completely Encased in Concrete. Publication 29-11, International Association for Bridge and Structural Engineering, pp. 67–100.

Dziewolski, R. (1972), Prestressed Composite Structures and Space Structures in Construction of Tall Buildings, *Proceedings of Planning and Design of Tall Buildings Conference*, Lehigh University, Vol. Ia, p. 617.

Emanuel, J. H. and Hulsbos, C. L. (1962), *Flexural Behavior of a Prestressed Welded Steel Beam*, Engineering Report No. 34, Iowa Engineering Experimental Station, Iowa.

Fan, H. M. and Heins, C. P. (1976), "Effective Width of Composite Bridges at Ultimate Load," *Journal of Structural Division*, Proc. ASCE, October.

Fintel, M. (1974), *Handbook of Reinforced Concrete Engineering*, Van Nostrand Reinhold Co., New York.

FIP-CACA (1970), *International Recommendations for the Design and Construction of Concrete Structures*, by the Comite Europeen du Beton, Federation Internationale de la Precontrainte, Cement and Concrete Association, London, England, pp. 39–40.

Fisher, J. W. (1970), Design of Composite Beams with Formed Metal Deck, *Eng. J., American Institute of Steel Construction*, Vol. 7, pp. 88–96, July.

Fraser, D. J. (1971), "The Effective Width of Simply Supported T and L Beams for the Calculation of Deflection," UNICIV Report No. R-66, Department of Civil Engineering, University of New South Wales, Kensington, Australia, June.

Furlong, R. W. (1968), Design of Steel-Encased Concrete Beam-Columns, *Proceedings of ASCE*, Structural Division, January.

Galambos, T. V., et al. (1970), Experiments on Composite Open-Web Steel Joists, *Research Report No. 13*, Structural Division, Civil and Environmental Engineering Department, Washington University, St. Louis, Mo., January.

Gardener, N. J. and Jacobson, E. R. (1967), Structural Behavior of Concrete Filled Steel Tubes, *Journal, American Concrete Institute*, Vol. 64, No. 7, July, pp. 404–413 (and Journal Supplement No. 2, Title No. 64-38).

Gillespie, P., MacKay, H. M., and Leluau, C. (1923), Report on the Strength of I-Beams Haunched in Concrete, *Engineering Journal (Montreal)*, p. 365.

Giriyappa, J. (1966), Behavior of Composite Castellated Hybrid Beams, thesis submitted to the University of Missouri, in partial fulfillment of the requirements for the degree of Master of Science.

Harstock, J. A. (1969), *Design of Foam-Filled Structures*, Technomic Publishing Co., Stamford, Connecticut.

Harstock, J. A. and Chong, K. P. (1976), Analysis of Sandwich Panels with Formed Faces, *Proceedings of ASCE*, ST4, April, p. 803.

Hasse, G. (1969), Effect of Creep on Statically Indeterminate Composite Beams and Frames of Variable Cross Section (in German), *Verein Deutscher Ingenieure, Berichte*, Part 4, No. 16, November, p. 111.

Hawkins, N. W. (1970), Strength of Concrete-Encased Steel Beams, *Proceedings of the American Society of Civil Engineers' Specialty Conference*, University of Missouri, Rolla, pp. 21–26.

Hisatoku, T. and Nishikawa, F. (1972), Mixed and Composite Concrete and Steel Systems, *Proceedings of Planning and Design of Tall Buildings Conference*, Lehigh University, Bethlehem, Penn., Vol. Ia, p. 501.

Idorn, G. M. and Fordos, Z. (1974), *Cement-Polymer Materials: Bibliography*, Technical Note, Cembureau, 2 Rue Saint Charles, Paris 15e, France, 28 pp.; also published in the 6th International Symposium on Chemistry of Cement, Moscow, September 1974.

Iwamoto, K. (1962), *On the Continuous Composite Girder*, Highway Res. Board Bulletin, 339. (Bridge deck design and loading studies p. 81.) National Academy of Sciences National Research Council, Washington, D.C.

Janss, J. (1972), *Research on Composite Structures Carried Out by the CRIF at the University of Liege*, (in French) Preliminary Report, International Association for Bridge and Structural Engineering, Ninth Congress, Amsterdam, pp. 125–132.

Johnson, R. P. (1970), Research on Steel-Concrete Composite Beams, *Proceedings of ASCE*, ST3, March, p. 445.

Johnson, R. P., Finlinson, J. C. H., and Heyman, J. (1966), A Plastic Composite Design, *Proc. Institute of Civil Engineers*, Vol. 32, pp. 198–209.

Johnson, R. P. and Cross, K. E. (1973), Design of Low-Cost Composite Structures for Car Parks, *Proc. Conference Multi-story and Underground Car Parks*, Institution of Structural Engineers, 81–87, May.

Johnson, R. P. (1975), *Composite Structures of Steel and Concrete*, Vol. 1–Beams, Columns, Frames and Applications in Buildings, Constrado Monographs, Halsted Press, John Wiley & Sons, New York.

Jones, R. M. (1972), *Mechanics of Composite Materials*, McGraw-Hill Publishing Co., Inc., New York.

Khan, F. R. (1974), Tubular Structures for Tall Buildings, in *Handbook of Concrete Engineering*, (Ed., M. Fintel), Van Nostrand Reinhold Co., New York.

Khan, F. R. and Amin, N. R. (1973), *Analysis and Design of Framed Tube Structures for Tall Concrete Buildings*, Sp-36, ACI Special Publication, pp. 39–49.

Knowles, P. R. (1973), *Composite Steel and Concrete Construction*, Halstead Press Book, John Wiley & Sons.

Krawinkler, H., Bertero, V. V., and Popov, B. (1975), Shear Behavior of Steel Frame Joints, *Proceedings of ASCE*, ST12, December.

Kuenzi, E. W. (1960), *Structural Sandwich Design Criteria*, National Academy of Sciences, N.R.C. Publication 798, Washington, D.C., pp. 9–18.

Larnach, W. J. and Park, R. (1964), The Behavior Under Load of Six Castellated Composite T-Beams, *Civil Engineering and Public Works Review*, March, pp. 339–343.

Lee, K. L., Adams, B. D., and Vagneron, J. J. (1975), *Reinforced Earth Retaining Walls*, *Proceedings of ASCE*, SM10, October, p. 745.

Lembeck, H. G., Jr. (1965), Composite Design of Open Web Steel Joists, thesis submitted to Washington University, St. Louis, Mo., in partial fullfillment of the requirements for the degree of Master of Science.

Lew, H. S. and Toprac, A. A. (1968), *The Static Strength of Hybrid Plate Girders*, Structural Research Laboratory Report No. P550-11, University of Texas, Austin, Texas, January.

Lew, H. S. (1970), *Effect of Shear Connector Spacing on the Ultimate Strength of Concrete-on-Steel Composite Beams*, National Bureau of Standards Report No. 10 246, United States Department of Commerce, August.

Mahin, S. A. and Bertero, V. V. (1976), Non-linear Seismic Response of Coupled Wall Systems, *Proceedings of ASCE*, ST9, September.

Mainz, B. and Wolff, H. J. (1972), Stress Analysis for Statically Indeterminate Composite Beams, (in German), *Der Stahlbau*, Vol. 41, No. 2, February, pp. 45–48.

McDermott, J. F. (1965), Tests Evaluating Punching Shear Resistance of Prefabricated Composite Bridge Units Made with Inverted Steel T-Beams, *Highway Research Record No. 103*, Highway Research Board, pp. 41–52.

McDermott, J. F. (1967), Structural Tests of a Composite Floor System, *Journal of the Structural Division*, ASCE, February, pp. 255–274.

McDevitt, J. and Viest, I. M. (1972), Interaction of Different Materials, *Proceedings of the Ninth Congress of IABSE*, Amsterdam, pp. 55–79.

McGarraugh, J. B. and Baldwin, J. W. (1971), Lightweight Concrete-on-Steel Composite Beams, *AISC Engineering Journal*, American Institute of Steel Construction, Vol. 8, No. 3, July, pp. 90–98.

Milek, W. A. (1970), Homogenous and Hybrid Girder Design in the 1969 AISC Specifications, *AISC Engineering Journal*, January.

Montreal (1971), *Proceedings of IAHS Conference on Panelized Construction in Housing Applications*, (Ed. Paul Fazio), Sir George Williams University, Montreal.

Nair, R. S. (1975), Overall Elastic Stability of Multistory Buildings, *Proceedings of ASCE*, ST12, December, pp. 2487–2503.

Naka, T., Wakabayashi, M., and Murata, J. (1972), *Steel Reinforced Concrete Construction*, Preliminary Report, International Association for Bridge and Structural Engineering, Ninth Congress, Amsterdam, May, pp. 165–172.

Newmark, N. M., Siess, C. P., and Viest, I. M. (1951), Tests and Analyses of Composite Beams with Incomplete Interaction, *Proc. Soc. for Experimental Stress Analysis.*

Nicholas, R. J. (1968), Development of the Preflexion of Bridges for Bridgeworks, *Proceedings of the Conference on Steel Bridges*, British Constructional Steelwork Association, June, pp. 123–132.

Okamoto (1973), *Earthquake Resistant Design of Concrete Structures*, University of Tokyo Press, Tokyo, and John Wiley & Sons, Halsted Press, New York.

PCA (1974), *Notes on Load Factor Design for Reinforced Concrete Bridge Structures with Design Applications*, Portland Cement Association, Skokie, Illinois.

Porter, M. L. and Ekberg, C. E. (1976), Design Recommendations for Steel Deck Floor Slabs, *Journal of the Structural Division*, ASCE, November, pp. 2121–2136.

Porter, M. L., Ekberg, C. E., Jr., Greimann, L. F., and Elleby, H. A. (1976), Shear-Bond Analysis of Steel-Deck-Reinforced Slabs, *Journal of the Structural Division*, Proc. ASCE, December, pp. 2255–2268.

Proctor, A. N. (1967), Full Size Tests Facilitate Derivation of Reliable Design Methods, *Consulting Engineer*, August.

Proctor, A. N. (1969), Composite Construction, *Consulting Engineer*, Vol. 33, No. 2, February, pp. 48.

Rao, J. K. S. (1971), *Composite Structural Systems in Buildings and Bridges: An Annotated Bibliography*, Report, NBS-CSSB Project, Department of Civil Engineering, Indian Institute of Technology, Kanpur, U.P., India, 1971. (A Report to the Structures Section, Center for Building Technology, National Bureau of Standards, Washington, D.C. of NBS PL-480 International Research Grant G-91).

Rao, J. K. S., ed. (1975), *Status Report on Housing and Construction Technology*, Section 3.b on Prefabrication and Industrialization of Buildings, National Committee on Science and Technology, Department of Science and Technology, New Delhi, June.

CTICM Recommendations (1965), Design of Composite Steel-Concrete Floors, (in French), *Construction Metallique*, Vol. 3.

Roderick, J. W. (1972), *Further Studies of Composite Steel and Concrete Structures*, Preliminary Report, International Association for Bridge and Structural Engineering, Ninth Congress, Amsterdam, May, pp. 165–172.

Roret, J. A. (1972), High Rise Construction in France, *Proceedings of the Conference on Planning and Design of Tall Buildings*, pp. 741–752.

Roll, F. (1971), Effects of Differential Shrinkage and Creep on a Composite Steel-Concrete Structure, *Designing for Effects of Creep, Shrinkage, Temperature in Concrete Structures*, SP-27, American Concrete Institute, Detroit.

Sabnis, G. M. and Aroni, S. (1971), Sandwich Construction in Housing, *Proceedings of Conference on Panelized Construction* (ed. Paul Fazio), Sir George Williams University, Montreal.

Saillard, Y. (1971), Determination of the Effective Width of the Compression Flange, *Reinforced Con-*

crete: An International Manual, by the Committee of Experts Commissioned by United Nations Economic, Social, and Cultural Organization, Butterworths, London, England, pp. 261-267.

Schuster, R. M. (1976), Composite Steel-Deck Concrete Floor Systems, *Journal of the Structural Division*, Proc. ASCE, May, pp. 899-917.

Schilling, C. G. (1968), Bending Behavior of Composite Hybrid Beams, *Proc. ASCE*, Vol. 94, No. ST8, August, pp. 1945-1964.

Scott, W. B. (1925), The Strength of Steel Joists Embedded in Concrete, *Structural Engineer (London)*, Vol. 26.

Shen, C. K., Romstad, K. M., and Herrman, L. R. (1976), Integrated Study of Reinforced Earth-II, *Proceedings ASCE*, SM6, June.

Siess, C. P. (1949), Composite Construction for I-Beam Bridges, *Transactions ASCE*, Vol. 114, pp. 1023-1045.

Simon, L. U. (1972), Tall Buildings in Australia—Construction Trends, *Proceedings of the Conference on Planning and Design of Tall Buildings*, p. 793.

Slutter, R. G. and Driscoll, G. (1965), Flexural Strength of Steel-Concrete Composite Beams, *Proceedings of ASCE*, Structural Division, April.

Slutter, R. G. (1974), Composite Steel-Concrete Members, Chapter 13 in *Structural Steel Design* (ed., L. Tall), Ronald Press Company, New York.

Smith, G. G. K. (1972), Construction Practices in the U.K., *Proceedings of the Conference on Planning and Design of Tall Buildings*, p. 816.

Soor, A. K. (1973), Composite Construction Brings Economy to Design of Large Canteen Building, *Indian Concrete Journal*, Bombay, India.

Stevens, R. F. (1965), Encased Stanchions, *The Structural Engineer*, Vol. 43, February, pp. 59-66.

Tall, L., editor (1972), *Structural Steel Design*, The Ronald Press Company, New York.

Taylor, E. E. F. (1972), Steel Construction, *Proceedings of the Conference on Planning and Design of Tall Buildings*, p. 767.

Tide, R. H. R. and Galambos, T. V. (1970), Composite Open-Web Joists, *AISC Engineering Journal*, American Institute of Steel Construction, Vol. 7, No. 1, January, pp. 27-36.

Toprac, A. A. and Eyre, D. G. (1967), Composite Beams with a Hybrid Tee Steel Section, *Proc. ASCE*, Vol. 93, No. ST5, October, pp. 309-322.

Toprac, A. A. (1965), Strength of Three New Types of Composite Beams, *Bulletin No. 2*, American Iron and Steel Institute, 29 p.

Trost, H. (1968), Design of Composite Steel Girders on the Basis of Recent Investigations of the Viscoelastic Behavior of Concrete, (in German), *Stahlbau*, Vol. 37, No. II, November.

Van Dalen, K. and Narasimhan, S. V. (1976), Shear Lag in Shallow Wide-Flanged Box Girders, *Journal of Structural Division*, Proc. ASCE, Paper 12449, October.

Varghese, P. C., Radhakrishnan, R., and Paramasivan, V. (1968), Flexural Strength of Encased Beams, *Indian Concrete Journal*, Vol. 42, No. 1, January, pp. 21-22, 27-28.

Vasiliev, A. P., et al. (1972), Prefabricated Reinforced Concrete Multistory Frame Buildings in the U.S.S.R., *Proceedings of the Conference on Planning and Design of Tall Buildings*, p. 489.

Verma, N. (1974), Prefabricated Brick Panels for Low Cost Rural Housing, *Proceedings of the Third IAHS Symposium on Low Cost Housing Problems* (ed. Paul Fazio), Concordia University, Montreal.

Viest, I. M. (1960), Review of Research on Composite Steel Concrete Beams, *Proceedings of ASCE*, ST6, June, p. 1-20.

Vincent, G. S. (1969), *Tentative Criteria for Load Factor Design of Steel Highway Bridges*, Steel Research for Variability of Strength, Construction Bulletin No. 15, American Iron and Steel Institute, March.

Von Karman, T. V. (1923), *Festschrift August Foppls*, p. 114. (See also, Timoshenko, S. and Goodier, J. N., *Theory of Elasticity*, 2nd Edition, McGraw-Hill, 1951, pp. 171-177.)

Vorlicek, M. and Tichy, M. (1969), *Statistical Aspects of Strength of Concrete Structures*, (in English), Czechoslavakian Academy of Sciences, Prague.

Wakabayashi, M., Naka, T., and Kato, B. (1972), *Elasto-Plastic Behavior of Encased Structures*, ASCE-IABSE International Conference on Planning and Design of Tall Buildings, August, Conference Reprints, Addendum, Vol. A, pp. 233-252.

Wang, P. C. and Kaley, D. J. (1967), Composite Action of Concrete Slab and Open Web Joist, *AISC Engineering Journal*, American Institute of Steel Construction, Vol. 4, No. 1, January, pp. 10-16.

Wang, C. K. and Salmon, C. G. (1973), *Design of Concrete Structures*, Intext Publishing Co., New York (Ch. 14).

Wilenko, L. K. (1969), The Application to Large Buildings of Structural Steelwork Prestressed during Erection, *Proc. Conference on Steel in Architecture*, British Constructional Steelwork Association, London.

Yam, L. C. P. and Chapman, J. C. (1968), Inelastic Behavior of Simply Supported Composite Beams of Steel and Concrete, *Proceedings, The Institution of Civil Engineers*, Vol. 41, December, pp. 651-683.

Zuk, W. (1965), Thermal Behavior of Composite Bridges, Insulated and Uninsulated, *Highway Research Record No. 76*, Highway Research Board, pp. 231-253.

2

Composite steel–concrete construction*

CHARLES G. SALMON, Ph.D., P.E.
Professor of Civil Engineering
University of Wisconsin
Madison, Wisconsin

JAMES M. FISHER, Ph.D., P.E.
Principal
Computerized Structural Design, Inc.
Milwaukee, Wisconsin

2.1 GENERAL

One of the most common applications of composite action between two materials occurring in structural engineering is that between steel beams and an overlaying concrete slab. In this chapter, various types of steel beams are considered acting compositely with a concrete slab. Actually, composite action between steel and concrete may be obtained by encasing the steel completely in concrete, by embedding only a small part such as the flange of a beam, or, as is commonly done, embedding only steel connectors to provide a shear transfer between the steel and concrete.

Steel beams encased in concrete were widely used from the early 1900s until the development of lightweight materials for fire protection in the past 25 years. Some such beams were designed compositely, and some were not. In the early 1930s, bridge construction began to use composite sections. Not until the early 1960s, however, was it economical to use composite construction for buildings. However, current practice† utilizes composite action in nearly all situations where concrete and steel are in contact, both on bridges and buildings.

An early study of composite structural action between steel and concrete was that of MacKay et al. (1923) involving steel beams encased in concrete. It was noted that such encased beams supporting a monolithically cast reinforced concrete slab had good interaction between the beam and the slab. In the early studies of encased beams, bond was depended upon to provide the interaction between steel and concrete. Composite beams, including encased beams as well as slabs on I-shaped beams, exhibited sufficient reserve strength so that Caughey (1929) recommended that they be designed on the basis of a homogeneous section wherein the concrete area is transformed into an equivalent area of steel.

Since stresses in a wide slab supported on a steel beam are not uniform across the slab width, the ordinary flexure formula, $f = Mc/I$, does not apply. Similarly to T-sections entirely of reinforced concrete, an equivalent width of the wide slab is used such that the flexure for-

*This chapter is an extension and adaptation of Chapter 16, Charles G. Salmon and John E. Johnson, *Steel Structures–Design and Behavior*, Harper & Row, 1971.

†Defined by AISC (1978), AASHTO (1973), interims 1974–1977, and ACI (1977). Throughout this chapter, these documents will be used as references.

mula may be applied to obtain the correct moment capacity. Theoretical studies to determine the correct effective width have been carried out by Von Karman (1924) and Reissner (1934), and an excellent summary of the subject has been presented by Brendel (1964).

Viest (1960), in his review of research, notes that the important factor in composite action is that the bond between concrete and steel remains unbroken. As designers began to place slabs on top of supporting steel beams, investigators began to study the behavior of mechanical shear connectors. The shear connectors provided the interaction necessary for the composite action between slab and steel I-shaped beams that had previously been supplied by bond for the fully encased beams. Studies of mechanical shear connectors began in the 1930s with the work of Voellmy reported by Ros (1934). Detailed information concerning the various types of shear connectors and their structural action is considered in the present volume in Chapters 1, 3, and 8. In what follows here, the general resulting behavior of shear connectors will be used in the form of specification-prescribed capacities for example problems.

A state-of-the-art survey by an ASCE committee (Viest 1974) provides an excellent summary of research relating to steel-concrete composite construction and contains an excellent bibliography.

2.2 COMPOSITE ACTION

Composite action is developed when two load-carrying structural members such as a concrete floor system and the supporting steel beams are integrally connected and deflect as a single unit. Typical examples of composite cross sections are shown in Fig. 2.1. The extent to which composite action is developed depends on the provisions made to insure a single linear strain from the top of the concrete slab to the bottom of the steel section.

In developing the concept of composite behavior, consider first the noncomposite beam of Fig. 2.2a, wherein if friction between the slab and beam is neglected, the beam and slab each carry separately a part of the load. This is further shown in Fig. 2.3a. When the slab deforms under vertical load, its lower suface is in tension and elongates, while the upper surface of the beam is in compression and shortens. Thus, a discontinuity will occur at the plane of contact. Since friction is neglected, only vertical internal forces act between the slab and beam.

When a system acts compositely (Fig. 2.2b and c), no relative slippage occurs between the slab and beam. Horizontal forces (shears) are developed, which, when acting on the lower surface of the slab compress and shorten it, while those acting on the upper surface of the beam elongate it.

By an examination of the strain distribution that occurs when there is no interaction between the concrete slab and the steel beam (Fig. 2.3a), it is seen that the total resisting moment is equal to

$$M = M_{\text{slab}} + M_{\text{beam}} \tag{2.1}$$

It is noted that for this case there are two neutral axes: one at the center of gravity of the slab and the other at the center of gravity of the beam. The horizontal slippage resulting from the bottom of the slab in tension and the top of the beam in compression is also indicated.

Consider next the case (Fig. 2.3b) where only partial interaction is present. The neutral axis of the slab is closer to the beam and that of the beam closer to the slab. Due to the partial interaction, the horizontal slippage has now decreased. The result of the partial interaction is the partial development of the compressive and tensile forces C' and T', the maximum capacities of the concrete slab and steel beam, respectively. The resisting moment of the section is now increased by the amount of $T'e'$ or $C'e'$.

When complete interaction between the slab and the beam is developed, no slippage occurs; the resulting strain diagram is shown in Fig. 2.3c. Under this condition, a single neutral axis occurs that lies below that of the slab and above that of the beam. In addition, the compressive and tensile forces C'' and T'' are larger than the C' and T' existing with partial interaction. The resisting moment of the fully developed section then becomes

$$M = T''e'' \text{ or } C''e'' \tag{2.2}$$

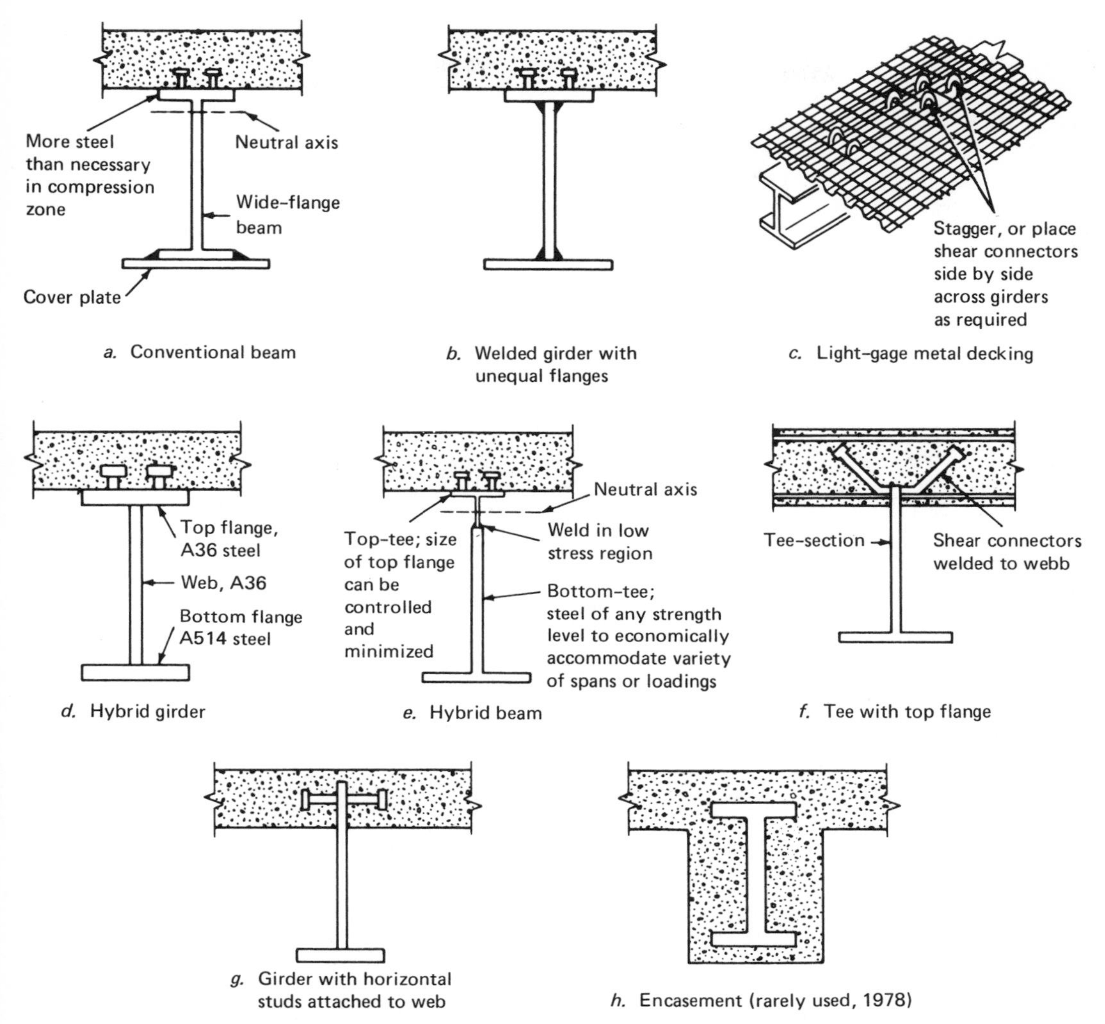

Fig. 2.1. Various types of composite steel concrete sections.

2.3 EFFECTIVE WIDTH

In order to compute in a practical way the section properties of a composite section, it is necessary to utilize the concept of effective width. Referring to Fig. 2.4, consider the composite section under stress in which the slab is infinitely wide. The intensity of the extreme fiber compressive stress σ_x, which is maximum over the steel beam, decreases nonlinearly as the distance from the supporting beam increases.

The effective width of a flange for a composite member can be taken as

$$b_E = b_f + 2b' \tag{2.3}$$

where $2b'$ times the maximum stress, $\sigma_{x\,\max}$ is equal to the area under the curves for σ_x. Various investigators, including Timoshenko and Goodier (1951) and von Kármán (1924), have derived expressions for the effective width of homogeneous beams with wide flanges; Johnson and Lewis (1966) have shown the expressions to be valid also for beams in which the flange and stem are of different materials.

The analysis for effective width involves theory of elasticity applied to plates, using an infinitely long continuous beam on equidistant supports, with an infinitely wide flange having a small thickness compared to the beam depth.

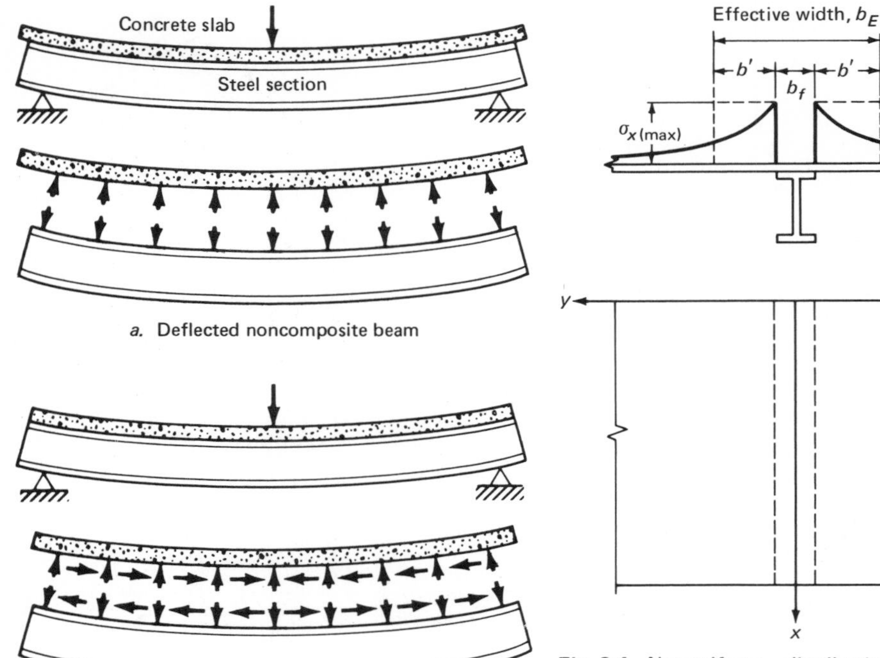

Fig. 2.2. Comparison of deflected beams with and without composite action.

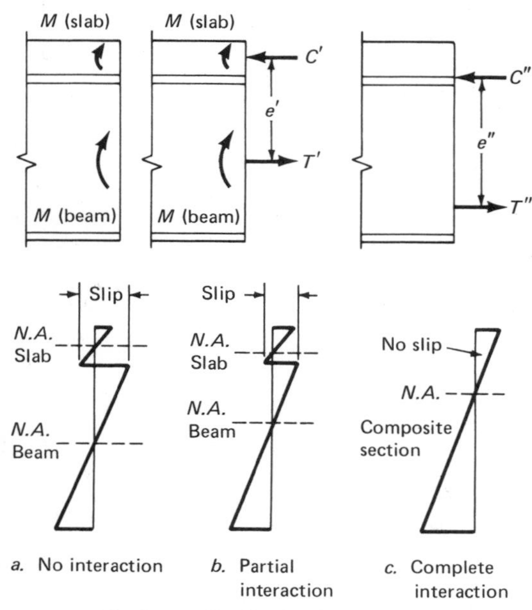

Fig. 2.3. Strain distribution in composite beams.

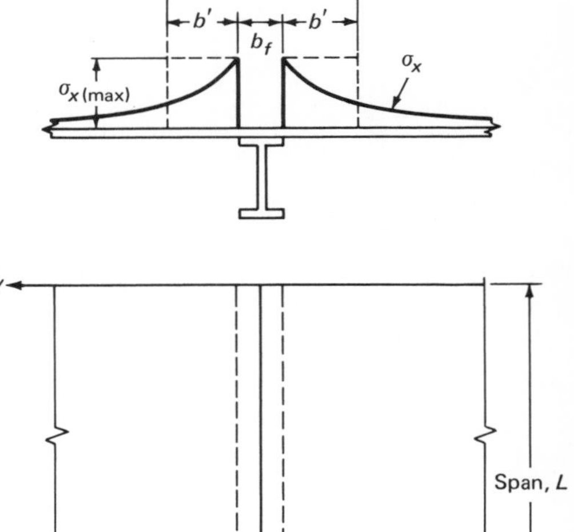

Fig. 2.4. Nonuniform distribution of compressive stress σ_x, and effective width b_E.

The total compression carried by the equivalent system is the same as that carried by the actual system. The value of b' depends on the span length and type of loading. According to Johnson and Lewis (1966), for loading that produces bending moment having a half-sine wave shape, the effective width is

$$b_E = b_f + \frac{2L}{\mu(3 + 2\mu - \mu^2)} \tag{2.4}$$

where

L = span length of beam
b_f = steel-beam flange width
μ = Poisson's ratio for the slab.

Assuming $\mu = 0.2$ for concrete,

$$b_E = b_f + 0.196L$$

As a simplification for design purposes, the American Institute of Steel Construction (AISC 1978) and the American Association of State Highway and Transportation Officials (AASHTO 1973) have adopted the same method of computing effective flange widths as used by

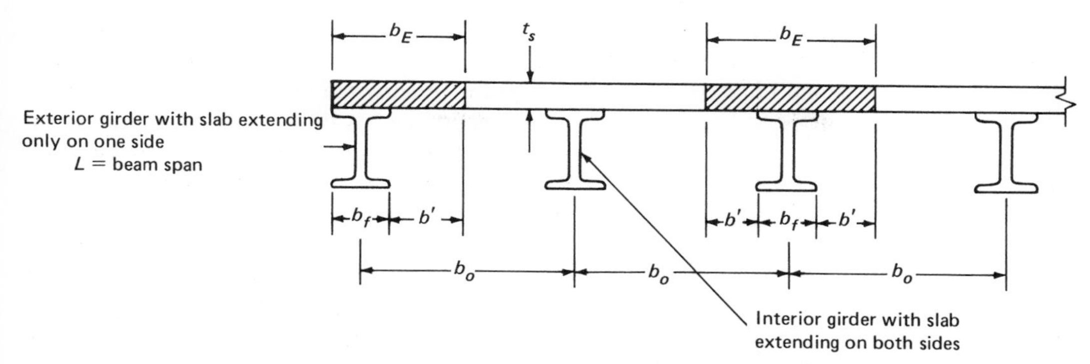

Fig. 2.5. Dimensions governing effective width, b_E, on composite steel-concrete beams.

the American Concrete Institute Building Code (ACI 1977) for reinforced concrete beams. Referring to Fig. 2.5, the maximum value of the effective width b_E permitted by the AISC-1.11.1 is the least value computed by the following relations:

1. For an interior girder with slab extending on both sides of girder:

   a. $b_E \leqslant L/4$ (2.5a)

   b. $b_E \leqslant b_o$ (for equal beam spacing) (2.5b)

   c. $b_E \leqslant b_f + 16t_s$ (2.5c)

2. For an exterior girder with slab extending only on one side:

   a. $b_E \leqslant L/12 + b_f$ (2.6a)

   b. $b_E \leqslant \frac{1}{2}(b_o + b_f)$ (2.6b)

   c. $b_E \leqslant b_f + 6t_s$ (2.6c)

Similarly, for highway bridge design, the effective width according to AASHTO-1.7.98 is identical with that given by AISC-1.11.1 except that Eq. 2.5c for an interior girder is replaced by

c. $b_E \leqslant 12t_s$ (2.7)

and for an exterior girder, Eqs. 2.6a and c are replaced by:

a. $b_E \leqslant L/12$ (2.8a)

c. $b_E \leqslant 6t_s$ (2.8b)

2.4 COMPUTATION OF SECTION PROPERTIES

The section properties of a composite section can be computed by the transformed area method. In contrast to reinforced concrete design, where the steel area is transformed into an equivalent concrete area, the concrete is transformed into equivalent steel. As a result, the concrete area is reduced by using a slab width equal to b_E/n, where n is the ratio of steel modulus of elasticity, E_s, to concrete modulus of elasticity, E_c.

2.4.1 Modular Ratio *n*

The modulus of elasticity of concrete in psi is generally taken as

$$E_c = w^{1.5} 33\sqrt{f_c'} \tag{2.9}$$

where w is weight of concrete in pcf, and f_c' is taken with the units of psi. For normal-weight concrete, weighing approximately 145 pcf, the value may be taken according to ACI (1977) as

$$E_c = 57{,}000\sqrt{f_c'} \tag{2.10}$$

Values of modulus of elasticity for various concrete strengths appear in Table 2.1. The minimum value for n permitted by the ACI Code and the AASHTO Specification is 6. For practical design purposes, the value of n from Table 2.2 is suggested, although the ACI Code recommends that modular ratio be taken as the nearest whole number.

Table 2.1. Values of modulus of elasticity for concrete (using $E_c = 33w^{1.5}\sqrt{f'_c}$ for normal-weight concrete[a] weighing 145 pcf).

f'_c (psi)	E_c (psi)	f'_c (N/mm^2)	E_c (N/mm^2)
3000	3,150,000	21[a]	21,700
3500	3,400,000	24	23,200
4000	3,640,000	28	25,000
4500	3,860,000	31	26,300
5000	4,070,000	35	28,000

[a]For normal-weight concrete, $E_c = 4730\sqrt{f'_c}$ for f'_c in N/mm^2.
[b]These SI values are rounded values approximating strengths in customary United States units.

Table 2.2. Practical design values for modular ratio n.

Concrete Strength (f'_c psi)	Modular Ratio ($n = E_s/E_c$)	f'_c (N/mm^2)
3000	9	21
3500	$8\frac{1}{2}$	24
4000	8	28
4500	$7\frac{1}{2}$	31
5000	7	35
6000	$6\frac{1}{2}$	42

2.4.2 Effective Section Modulus

A complete beam may be considered as a steel member to which a cover plate has been added on the top flange. This "cover plate," being concrete, is considered to be effective only when the top flange is in compression. In continuous beams, the concrete slab is usually ignored in regions of negative moment. If the neutral axis falls within the concrete slab, present practice is to consider only that portion of the concrete slab that is in compression. AISC-1.11.2.2 permits reinforcement parallel to the steel beam and lying within the effective slab width to be included in computing properties of composite sections. These reinforcing bars usually make little difference to the composite section modulus and are frequently neglected.

EXAMPLE 2.1

Compute the section properties of the composite section shown in Fig. 2.6 according to AISC Specification, assuming $f'_c = 3000$ psi and $n = 9$.

Solution

First, determine effective width:

$$b_E = (0.25)(\text{span length}) = 0.25(30)12 = 90 \text{ in.}$$

$$b_E = b_o = 8(12) = 96 \text{ in.}$$

$$b_E = b_f + 2(8)t_s = 8.24 + 64$$

$$= \underline{72.24} \text{ in. (controls)}$$

The width of equivalent steel is $b_E/n = 8.03$ in. The computation of centroid and moment of inertia is shown in Table 2.3.

Table 2.3

Element	Transformed Area (in.^{2})	Moment Arm from Centroid (y)	Ay	Ay^2	I_o
Slab	32.1	+12.495	+401	5012	43
W21 × 62	18.3	0	0	0	1330
Cover plate	7.0	10.995	−77	846	
Total	57.4		+324	5858	1373
$Ay^2 + I_o = I_x$					7231 in.^{4}

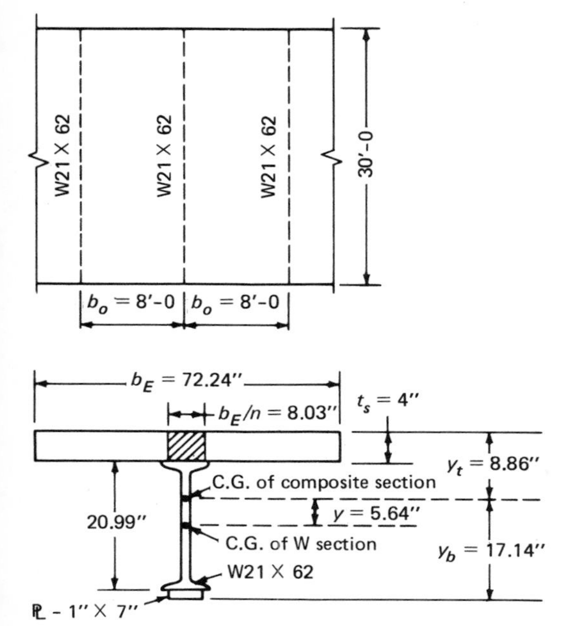

Fig. 2.6. Composite section for Ex. 2.1.

$$y = \frac{+324}{57.4} = +5.64 \text{ in.}$$

$$I = I_x - Ay^2 = 7231 - 57.4(5.64)^2$$

$$= 5402 \text{ in.}^4 (22.5 \times 10^6 \text{cm}^4)$$

$$y_t = 10.50 - 5.64 + 4.0 = 8.86 \text{ in.}$$

$$y_b = 10.50 + 5.64 + 1.0 = 17.14 \text{ in.}$$

The section modulus with respect to the top fiber, S_t, is

$$S_t = I/y_t = 5402/8.86 = 610 \text{ in.}^3 \ (10{,}000 \text{ cm}^3)$$

and the section modulus with respect to the bottom fiber, S_b, is

$$S_b = I/y_b = 5402/17.14 = 315 \text{ in.}^3 (5160 \text{ cm}^3)$$

2.5 SERVICE-LOAD STRESSES WITH AND WITHOUT SHORING

The actual stresses that result due to a given loading on a composite member are dependent upon the manner of construction.

The simplest construction occurs when the steel beams are placed first and used to support the concrete slab formwork. In this case, the steel beam acting noncompositely (i.e., by itself) supports the weight of the forms, the wet concrete, and its own weight. Once forms are removed and concrete has cured, the section will act compositely to resist all dead and live loads placed after the curing of concrete. Such construction is said to be without temporary shoring (i.e., unshored).

Alternatively, to reduce the service load stresses, the steel beam may be supported on temporary shoring, in which case the steel beam, forms, and wet concrete are carried by the shores. After curing of the concrete, the shores are removed, and the section acts compositely to resist all loads. This system is called *shored* construction.

The following example illustrates the difference in service-load stresses under the two systems of construction.

EXAMPLE 2.2

For the steel W21 X 62 with the 1 X 7 plate of Fig. 2.6, determine the service-load stresses considering that (1) construction is without temporary shoring, and (2) construction uses temporary shores. The dead- and live-load moment to be superimposed on the system after the concrete has cured is 560-ft-kips (760 kN-m).

Solution

The composite section properties as computed in Example 2.1 are

$$S_{top} = 610 \text{ in.}^3 (\text{top of concrete})$$

$$(10{,}000 \text{ cm}^3)$$

$$S_{bottom} = S_{tr} (\text{AISC-1969})$$

$$= 315 \text{ in.}^3 (\text{bottom of steel}) (5160 \text{ cm}^3)$$

From *AISC Manual** tables, "Composite Design", the noncomposite properties may be obtained or computed as follows. For the steel section alone, see Fig. 2.7.

**Manual of Steel Construction*, 7th Edition, American Institute of Steel Construction, New York, 1970.

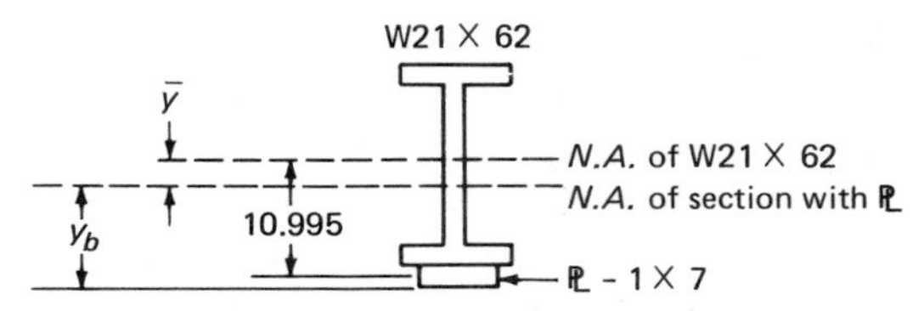

Fig. 2.7. Steel section for Ex. 2.2.

$$\bar{y} = \frac{7.0(10.995)}{7.0 + 18.3} = 3.04 \text{ in.}$$

$$y_b = 10.495 - 3.04 + 1.00 = 8.45 \text{ in.}$$

$$I = I_o(\text{W21} \times 62) + A_p y^2 - A\bar{y}^2$$

$$= 1330 + 7.0(10.995)^2 - 25.3(3.04)^2$$

$$= 1330 + 846 - 234$$

$$= 1940 \text{ in.}^4 (52{,}000 \text{ cm}^4)$$

$$S_{st} = \frac{1940}{13.55} = 143 \text{ in.}^3 (\text{top})(2340 \text{ cm}^3)$$

$$S_{sb} = \frac{1940}{8.45} = 230 \text{ in.}^3 (\text{bottom})(3740 \text{ cm}^3)$$

1. *Without temporary shores.* Weight due to the concrete slab and steel beam,

$$w(\text{concrete slab}) = \tfrac{4}{12}(8)(0.15) = 0.40$$

$$w(\text{steel beam}) = \underline{0.06}$$

$$0.46 \text{ kip/ft}$$

M(DL on noncomposite)

$$= \tfrac{1}{8}(0.46)(30)^2$$

$$= 51.8 \text{ ft-kips } (70.3 \text{ kN-m})$$

$$f_{\text{top}} = \frac{M_{DL}}{S_{st}(\text{steel section})} = \frac{51.8(12)}{143}$$

$$= 4.3 \text{ ksi } (30 \text{ N/mm}^2)$$

$$f_{\text{bottom}} = \frac{M_{DL}}{S_{sb}(\text{steel section})} = \frac{51.8(12)}{230}$$

$$= 2.7 \text{ ksi } (19 \text{ N/mm}^2)$$

The additional stresses after the concrete has cured are

$$f_{\text{top}} = \frac{M_{LL}}{S_{\text{top}}(\text{concrete})} = \frac{560(12)}{610(9)}$$

$$= 1.22 \text{ ksi (concrete stress)}(8.4 \text{ N/mm}^2)$$

where the stress in the concrete is $1/n$ times the stress on equivalent steel (transformed section).

$$f_{\text{bottom}} = \frac{M_{LL}}{S_{tr}} = \frac{560(12)}{315}$$

$$= 21.3 \text{ ksi } (147 \text{ N/mm}^2)$$

The total maximum tensile stress in the steel is

$$f = f(\text{noncomposite}) + f(\text{composite})$$

$$= 2.7 + 21.3 = 24.0 \text{ ksi } (165 \text{ N/mm}^2)$$

2. *With temporary shores.* Under this condition, all loads are resisted by the composite section.

$$f_{\text{top}} = \frac{M_{DL} + M_{LL}}{S_{\text{top}}(\text{composite})} = \frac{(560 + 51.8)12}{610(9)}$$

$$= 1.34 \text{ ksi on concrete } (9.2 \text{ N/mm}^2)$$

$$f_{\text{bottom}} = \frac{M_{DL} + M_{LL}}{S_{tr}} = \frac{(560 + 51.8)(12)}{315}$$

$$= 23.3 \text{ ksi } (161 \text{ N/mm}^2)$$

Stress distributions for both with and without shores are given in Fig. 2.8. Since the dead load

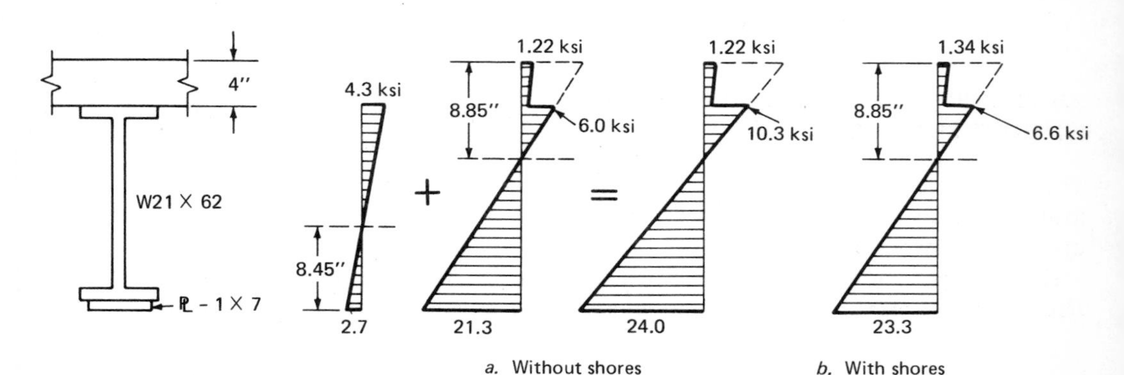

Fig. 2.8. Service-load stresses for Ex. 2.2.

was small in this example, the use of shores gave insignificant reduction in service-load stress. Where thicker slabs are used, the dead-load stresses carried by the noncomposite section may become as high as 30%, in which case using or not using shores will make a significant difference in the *service-load stress.* In the next section, it will be shown that the *strength* of the section is not affected by the service-load stress level.

2.6 ULTIMATE STRENGTH OF FULLY COMPOSITE SECTIONS

The ultimate strength in terms of an ultimate depends on the yield strength and section properties of the steel beam, the concrete slab strength, and the interaction capacity of the shear connectors joining the slab to the beam.

The provisions of AISC-1.11 (1978) are nearly all based on ultimate-strength behavior, even though all relationships are adjusted to be in the service-load range. These ultimate-strength concepts were applied to design practice as recommended by the ASCE-ACI Joint Committee on Composite Construction (1960) and further modified as a result of research by Slutter and Driscoll (1965).

The ultimate strength in terms of an ultimate moment capacity gives a clearer understanding of composite behavior as well as providing a more accurate measure of the true factor of safety. The true factor of safety is the ratio of the ultimate moment capacity to the actually applied moment. In both cases, whether the *slab* is termed "adequate" or "inadequate" compared to the tensile yield capacity of the beam, the *connection* between the slab and beam is considered adequate in the following development. Complete shear transfer at the steel-concrete interface is assumed.

In determining the ultimate moment capacity, the concrete is assumed to take only compressive stress. Although concrete is able to sustain a limited amount of tensile stress, the tensile stress at the strains occurring during the development of the ultimate moment capacity is negligible.

The procedure for determining the ultimate moment capacity depends on whether the neutral axis occurs within the concrete slab or within the steel beam. If the neutral axis occurs within the slab, the slab is said to be adequate—i.e., the slab is capable of resisting the total compressive force. If the neutral axis falls within the steel beam, the slab is considered inadequate—i.e., the slab is able to resist only a portion of the compressive force, the remainder being taken by the steel beam. Figure 2.9 shows the stress distribution for these two cases.

Case 1: Slab Adequate

Referring to Fig. 2.9b and assuming the Whitney rectangular stress block* (uniform stress of $0.85f_c'$ acting over a depth a), the ultimate compressive force C is

*For the development of the concept of replacing the true distribution of compressive stress in concrete by a rectangular stress distribution, see Chu-Kia Wang and Charles G. Salmon, *Reinforced Concrete Design*, 3rd Edition, Harper & Row, New York, 1978, Chapter 3, or any other appropriate text on concrete design.

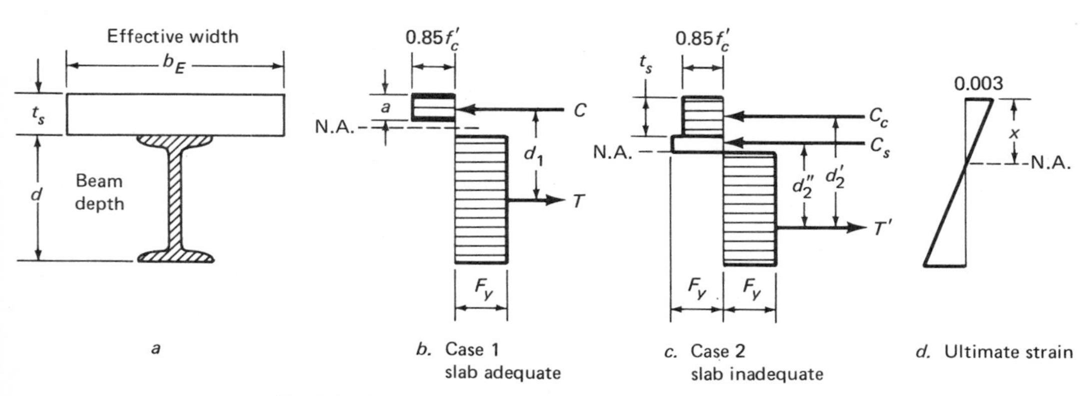

Fig. 2.9. Stress distributions at ultimate-moment capacity.

$$C = 0.85 f_c' a b_E \tag{2.11}$$

The ultimate tensile force T is the yield strength of the beam times its area:

$$T = A_s F_y \tag{2.12}$$

Equating the ultimate compressive force C to the ultimate tensile force T gives

$$a = \frac{A_s F_y}{0.85 f_c' b_E} \tag{2.13}$$

According to the ACI-accepted rectangular stress block approach, the neutral axis distance x, as shown in Fig. 2.9, equals a/β_1, where $\beta_1 = 0.85$ for $f_c' = 4000$ psi. The ultimate moment capacity M_u becomes

$$M_u = C d_1 \quad \text{or} \quad T d_1 \tag{2.14}$$

Since the slab is assumed adequate, it is capable of developing a compressive force equal to the full yield capacity of the steel beam. Expressing the ultimate moment in terms of the steel force gives

$$M_u = A_s F_y \left(\frac{d}{2} + t_s - \frac{a}{2} \right) \tag{2.15}$$

The usual procedure is to determine the depth of the stress block a by Eq. 2.13 and, if a is less than the slab thickness t_s, to determine the ultimate moment capacity by Eq. 2.15.

Case 2: Slab Inadequate

If the depth a of the stress block as determined in Eq. 2.13 exceeds the slab thickness, the stress distribution will be shown in Fig. 2.9c. The ultimate compressive force C_c in the slab is

$$C_c = 0.85 f_c' b_E t_s \tag{2.16}$$

The compressive force in the steel beam resulting from the portion of the beam above the neutral axis is shown in Fig. 2.9c as C_s.

The ultimate tensile force T' which is now less than $A_s F_y$ must equal the sum of the compressive forces:

$$T' = C_c + C_s \tag{2.17}$$

Also,

$$T' = A_s F_y - C_s \tag{2.18}$$

Equating Eqs. 2.17 and 2.18, C_s becomes

$$C_s = \frac{A_s F_y - C_c}{2}$$

or

$$C_s = \frac{A_s F_y - 0.85 f_c' b_E t_s}{2} \tag{2.19}$$

Considering the compressive forces C_c and C_s, the ultimate moment capacity M_u for Case 2 is

$$M_u = C_c d_2' + C_s d_2'' \tag{2.20}$$

the moment arms d_2' and d_2'' are shown in Fig. 2.9c.

Whenever the Case 2 situation occurs, the steel beam is assumed to accommodate plastic strain in both tension and compression at ultimate strength. Certainly, it is implied that such a steel section satisfy the requirements of "compact" sections; that is, it should have proportions that insure its ability to develop its plastic moment capacity. Little research has been performed on Case 2 situations because they rarely occur in practice.

EXAMPLE 2.3

Determine the ultimate moment capacity of the composite section shown in Fig. 2.10. Assume A35 steel, $f_c' = 3000$ psi, and $n = 9$.

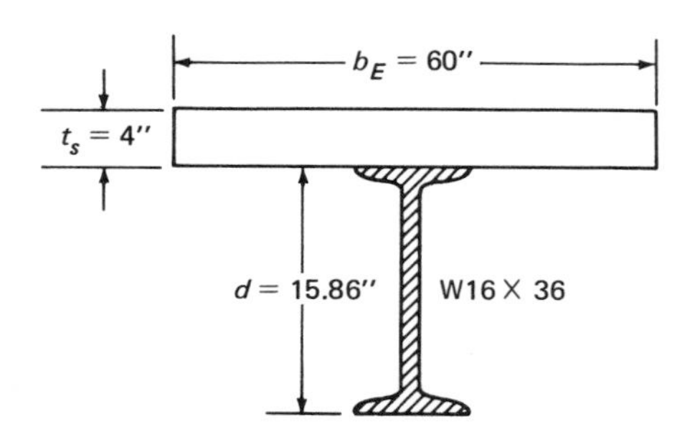

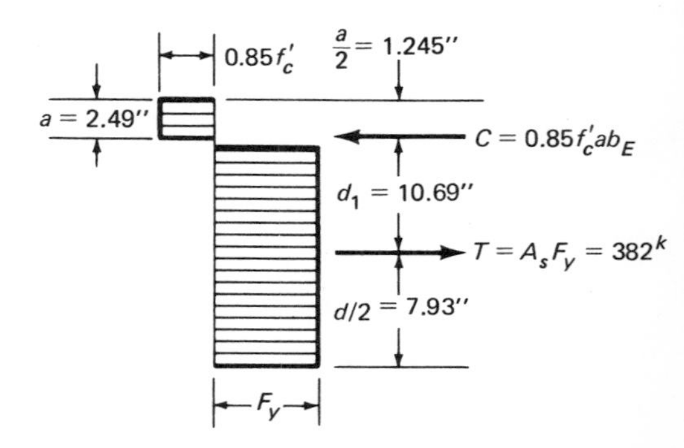

Fig. 2.10. Example 2.3.

Referring to Fig. 2.10, determine that the concrete slab is fully adequate, i.e., Case 1.

$$a = \frac{A_s F_y}{0.85 f_c' b_E} = \frac{10.6(36)}{0.85(3)60} = 2.49 \text{ in.} < t_s \text{ OK}$$

$$C = 0.85 f_c' a b_E = 0.85(3)(2.49)(60)$$

$$= 382 \text{ kips } (1700 \text{ kN})$$

$$T = A_s F_y = 10.6(36) = 382 \text{ kips (checks)}$$

$$\text{Arm } d_1 = \frac{d}{2} + t - \frac{a}{2} = 7.93 + 4.0 - 1.245$$

$$= 10.69 \text{ in. } (271 \text{ mm})$$

Ultimate composite moment capacity,

$$M_u = Cd_1 = Td_1 = 382(10.69)/12$$

$$= 340 \text{ ft-kips } (461 \text{ kN-m})$$

EXAMPLE 2.4

Determine the ultimate moment capacity of the composite section shown in Fig. 2.11. Assume A36 steel, $f_c' = 3000$ psi, and $n = 9$.

Solution

Referring to Fig. 2.11, determine whether or not the slab is adequate. Assuming the slab is adequate to balance tensile capacity of the steel section (i.e., Case 1),

$$a = \frac{A_s F_y}{0.85 f_c' b_E} = \frac{47.0(36)}{0.85(3)(72)}$$

$$= 9.22 \text{ in.} > t_s = 7 \text{ in. NG}$$

Since the concrete slab is only 7 in. thick, the slab is inadequate to carry a compressive force equal to the tensile force that can be developed by the W36 X 160. Thus, Case 2 applies. Using Eq. 2.16,

$$C_c = 0.85 f_c' b_E t_s = 0.85(3)72(7)$$

$$= 1285 \text{ kips } (5720 \text{ kN})$$

Using Eq. 2.19,

$$C_s = \frac{A_s F_y - 0.85 f_c' b_E t_s}{2} = \frac{47.0(36) - 1285}{2}$$

$$= 204 \text{ kips } (905 \text{ kN})$$

Assuming that only the flange of the W36 X 160 ($b_f = 12.00$ in.) is in compression, the portion of the flange d_f to the neutral axis is

$$d_f = \frac{204}{36(12.00)} = 0.472 \text{ in.}$$

The location of the centroid of the tension portion of the steel beam from the bottom is

$$\bar{y} = \frac{47.0(18) - 0.472(12)35.77}{47.0 - 0.472(12)}$$

$$= 15.57 \text{ in. } (39.5 \text{ mm})$$

Referring to Fig. 2.11, the ultimate composite moment capacity from Eq. 2.20 is

$$M_u = C_c d_2' + C_s d_2''$$

$$= [1285(23.94) + 204(20.21)]/12$$

$$= 2910 \text{ ft-kips } (3950 \text{ kN-m})$$

In the development of the ultimate strength of composite sections, it has been implicitly assumed that sufficient interaction between the concrete slab and the steel beam existed. Such interaction is usually assured by providing a sufficient number of "shear connectors." This matter is treated in the next section.

The foregoing development and examples have emphasized the computation of ultimate moment capacity. Tests have verified that such capacities are achieved. Furthermore, whether

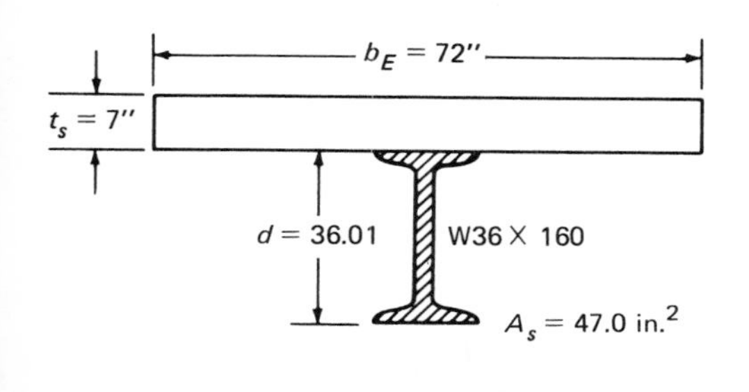

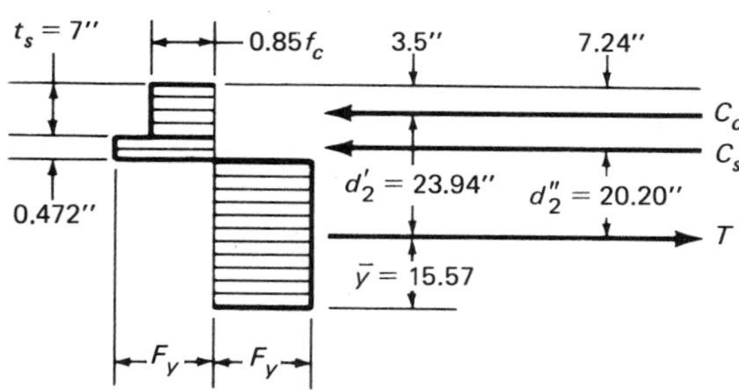

Fig. 2.11. Example 2.4.

the steel beam is *shored* or *unshored* during construction, the ultimate strength is identical. Even though service-load stresses, as discussed in Section 2.5, are lower when temporary shoring is used to support the beam during construction than when such shoring is omitted, the same ultimate moment capacity is achieved. Current AISC design practice, as discussed in Sections 2.7 and 2.8, uses this ultimate-strength concept as the basis for its working stress method.

2.7 SHEAR CONNECTORS

The horizontal shear that develops between the concrete slab and the steel beam during loading must be resisted so that the composite section acts monolithically. Although the bond developed between the slab and the steel beam may be significantly high, it cannot be depended upon to provide the required interaction. Neither can the frictional force developed between the slab and the steel beam.

Instead, mechanical shear connectors attached to the top of the beam must be provided. Typical shear connectors are shown in Fig. 2.12.

Ideally, the shear connectors should be stiff enough to provide the complete interaction shown in Fig. 2.3c. This, however, would require that the stiffeners be infinitely rigid.

Headed stud
Hooked stud
h d_s
a. Stud connectors

W t h
flexible channel
b. Channel connectors

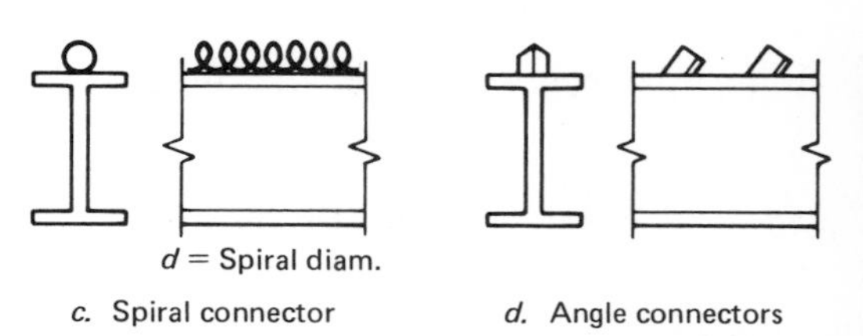

Fig. 2.12. Typical shear connectors.

Also, by referring to the shear diagram of a uniformly loaded beam as shown in Fig. 2.13, it would be inferred, theoretically at least, that more shear connectors would be required near the ends of the span than at the midspan. Consider the shear-stress distribution of Fig. 2.13b wherein the stress v_1 must be developed by the connection between the slab and beam.

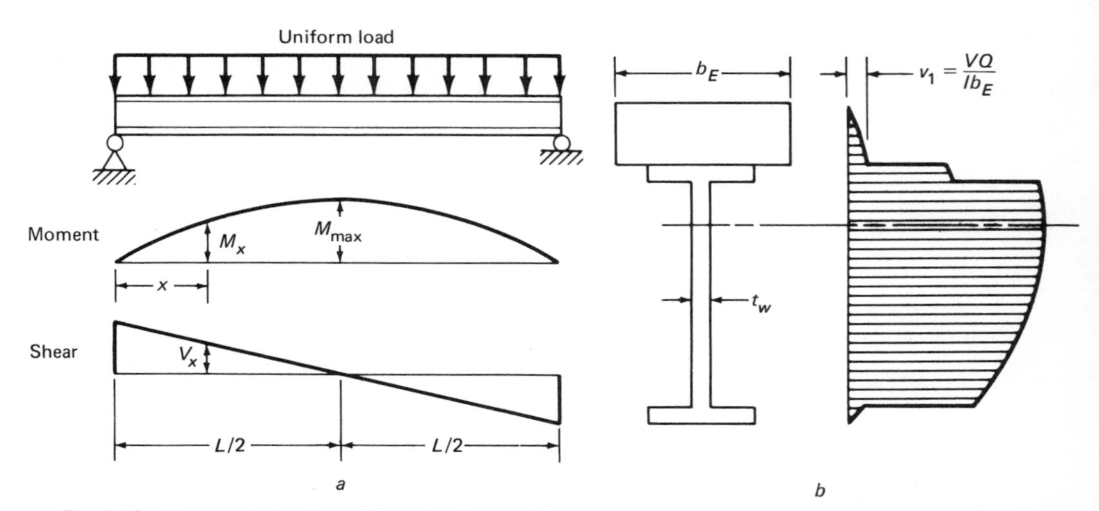

Fig. 2.13. Shear variation for uniform loading and distribution across a steel-concrete composite section.

Under service load, the stress on the beam of Fig. 2.13 varies from zero at midspan to a maximum at the support. Next, examine the equilibrium of an elemental slice of the beam, as in Fig. 2.14. The shear force per unit distance along the span is $dC/dx = v_1 b_E = VQ/I$. Thus, if a given connector has an allowable capacity of q kips, the maximum spacing p to provide the required capacity is

$$p = \frac{q}{VQ/I} \tag{2.21}$$

Until recent years, composite design has used Eq. 2.21 to space connectors. AASHTO-1.7.100(A) (1973) requires using Eq. 2.21 to design for fatigue; then a check is required for ultimate strength.

If one uses an ultimate-strength concept, the shear connectors under ultimate bending moment share equally in carrying the total maximum compressive force developed in the concrete slab. This would mean, referring to Fig. 2.12a, that shear connection is required to transfer the compressive force developed in the slab at midspan to the steel beam in the distance $L/2$, since no compressive force can exist in the slab at the end of the span where zero moment exists. The ultimate compressive force to be resisted could not exceed that which the concrete can carry:

$$C_{\max} = 0.85 f'_c b_E t_s \tag{2.22}$$

or, if the ultimate tensile force below the bottom of the slab is less than $C_{\max}$,

$$T_{\max} = A_s F_y \tag{2.23}$$

Thus, if a given connector has an ultimate capacity q_{ult}, the total number of connectors N required between the points of maximum and zero bending moment is

$$N = \frac{C_{\max}}{q_{\text{ult}}} \text{ or } \frac{T_{\max}}{q_{\text{ult}}} \tag{2.24}$$

whichever is smaller. Under the ultimate-strength approach, the total required number of shear connectors are distributed over the region of the beam between maximum and zero bending moment.

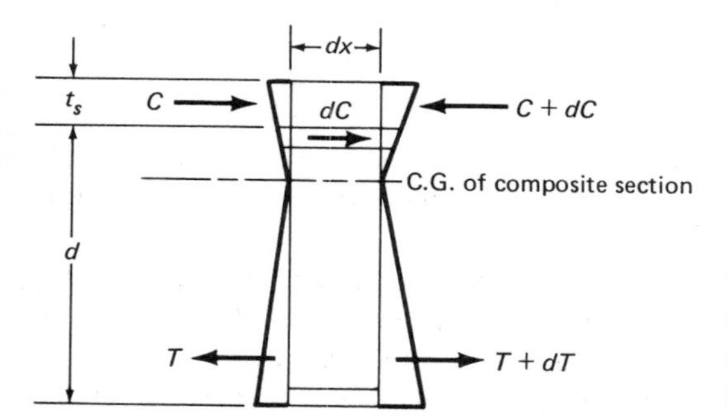

Fig. 2.14. Force required of shear connectors at service loads.

The determination of the connector capacity analytically is complex, since the shear connector deforms under load and the concrete that surrounds it is also a deformable material. Moreover, the amount of deformation a shear connector undergoes is dependent upon factors such as its own shape and size, its location along the beam, the location of the maximum moment, and the manner in which it is attached to the top flange of the steel beam. In addition, any particular shear connector may yield sufficiently to cause a slipping between the beam and the slab, in which case the adjacent shear connectors pick up the additional shear.

As a result of the extremely complex behavior of shear connectors, their capacities are not based solely on a theoretical analysis. In order to develop a rational approach, a number of research programs discussed by Viest (1974), Lew (1969), and Sabnis and Reddy (1977) have been undertaken to develop the strength of the various types of shear connectors.

Investigators determined that shear connectors will not fail if the average load per connector is kept below that load which causes 0.003-in. (0.076-mm) residual slip between concrete and steel. The slippage is also a function of the strength of the concrete that surrounds the shear connector and will be different for lightweight aggregate concrete than for normal-weight stone aggregate concrete. Relating connector capacity to a specified slip may be realistic for bridge design where fatigue strength is important, but it is very conservative with respect to failure loads. So-called ultimate capacities used prior to 1965 were based on slip limitation, giving values about one-third of the ultimate strengths obtained when actual failure of a connector is the criterion. Recently, Sabnis and Dabholkar (1977) investigated the low-cycle fatigue strength of shear studs, the phenomenon commonly experienced by structures subjected to earthquake-type loading.

When ultimate flexural capacity of the composite section is the basis for design, the connectors must be adequate to satisfy the equilibrium of the concrete slab between the points of maximum and zero moment, as discussed in the development of Eqs. 2.22 through 2.24. Slip is not a criterion for this equilibrium requirement. As stated by Slutter and Driscoll (1965), "the magnitude of slip will not reduce the ultimate moment provided that (1) the equilibrium condition is satisfied, and (2) the magnitude of slip is no greater than the lowest value of slip at which an individual connector might fail." More recent studies by Ollgaard et al. (1971) at Lehigh University and by McGarraugh and Baldwin (1971) at the University of Missouri included the effect of lightweight concrete on stud connector capacity.

Two currently accepted expressions by AASHTO Specifications (1973) for ultimate connector capacity are as follows:

1. Hooked or headed shear stud connectors welded to flange (Fig. 2.13a). The 1974 Interim AASHTO Specifications (AASHTO 1973) give essentially the same expression developed at Lehigh by Ollgaard et al. (1971):

$$q_{\text{ult}} = 0.4 d_s^2 \sqrt{f'_c E_c} \text{ for } H/d_s \geqslant 4 \tag{2.25}$$

where

d_s = stud diameter (in.)
q_{ult} = connector capacity for one stud (lb)
f'_c = 28-day compressive strength of concrete (psi)
E_c = modulus of elasticity of concrete (psi); $E_c = 33w^{1.5}\sqrt{f'_c}$ where w is the unit weight of concrete; for normal-weight concrete of 145 pcf, $E_c = 57{,}600\sqrt{f'_c}$.

In metric units, Eq. 2.25 becomes

$$q_{\text{ult}} = 0.0004 d_s^2 \sqrt{f'_c E_c} \tag{2.25a}$$

with d_s, mm; f'_c and E_c, N/mm^2; and q, kN. For this,

$$E_c = w^{1.5}(0.0426)\sqrt{f'_c}$$

with w, kg/m^3.

2. Channel connectors (Fig. 2.13b). The AASHTO Specifications give

$$q_{\text{ult}} = 550(h + 0.5t)W\sqrt{f'_c} \tag{2.26}$$

where

h = average thickness of channel flange (in.)
t = thickness of channel web (in.)
W = length of channel shear connector (in.).

In SI units, Eq. 2.26 becomes

$$q_{\text{ult}} = 0.588(h + 0.5t)W\sqrt{f'_c} \tag{2.26a}$$

with h, t, and W expressed in mm, f'_c in N/mm^2, and q in kN.

2.7.1 Connector Design—Ultimate Strength Concept

It may be noted that the connection and the beam must resist the same ultimate load. However, under service loads the beam resists dead and live loads, but unless shores are used, the connectors resist essentially only the live load. Working stress method might design the connection only for live load; however, an increased factor of safety should be used, since the ultimate capacity would otherwise be inadequate.

AISC-1.11 (1978) also uses an ultimate-strength concept but converts both the forces to be designed for and the connector capacities into the service-load range by dividing them by a suitable safety factor, with a nominal value of 2. Thus, for design under service loads,

$$V_h = \frac{C_{\max}}{2} = \frac{0.85 f'_c A_c}{2} \tag{2.27}$$

which is AISC (1978) Formula 1.11-3, where $A_c = b_E t_s$, the effective concrete area. Equation 2.23 divided by 2 becomes

$$V_h = \frac{T_{\max}}{2} = \frac{A_s F_y}{2} \tag{2.28}$$

which is AISC Formula 1.11-4. In Eqs. 2.27 and 2.28, V_h is the horizontal shear to be resisted between the points of maximum positive moment and zero moment; the smaller of Eq. 2.27 or Eq. 2.28 is to be used. The other terms are the same as defined previously.

The connector ultimate capacities must also be divided by factors to give "allowable values" for the working stress method. The AISC allowable values are obtained by dividing ultimate capacities from Eqs. 2.25 and 2.26 by a factor of safety of approximately 2.0. AISC allowable values are given in Table 2.4 for hooked or headed studs and channels. Since the ultimate capacity expressions for hooked or headed studs are valid for $H/d_s \geqslant 4$, the values in Table 2.4 are also applicable to studs longer than the lengths indicated in the table.

When lightweight aggregate concrete is used, the connector values in Table 2.4 are to be multiplied by the coefficients given in Table 2.5.

The number N_1 of connectors required is obtained by dividing the smaller value of V_h by the allowable shear per connector:

$$N_1 = \frac{\text{Smaller } V_h}{q} \tag{2.29}$$

where

q = allowable load from Table 2.4.

The smaller value of V_h, determined by Eqs. 2.27 or 2.28, is used since it represents the maximum force to give equilibrium at ultimate strength, as discussed in developing Eqs. 2.22 and 2.23. It would be needless to provide more shear resistance than either the concrete slab or the steel beam could develop. Also, it is doubtful that an excessive number of shear connectors would perceptibly reduce deflection.

AASHTO-1.7.100 (1973 and 1974 Interim) uses the ultimate strength concept directly (i.e., without dividing by a factor to convert the computation into a nominal service-load comparison). However, the strength calculation is not used as the sole design procedure but rather as an additional check after determining the shear connectors required for the fatigue criterion. The fatigue requirement is an elastic procedure based on a slip limitation.

EXAMPLE 2.4

Determine the number of $\frac{3}{4}$-in. diameter × 3-in. shear stud connectors required according to AISC Specifications for the composite section shown in Fig. 2.15. Assume a uniform loading and simple beam supports, F_y = 36 ksi and f'_c = 3 ksi.

Solution

Using Eqs. 2.27 and 2.28,

$$V_h = \frac{0.85 f'_c A_c}{2} = \frac{0.85(3.0)72(7)}{2}$$

$$= 643 \text{ kips } (2860 \text{ kN})$$

Table 2.4. AISC design capacities for shear connectors (from AISC-1.11.4).[a]

Shear Connectors	Allowable Horizontal Shear Load, q (kips)		
	f'_c = 3000 psi	f'_c = 3500 psi	$f'_c \geqslant$ 4000 psi
$\frac{1}{2}$-in. diam, 2-in. hooked or headed stud	5.1	5.5	5.9
$\frac{5}{8}$-in. diam, $2\frac{1}{2}$-in. hooked or headed stud	8.0	8.6	9.2
$\frac{3}{4}$-in. diam, 3-in. hooked or headed stud	11.5	12.5	13.3
$\frac{7}{8}$-in. diam, $3\frac{1}{2}$-in. hooked or headed stud	15.6	16.8	18.0
3-in. channel, 4.1 lb	4.3 w[b]	4.7 w	5.0 w
4-in. channel, 5.4 lb	4.6 w	5.0 w	5.3 w
5-in. channel, 6.7 lb	4.9 w	5.3 w	5.6 w

[a] Applicable only to normal-weight concrete.
[b] w = length of channel (in.).

Table 2.5. Reduction factors for connector capacities when using lightweight aggregate concrete (from AISC-1.11.4).

Air dry unit weight, pcf	90	95	100	105	110	115	120
Coefficient, for $f'_c \leqslant$ 4.0 ksi	0.73	0.76	0.78	0.81	0.83	0.86	0.88
Coefficient, for $f'_c \geqslant$ 5.0 ksi	0.82	0.85	0.87	0.91	0.93	0.96	0.99

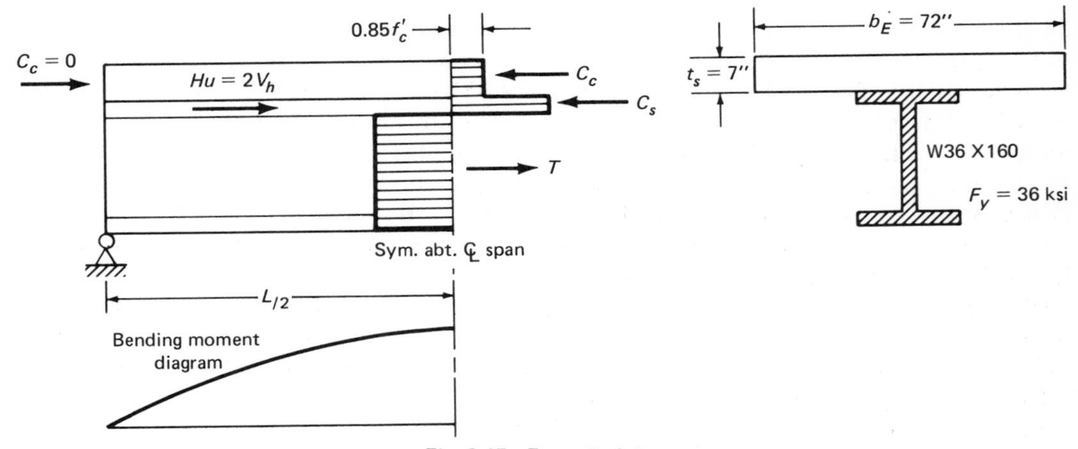

Fig. 2.15. Example 2.4.

or

$$V_h = \frac{A_s F_y}{2} = \frac{47.0(36)}{2} = 846 \text{ kips } (3760 \text{ kN})$$

From Table 2.4, the allowable shear per connector is 11.5 kips. Taking the smaller value of V_h, the number of shear connectors, N, required for each half of the span is therefore

$$N = \frac{643}{11.5} = 56$$

Use 56, $\frac{3}{4}$-in. diameter × 3-in. studs per half-span.

Ordinarily, AISC design would not involve an ultimate-strength analysis that would permit direct computation of H_u (Fig. 2.15) equal to C_c as was done in Ex. 2.4. In lieu of such computation, the V_h two-formula procedure is necessary.

Adequately anchored, longitudinal reinforcing bar steel within the effective width of the concrete slab may be assumed to act compositely with the steel beam. The total horizontal shear to be resisted by the shear connectors between an interior support on a continuous beam and each adjacent point of contraflexure equals the maximum tensile force that can be developed in the reinforced concrete slab, i.e., neglecting tensile capacity of the concrete.

$$T_{\text{slab}} = A_{sr} F_{yr} \tag{2.30}$$

where

A_{sr} = total area of longitudinal reinforcing bar steel at the interior support located within the effective flange width

F_{yr} = specified minimum yield strength of the longitudinal reinforcing steel.

In working stress design, the ultimate shear force, T_{slab}, developed between maximum negative moment and point of contraflexure, is divided by 2 to bring it into the service-load range.

$$V_h = \frac{T_{\text{slab}}}{2} = \frac{A_{sr} F_{yr}}{2} \tag{2.31}$$

It is logically presumed that the tensile capacity of the reinforced concrete slab will be less than the tensile capacity of the steel beam, so that for negative moment, only Eq. 2.31 is used.

2.7.2 Connector Design—Elastic Concept for Static Loads

In the elastic approach to shear connector design, the connectors are distributed according to the variation in horizontal shear between the slab and the steel beam. The connector capacities would be based on a limitation on slip. Formerly, design rules for both buildings and bridges used this approach with Eq. 2.21. When this elastic method was used where slip was limited, fatigue was not a controlling factor. The number of connectors required was conservatively large, indicating more than enough to develop the ultimate flexural strength of the composite member. If the shear connector requirements are reduced to the number required to just develop the ultimate flexural strength, fatigue failure then may become the governing factor. Present bridge design considers both fatigue strength as well as ultimate strength.

2.7.3 Connector Design—Elastic Concept for Fatigue Strength

The 1973 AASHTO specification requirements for fatigue are based largely on the work of Slutter and Fisher (1966). For fatigue, the *range* of stress rather than the magnitude is the major variable influencing fatigue strength. Fatigue strength may be expressed as

$$\log N = A + BS_r \tag{2.32}$$

where

S_r = range of horizontal shear stress
N = number of cycles to failure
A and B = empirical constants.

The equation used for design is shown in Fig. 2.16.

Since the magnitude of shear force transmitted by individual connectors when service loads act on a structure agrees well with that predicted by elastic theory, the horizontal shear may be calculated as VQ/I. Fatigue is critical under repeated applications of service load; thus, it is reasonable to determine variation in shear stress using elastic theory. Using Eq. 2.21, for static load

$$\frac{VQ}{I} = \frac{\text{Allowable load}, q}{p} \tag{2.33}$$

For cyclic load,

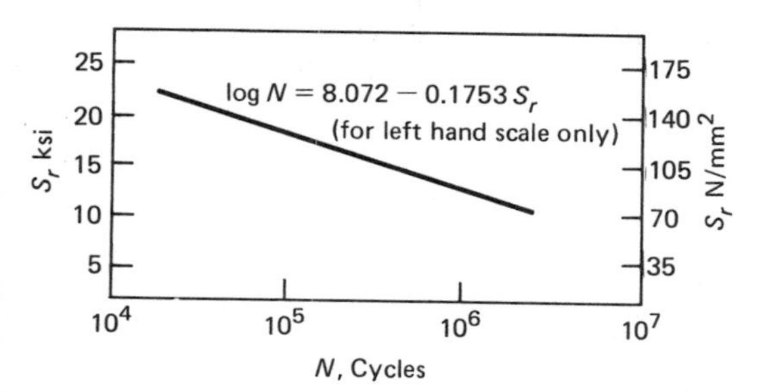

Fig. 2.16. Fatigue strength of stud shear connectors. (*From Slutter and Fisher 1966.*)

$$\frac{(V_{max} - V_{min})Q}{I} = \frac{\text{Allowable range}, Z_r}{p} \tag{2.34}$$

where

p = connector spacing.

AASHTO-1.7.100(A) (1973) gives Eq. 2.34 as

$$S_r = \frac{V_r Q}{I} \leqslant Z_r \tag{2.35}$$

where

$V_r = V_{max} - V_{min}$
$Z_r = \alpha d_s^2$ for welded studs
α = 13,000 for 100,000 cycles
10,600 for 500,000 cycles
7,850 for 2,000,000 cycles.

EXAMPLE 2.5

Redesign the shear connectors for the beam of Ex. 2.4 (Fig. 2.15) using the working stress fatigue requirement of AASHTO with $\frac{3}{4}$-in. diameter × 3-in. stud connectors. Assume that the 500,000 cycles loading case for the live load is to be designed for whether or not the beam is shored. Only the live load is involved in the cyclical load. Spacing of beams = 7 ft, F_y = 36 ksi, and f'_c = 3 ksi. Use uniform live load of 3.5 kips/ft and a beam span of 45 ft.

Solution

1. Loads and shears: For the fatigue requirement in AASHTO-1.7.100(A) (1), only the *range* of live-load shear is needed.

$$V_{max}(\text{support}) = \frac{3.5(45)}{2} = 78.8 \text{ kips}$$

Considering partial span loading of live load,

$$V_{\frac{1}{4}\text{ span}} = 3.5(45)(0.75)(0.375) = 44.3 \text{ kips}$$

$$V_{\text{midspan}} = \tfrac{1}{8}(3.5)45 = 19.7 \text{ kips}$$

The envelope showing the *range* of live-load shear is given in Fig. 2.17. Inclusion of dead-load shear would change both V_{max} and V_{min} by the same amount at any section along the beam; however, $V_{max} - V_{min}$, the range (V_r) would not be affected.

2. Compute composite section properties (n = 9) (see Fig. 2.15).

Element	*Effective Area (A)*	*Arm from C.G. of steel beam, y(in.)*	*Ay*	*Ay²*	I_o
Slab, 72(7)/9	56.0	21.5	1204	25,900	230
W36 × 160	47.0	—	—	—	9750
	103.0 in.²		1204	25,900 in.⁴	9980 in.⁴

$$I_x = Ay^2 + I_o = 25{,}900 + 9980 = 35{,}900 \text{ in.}^4$$

$$\bar{y} = \frac{1204}{103.0} = 11.69 \text{ in.}$$

$$I = 35{,}900 - 103.0(11.69)^2 = 21{,}800 \text{ in.}^4$$

$$y_t = 18.0 + 7.0 - 11.69 = 13.31 \text{ in.}$$

$$y_b = 18.0 + 11.69 = 29.69 \text{ in.}$$

$$S_t = \frac{21{,}800}{13.31} = 1640 \text{ in.}^3$$

$$S_b = \frac{21{,}800}{29.69} = 734 \text{ in.}^3$$

Determine the static moment of the effective concrete area about the centroid of the composite section,

$$Q = 56.0(y_t - 3.5) = 56.0(9.81) = 549 \text{ in.}^3$$

3. Determine the allowable load for $\frac{3}{4}$-in.-diameter × 3-in. stud connectors. AASHTO-1.7.100 gives an allowable service-load capacity based on fatigue for 500,000 cycles of loading as

$$\text{Allowable } S_r = 10.6 d_s^2$$
$$= 10.6(0.75)^2 = 5.96 \text{ kips}$$

AASHTO (1973) allowable values are higher than those of earlier editions but are still lower than AISC values because the AASHTO values are related to a slip limitation and a *range* of stress. This is appropriate whenever fatigue loading may occur.

4. Determine spacing of connectors. Use four studs across the beam flange width at each location:

$$S_r \text{ for four studs} = 4(5.96) = 23.8 \text{ kips}$$

Using Eq. 2.35,

$$p = \frac{S_r}{V_r Q/I} = \frac{S_r I}{V_r Q}$$

where

$$I/Q = 21{,}800/549 = 39.6 \text{ in.}$$

$$p = \frac{23.8(39.6)}{V_r} \quad \frac{943}{V_r(\text{kips})}$$

The values are computed in the table below, and the spacing is determined graphically on the shear diagram of Fig. 2.17.

p (in.)	V_r *(kips)*
12	79
15	63
18	52
21	45
24	39

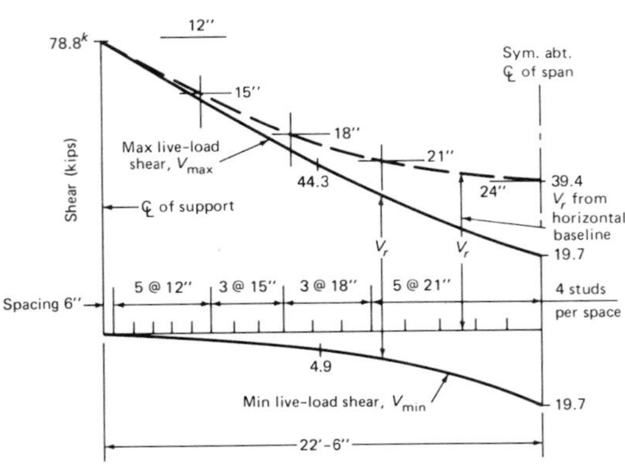

Fig. 2.17. Shear range diagram and stud spacing according to elastic fatigue theory used by AASHTO.

The fatigue service-load criterion requires nearly 18% more connectors (66 vs. 56 per half-span) than does the procedure based on the ultimate strength concept.

2.8 HYBRID COMPOSITE GIRDERS

The preceding discussion in this chapter relates specifically to the steel section being of one grade of steel. If only one grade of steel is involved, the behavior is the same whether the steel section is a rolled shape, with or without cover plates, or a built-up shape consisting of plates welded together. The welded built-up shape, known as a plate girder, has certain design considerations that are unrelated to composite construction. The general behavior of plate girders is outside the scope of this book and may be found in texts on steel structures.*

Of particular interest here are the special behavioral features of the hybrid plate girder and its use in composite construction. A *hybrid plate girder* is one that has either the tension flange or both flanges of the steel section made with a higher strength grade of steel than used for the web (see Fig. 2.1d and e). There are economic advantages of this in composite construction where the concrete slab provides a large compression capacity. This causes the neutral axis of the composite section to lie near the compression face of the composite girder and make the higher stressed tension flange become the controlling element determining the strength. The use of a higher strength tension flange permits using a smaller more economical plate.

*See, for example, Charles G. Salmon and John E. Johnson, *Steel Structures: Design and Behavior*, Harper & Row, New York, 1971, Chapter 11.

Currently, the procedures used to design hybrid steel girders are primarily those recommended by Subcommittee 1 of the ASCE-AASHTO Joint Committee (1968). Frost and Schilling (1964) have explained the behavior of noncomposite hybrid beams under static loads. Toprac (1965a, 1965b) and Toprac and Eyre (1967) studied hybrid tee steel sections (similar to Fig. 2.1g) having no top flange, A36 web, and A36 or A514 steel bottom flange. Schilling (1968) presented the theoretical aspects of the bending behavior of composite hybrid beams.

Although an accurate analysis of a hybrid girder is complicated, the primary behavioral feature of practical concern to the designer is the yielding of the web prior to reaching maximum strength in the flanges.

EXAMPLE 2.6

For the section given in Fig. 2.18 whose flexural properties are $I_x = 13{,}640$ in.^{4} and $S_x = 910$ in.^{3}, determine the moment-rotation characteristics (1) for the section as a homogeneous one of A514 steel and (2) for the section as a hybrid A514/A36 beam.

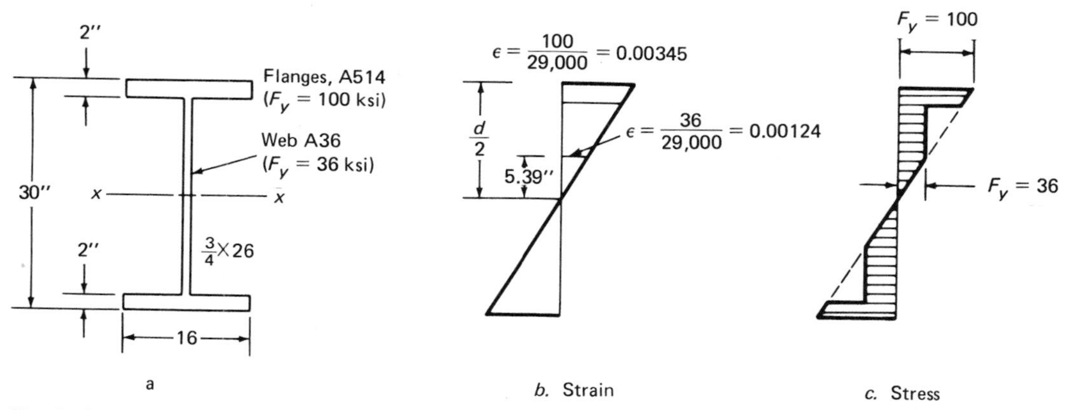

Fig. 2.18. Section for Ex. 2.6, showing strain and stress on hybrid section when F_y is reached at extreme fiber of section.

Solution

1. Homogeneous A514 (F_y = 100 ksi) section. For strain $\epsilon = F_y/E_s$ at extreme fiber of section:

$$M_y = S_x F_y = \frac{910(100)}{12}$$

$$= 7580 \text{ ft-kips } (10{,}300 \text{ kN-m})$$

$$\frac{d}{2}\phi = \frac{F_y}{E_s} = \frac{100}{29{,}000} = 0.00345$$

For plastic moment (i.e., ultimate strength):

$$Z_x = [16(2)(14) + 13(0.75)(6.5)]\,2 = 1020 \text{ in.}^3$$

$$M_p = Z_x F_y = \frac{1020(100)}{12}$$

$$= 8520 \text{ ft-kips } (11{,}600 \text{ kN-m})$$

2. Hybrid A514/A36 section. For strain $\epsilon = F_y/E_s$ at extreme fiber of *web*:

$$f \text{ at extreme fiber} = 36\left(\frac{15}{13}\right) = 41.5 \text{ ksi}$$

$$M_{yw} = S_x(41.5) = \frac{910(41.5)}{12}$$

$$= 3150 \text{ ft-kips } (4270 \text{ kN-m})$$

$$\frac{d}{2}\phi = \frac{41.5}{29{,}000} = 0.00143$$

For strain $\epsilon = F_y/E_s$ at extreme fiber of section (Fig. 2.18b and c): In this condition, the web is partially plastic while the flanges are at initial yielding.

$$M_y = \left\{100\left[\frac{2(32)(14)^2}{15}\right] + 36\left[\frac{1}{6}\left(\frac{3}{4}\right)(10.78)^2\right] + 36\left[(13 - 5.39)\left(\frac{3}{4}\right)2\left(\frac{13-5.39}{2} + 5.39\right)\right]\right\}\frac{1}{12}$$

$$= 6968 + 44 + 315 = 7330 \text{ ft-kips } (9950 \text{ kN-m})$$

For fully plastic flanges but partially plastic web on hybrid section, the strain at the extreme fiber of the *web* will be (100/29,000) = 0.00345. The distance is 4.67 in. from the neutral axis to the point where the stress on the web is 36 ksi.

$$M_{pf} = \left\{100(32)(15)2 + 36\left[\frac{1}{6}\left(\frac{3}{4}\right)(9.34)^2\right] + 36\ (13 - 4.67)\left(\frac{3}{4}\right)(2)\left(\frac{13-4.67}{2} + 4.67\right)\right]\right\}\frac{1}{12} = 8000 + 33 + 331$$

$$= 8360 \text{ ft-kips } (11{,}300 \text{ kN-m})$$

For the fully plastic hybrid section,

$$M_p = [\underset{\text{flanges}}{100(32)(15)(2)} + \underset{\text{web}}{36(13)(0.75)(6.5)(2)}]\frac{1}{12} = 8000 + 380$$

$$= 8380 \text{ ft-kips } (11{,}400 \text{ kN-m})$$

The results are shown in Fig. 2.19, which compares the behavior of the hybrid and the homogeneous beam.

There are two principal effects apparent from Fig. 2.19. First, the onset of yielding of the web at 38% of the strength based on yielding of the flanges means that, even at service load, inelastic behavior of the web is to be expected. Second, the strength of the section computed when the flanges have entirely yielded but the web is still inelastic gives a value not significantly different than obtained if the full plastic strength is used. In this example, the ratio A_w/A_f, web area to flange area, was only 0.61. For higher ratios, the effect of the web increases but, for practical purposes, it still does not greatly change the behavior from that of a homogeneous member. Any yielding occurring in the web is restricted by the elastic flanges.

For design of hybrid beams, it is recommended by Subcommittee 1 of the ASCE-AASHTO Joint Committee (1968) that, to account for the small effect on capacity for a hybrid beam with a low yield strength web, either of the following procedures be used:

1. An allowable moment be determined by taking the flange-yield moment (point *B* on Fig. 2.19) and dividing by a factor of safety, or

2. An allowable moment be calculated as the elastic section modulus of the full section mul-

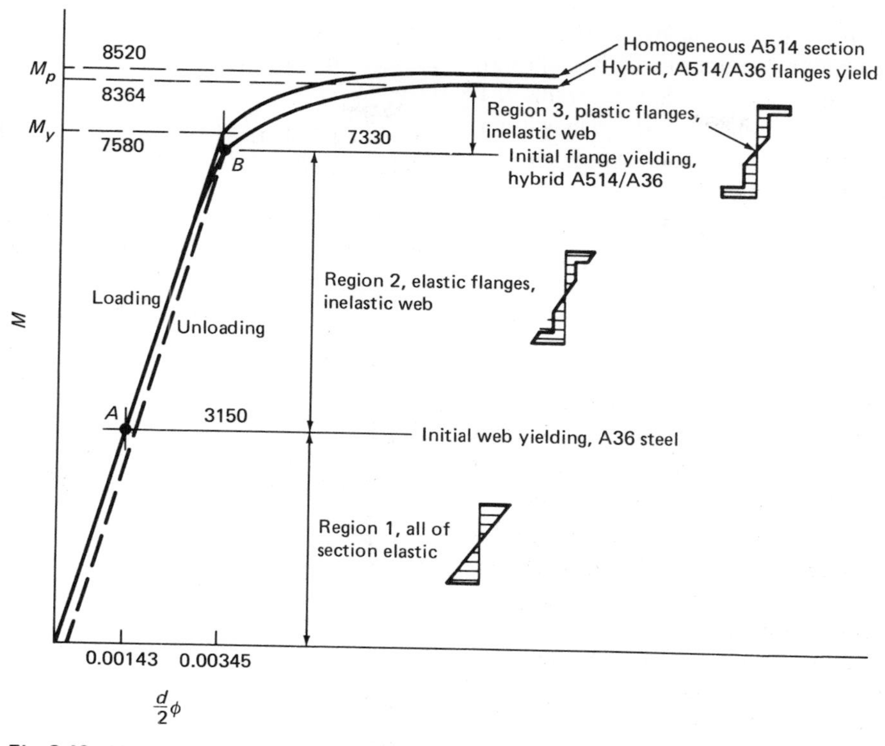

Fig. 2.19. Moment-rotation relationships for section of Fig. 2.18. Assumes no residual stress.

tiplied by a reduced allowable flange extreme fiber stress.

Both AISC-1.10.6 and AASHTO-1.7.111 use the latter approach. The allowable extreme fiber stress for a homogeneous beam based on flexural strength (including lateral-torsional buckling) is then multiplied by a reduction factor to account for the lower strength web.

AISC-1.10.6, Formula (1.10-6) is

$$F_b' \leqslant F_b \left[\frac{12 + \beta(3\alpha - \alpha^3)}{12 + 2\beta}\right] \tag{2.36}$$

where

$\beta = A_w/A_f$ = ratio of cross-sectional area of the web to the cross-sectional area of one flange

$\alpha = F_y(\text{web})/F_y(\text{flange})$ = ratio of the yield strength of the web steel to the yield strength of the flange steel

F_b = allowable stress, including consideration of lateral-torsional buckling, assuming entire section consists of the steel in the flanges

F_b' = reduced allowable stress to account for lower strength web steel.

When extending the hybrid-beam concept to composite steel-concrete members, the behavior is essentially the same. The theoretical analysis of such sections has been presented in detail by Schilling (1968). The additional complicating factors are that (1) the neutral axis of the composite section is not at middepth, requiring an evaluation of an unsymmetrical hybrid member, and (2) the relative stiffness of the concrete deck and the steel section are continuously changing as the yielding progresses. In the composite hybrid section, the principal concern is with the tension flange (bottom flange in positive moment zones). Since a greater percentage of the depth of web is located below the neutral axis (tension side), the early yielding of the low-strength web means a greater reduction in strength for a hybrid beam in composite construction than for a noncomposite symmetrical beam.

To account for a variable distance from the

tension flange to the neutral axis, ASCE-AASHTO (1968) recommended the following equation as applicable "to hybrid beams that support the dead weight of the slab without composite action but act compositely with the slab in support of live load."

$$F_b' = F_b \left[1 - \frac{\beta\psi(1-\alpha)^2(3-\psi+\alpha\psi)}{6+\beta\psi(3-\psi)} \right] \tag{2.37}$$

where

ψ = ratio of the distance from the bottom of the beam to the neutral axis of the transformed section (composite section) to the overall depth of the steel section.

All other variables are as indicated in regard to Eq. 2.36. *Equation 2.37 is not to be used if the top flange has a higher yield strength or larger area than the bottom flange.*

AASHTO-1.7.111 or 1.7.133 uses Eq. 2.37 for both symmetrical noncomposite hybrid girders and composite hybrid girders. AISC-1.11 makes no reference to hybrid composite construction, but presumably it is permitted. Presumably, Eq. 2.36 would have to be used to obtain the allowable stress reduction. When the section is symmetrical and $\psi = 0.5$ for Eq. (2.37), the value of the reduction factor is about the same, whichever formula is used.

The reduction factors given by Eqs. 2.36 and 2.37 are in Table 2.6.

2.9 COMPOSITE STEEL BEAMS WITH METAL DECK

A very common type of composite construction is shown in Fig. 2.20. The composite sec-

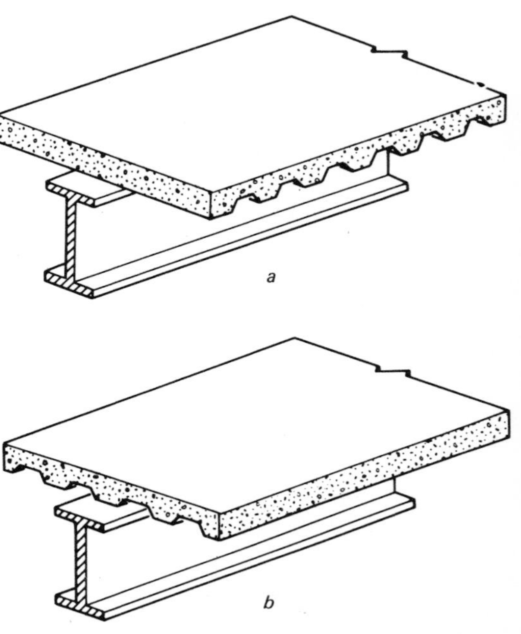

Fig. 2.20. Composite beams with metal deck.

Table 2.6. Multipliers to reduce allowable bending stress for hybrid composite and noncomposite girders.

α \ β	0.50	1.0	2.0	3.0	4.0	
ψ = 0.5 (Neutral axis at mid-depth)						
0.36	0.963	0.931	0.879	0.839	0.807	AISC Eq. 2.36
0.50	0.976	0.955	0.922	0.896	0.875	
0.72	0.992	0.985	0.973	0.964	0.957	
0.36	0.959	0.924	0.871	0.831	0.800	AASHTO Eq. 2.37
0.50	0.974	0.953	0.919	0.894	0.857	
0.72	0.992	0.985	0.974	0.965	0.959	
ψ = 0.75						
0.36	0.943	0.899	0.835	0.790	0.757	AASHTO
0.50	0.964	0.936	0.895	0.866	0.846	
0.72	0.988	0.979	0.965	0.956	0.949	
ψ = 1.00						
0.36	0.931	0.879	0.807	0.758	0.724	AASHTO
0.50	0.955	0.922	0.875	0.844	0.821	
0.72	0.985	0.973	0.957	0.947	0.939	

tion consists of a steel beam, metal deck, and concrete slab. The metal deck can be oriented either parallel or perpendicular to the steel beam. The metal deck can also be acting compositely or noncompositely with the concrete slab. The composite behavior of metal deck is treated in Chapter 3. The behavior of composite beams presented in this chapter is directly applicable to composite-beam–metal-deck systems with a few exceptions. These exceptions are:

1. The effective slab thickness of the composite system is normally calculated based on the solid portion of the concrete slab.
2. The capacity of the shear connectors is influenced by the geometry of the metal deck. In the case of the metal deck spanning parallel to the composite beam, no reduction in shear connector capacity is required; however, shear connector capacities may be less than their maximum strength when the metal deck is spanning perpendicular to the composite beams. The behavior of such shear connectors is treated in Chapter 4.

The following example is presented to illustrate the design procedure.

EXAMPLE 2.7

Design a composite beam to support a total loading of 130 psf, including live load (60 psf), partition load (20 psf), ceiling (5 psf), and $5\frac{1}{2}$-in. floor slab on 3-in. deck (45 psf). Assume the steel beams are of A36 steel, on a span of 30 ft spaced 10 ft center-to-center. The concrete is normal weight, with $f'_c = 3000$ psi ($n = 9$). No shoring is to be used.

Solution

1. Loading and properties. Load on beam, $0.130(10) + 0.030 = 1.33$ kips/ft (assuming 30 plf beam)

$$M = \frac{wL^2}{8} = \frac{1.33(30)^2}{8} = 150 \text{ ft-kips}$$

$$S_{tr} = \frac{M}{F_b} = \frac{12(150)}{24} = 74.8 \text{ in.}^3$$

Try W14 × 30:

$$b_E = 16t + b_f = 16(5.5) + 6.73 = 94.7 \text{ in.}$$
$$= L/4 = 0.25(30)(12) = 90 \text{ in. (controls)}$$
$$= \text{Spacing} = 120 \text{ in.}$$

$$A_c = \frac{90}{9}(2.5) = 25 \text{ in.}^2$$

$$I_c = \frac{bh^3}{12} = \frac{10(2.5)^3}{12} = 13.02 \text{ in.}^4$$

Part	A	y	Ay	h	Ah^2	I_o
Concrete	25	18.09	452.2	2.92	213	13.02
W14 × 30	8.85	6.92	61.2	8.25	602	291
	33.85 in.^{2}		513.4 in.^{3}		815 in.^{4}	304 in.^{4}

$$y_b = \frac{Ay}{A} = \frac{513.4}{33.85} = 15.17 \text{ in.}$$

$$h_c = 18.09 - 15.17 = 2.92 \text{ in.}$$

$$h_s = 15.17 - 6.92 = 8.25 \text{ in.}$$

$$I_{tr} = Ah^2 + I_o = 815 + 304 = 1119 \text{ in.}^4$$

$$S_{tr} = I_{tr}/y_b = 1119/15.17$$
$$= 73.8 \text{ in.}^3 \approx 74.8 \text{ in.}^3 \quad \text{OK}$$

$$S_{top} = \frac{1119}{4.18} = 268 \text{ in.}^3$$

2. Check stresses.

$$\text{Concrete stress} = \frac{M}{nS_{top}} = \frac{150(12)}{9(268)}$$
$$= 0.75 \text{ ksi} < 1.35 \text{ ksi} \quad \text{OK}$$

$$\text{Maximum } S_{tr} = \left(1.35 + 0.35\frac{M_L}{M_D}\right)S_s$$
$$= \left[1.35 + 0.35\left(\frac{85}{45}\right)\right]42.0$$
$$= 84.5 > 73.7 \text{ in.}^3 \quad \text{OK}$$

No shoring required.

3. Determine shear connector requirements.

$$V_h = \frac{0.85 f'_c A_c}{2} = \frac{0.85(3)(2.5)(90)}{2} = 287 \text{ kips}$$

$$V_h = \frac{A_s F_y}{2} = \frac{8.85(36)}{2} = 159 \text{ kips (controls)}$$

Using $\frac{3}{4}$-in.-diameter × 3-in. headed stud and assuming full connector capacity* ($q = 11.5$ kips/stud), determine the number of studs required.

$$N = \frac{159(2)}{11.5} = 28$$

Use 28, $\frac{3}{4}$-in.-diameter × 3-in. studs over the total beam length.

4. Deflection calculations.

Dead load:

$$\Delta_{DL} = \frac{5wL^4}{384EI} = \frac{5(450)(30)^4 1728}{384(29.5 \times 10^6)(291)}$$

$$= 0.96 \text{ in.} \qquad \text{Assume OK}$$

Live load:

$$\Delta_{LL} = \frac{5(850)(30)^4 1728}{384(29.5 \times 10^6)(1119)}$$

$$= 0.47 \text{ in.} < \frac{L}{360} = 1.0 \text{ in.}$$

2.10 AISC DESIGN PROCEDURE

As discussed in Section 2.5, the actual stresses that occur under service load in a given composite member depend on the manner of construction. The slab formwork must be supported by either the steel beam acting alone or by temporary shoring that also would support the beam. When temporary shoring is used, service-load stresses will be lower than when such shoring is not used, since *all* loads will be supported by the composite section. If the system is built without temporary shoring, the steel beam alone must support itself and the slab without benefit of composite action.

For economical construction, it is desirable to avoid use of shoring wherever possible. In Section 2.6, it was shown that no matter which construction system is used, the ultimate moment capacity is identical. It is a simple procedure, therefore, to design as if the entire load is to be carried compositely (i.e., assume shores are used) *even when shores are not to be used.* Strength is assured; however, it is necessary to insure that the stress in the steel beam does not approach too closely the yield stress under service-load conditions.

In order to resist loads compositely, the concrete strength must be adequately developed. AISC-1.11.2.2 requires that 75% of the compressive strength f'_c of the concrete must be developed before composite action may be assumed.

The AISC (1978) design procedure for flexure may be summarized by the following steps:

1. *Select section as if shores are to be used.* The required composite section modulus S_{tr} with reference to the tension fiber is

$$S_{tr}(\text{reqd}) = \frac{M_D + M_L}{F_b} \tag{2.38}$$

where

M_D = the service-load moment caused by loads applied *prior* to the time the concrete has reached 75% of its required strength

M_L = the service-load moment caused by loads applied *after* the concrete has reached 75% of its required strength

F_b = allowable service-load stress, 0.66 F_y for positive moment regions (where sections are exempt from the "compactness" requirements of AISC-1.5.1.4.1).

Lateral support is assumed to be adequately provided by the concrete slab and its shear connector attachments.

2. *Check AISC Formula 1.11-2.* When shores are actually not to be used, service-load stress on the steel section must be assured of being less than yield stress. AISC-1.11.2.2 uses an indirect procedure for checking this. The section modulus of the composite section S_{tr} may

*At the time of the design, the engineer will probably not know what manufacturer will be supplying the metal deck; thus, the authors recommend that the engineer base the calculations on full connector capacity. Once the deck geometry is known, the appropriate reductions in connector capacity can be made.

not exceed (or be considered more effective than) the following:

$$S_{tr}\text{ (effective)} \leqslant \left(1.35 + 0.35\frac{M_L}{M_D}\right)S_s \tag{2.39}$$

which is AISC Formula 1.11.2. In this formula, M_L refers to the moment caused by loads that are to be carried compositely, and M_D refers to loads carried by the steel beam.

To understand the development of Eq. 2.39, the reader is referred back to Section 2.5 where service-load stresses are computed for construction with and without shores. Service-load tension stresses on the steel beam may be expressed in general as

$$f_b = \frac{M_D}{S_s} + \frac{M_L}{S_{tr}} \leqslant k_1 F_y \text{ without shores} \quad \text{(a)}$$

$$f_b = \frac{M_D + M_L}{S_{tr}} \leqslant k_2 F_y \text{ with shores} \quad \text{(b)}$$

where

S_s = section modulus of the steel beam referred to its bottom flange (tension flange)

S_{tr} = section modulus of composite section referred to its bottom flange (tension flange)

k_1, k_2 = constants to obtain the allowable stresses in tension without shores and with shores, respectively.

Divide Eq. (a) by Eq. (b), letting $kS_s = S_{tr}$:

$$\frac{k_1}{k_2} \geqslant \frac{\dfrac{M_D}{S_s} + \dfrac{M_L}{kS_s}}{\dfrac{M_D + M_L}{kS_s}} = \frac{kM_D + M_L}{M_D + M_L} \quad \text{(c)}$$

$$\frac{k_1}{k_2}(M_D + M_L) - M_L \geqslant kM_D \quad \text{(d)}$$

Divide by M_D,

$$k \leqslant \frac{k_1}{k_2}\left(1 + \frac{M_L}{M_D}\right) - \frac{M_L}{M_D} \quad \text{(e)}$$

Replacing k by S_{tr}/S_s gives the AISC formula in general terms.

$$S_{tr} \leqslant \left[\frac{k_1}{k_2} + \frac{M_L}{M_D}\left(\frac{k_1}{k_2} - 1\right)\right] S_s \quad \text{(f)}$$

The AISC value of $k_1/k_2 = 1.35$ is obtained if a compact section ($F_b = 0.66F_y$) is allowed to reach a service-load stress of $0.89F_y$ ($0.89/0.66 = 1.35$). As is seen from Eq. (f), this limitation of stress is valid no matter what ratio of M_L to M_D is used.

3. *Check stress on steel beam* supporting the loads acting before concrete has hardened.

$$\text{Required } S_s = \frac{M_D}{F_b} \tag{2.40}$$

where F_b may be $0.66F_y$, $0.60F_y$, or some lower value if adequate lateral support is not provided. It is to be noted that Eq. 2.40 is frequently controlling on the compression fiber (top in positive moment zone), particularly if a steel cover plate is used on the bottom.

4. *Partial composite action.* When fewer connectors are used than necessary to develop full composite action, an effective section modulus may be obtained by linear interpolation. AISC-1.11.2.2 allows

$$S_{\text{eff}} = S_s + \sqrt{\frac{V_h'}{V_h}}(S_{tr} - S_s) \tag{2.41}$$

where

V_h = design horizontal shear for full composite action

V_h' = actual capacity of connectors used; less than V_h

S_s and S_{tr} as defined for Eqs. a and b.

For this case, S_{eff} is used to design calculations in place of that computed from beam dimensions, and is the quantity that may not exceed the value given by Eq. 2.39.

2.11. EXAMPLES—SIMPLY SUPPORTED BEAMS

EXAMPLE 2.8

Design an interior member of the floor shown in Fig. 2.21, assuming it is constructed without temporary shoring. Assume $F_y = 36$ ksi, $n = 9$, $f_c' = 3000$ psi, $f_c = 1350$ psi, and a 4-in. slab. Use AISC Specification.

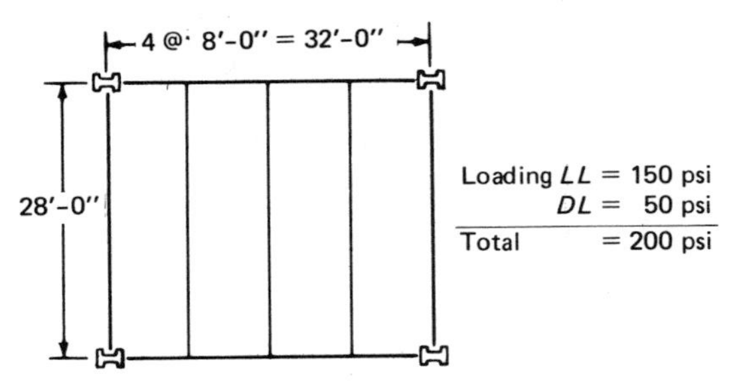

Fig. 2.21. Example 2.8.

Solution

1. Loads and bending moments. Load carried on steel beam,

concrete slab, $\frac{4}{12}(0.15)8 = 0.40$ kip/ft

beam-weight (estimated) $= \underline{0.04}$

0.44 kip/ft

$$M_D = \frac{0.44(28)^2}{8} = 43.1 \text{ ft-kips } (58.5 \text{ kN-m})$$

Load carried by composite section, live load, $0.15(8) = 1.20$ kip/ft

$$M_L = \frac{1.20(28)^2}{2} = 118 \text{ ft-kips } (160 \text{ kN-m})$$

2. Select beam as if shores were to be used. For $M_D + M_L$, the allowable stress is $0.66F_y$ on the composite section,

$$\text{Required } S_{tr} = \frac{(118 + 43)12}{24}$$

$$= 80.5 \text{ in.}^3 \ (1320 \text{ cm}^3)$$

For M_D acting on the steel section alone, the allowable stress is at least $0.60F_y$ if adequate lateral support is provided.

$$\text{Required } S_s = \frac{M_D}{0.60F_y} = \frac{43(12)}{22}$$

$$= 23.4 \text{ in.}^3 \ (383 \text{ cm}^3)$$

Enter *AISC Manual*, "Composite Beam Selection Table," and find W16 X 36 with no cover plate.

Try W16 X 36:

For properties, see *AISC Manual*, "Composite Design, Properties of Composite Beams," 4-in. slab, no cover plate.

Steel section alone:

$A = 10.6$ in.^{2}; $I_x = 447$ in.^{4}; $S_x = 56.5$ in.^{3}; $b_f = 6.99$ in. ($A = 68.4$ cm^2; $I_x = 18{,}600$ cm^4; $S_x = 926$ cm^3; $b_f = 178$ mm). Determine effective width (see Fig. 2.22).

$$b_E = \frac{1}{4} \text{ of span} = \frac{28(12)}{4} = 84 \text{ in.}$$

$$b_E = \text{beam spacing} = 96 \text{ in.}$$

$$b_E = 16 \text{ (thickness of slab)} + b_f$$

$$= 16(4) + 6.99 = 70.99$$

$$= 71 \text{ in. (controls)}$$

$$y_b = \frac{10.6\left(\frac{15.85}{2}\right) + \left(\frac{71(4)}{9}\right) 17.85}{10.6 + \frac{(71)4}{9}} = 15.35 \text{ in.}$$

$$I_{\text{comp}} = 447 + 10.6(7.42)^2 + \frac{1}{12}\left(\frac{71}{9}\right)(4)^3 + \frac{71(4)}{9}(2.50)^2$$

$$= 1270 \text{ in.}^4 \text{ (checks with AISC tables)*}$$

$$S_{tr} = \frac{1270}{15.35}$$

$$= 82.6 \text{ in.}^3 \text{ (for bottom of steel beam)}$$

$$S_{\text{top}} = \frac{1270}{4.50} = 282 \text{ in.}^3 \text{ (for top of concrete)}$$

Recomputing moments, $w = 0.05(8) + 0.036 = 0.436$ kips/ft

$$M_D = \frac{0.436(28)^2}{8} = 42.8 \text{ ft-kips } (58.1 \text{ kN-m})$$

*For depths of 4, $4\frac{1}{2}$, 5, and $5\frac{1}{2}$ in., the AISC *Manual of Steel Construction* gives tables for various properties of composite sections on pp. 2-152 through 2-191.

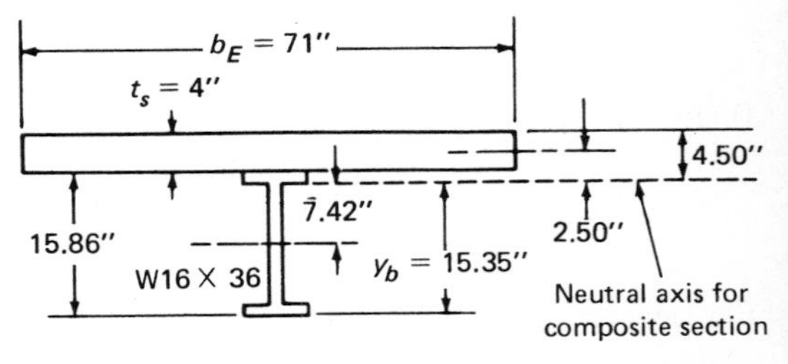

Fig. 2.22.

$$M_L = \frac{0.150(8)(28)^2}{8} = 118 \text{ ft-kips (160 kN-m)}$$

Total 161 ft-kips (218 kN-m)

Check stresses:

At top of concrete slab; allowable $f_c = 0.45f'_c = 1.35$ ksi

$$f_c = \frac{161(12)}{9(282)} = 0.76 \text{ ksi} < 1.35 \text{ ksi} \qquad \text{OK}$$

At bottom of steel beam; allowable $F_b = 0.66F_y = 24$ ksi

$$f_b = \frac{161(12)}{82.6} = 23.4 \text{ ksi} < 24.0 \text{ ksi} \qquad \text{OK}$$
$$(161 \text{ N/mm}^2)$$

3. Check Eq. 2.39 to determine maximum transformed section modulus S_{tr} that can be used.

$$S_{tr} = \left(1.35 + 0.35 \frac{118}{42.8}\right) 56.5$$

$$= 131 \text{ in.}^3 > 82.6 \text{ in.}^3 \qquad \text{OK}$$

Thus, shores need not be used.

In *AISC Manual*, it is stated that all beams included in the composite design tables in Part 2 of the manual have $S_{tr} \leqslant 2.31S_s$. Figure 2.39 then indicates

$$2.31S_s = \left(1.35 + 0.35 \frac{M_L}{M_D}\right) S_s$$

$$\frac{M_L}{M_D} = 2.74$$

Whenever M_L/M_D exceeds 2.74, a large enough proportion of the total load is applied after the composite section is effective so that there is no danger of reaching yield stress on the steel section acting along to carry dead load. In other words, if M_L/M_D exceeds 2.74, AISC Formula 1.11-2 is automatically satisfied. Most of the time, the ratio M_L/M_D is less than 2.74 and AISC Formula 1.11-2 must be checked.

4. Check steel stress for loads carried noncompositely, using Eq. 2.40,

$$f_b = \frac{M_D}{S_s} = \frac{42.8(12)}{56.5} = 9.1 \text{ ksi}$$

$$< 0.60F_y \ (62.7 \text{ N/mm}^2)$$

The preceding equation assumes adequate lateral support during construction so that the laterally unbraced length is less than L_u, based on either $\sqrt{102{,}000C_b r_T^2/F_Y}$ or $20{,}000C_b/[(d/A_f)F_y]$ as given in AISC-1.5.1.4.6a. Stress on the steel section resulting from noncomposite loads is more likely to govern when a cover plate is used on the bottom than when such a plate is not used. The beam section selected is therefore satisfactory.

Use W16 X 36.

4. Design shear connectors:
From Eq. 2.27,

$$V_h = \frac{0.85(3)71(4)}{2} = 362 \text{ kips}$$

From Eq. 2.28,

$$V_h = \frac{10.6(36)}{2} = 191 \text{ kips}$$

From Table 2.4, $\frac{5}{8}$-in.-diameter X $2\frac{1}{2}$-in. headed stud, $q = 8.0$ kips/stud

$$N = \frac{V_h}{q} = \frac{191}{8.0} = 23.8, \text{ say } 24.$$

Use 24 shear connectors on each side of the centerline at midspan. Use a uniform spacing with two studs at a section across the beam width:

$$s = \frac{L/2}{N/2} = \frac{28(12)}{24} = 14.1 \text{ in. (35.8 cm)}$$

Use a 14-in. spacing for the pairs of stud connectors, starting at the support.

EXAMPLE 2.9

Design a composite section, without shores, for use as an interior floor beam of an office building. Use AISC Specifications. $f'_c = 3000$ psi; $n = 9$, $F_y = 36$ ksi ($f'_c = 21$ N/mm^2 and $F_y = 250$ N/mm^2)

Span = 30 ft.	Live load = 150 psf
Beam spacing = 8 ft	Partitions = 25 psf
Slab thickness = 5 in.	Ceiling = 7 psf

Solution

1. Determine moments:

$$\text{5-in. slab, } \tfrac{5}{12}(8)0.15 = 0.50 \text{ kips/ft}$$

$$\text{Steel beam (assumed)} = \frac{0.03}{0.53 \text{ kips/ft}}$$

$$M_D = \tfrac{1}{8}(0.53)(30)^2 = 60 \text{ ft-kips}$$

Live load 0.15(8) = 1.2 kips/ft

Partitions 0.025(8) = 0.2

$$\text{Ceiling } 0.007(8) = \frac{0.05}{1.45 \text{ kips/ft}}$$

$$M_L = \tfrac{1}{8}(1.45)(30)^2$$
$$= 163 \text{ ft kips (221 kN-m)}$$

2. Select section

$$\text{Required } S_{tr} = \frac{M_D + M_L}{0.66F_y} = \frac{223(12)}{24}$$
$$= 112 \text{ in.}^3 \text{ (1830 cm}^3\text{)}$$

If shores are not used, stress on the steel section prior to developing composite action must not be excessive. Assuming adequate lateral bracing such that unbraced length $L < L_u$,

$$\text{Required } S_s = \frac{M_D}{0.60F_y} = \frac{60(12)}{22}$$
$$= 32.7 \text{ in.}^3 \text{ (536 cm}^3\text{)}$$

If the unbraced length $L < L_c$ and the section is "compact" for local buckling, $0.66F_y$ could be used for the allowable stress. Use *AISC Manual* "Composite Beam Selection Table" for 5-in. slab to select this member. Find W14 X 22 with 4 in.^{2} cover plate, whose properties are given in *AISC Manual*, "Properties of Composite Beams."

Composite section properties:

$$S_{top} = 361 \text{ in.}^3 \text{ (5920 cm}^3\text{)}$$

$$S_{bottom} = 113 \text{ in.}^3 \text{ (1850 cm}^3\text{)}$$

Steel section alone:

$$S_{top} = 34.3 \text{ in.}^3 \text{ (562 cm}^3\text{)}$$

$$S_{bottom} = 65.6 \text{ in.}^3 \text{ (1075 cm}^3\text{)}$$

Further checking is required, even for the properties, since the effective width of concrete slab b_E may not be the same as that used for computing the properties in the *AISC Manual*.

3. Check effective width, and compute properties. Effective width:

$$b_E \leqslant L/4 = 90 \text{ in.}$$
$$\leqslant 16t_s + b_f = 16(5) + 5.0$$
$$= 85.0 \text{ in. (governs)}$$
$$\leqslant \text{spacing of beams} = 96 \text{ in.}$$

Since $b_E = 85$ in. was used to obtain the AISC Manual properties, the values are correct and are not rechecked here.

4. Check stress under $M_D + M_L$ (ultimate strength concept; nominal working-stress procedure). Since girder weight is 32 lb/ft ≈ 30 as assumed, original moments are used:

$$f_b = \frac{M_D + M_L}{S_{tr}} = \frac{223(12)}{113}$$
$$= 23.7 \text{ ksi} < 24 \text{ ksi} \quad \text{OK}$$

5. Check AISC Formula 1.11-2:

$$S_{tr} = \left(1.35 + 0.35\frac{M_L}{M_D}\right)S_s$$
$$= \left(1.35 + 0.35\frac{163}{60}\right)65.8$$
$$= 2.30(65.8) = 151 \text{ in.}^3$$

Since actual $S_{tr} = S_{bottom} = 113$ does not exceed the upper limit of 151, shores are not required to prevent service-load stresses from becoming too close to the yield stress. As discussed earlier, the same conclusion could be reached if actual service-load stresses were computed as follows for no shoring:

$$f_b = \frac{M_D}{S_s} + \frac{M_L}{S_{tr}}$$
$$= \frac{60(12)}{65.6} + \frac{163(12)}{113} = 11.0 + 17.3$$
$$= 28.3 \text{ ksi (195 N/mm}^2\text{)}$$

which is acceptable, since it does not exceed $0.89F_y = 32$ ksi at service load. AISC requires the formula check, instead of the stress check, because without the understanding of ultimate strength, the $0.89F_y$ might appear to be an unsafe value.

6. Check stress on steel beam prior to development of composite action for the unshored system. The maximum stress occurs in compression at the top of the beam:

$$f_b = \frac{M_D}{S_s} = \frac{60(12)}{34.4} = 21.0 \text{ ksi}$$

$$< F_b = 0.60F_y \text{ (145 N/mm}^2\text{)}$$

Maximum laterally unsupported length during construction is $L = L_u = 5.7$ ft. Use W14 X 22 with $\frac{1}{2}$ X 8 cover plate. The section with service-load stresses is given in Fig. 2.23. Design of connectors is not illustrated since no new principles are involved.

EXAMPLE 2.10

Design a composite section, without shores, for the same loading conditions as in Ex. 2.9, except use F_y = 50 ksi steel.

1. Select a section using no cover plate and with the minimum number of $\frac{3}{4}$-diam. by 3-in. stud shear connectors.

2. Compare with a section having a cover plate. Determine the length and connection for the cover plate.

Solution

1. Loads and moments (from Ex. 2.9)

M_D = 60 ft-kips (81 kN-m)

(assumes 30 lb/ft beam)

M_L = 163 ft-kips (221 kN-m)

2. Select section. $F_b = 0.66F_y$ = 35 ksi (228 N/mm²)

$$\text{Required } S_{tr} = \frac{M_D + M_L}{0.66F_y} = \frac{223(12)}{33}$$

$$= 81.1 \text{ in.}^3 \text{ (1530 cm}^3\text{)}$$

For the steel section alone,

$$\text{Required } S_s = \frac{M_D}{0.60F_y} = \frac{60(12)}{30}$$

$$= 24 \text{ in.}^3 \text{ (393 cm}^3\text{)}$$

Select W18 X 35 (S_{tr} = 93.6 in.³) (from "Composite Design Selection Table," *AISC Manual*).

The properties are:

Composite section properties:

$$S_{top} = 379 \text{ in.}^3$$

$$S_{bottom} = 93.6 \text{ in.}^3$$

Steel section alone:

$$S_{top} = 57.6 \text{ in.}^3$$

$$S_{bottom} = 57.6 \text{ in.}^3$$

Controlling $b_E = 16t + b_f$
$= 16(5) + 6.00 = 86$ in.

3. Check AISC Formula 1.11-2.

$$\text{Max } S_{tr} = \left(1.35 + 0.35 \frac{M_L}{M_D}\right) S_s$$

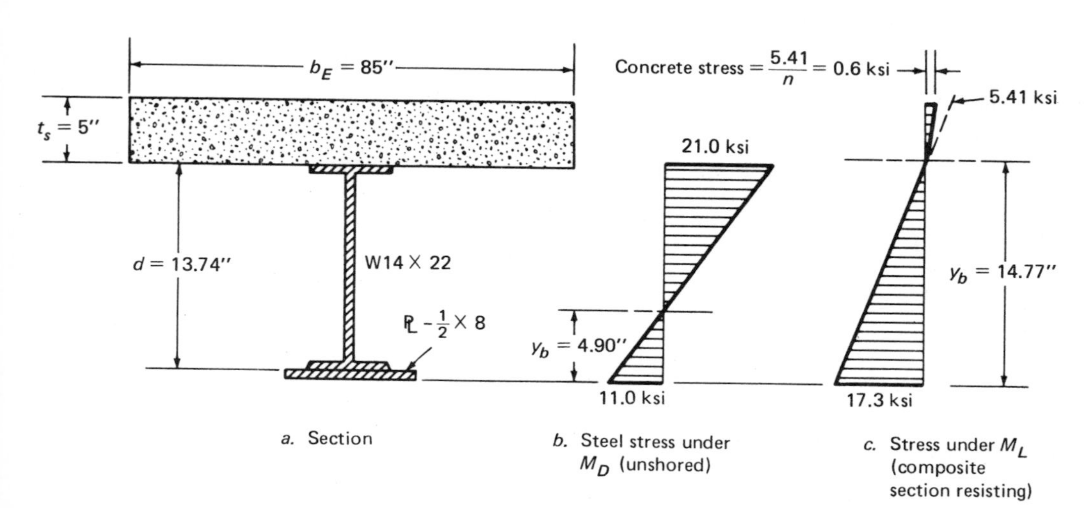

Fig. 2.23. Solution for Ex. 2.9 showing stress under service loads.

$$= \left(1.35 + 0.35 \frac{163}{60}\right) 57.6$$

$$= 2.30(57.6) = 132 \text{ in.}^3 > 93.6 \text{ in.}^3 \quad \text{OK}$$

Use W18 X 35.

4. Determine the minimum number of $\frac{3}{4}$-in.-diameter X 3-in. shear studs required. Sometimes savings can be made by using less than full shear transfer between the concrete slab and the steel beam. For partial composite action, the section modulus used is obtained from Eq. 2.41.

From Eqs. 2.27 and 2.28,

$$V_h = \frac{A_s F_y}{2} = \frac{(10.3)50}{2} = 258 \text{ kips (controls)}$$

or

$$V_h = \frac{0.85 f_c' A_c}{2} = \frac{0.85(3)(86)5}{2} = 548 \text{ kips}$$

Solving Eq. 2.41 for V_h',

$$V_h' = \left[\frac{(S_{\text{eff}} - S_s)}{(S_{tr} - S_s)}\right]^2 V_h$$

$$= \left[\frac{(81.1 - 57.6)}{(93.6 - 57.6)}\right]^2 258 = 110 \text{ kips}$$

The number of connectors required between midspan (point of maximum moment) and the end of the beam is

$$N_1 = \frac{V_h'}{q} = \frac{110}{11.5} = 9.6, \text{ say } 10 \text{ (20 per span)}$$

If full composite action were developed, the effective S would be 93.6 in.^{3}, and the number of connectors would be

$$N_1 = \frac{V_h}{q} = \frac{258}{11.5} = 22.4, \text{ say } 24 \text{ (48 per span)}$$

The spacing required for 20 studs per span (10 pairs) is

$$\text{Spacing} = \frac{30(12)}{10} = 36 \text{ in.}$$

Maximum spacing (AISC-1.11.4)

$$= 8t = 8(5) = 40 \text{ in.} > 36 \text{ in.} \quad \text{OK}$$

Use 20, $\frac{3}{4}$-in.-diameter X 3-in. studs per beam.

5. Alternate design with cover plate. If a cover plate has been used, a W12 X 19 with 1 X 3 cover plate would be selected. The S_{tr} provided (83.3) would have been only slightly higher than that required (81.1); thus, nearly full composite action would have to be developed.

$$V_h = \frac{(5.59 + 3.0)50}{2} = 215 \text{ kips}$$

$$V_h' = \left(\frac{81.1 - 44.9}{83.3 - 44.9}\right)^2 215 = 191 \text{ kips}$$

$$N_1 = \frac{191}{11.5} = 16.6, \text{ say } 17 \text{ (34 per span)}$$

It is problematical whether the W18 X 35 without cover plate with 20 studs, or the W12 X 19 with a 1 X 3 cover plate with 34 studs, is the better choice.

6. Determine the cover-plate length, and specify its connection.

Provided S_x (with 1 X 3 cover) = 83.3 in.^{3}
Provided S_x (without cover) = 41.1 in.^{3}
From Fig. 2.24, the cover plate is required over the distance L_1 between points A and B.

$$\frac{\left(\frac{L_1}{2}\right)^2}{\left(\frac{L}{2}\right)^2} = \frac{223 - 113}{223} = 0.493$$

$$L_1^2 = 0.493 L^2$$

$$L_1 = 0.702 L$$

The *AISC Manual*, "Properties of Composite Beams," for a 5-in. slab also gives L, in terms of KL, where K is given as 0.71.

Under provisions of AISC-1.10.4, the cover plate must develop the cover plate's proportion of the flexural stresses in the beam at the theoretical cutoff point.

Stress at middle of cover plate,

$$f = \frac{M_D + M_L}{I_{tr}} (y_b - 0.5)$$

$$= \frac{113(12)(14.03 - 0.5)}{1170} = 15.7 \text{ ksi}$$

Force in cover plate,

$$F = f A_{pl} = 15.7(3.0) = 47.1 \text{ kips}$$

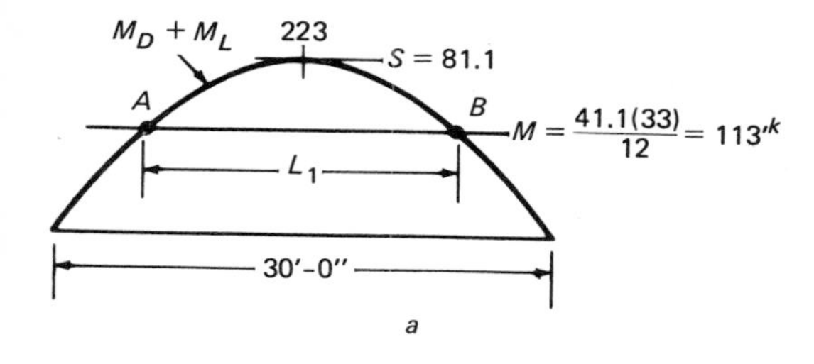

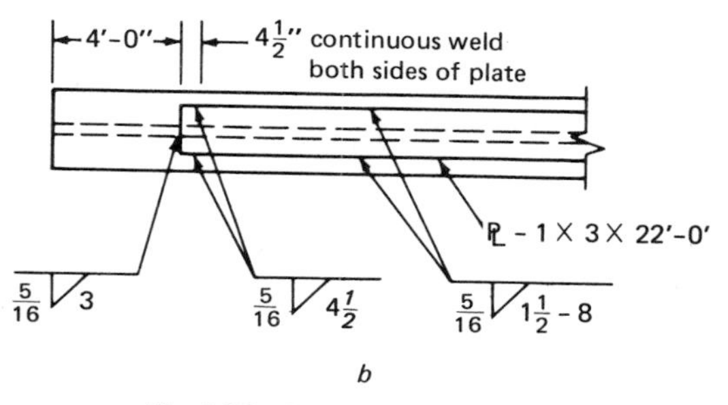

Fig. 2.24. Cover plate for Ex. 2.10.

If one observes that the static moment of area $Q = A_{pl}(y_b - 0.5)$, then

$$F = \frac{MQ}{I}$$

$$F = M\frac{(12\,Q)}{I} \text{ if } M \text{ is in ft-kips}$$

The quantity $12\,Q/I$ is given in *AISC Manual*, "Properties of Composite Beams," for a 5-in. slab. For this problem, $12\,Q/I = 0.42$.

$$F = 113(0.42) = 47.4 \text{ kips}$$

Length of plate required beyond theoretical termination point:

Minimum weld size = $\frac{5}{16}$ in. (AISC Table 1.17.2 based on the thicker of the cover plate or the flange of W12 X 19).

Weld capacity per inch, $R_W = \frac{5}{16}(0.707)21 = 4.64$ kips/in. for E70 electrodes).

$$\text{Weld length} = \frac{47}{4.64} = 10.1 \text{ in.}$$

Try using weld along the end (3 in.), and use 4 in. along each side. Check AISC-1.10.4, Case 2,

Minimum length beyond theoretical cutoff

$$= 1\tfrac{1}{2} \text{ times plate width}$$

$$= 1\tfrac{1}{2}(3) = 4.5 \text{ in.}$$

$$\text{Cover-plate length} = 0.702L + 2\frac{(4.5)}{12}$$

$$= 0.702(30) + 0.75 = 21.8 \text{ ft.}$$

Use Bottom cover plate 1 X 3 X 22'-0",welded and positioned as shown in Fig. 2.24.

Except for the first 4.5 in. on each side of the cover plate, the remainder of the connection can be welded with intermittent welds.

Minimum weld size = $\frac{5}{16}$-in.

(based on AISC-1.17.2 for maximum thickness over $\frac{3}{4}$-in.)

For a minimum segment length of 1.5 in.,

$$\text{Segment capacity} = 1.5(4.64) = 6.96 \text{ kips}$$

The maximum horizontal shear to be transferred occurs at the cutoff location. Neglecting partial span loading, the shear is

$$V = \frac{W_{D+L}L}{2} - W_{D+L}(4.0)$$

$$= \frac{(0.53 + 1.45)(30)}{2} - 1.98(4) = 21.8 \text{ kips}$$

$$\frac{VQ}{I} = 21.8\left(\frac{0.42}{12}\right) = 0.76 \text{ kips/in.}$$

$$\text{Required spacing} = \frac{2(6.96)}{VQ/I} = \frac{2(6.96)}{0.76} = 18.4 \text{ in.}$$

Maximum spacing permitted = $24t$ = 24 (0.349) = 8.4 in. (governs) (AISC 1.18.3.1) and not more then 24 in. in any case.

Use Intermittent $\frac{5}{16}$-in. fillet welds, $1\frac{1}{2}$-in. segments @ 8-in. pitch, except for the first 4.5 in. at each end of plate where continuous weld is to be used. See Fig. 2.24b.

Comparison:

1. W18 X 35 with 32 studs — 35 lb/ft
2. W12 X 19 with 1 X 3 X 22'-0" plate with 36 studs — 26.3 lb/ft

It is likely that the economical choice is to use the cover-plated beam, although it is a borderline decision. If less than 7 or 8 lb/ft is saved by using a cover plate, the plate should not be used.

EXAMPLE 2.11

Design a composite hybrid beam to carry M_D = 90 ft-kips and M_L = 220 ft-kips. Use an A36 web and A514 (F_y = 100 ksi) steel for either the tension flange only or both flanges. The span is 30 ft, beam spacing is 8 ft, and a 5-in. slab (f_c' = 3000 psi, n = 9) is used.

Solution

1. Moments and required section modulus values.

$$M_D = 90 \text{ ft-kips (122 kN-m)};$$

$$M_L = 210 \text{ ft-kips (285 kN-m)}$$

$$\text{Required } S_{tr} = \frac{M_D + M_L}{0.60F_y} = \frac{300(12)}{60}$$

$$= 60 \text{ in.}^3 \text{ (983 cm}^3\text{)}$$

A514 steel is not permitted to be used as a "compact" section under AISC-1.5.1.4.1; thus, the maximum allowable stress is $0.60F_y$. The use of an A36 web will reduce the allowable below $0.60F_y$ in accordance with Eq. 2.36 (AISC-1.10.6).

For the steel section alone.

$$\text{Required } S_s = \frac{M_D}{0.60F_y} = \frac{90(12)}{60}$$

$$= 18 \text{ in.}^3 \text{ (295 cm}^3\text{)}$$

The use of $0.60F_y$ for the steel section acting noncompositely assumes lateral support at spacing no greater than $32r_T\sqrt{C_b}$ (AISC-1.5.1.4.6a). Note that the AISC Formula (1.5-7) does *not* apply to hybrid plate girders.

2. Select trial section. As a guideline for establishing depth, use L/d of about 20 for the steel section alone for situations where deflection control is an important consideration. (See Section 2.12 for details regarding deflection.)

$$d = \frac{L}{20} = \frac{30(12)}{20} = 18 \text{ in.}$$

In this case, the slab is relatively stiff, and the steel beam will represent a smaller than usual proportion of the total effective area; thus, a shallower than 18-in. section may be acceptable. The AISC "Composite Beam Selection Table" for a 5-in. slab indicates a very light steel section in the range of 14- to 16-in. depth. Try a 14-in. deep and minimum $\frac{1}{4}$-in. thick web plate.

Assuming a symmetrical section,

$$S_s = \frac{I}{d/2} \approx \frac{2A_f(d/2)^2 + t_w d^3/12}{d/2}$$

$$= A_f d + t_w d^2/6 = A_f d + A_w d/6$$

$$\text{Required } A_f = \frac{\text{Reqd } S_s - A_w d/6}{d}$$

$$= \frac{18}{14} - \frac{0.25(14)}{6} = 0.70 \text{ in.}^2$$

for the steel section alone. An unsymmetrical section will likely give the most economical arrangement, but, with the same flange area required, minimum-size plates will be necessary so that little advantage will accrue to an unsymmetrical section for this problem.

Try flanges, $\frac{1}{4} \times 3$, $A_f = 0.75$ in.^{2} with $\frac{1}{4} \times$ 14 web. Properties of the steel section:

Flanges $2(0.75)(7.125)^2$ =	76.1
Web $0.25(14)^3/12$ =	57.2
I =	133.3 in.^{4}

$$\text{Area} = 2(0.75) + 0.25(14) = 5.0 \text{ in.}^2$$

$$S_s = \frac{133.3}{7.25} = 18.4 \text{ in.}^3 \text{ (302 cm}^3\text{)}$$

Properties of the composite section:

$$b_E = 16t + b_f = 16(5) + 3 = 83 \text{ in.}$$

$$S_{tr} = 39 \text{ in.}^3 < 60 \text{ in.}^3 \text{ required.}$$

Composite section modulus governs! Increase section to $\frac{1}{4} \times 16$ web, $\frac{1}{4} \times 4$ top flange, $\frac{3}{8} \times 6$ bottom flange.

Properties of the steel section:

Element	Area (A)	Moment Arm from Top of Slab (y)	Ay	Ay²	I_o
Top flange	1.0	0.5	0.50	0.25	-
Web	4.0	8.25	33.0	272	85.3
Bottom flange	2.25	16.44	37.0	608	-
	7.25 in.²		70.5	880	85
				85	
			I_{top} =	965 in.⁴	

$$y_{top} = \frac{70.5}{7.25} = 9.72 \text{ in.}$$

$$I_{cg} = 965 - 7.25(9.72)^2 = 280 \text{ in.}^4$$

$$S_s(\text{bottom}) = \frac{280}{16.63 - 9.72} = 40.6 \text{ in.}^3 \ (665 \text{ cm}^3)$$

$$S_s(\text{top}) = \frac{280}{9.72} = 28.8 \text{ in.}^3 \ (472 \text{ cm}^3)$$

Properties of the composite section:

Element	Area (A)	Moment Arm (y)	Ay	Ay²	I_o
Slab	46.1	2.5	115.3	288	96
Steel section	7.25	14.72	106.7	1571	280
	53.35 in.²		222.0	1859	376
				376	
			I_{top} =	2235 in.⁴	

$$y_{top} = \frac{222}{53.35} = 4.16 \text{ in.}$$

$$I_{cg} = 2235 - 53.35(4.16)^2 = 1312 \text{ in.}^4$$

Note that 0.84 in. of concrete slab near the neutral axis is in tension and was considered effective in computing the properties. AISC-1.11.2.2 says that concrete tension stresses shall be neglected. It makes little difference here and simplifies computing properties (S = 74.7 "exact" vs. 75.1 as computed).

$$S_{tr} \text{ (tension flange)} = \frac{1312}{21.63 - 4.16}$$

$$= 75.1 \text{ in.}^3 \ (1230 \text{ cm}^3)$$

3. Check stresses on section. Because of the hybrid section, the A36 web will yield before the strength of the A514 flanges has been developed. The flange allowable stress is reduced according to AISC-1.10.6 to account for this (see Section 2.8). Using Eq. 2.36 (see also Table 2.6),

$$F_b' < F_b \left[\frac{12 + \beta(3\alpha - \alpha^3)}{12 + 2\beta}\right]$$

$$\beta = \frac{A_w}{A_f} = \frac{4.0}{2.25} = 1.78 \text{ for tension flange}$$

$$\alpha = \frac{\text{Web } F_y}{\text{Flange } F_y} = \frac{36}{100} = 0.36$$

$$F_b' = 60 \left\{\frac{12 + 1.78[3(0.36) - (0.36)^3]}{12 + 2(1.78)}\right\}$$

$$= 60(0.89) = 53.4 \text{ ksi } (368 \text{ N/mm}^2)$$

$$f_b = \frac{M_D + M_L}{S_{tr}} = \frac{300(12)}{75.1}$$

$$= 47.9 \text{ ksi} < 53.4 \text{ ksi } (330 \text{ N/mm}^2)$$

For the noncomposite loading, the compression stress on the steel beam alone controls. Since the compression flange has less area than the tension flange, the allowable stress F_b' will be different.

$$\beta = \frac{A_w}{A_f} = \frac{4.0}{1.0} = 4.0 \text{ for compression flange}$$

$$F_b' = 0.807(60) = 48.4 \text{ ksi}$$

(from Table 2.18 for $\psi = 0.5$ and $\beta = 4$)

It is to be noted that the AISC formula does not account for the unsymmetrical section.

$$f_b = \frac{M_D}{S_s \text{ (top)}} = \frac{90(12)}{28.8}$$

$$= 37.5 \text{ ksi} < 48.4 \text{ ksi} \quad \text{OK}$$

The stresses are somewhat low. Try reducing the tension flange to $\frac{5}{16} \times 6$. Retain $\frac{1}{4} \times 16$ web and $\frac{1}{4} \times 4$ top flange.
Steel section properties:

$A = 6.88$ sq in.^{2}; $\quad y_{top} = 9.35$ in.;

$I = 262$ in.^{4}; $\quad S_s(\text{top}) = 28.0$ in.^{3}

Composite section properties:

$A = 52.98$ in.^{2}; $\quad y_{top} = 4.04$ in.;

$I = 1198$ in.^{4}; $\quad S_{tr} = 68.4$ in.^{3}

For tension flange, $\beta = A_w/A_f = 2.13$

$$F_b' = 52.4 \text{ ksi } (361 \text{ N/mm}^2)$$

$$f_b = \frac{M_D + M_L}{S_{tr}} = \frac{300(12)}{68.4}$$

$$= 52.6 \text{ ksi} \approx F_b' = 52.4 \text{ ksi} \quad \text{OK}$$

On steel section, for top flange,

$$f_b = \frac{M_D}{S_s} = \frac{90(12)}{28.0} = 28.6 \text{ ksi} < F_b' = 48.4 \text{ ksi } (197 \text{ N/mm}^2)$$

Use ℞ - $\frac{1}{4} \times 16$ (A36) for web; ℞ - $\frac{1}{4} \times 4$ (A514) for top flange; ℞ - $\frac{5}{16} \times 6$ (A514) for bottom flange. See Fig. 2.25.

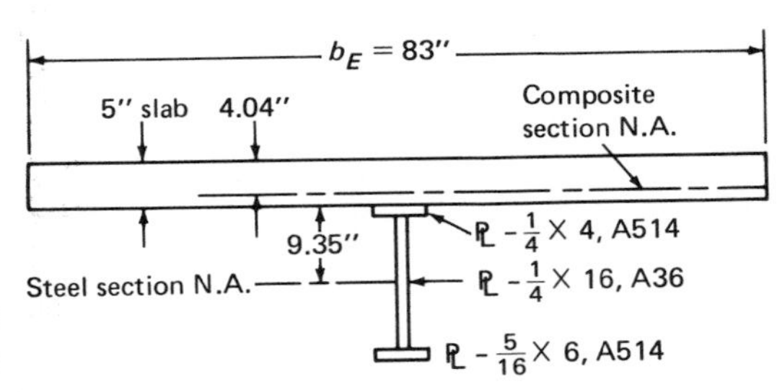

Fig. 2.25. Hybrid composite section of Ex. 2.11.

The design of shear connectors for this girder requires no new concepts and is not illustrated.

2.12 DEFLECTIONS

In order to accurately determine the deflections of composite members, a number of factors must be taken into account that are not normally considered. These are: the method of construction, the separation of the live-load and dead-load moments, and the effect of creep and shrinkage in the concrete slab.

Limits on deflections are not well defined. Most engineers limit live-load deflections to $L/360$ for both shored and unshored construction. A well-defined limit on dead-load deflection for unshored construction does not exist. Some engineers limit the dead-load deflection to 1-in. (25.4 mm) so as to prevent a ponding problem on the unshored steel beam during the placing of the concrete.

If the construction is without shoring, the total deflection will be the sum of the dead-load deflection of the steel beam and the live-load deflection of the composite section. If shoring is used, then the total deflection will result from the dead and live loads on the composite section.

EXAMPLE 2.12

Determine the total deflection of the composite section in Ex. 2.8, and check against maximum deflection permitted by AISC if no shoring is used (see Fig. 2.26).

Solution

Dead-load deflection

$$\Delta_{DL} = \frac{5wL^4}{384EI} = \frac{5(0.40 + 0.036)(28)^4(12)^3}{384(29{,}000)447}$$

$$= 0.46 \text{ in. } (11.7 \text{ mm})$$

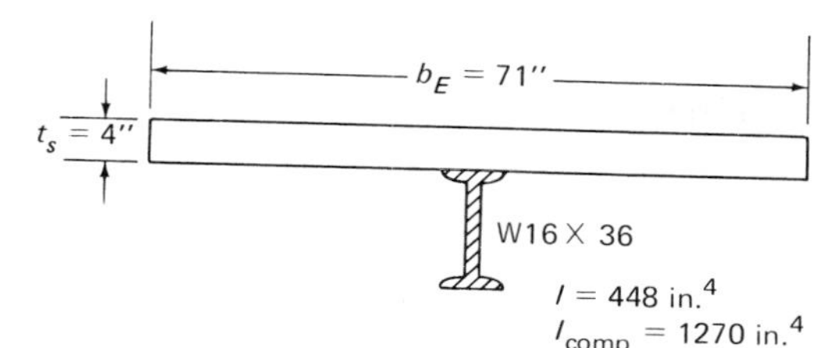

Fig. 2.26. Example 2.12.

Live-load deflection

$$\Delta_{LL} = \frac{5wL^4}{384EI_{comp}} = \frac{5(1.2)(28)^4(12)^3}{384(29{,}000)1270}$$

$$= 0.45 \text{ in. } (11.4 \text{ mm})$$

Total deflection

$$\Delta = \Delta_{DL} + \Delta_{LL} = 0.46 + 0.45$$

$$= 0.91 \text{ in. } (23.1 \text{ mm})$$

Check maximum deflection permitted

$$\Delta_{max} = \frac{L}{360} = \frac{28(12)}{360} = 0.93 \text{ in. } (23.6 \text{ mm})$$

Since $\Delta_{actual} < 0.93$ in., the deflection criterion is satisfied.

If shoring provides support during the hardening of the concrete, the total deflection will be a function of the total composite section. Account must be taken of the fact that concrete is subject to creep under long-time loadings and that shrinkage will occur. This inelastic behavior may be approximated by multiplying the modular ratio n by a factor to reduce the net effective width. The result is a reduced moment of inertia for the composite section that is used in computing the dead-load deflection. The live-load deflection is then usually computed on the basis of the elastic composite moment of inertia. Occasionally, the conservative approach is to use the reduced composite moment inertia when the live loads are expected to remain for extended periods of time.

Because the concrete slab in building construction is normally not too thick (e.g., $t \leqslant 5$ in.), creep deflection is not considered to be a problem. AISC gives no indication that one need be concerned with anything but live-load short-time deflection. ASCE-ACI (1960) recommends using one-half the concrete modulus of elasticity $E_c/2$ instead of E_c when computing sustained load creep deflection. AASHTO (1973) uses $E_c/3$ instead of E_c. Such arbitrary procedures can at best give an estimate of creep effects, probably not better than ±30%. The steel section, exhibiting no creep, and representing the principal carrying element, insures that creep problems will usually be minimal.

More accurate procedures for computing deflections to account for creep and shrinkage on composite steel-concrete beams have been studied by Roll (1971) and in detail by Branson (1968, 1977).

EXAMPLE 2.13

Determine the total deflection of the composite section in Ex. 2.8 if temporary shoring is provided. Assume a value of $3n$ for the modular ratio, i.e., $E_c' = E_c/3$ (see Fig. 2.27).

Solution

Properties for composite section using $3n$ instead of n:

$$y_b = \frac{10.6\left(\frac{15.85}{2}\right) + 4(2.63)17.85}{10.6 + 4(2.63)} = 12.87 \text{ in.}$$

$$I_{comp} = 447 + 10.6(4.94)^2 + \frac{1}{12}(2.63)(4)^3$$

$$+ 2.63(4)(4.98)^2 = 981 \text{ in.}^4$$

Dead-load deflection:

$$\Delta_{DL} = \frac{5(0.4 + 0.036)(28)^4(12)^3}{384(29{,}000)981}$$

$$= 0.21 \text{ in. } (5.3 \text{ mm})$$

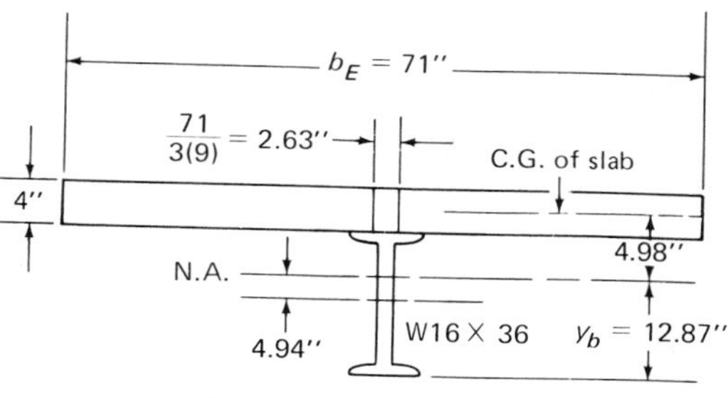

Fig. 2.27. Example 2.13.

Live-load deflection, using elastic I from Ex. 2.12

$$\Delta_{LL} = 0.45 \text{ in. } (11.4 \text{ mm})$$

Total deflection, $\Delta = \Delta_{DL} + \Delta_{LL}$:

$$\Delta = 0.21 + 0.45 = 0.66 \text{ in. } (16.8 \text{ mm})$$

It should be noted that the difference in the computed deflection of the two previous examples was only about $\frac{5}{16}$ in. It is doubtful that this difference would be worth the extra expense of shoring. In general, the advantage of providing shoring during the hardening of the concrete is minimal, and therefore, such shoring is not recommended unless deflection criteria govern.

2.13 CONTINUOUS BEAMS

It has been traditional to design the positive moment region on continuous beams as a composite section and the negative moment region as a noncomposite section. However, some composite action has been known to exist. Significant contribution to the knowledge about the strength of continuous composite beams has been made by Barnard and Johnson (1965), Park (1967), Daniels and Fisher (1967), Johnson et al. (1967), and more recently by Hamada and Longworth (1974, 1976).

According to both AISC-1.11.2.2 and AASHTO-1.7.99 and 1.7.127, the steel reinforcement that extends parallel to the beam span and that is contained within the concrete slab effective width b_E *may be used* as part of the effective composite section. This is true for both positive and negative bending regions. The inclusion of such steel reinforcement has little effect in positive moment regions but can help in negative moment regions. In the negative moment region, the concrete is ordinarily all in tension and is therefore not considered effective (AISC-1.11.2.2 and AASHTO-1.7.96 and 1.7.128).

When the reinforcing bars in the concrete slab are utilized as part of the composite section, the force developed by them must be transferred in shear by the use of mechanical shear connectors. The ultimate force developed would be

$$T \text{ (for } -\text{M region)} = A_{sr}F_{yr} \qquad (2.42)$$

$$C \text{ (for } +\text{M region)} = A_s'F_{yr} \qquad (2.43)$$

where

A_{sr} = total area of longitudinal reinforcing steel at the interior support located within the effective flange width b_E.

A_s' = total area of longitudinal compression steel acting with the concrete slab at the location of maximum positive moment and lying within the effective width b_E.

F_{yr} = specified minimum yield stress of the longitudinal reinforcing steel.

In order to use a working stress method, division by a factor of 2 is used to reduce these forces into the service-load range (see Section 2.7.1, p. 54). Thus, the service-load horizontal shear force to be designed for in the *negative moment zone* is

$$V_h = \frac{A_{sr}F_{yr}}{2} \qquad (2.44)$$

In the *positive-moment zone*, when the compression steel is included in computing the composite section properties, Eq. 2.43 divided by 2 is added to Eq. 2.27. Thus,

$$V_h = \frac{0.85f_c'A_c}{2} + \frac{A_s'F_{yr}}{2} \qquad (2.45)$$

Equations 2.44 and 2.45 are as prescribed by AISC-1.11.4. AASHTO-1.7.100 uses Eq. 2.44 but makes no reference to using the compression steel in positive-moment regions; however, the AASHTO load factor provisions of 1.7.127 do utilize the concept of Eq. 2.45.

Under both AISC and AASHTO, the inclusion of the longitudinal reinforcing bars A_{sr} appears optional. If A_{sr} is included in computing properties, the horizontal shear V_h produced by such bars must be developed by shear connectors. Under AASHTO-1.7.100(3), there are additional shear connectors required at points of contraflexure when A_{sr} is *not* utilized in computing section properties. The minimum number N_e of added connectors for this fatigue requirement is

$$N_e = \frac{A_{sr}f_r}{Z_r} \qquad (2.46)$$

where

f_r = range of stress due to live load plus impact in the slab reinforcement over the support (in lieu of more accurate computations, f_r, may be taken as equal to 10,000 psi)

Z_r = allowable range of horizontal shear on an individual shear connector (see Eq. 2.35).

As discussed in Section 2.6, the one failure mode in the positive-moment region is crushing of the concrete slab. This assumes no shear connector failure and no longitudinal splitting or shear failure in the concrete slab. In the negative-moment region, the usual failure mode is local flange buckling according to Hamada and Longworth (1976).

Under current AISC and AASHTO provisions, the usual lateral buckling provisions for noncomposite steel sections apply to the negative moment region of continuous composite beams. In the use of the lateral-torsional buckling formulas of AISC-1.5.1.4.6a and AASHTO-1.7.1 or 1.7.124(D), the point of contraflexure is generally treated as a braced point. Local buckling limitations for the flange and web also apply (AISC-1.5.1.4.1 and 1.9; AASHTO-1.7.69 or 1.7.124).

The work of Hamada and Longworth (1974) indicates that the negative region of composite continuous beams has measurably greater resistance to lateral buckling than a noncomposite steel section without the concrete slab attached to its top flange. They conclude that "the ultimate moment capacity of composite beams in negative bending is affected by local flange buckling unless the compression flange is stiffened by a cover plate." The cover plate increases the torsional rigidity and may cause lateral buckling to be more critical than local buckling. The most recent local buckling recommendations of Hamada and Longworth (1974) are summarized as follows for cases where the compression flange is a single plate element:

For $A_{sr}/A_w \leqslant 1.0$,

$$\frac{b_f}{2t_f} < \frac{54}{\sqrt{F_y}} \tag{2.47}$$

For $1.0 < A_{sr}/A_w \leqslant 2.0$,

$$\frac{b_f}{2t_f} \leqslant \frac{49}{\sqrt{F_y}} \tag{2.48}$$

These local buckling limitations for the compression flange are more conservative than required under AISC or AASHTO even for "compact" sections. (AISC-1.5.1.4.1b uses $b_f/(2t_f) \leqslant 65/\sqrt{F_y}$.) It would appear that present design procedures are conservative with regard to lateral-torsional buckling in the negative region of composite continuous beams but may not be conservative with regard to local buckling.

EXAMPLE 2.14

Investigate the section of Fig. 2.28 subject to a negative bending moment of 135 ft-kips acting as a composite section under the AISC Specifications. The reinforcing bars have F_y = 50 ksi, and the W12 X 26 has F_y = 60 ksi.

Solution

1. Compute section properties under negative bending. The concrete slab is not considered to participate since it is on the tension side of the neutral axis.

	Area (A)	*Moment arm from top* (y)	Ay	Ay^2	I_o
#5 bars, A_s = 0.31	3.10	2.0	6.2	12	—
W12 X 26	7.65	10.11	77.3	782	204
	10.75 in.^{2}		83.5	794	204
				204	
			I_{top} = 998 in.^{4}		

$$y_{top} = \frac{83.5}{10.75} = 7.77 \text{ in.}$$

$$I_{cg} = 998 - 10.75(7.77)^2 = 349 \text{ in.}^4$$

$$S_{tr}(\text{bottom}) = \frac{349}{8.22} = 41.3 \text{ in.}^3 \ (677 \text{ cm}^3)$$

$$S_{tr}(\text{at \#5 bars}) = \frac{349}{5.77} = 60.5 \text{ in.}^3 \ (991 \text{ cm}^3)$$

2. Check stresses.

On the composite section,

$$f_b = \frac{M_{D+L}}{S_{tr}} = \frac{135(12)}{41.3} = 39.2 \text{ ksi } (270 \text{ N/mm}^2)$$

Assuming that the distance of lateral unsupport does not exceed $L_c = 9.8 b_f = 5.3$ ft (AISC-1.5.1.4.1), the section may be considered "compact" and the allowable stress is $0.66F_y$, since the W12 X 26 satisfies the local buckling requirements ($b_f/2t_f \leqslant 8.4$ and $d/t \leqslant 82.6$).

$$f_b = 39.2 \text{ ksi} < F_b = 39.6 \text{ ksi} \quad \text{OK}$$
$$(270 \text{ N/mm}^2) \quad (273 \text{ N/mm}^2)$$

If the section had been required to satisfy the more conservative requirements of Eq. 2.47 or 2.48 for local flange buckling, the section would not have qualified. Thus,

$$\text{W12} \times 26, \frac{b_f}{2t_f} = 8.54 > \frac{54}{\sqrt{F_y}} = 7.0 \quad \text{NG}$$

3. Shear connectors. Use $\frac{5}{8}$-in.-diameter X $2\frac{1}{2}$-in. studs: $q = 8.0$ for $f'_c = 3000$ psi concrete.

$$V_h = \frac{A_{sr}F}{2} = \frac{10(0.31)50}{2} = 77.5 \text{ kips}$$

The number N_1 of studs required between the maximum negative-moment location (the support) and the point of contraflexure is

$$N_1 = \frac{V_h}{q} = \frac{77.5}{8.0} = 9.7, \text{ say } 10.$$

REFERENCES

AASHTO (1973), *Standard Specifications for Highway Bridges*, 11th Edition, American Association of State Highway and Transportation Officials, Washington D.C., 1973. Also, 1974–77 *Interim Specifications Bridges.*

ACI (1977), ACI Committee 318, *Building Code Requirements for Reinforced Concrete*, American Concrete Institute, Detroit, Michigan.

AISC (1978), *Specification for the Design, Fabrication and Erection of Structural Steel for Buildings*, American Institute of Steel Construction, New York.

ASCE-AASHTO (1968), Report of Subcommittee 1 on Hybrid Beam and Girders, Joint ASCE-AASHTO Committee on Flexural Members, C. G. Schilling, Chmn., Design of Hybrid Steel Beams, *J. Structural Division*, ASCE, Vol. 94, No. ST6 (June), pp. 1397–1426.

ASCE-ACI (1960), Joint ASCE-ACI Committee on Composite Construction, "Tentative Recommendations for the Design and Construction of Composite Beams and Girders for Buildings," *J. Structural Division*, ASCE, Vol. 68, No. ST12 (December), pp. 73–92.

Barnard, P. R. and Johnson, R. P. (1965), Plastic Be-

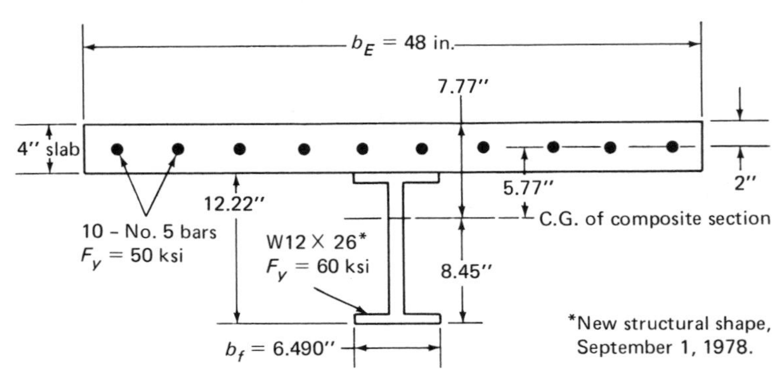

Fig. 2.28. Composite section for negative bending of Ex. 2.14.

havior of Continuous Composite Beams, *Proceedings, Institution of Civil Engineers*, October.

Branson, D. E. (1977), *Deformation of Concrete Structures*, McGraw-Hill, Dusseldorf and New York. pany, Center for Advanced Publishing, Dusseldorf and New York.

Branson, D. E. (1968), Design Procedures for Computing Deflections, *ACI Journal, Proceedings*, Vol. 65, September, pp. 703–742.

Brendel, G. (1964), Strength of the Compression Slab of T-Beams Subject to Simple Bending, *ACI Journal, Proceedings*, Vol. 61, January, pp. 57–76.

Caughey, R. A. (1929), Composite Beams of Concrete and Structural Steel, *Proceedings*, 41st Annual Meeting, Iowa Engineering Society.

Daniels, J. H. and Fisher, J. W. (1967), *Static Behavior of Continuous Composite Beams*, Fritz Engineering Laboratory Report No. 324.2., Lehigh University, Bethlehem, Pennsylvania, March.

Frost, R. W. and Schilling, C. G. (1964), Behavior of Hybrid Beams Subjected to Static Loads, *J. Structural Division*, ASCE, Vol. 90, No. ST3 (June), pp. 55–88.

Hamada, S. and Longworth, J. (1974), Buckling of Composite Beams in Negative Bending, *J. Structural Division*, ASCE, Vol. 100, No. ST11 (November), pp. 2205–2222.

Hamada, S. and Longworth, J. (1976), Ultimate Strength of Continuous Composite Beams, *J. Structural Division*, ASCE Vol. 102, No. ST7 (July), pp. 1463–1478.

Johnson, J. E. and Lewis, A. D. M. (1966), Structural Behavior in a Gypsum Roof-Deck System, *J. Structural Division*, ASCE Vol. 92, No. ST2 (April), pp. 283–296.

Johnson, R. P., Van Dalen, K., and Kemp, A. R. (1967), Ultimate Strength of Continuous Composite Beams, *Proceedings of the Conference on Structural Steelwork*, British Constructional Steelwork Association, November.

Lew, H. S. (1970), *Effect of Shear Connector Spacing on the Ultimate Strength of Concrete-on-Steel Composite Beams*, National Bureau of Standards Report No. 10 246, United States Department of Commerce, August.

MacKay, H. M., Gillespie, P., and Leluau, C. (1923), Report on the Strength of Steel I-Beams Haunched with Concrete, *Engineering Journal*, Engineering Institute of Canada, Vol. 6, No. 8, pp. 365–369.

McGarraugh, J. B. and Baldwin, J. W., Jr. (1971), Lightweight Concrete-on-Steel Composite Beams, *Engineering Journal*, AISC, Vol. 8, No. 3, July, pp. 90–98.

Ollgaard, J. G., Slutter, R. G., and Fisher, J. W. (1971), Shear Strength of Stud Connectors in Lightweight and Normal-Weight Concrete, *Engineering Journal*, American Institute of Steel Construction, Vol. 8, No. 2 (April), pp. 55–64.

Park, R. (1967), The Ultimate Strength of Continuous Composite Beams, *Civil Engineering Transactions*, Australia, Vol. CE9, October.

Reissner, E. (1934), Uber die Berechnung von Plattenbalkan, *Der Stahlbau*, Vol. 26, December.

Roll, F. (1971), Effects of Differential Shrinkage and Creep on a Composite Steel-Concrete Structure, *Designing for Effects of Creep, Shrinkage, Temperature in Concrete Structures*, SP-27, American Concrete Institute, Detroit.

Ros, M. (1934), Les constructions acier-beton, system alpha, *L'Ossature Metallique* (Bruxelle), Vol. 3, No. 4, pp. 195–208.

Sabnis, G. M. and Reddy, G. (1977), *Some Aspects of Shear Strength of Steel-Concrete Composite Beams*, Department of Civil Engineering, Howard University, Washington, D.C., NBS Contract No. B-63-291-1, September, 173 pp. (Submitted to the National Bureau of Standards, Gaithersburg, Maryland.)

Sabnis, G. M. and Dabholkar, A. Y. (1977), *Low Cycle Fatigue Strength of Shear Connectors*, Department of Civil Engineering, Howard University, Washington, D.C., NBS Contract No. B-63-291-2, September, 77 pp. (Submitted to the National Bureau of Standards, Gaithersburg, Maryland.)

Schilling, C. G. (1968), Bending Behavior of Composite Hybrid Beams, *J. Structural Division*, ASCE, Vol. 94, No. ST8 (August), pp. 1945–1964.

Slutter, R. G. and Fisher, J. W. (1966), Fatigue of Shear Connectors, *Highway Research Record No. 147*, Highway Research Board, pp. 65–88.

Slutter, R. G. and Driscoll, G. C. (1965), Flexural Strength of Steel-Concrete Composite Beams, *J. Structural Division*, ASCE, Vol. 91, No. ST2 (April), pp. 71–99.

Timoshenko, S. and Goodier, J. (1951), *Theory of Elasticity*, 2nd Ed., McGraw-Hill Book Company, Chapter 6.

Toprac, A. A. (1965a), Strength of Three New Types of Composite Beams, *Bulletin No. 2*, American Iron and Steel Institute, October (contains test results not in Toprac 1965b).

Toprac, A. A. (1965b), Strength of Three New Types of Composite Beams, *Engineering Journal*, American Institute of Steel Construction, Vol. 2, No. 1 (January), pp. 21–30.

Toprac, A. A. and Eyre, D. G. (1967), Composite Beams with a Hybrid Tee Steel Section, *J. Structural Division*, ASCE, Vol. 93, No. ST5 (October), pp. 309–322.

Viest, I. M. (Chairman) (1974), Composite Steel-Concrete Construction, Report of the Subcommittee on the State-of-the-Art Survey of the Task Committee on Composite Construction of the Committee on Metals of the Structural Division, *J. Structural Division*, ASCE, Vol. 100, No. ST5 (May), pp. 1085–1139.

Viest, I. M. (1960), Review of Research on Composite Steel-Concrete Beams, *J. Structural Division*, ASCE, Vol. 86, No. ST6 (June), pp. 1–21.

Von Kármán, T. (1924), Die Mittragende Breitte, Festschrift. August Föppls. (See also *Collected Works of Theodore von Kármán*, Volume II, p. 176.)

3

Applications of light-gauge steel in composite construction

JAMES M. FISHER, Ph.D., P.E.
Principal
Computerized Structural Design, Inc.
Milwaukee, Wisconsin

DONALD R. BUETTNER, Ph.D., P.E.
Principal
Computerized Structural Design, Inc.
Milwaukee, Wisconsin

3.1 INTRODUCTION

Light-gauge steel-concrete composite floor systems are in common use. Generally, such systems consist of a concrete slab on some type of cold-formed corrugated and/or ribbed decking. When composite construction is discussed, it is the use of metal deck that is commonly considered. There are, however, other uses for the deck:

1. form for wet concrete
2. working platform during construction
3. diaphragm for the transfer of lateral loads
4. part of a composite beam system.

This chapter has a twofold objective. First, a general overview of the background, research, and resulting design procedures for deck-concrete composite systems are presented. Second, examples are given to illustrate how composite systems can be designed. It is presumed that the reader is experienced in the basic rudiments of structural analysis and design.

3.2 BACKGROUND

The advantages of combining the structural properties of cold-formed light-gauge steel deck and concrete for use in floor systems for buildings were recognized many years ago. The most significant advantage was the reduced cost of the structure and foundations. Furthermore, the use of the deck, both as a platform for construction operations and as a form for the concrete, replaced the expensive conventional forming systems used previously. The deck, even in the early years, was recognized for its potential in channeling electrical and communications wiring through cellular construction. The steel ceiling formed by the underside of the deck facilitates the attachment of hanger supports for piping, ductwork, and suspended ceilings. Also, since the metal deck acts as the form for the wet concrete and generally does not require shoring, the time of construction for a structure may be reduced since each floor is independent and one need not wait for concrete to gain strength to support superimposed shoring as in cast-in-place systems. In addition, time is not required to remove shoring. The use of steel deck is accompanied by a nominal amount of temperature and shrinkage reinforcement, and the deck itself serves as the positive reinforcement once the concrete has hardened. (In building construction, top reinforcement is generally provided only when continuity is required.) Not withstanding all of these advantages, the earliest use of light-gauge decks was in *non*composite applications.

In order to present an unbiased treatment of the subject, it is reasonable to discuss the disadvantages of steel-deck composite construction. Such disadvantages seem centered on the characteristics of the deck. To insure good bond, the deck requires cleaning prior to placing concrete. Removal of oil and sawdust can be a problem. Furthermore, oil and water create a slippery and potentially dangerous working surface, tending to nullify one of the principal advantages. Extensive (and expensive) fire testing is required to obtain the fire ratings called for by many national and local codes. Finally, the advantage of the availability of the deck to act as a working platform can become a disadvantage if the deck is damaged by the temporary storage of heavy, concentrated loads, such as brick pallets.

Composite construction utilizing light-gauge deck began with the Granco Steel Products Co. "Cofar" deck. Rapid development of other deck systems followed. Much of the early work done with deck systems in terms of analysis, research, testing, and development of design procedures was on a proprietary basis, with each deck manufacturer working essentially independently.

The early engineering studies on composite concrete metal-deck systems were based on working stress design concepts. The independent nature of the early research was prompted by the need to obtain approvals of various code authorities. Since no unified design theory had evolved, such independent research and testing were the only methods by which such approvals could be obtained. This is understandable because of the highly competitive character of the industry.

Early studies, primarily of Friberg (1954) and Bryl (1967), examined the mechanism of composite construction. Friberg studied the "Cofar" system, and Bryl examined several deck systems. Both agreed on the cost-saving potential of this type of construction. Early studies concluded that shear transfer devices between the deck and slab are required to prevent brittle slab failures, and that slabs with shear transfer devices can accommodate large plastic deformations and develop substantial increases in load-carrying capacity.

In the early development of composite design with light-gauge steel-concrete systems, the competitive nature of the industry was significant. Since the properties and configurations of the decks of the various manufacturers were different, little exchange of information occurred. Eventually, this had a negative effect on the entire market for these systems.

Consequently, in 1967, the American Iron and Steel Institute (AISI) sponsored a research project at Iowa State University, with the principal objective of developing unified design criteria for steel deck-concrete composite floor systems. Numerous research publications and technical papers (e.g., Porter 1968, Porter and Eckberg 1971, and Schuster 1972) have evolved as a result of that investigation.

The Iowa State research was directed primarily at the development of an ultimate strength-design approach. The initial direction of the program involved extensive pushout tests. However, it was shown that many deck systems were not compatible with pushout testing, and it was ultimately concluded that only beam tests could be used to predict capacity based on the ultimate strength design approach. The beam-test data clearly revealed that failures were related to a horizontal shear (separation) at the concrete-deck interface. Principal variables affecting capacity were shown to be shear span, steel percentage, and the shear-transfer characteristics of the particular deck. Only in rare instances did flexural failures occur in these tests. The results of the Iowa State research were presented at the Internation Association of Bridge and Structural Engineers (IABSE) Congress by Eckberg and Schuster (1968). Results of studies relating to the ultimate-strength characteristics of composite systems were presented in the later IABSE Congress by Porter and Eckberg (1972). An excellent summary of pertinent literature dealing with composite design may be found in a recent paper by Schuster (1976).

This text does not have a basic objective of examining the literature in detail. However, the significant contributions developed from these various investigations did yield the following generalized conclusions:

1. Maximum capacity of composite systems depends upon shear bond strength (Eckberg and Schuster 1968, Porter and Eckberg 1972).
2. The steel deck may be designed to carry the wet weight of concrete using elastic design methods (Porter and Eckberg 1976).
3. Design for deflection limitations of the composite system can be based on ACI (1971) Building Code deflection design criteria.
4. Shallow deck profiles, while limited in form span capability, exhibit better composite action than deep deck sections (Luttrell and Division 1971).

If there is a general gap in the state of knowledge of the behavior and design procedures at the present time, it would have to center on the areas of dynamic and cyclic (repeated) loadings. While some work has been done by Sayed et al. (1974), the subject has not been studied exhaustively as with other aspects. As a result, metal deck-composite slabs have not experienced a wide application in bridge decks and parking structures where such loadings exist.

3.3 STEEL DECKS

As stated previously, the exact configuration of the various decks is proprietary with the various manufacturers. Widths of decks extend up to 30 in. (75 cm), and depths generally vary from $1\frac{1}{2}$ to 3 in. (38 to 76 mm). The heavy-gauge 0.07-in. (2 mm) decks weigh up to 8 psf (380 N/m^2). Figures 3.1 through 3.4 show typical examples of various decks.

It is important to note the various ways in which the different deck units achieve composite action. There are basically two ways in which such composite behavior is accomplished. First, many of the decks have ridges, corrugations, or other local "deformations" that act as shear-transfer devices. These "shear connectors" act in a mechanical manner to "lock" the slab to the deck. The second class of decks have no deformations and are classified as "smooth" decks, and composite action is accomplished only through the chemical bond that is achieved.

3.4 COMPOSITE BEAMS AND JOISTS

One of the most recent applications of composite construction using light-gauge steel deck is to use the deck as the form onto which the slab is cast, and to use the resulting combination to act compositely with beams, girders, and open-web joists. Significant research activities of Fisher (1970) and Robinson (1969) contributed heavily to current knowledge of the behavior of these members. Essentially, the composite slab-deck acts with the beam or joist to create T-beam action, stiffening the floor for a given design or permitting a lighter beam section where stiffness is not a consideration. Shown in Fig. 3.5 is a composite steel beam.

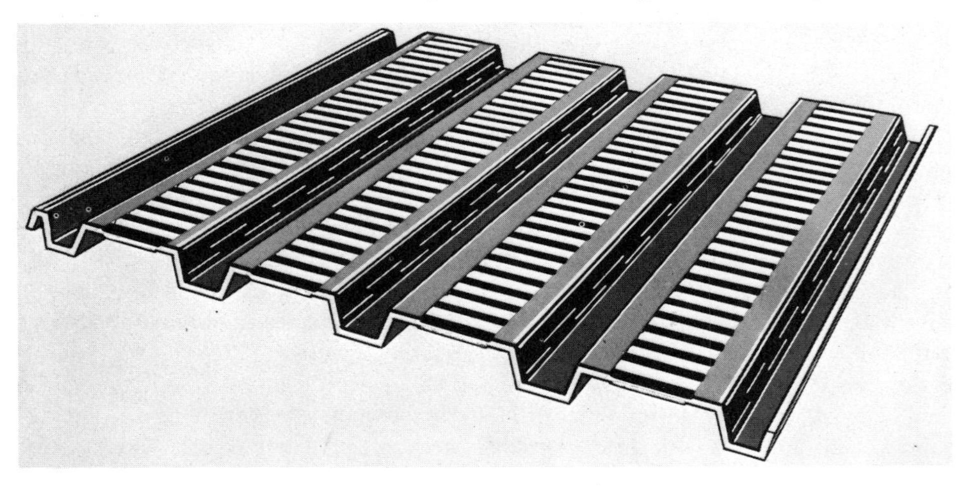

Fig. 3.1. Thirty-inch coverage, $1\frac{1}{2}$-in. deck.

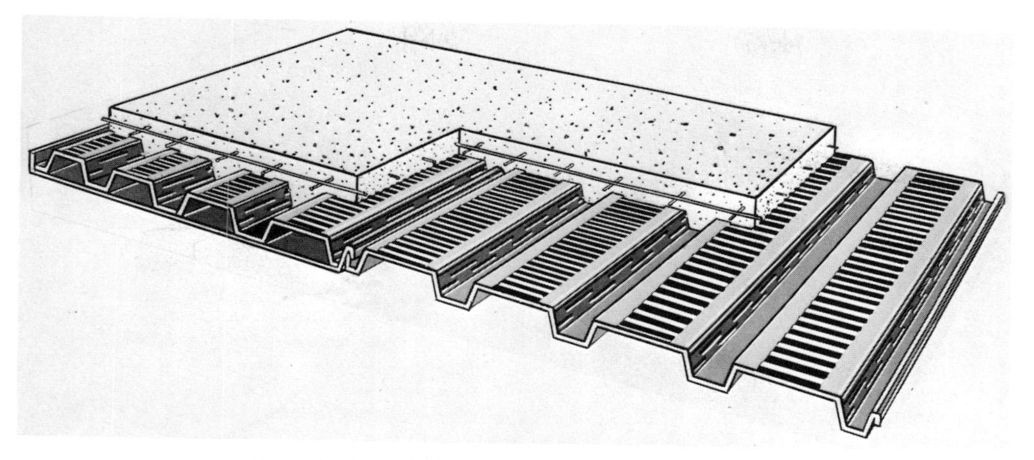

Fig. 3.2. Blended (cellular and noncellular) $1\frac{1}{2}$-in. deck. (*Courtesy RollForm Products Corp., Boston, Massachusetts.*)

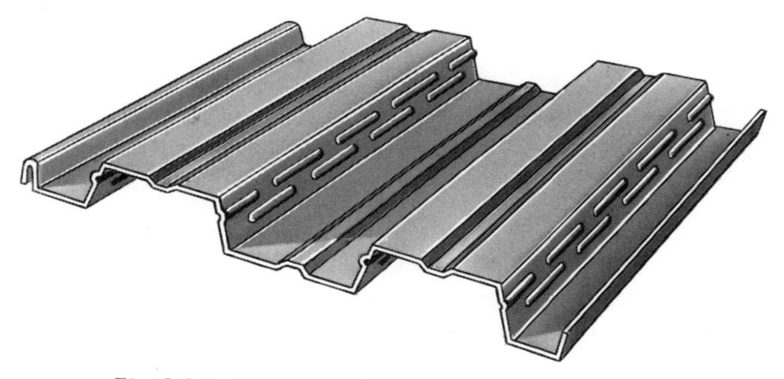

Fig. 3.3. Twenty-four-inch coverage, 2-in. deck.

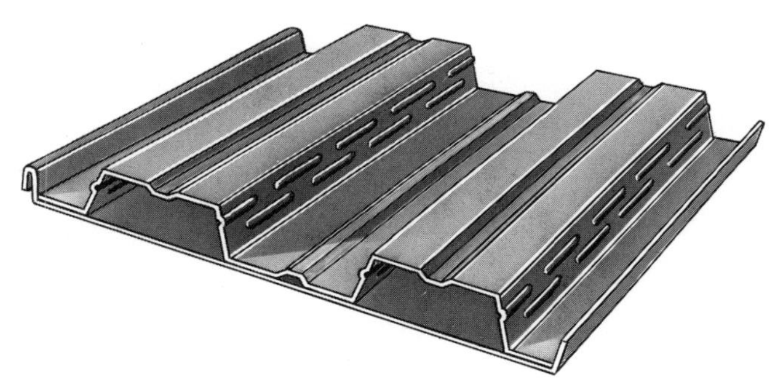

Fig. 3.4. Twenty-four-inch coverage, 2-in. cellular deck.

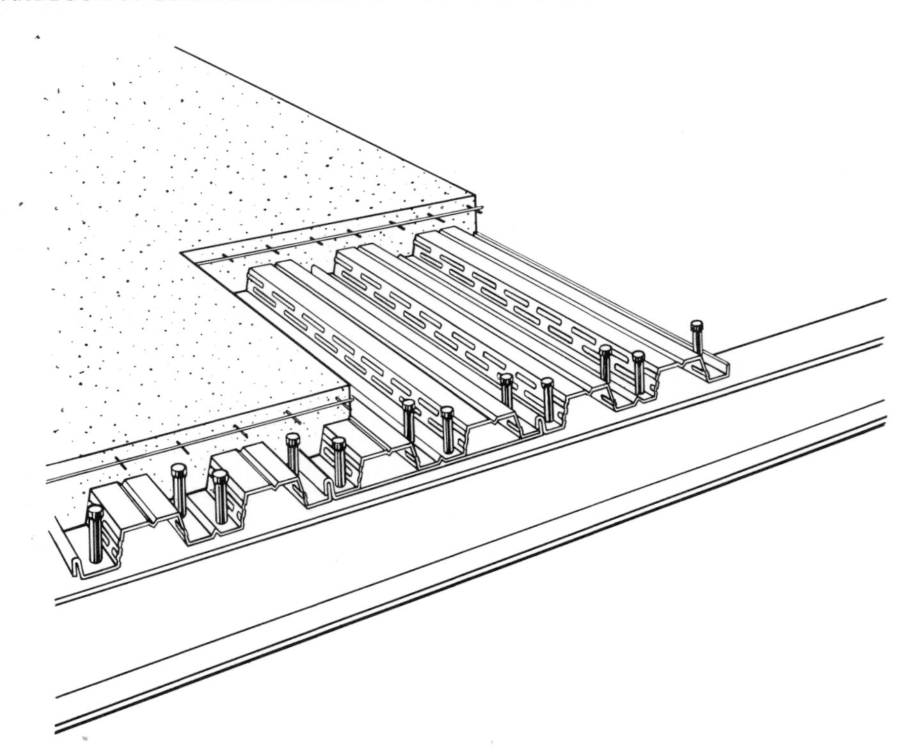

Fig. 3.5. Details of combined slab, deck, and beam composite system with shear connectors.

Shear connectors* are required to develop composite action between beam (or joist) and slab.

There are many devices that could be used as shear connectors. Certainly, *the* common connector in use today is the stud. The expanded use of the stud as the dominant shear connector relates to the ease of installation (directly through several layers of deck) by use of a welding gun.

Due to the limited capacity of the stud, large numbers of studs are often required on any given beam, and the flexural capacity of a beam may in fact be limited by the minimum spacing requirement for the studs. In addition, the studs on a beam present a hazard to workers, and so many unions in the United States have refused to erect steel with shop-fabricated studs.

There are a number of other types of shear connectors, which are used primarily in Europe. These include tees, angles, straps, and dowels. Their use involves extensive field welding and, as a result of the accompanying higher cost, has not and will not see extensive application in the United States.

*See Chapter 2 for a more detailed treatment of this topic.

3.5 DIAPHRAGM BEHAVIOR

Composite metal-deck systems are often used to resist lateral loads and to brace other structural members. Pertinent diaphragm nomenclature is shown in Fig. 3.6. Prior to discussing composite metal deck diaphragm behavior, a short review of general diaphragm behavior is in order. The pertinent features of a diaphragm are:

1. A diaphragm is primarily used to resist in-place forces.
2. Shear stresses and deflections are normally dominant as compared with flexural behavior.
3. For analysis purposes, a diaphragm can normally be considered analogous to the web of a plate girder; that is, its main function is the transmission of shear forces. The perimeter members thus serve as the flanges of the plate girder.

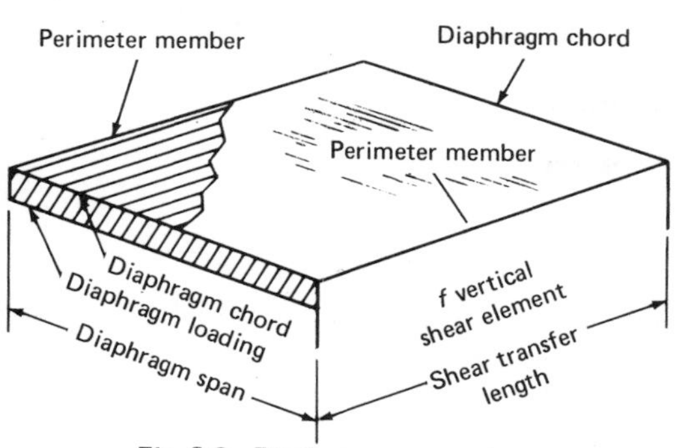

Fig. 3.6. Diaphragm nomenclature.

Some important conclusions relative to the performance of diaphragms are as follows:

1. Shear strength is approximately a linear function of the thickness of the panel material.
2. No definite relationship has been found to exist between shear rigidity and material thickness.
3. Shear strength per unit depth of diaphragm is relatively independent of panel length (dimension parallel to corrugations), provided that a "regular fastener arrangement" is used.
4. Shear rigidity is dependent (among other factors) upon panel length.
5. Frame size and stiffness have a very moderate influence on strength and stiffness. Perimeter members ordinarily used in practice are generally sufficient for an adequate performance, and no minimum requirements have been established.
6. The general behavior of diaphragms fabricated with *high-strength steel* panels is similar in nature to that of diaphragms made of *mild steel* panels. No direct correspondence exists between material strength and diaphragm strength; for example, an increase of 40% in material strength results in about 10% increase in strength and stiffness of diaphragm.
7. Sidelap, panel-to-panel connections are very influential on strength and stiffness. Because connections are a source of increased flexibility and reduced strength, the greater the cover width of a single panel, the better is the diaphragm performance. On the other hand, diaphragms with no sidelap fasteners at all are weak and extremely flexible.
8. In general, fastener type, spacing pattern, and configuration are very important factors in diaphragm performance. The shear rigidity especially is sensitive to panel configuration and spacing of end fasteners.

With respect to composite concrete-steel diaphragms, all of the preceding criteria exist, with the obvious additions that diaphragm strengths and stiffnesses increase with greater depths of concrete slab. In addition, normal-weight concrete slabs provide greater diaphragm strengths than lightweight slabs.

In most cases, engineers can and do design the *diaphragms* correctly. Far too often, however, the *details* concerning how lateral loads are transmitted into and out of the diaphragm system are overlooked. Proper design of welds or screw connectors to perimeter members can be obtained if the engineer *remembers* to follow the recommendations of the various reference manuals, such as those published by SDI (1973), H. H. Robertson Company (1970), and INRYCO, Inc. (1975).

EXAMPLE 3.1

A 2-in. (5-cm) composite floor deck (20 gauge) (0.91 mm thick) is used to span 8 ft, 0 in. (2.4 m) between beams in a braced frame. A $4\frac{1}{2}$-in. (12-cm) lightweight concrete slab is used. Determine if the composite floor deck can support a load of 1000 lb/ft. (14,590 N/m) on the span shown. Design the attachments to the perimeter members.

Solution

$$\text{Shear/ft} = 1000\,\frac{(100)}{100}$$

$$= 1000 \text{ lb/ft } (14{,}590 \text{ N/m})$$

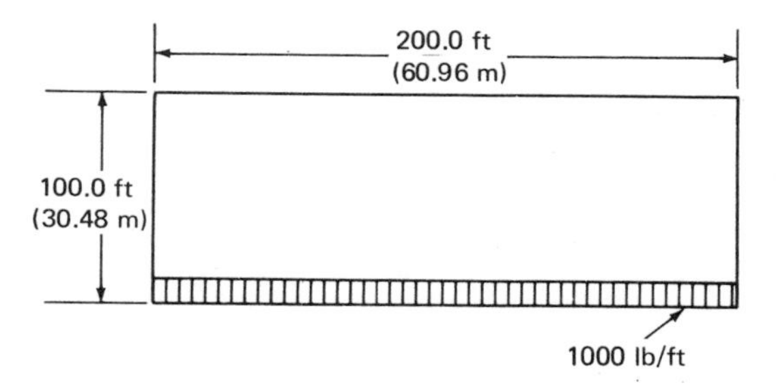

Fig. 3.7.

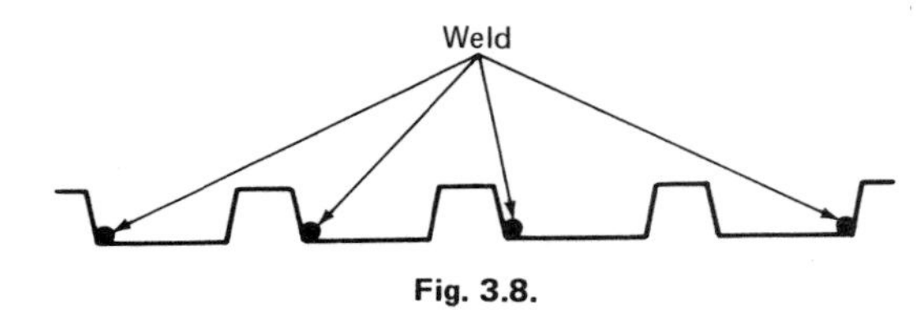

Fig. 3.8.

Based on the INRYCO manual, use a 2-in., 20-gauge deck with four puddle welds, as illustrated in Fig. 3.8 (attaching deck to joists or beams).

The allowable shear/ft of the composite panel is 1045 lb/ft (15,250 N/m). Therefore, the deck is adequate. Using the same reference, the allowable shear on a $\frac{1}{2}$-in. (12-mm) diameter weld in 20-gauge material is given as 1150 lb (5100 N). Therefore, the deck should be attached to the perimeter members using $\frac{1}{2}$-in. (12-mm) diameter welds at 12 in. (30 cm) on center $(1150 \div 1000) \simeq 12$ in.

It should be noted that in certain cases, the design of a composite steel deck concrete diaphragms, using this approach, could lead to serious consequences. One such case occurs when the lateral loads are transferred into the concrete slab via shear connectors or other means. In this case, the designer must make provisions to remove the diaphragm forces from the *concrete* at the *perimeter* members. A reasonable and commonly used solution to this transfer problem is to use shear studs spaced at appropriate intervals along the perimeter members. If such connectors are not used, then the only means for loads to transfer out of the system is for the shear to transfer from the concrete to the metal deck via the natural bonding mechanism between the deck and the concrete. In many cases, this transfer is difficult to obtain.

3.6 DESIGN OF DECK AS A FORM

Most of the deck manufacturers provide tables, which indicate the section properties at their various deck profiles. These properties can be used for the design of the deck to support the wet concrete load and construction loads. The design of the deck for this use is based on elastic behavior and a working-stress design approach. Deflections should normally be checked and never be allowed to exceed the span length ÷ 180. A typical table is shown as Table 3.1.

3.7 DESIGN OF COMPOSITE DECK

As in the design of any structural system, the design equations used for steel-deck composite systems are based on the various failure modes that may exist for the system. As stated previously, two primary failure modes exist for composite metal deck systems:

1. shear-bond failure
2. flexural failure.

The shear-bond failure is by far the most prevalent type of failure. It is characterized by the formation of diagonal cracks in the concrete slab in conjunction with bond slip between the metal deck and the concrete. As

Table 3.1. Section properties, $1\frac{1}{2}$-in. deck. (Courtesy of Roll Form Products, Boston, Massachusetts)

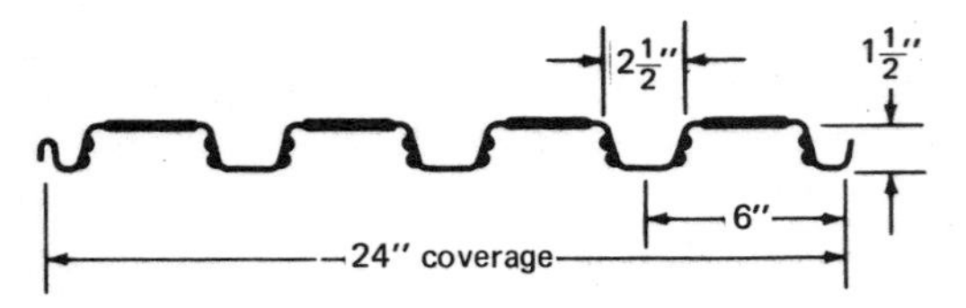

Gage*	Weight–psf*	S_p–in.^{3}	S_n–in.^{3}	I–in.^{4}
22	1.90	.21	.21	.18
20	2.26	.26	.28	.22
18	2.98	.35	.36	.32
16	3.69	.44	.45	.42

soon as slippage occurs, a significant drop in the load-carrying capacity results.

Flexural failures are identical to those of ordinary reinforced concrete flexural members. These failures can be either of the underreinforced (steel yielding prior to concrete-crushing) or overreinforced type (concrete-crushing prior to steel yielding).

Traditionally, design engineers have had to rely on experimental test results in order to determine ultimate shear and flexural capacities of steel-deck composite systems. The design procedure has typically been one of picking appropriate concrete thicknesses and steel-deck type and thickness from load tables provided by the various metal-deck manufacturers. Researchers have not attempted to derive generalized design equations. However, tentative design recommendations are currently being established by the American Iron and Steel Institute, based on extensive experimental investigations at several major universities. The results of the experimental investigations will be summarized here since presently they are the only existing design criteria available to the design engineer.

3.7.1 Ultimate-Strength Equations

3.7.1.1 Ultimate-moment equations

Two ultimate-moment equations exist, one for overreinforced beams and one for underreinforced beams. It is the authors' experience that ultimate moment seldom controls the design or is the mode of failure in metal-deck systems. Based on strain compatibility and internal equilibrium, the balanced steel ratio ρ_b is defined as:

$$\rho_b = \frac{0.85\beta_1 f'_c}{F_y} \frac{87{,}000(D - d_d)}{(87{,}000 + F_y)d} \quad (3.1)$$

where

β_1 = 0.85 for concrete with $f'_c \leqslant 4000$ psi and is reduced at the rate of 0.05 for each 1000-psi strength increase above 4000 psi; (β_1 minimum = 0.65)
f'_c = compressive concrete strength, psi
F_y = steel-deck yield strength, psi
d_d = depth of the steel deck, in.
D = nominal out-to-out slab depth, in.
d = effective slab depth (distance from the extreme concrete compression fiber to the centroid of the steel), in.

The reader who is familiar with design of reinforced concrete will recognize ρ_b as the same balanced reinforcement ratio used for that system. For values of $\rho = (A_s/bd)$ less than ρ_b (underreinforced sections), the ultimate moment is:

$$M_u = \Phi \frac{A_s F_y}{b}\left(d - \frac{a}{2}\right) \quad (3.2)$$

This equation is the conventional underreinforced concrete beam equation where

$\Phi = 0.90$
A_s = the net area of steel deck/unit width
$a = A_s F_y/(0.85 f'_c b)$, the depth of the equivalent rectangular stress block of width b.

In American usage, Eq. 3.2 may be written as:

$$M_u = \Phi \frac{A_s F_y}{12}\left(d - \frac{a}{2}\right) \quad (3.2a)$$

in which M_u is ft-lb/ft, A_s = net area of steel deck per ft width, and $a = A_s F_y/0.85 f'_c \times 12$, the depth of equivalent rectangular stress block of width b = 12 in.

It is assumed in Eq. 3.2 that the only reinforcing steel present is the steel deck itself. In addition, the normal assumption is made that the concrete reaches a strain of 0.003 in./in. (mm/mm) at the outermost fiber. Caution should be used in applying this equation to steel deck with a depth greater than 3 in. (7.5 cm) because the steel deck may not have sufficient ductility to reach full yield at all fibers or it may simply fracture prior to the concrete reaching a strain of 0.003 in./in.

For overreinforced systems, which are extremely rare, the ultimate moment (ft-lb/ft) from Eckberg and Porter (1976) is

$$M_u = \Phi \frac{0.85\beta_1 f'_c b d^2 k_u}{12}(1 - \beta_2 k_u) \quad (3.3)$$

where

Φ = 0.75 is recommended
$k_u = \sqrt{\rho\lambda + (\rho\lambda/2)^2} - \rho\lambda/2$
$\lambda = E_s \epsilon_u/(0.85\beta_1 f'_c)$

ϵ_u = 0.003 in./in.
E_s = 29,500,000 psi
β_2 = 0.425 for $f_c' \leqslant 4000$ psi and is reduced at the rate of 0.025 for each 1000-psi strength increase over 4000 psi f_c', d, and β_1 as defined previously.

More general equations for M_u may be written with suitable modifications in Eq. 3.3.

3.7.1.2 Shear-Bond Equation

The ultimate shear capacity of a composite metal deck system can be determined from the following empirical equation (Eckberg and Porter 1976):

$$V_u = \frac{bd}{S}\left(\frac{m\rho d}{L'} + kf_c'\right) \tag{3.4}$$

V_u = the ultimate shear (experimentally obtained)
s = a term that accounts for the spacing of the shear-transfer devices in the deck. For all decks produced within the United States in which shear transfer is based on surface bond and the deck profile, s is to be taken as unity; however, for steel deck in which shear transfer is provided by transverse wires welded to the deck or by holes or welded nobs, s should be taken as the spacing of these devices in inches.
L' = distance from the end reaction to the concentrated load point for a particular test setup (see Fig. 3.9).

Equation 3.4 is empirical and is based on the results of more than 400 tests. The values of constants m and k must be empirically obtained for each metal deck profile since each has its own shear-transfer characteristics. A minimum of four tests should be made (for each type of deck) to determine m and k for a specific deck profile, gauge, surface coating, and each concrete type (lightweight and normal-weight). Shown in Fig. 3.10 is a typical plot used to arrive at m and k values. The ordinate is $V_e s/bd\sqrt{f_c'}$, and the abscissa is $d/L'\sqrt{f_c'}$. A linear regression line is drawn between the test points (a minimum of two points is recommended at extreme $\rho d/L^1\sqrt{f_c'}$ values). Once the data are

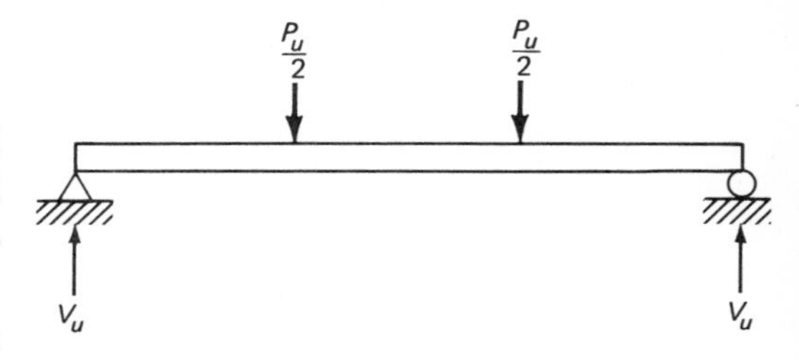

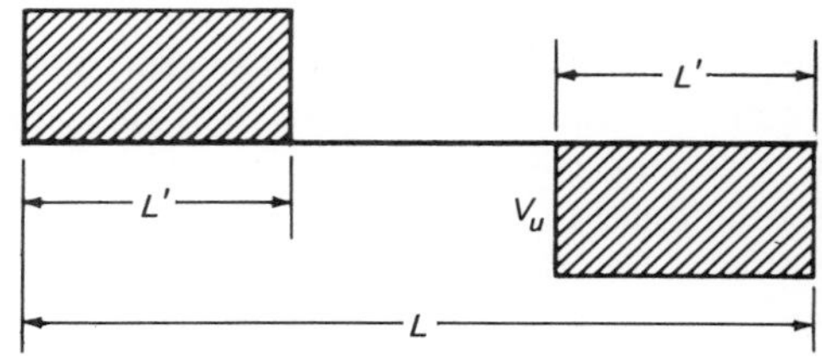

Fig. 3.9. Test configuration.

plotted and the regression line established, an "adjusted" line with a 15% reduction (as recommended by Porter and Eckberg 1976) is constructed, from which m and k are determined. This reduction may be looked upon as a safety recommendation.

3.7.2 Design Approach

For the most part, design engineers will not normally use the equations just presented due to the uniformly distributed load used in their

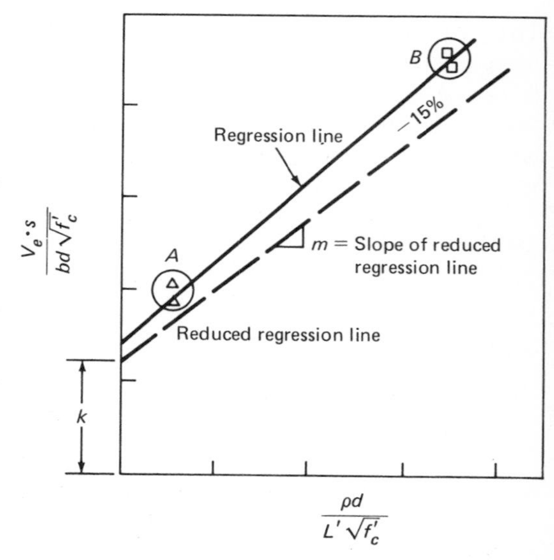

Fig. 3.10. Shear-bond failure relationship (constant steel thickness).

design. In the majority of designs, engineers will continue to select composite metal deck systems from design aids and load tables (for uniform loads) provided by the manufacturers of metal deck. However, in designs where *concentrated load* exists, the designer can use the equations in lieu of expensive testing in order to determine the ultimate shear and moment capacities of a composite deck system.

The ultimate moment equations can be used directly in a design application; however, the ultimate shear equation requires certain modifications. For ease of use, the shear-bond equation should be in convenient units (i.e., in capacity per-foot width). In addition, the effects of shoring must be considered.

The test specimens used in developing Eq. 3.4 were generally formed and concrete placed without the use of shoring. If shoring is used, Eq. 3.4 can be modified to take the shoring effect into account. According to Porter and Eckberg (1976), the ultimate shear capacity V_u in lb/ft of width of slab for a simply supported span L is:

$$V_u = \Phi\left[\frac{12d}{s}\left(\frac{m\rho d}{L'} + k\sqrt{f'_c}\right) + \frac{\gamma W_1 L}{2}\right] \tag{3.5}$$

where

$\Phi = 0.80$

W_1 = weight of the concrete slab, psf

γ = a coefficient used to proportion the amount of dead load added when the shore is removed.

For slabs of uniform thickness, $\gamma = 1.0$ for complete shoring, 0 for unshored construction, and 0.625 for center shoring.

In actual design problems, L', the shear span, must be adjusted in order to correlate with the test procedures used in determining Eq. 3.5. For simple span conditions, L' can be determined from the following:

$$L' = \frac{M}{V}$$

Some common values of L' are summarized below:

P, L/2, L

W, L

$$L' = \frac{M}{V} = \frac{\frac{PL}{4}}{P/2} = L/2$$

$$L' = \frac{M}{V} = \frac{\frac{1}{8}wL^2}{\frac{wL}{2}} = L/4$$

For unsymmetrical loading, the largest value of L' would be used.

P, L/3, 2L/3

$$L' = \frac{M}{V} = \frac{\frac{P}{L} \times \frac{L}{3} \times \frac{2L}{3}}{P/3} = \frac{2L}{3}$$

The reader should notice that for all cases of concentrated loads, L' is the largest distance from the load to the support.

Since the uniformly loaded slab is by far the most common case, Eq. 3.5 can be rewritten as:

$$V_u = \Phi\left[\frac{d}{s}\left(\frac{4m\rho d}{L} + 12k\sqrt{f'_c}\right) + \frac{\gamma W_1 L}{2}\right] \tag{3.6}$$

3.7.3 Deflection Calculations

For long-span conditions, deflection calculations should be made. Deflections of the metal deck due to the weight of the wet concrete should be limited to $\frac{3}{4}$ in. (20 mm) or L/180, whichever is smaller. Deflections equaling L/180 are noticeable. If deflections exceeding these limitations are allowed, the possibilities of failure of the metal deck due to "ponding" are significantly increased. A ponding failure occurs when, as concrete is placed, the deck excessively deflects under the dead load. The deflection of the deck creates an obvious "dip" in the surface, which requires more concrete. This process can continue until failure results. Deflection information can normally be obtained from the various metal deck manufacturers; however, the engineer may want to check these deflections. Moments of inertia for the plain metal deck should be calculated based on *Specification for the Design of Cold-Formed Steel Structural Members* (AISI 1968a). A companion publication entitled Design Examples (AISI 1968b) illustrates how the moment of inertia is calculated.

Deflections due to superimposed live and dead loads should also be checked when longer spans are used. The criterion suggested by several investigators is to base the composite deflections on the average moment of inertia of the uncracked and cracked sections, i.e., $\frac{1}{2}(I_u + I_c)$. The cracked moment of inertia may be written as

$$I_c = \frac{b}{3}(Y_{cc})^3 + nA_s(Y_{cs})^2 + nI_{sf} \quad (3.7)$$

where

$$Y_{cc} = d\{[2\rho n + (\rho n)^2]^{1/2} - \rho n\} \text{(see Fig. 3.11)}$$

$$\rho = A_s/bd$$
$$n = E_s/E_c$$

If $Y_{cc} > t_c$, then use $Y_{cc} = t_c$.

The uncracked moment of inertia is

$$I_u = \frac{bt_c^3}{12} + bt_c(Y_{cc} - 0.5t_c)^2 + nI_{sf} + nA_sY_{cs}^2 + \frac{W_r bd_d}{C_s}\frac{d_d^2}{12} + (D - Y_{cc} - 0.5d_d)^2 \quad (3.8)$$

where

$$Y_{cc} = \frac{0.5bD^2 + nA_sd - (C_s - W_r)\dfrac{bd_d}{C_s}(D - 0.5d_d)}{bD + nA_s - \dfrac{b}{C_s}d_d(C_s - W_r)}$$

C_s = cell spacing, in.
W_r = average rib width, in.
$d = D - Y_{sb}$

In many cases, deflections need not be checked since experience and depth-span ratios will give a good indication of whether a potential deflection problem exists. A span-to-depth ratio of 22 is recommended for simply supported spans. A quick check of deflection can be obtained by using the gross moment of inertia for the composite section. Using the equation

$$I_{gross} = \frac{1}{12}bh^3$$

where

$h = D - Y_{sb}$

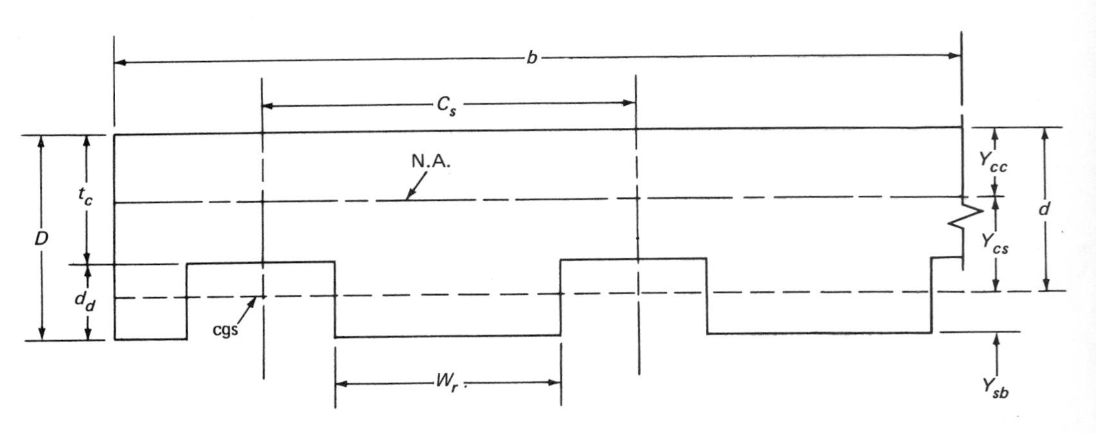

Fig. 3.11. Transformed composite section.

will normally result in a conservative approximation for the short-term deflection of a simple span system. Long-term deflection considerations can be accounted for by the equation

$$\Delta_{LT} = \Delta'_{ST}[2 - 1.2(A'_s/A_s)]$$

where

Δ'_{ST} = short-term deflection due to sustained load
A'_s = area of compressive reinforcement, in.²/ft of width.

The term within the brackets is a suggested multiplier used to account for the long-term effects and should be equal to or greater than 0.6. In general, assuming Δ_{LT} equal to twice the short-term deflection is conservative.

EXAMPLE 3.2

Determine if the $1\frac{1}{2}$-in. (38-mm) composite metal deck shown in Fig. 3.12 has adequate capacity to support the applied loads. The deck consists of 20-gauge (0.91 mm thick) material. The manufacturer of the deck has published k and m values for this deck as 0.113 and 17,300, respectively. A 4-in. (10-cm) normal-weight concrete slab is assumed.

Deck Properties

$S^+ = 0.26$ in.³ (4260 mm³)
$S^- = 0.28$ in.³ (4588 mm³)
$I = 0.22$ in.⁴ (91,570 mm²)
$C_s = 6$ in. (153 mm)
$W_r = 1.95$ in. (50 mm)
$Y_{sb} = 1.0$ in. (25 mm)
$f'_c = 3{,}000$ psi (21 N/mm²)
$A_s = 0.665$ in. (17 mm)
$d = 3.00$ in. (7.5 cm)
$F_y = 33{,}000$ psi (231 N/mm²)

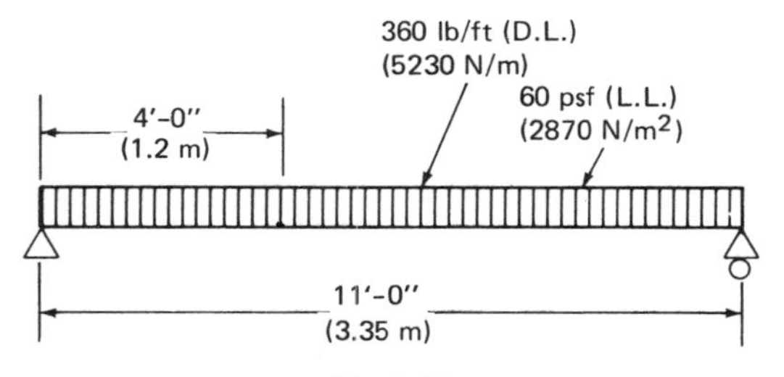

Fig. 3.12.

For completeness, the metal deck stresses due to the construction loads will be calculated.*

Check the deck for a 20-psf (960-N/m²) uniform construction load and a 150-lb (660-N) concentrated construction load.

Check:

1. deflection of deck
2. maximum positive-moment stresses
3. maximum negative-moment stresses
4. ultimate moment condition
5. shear
6. deflection of composite system.

1. *Check deflection of deck.* The maximum deck deflection will occur when construction loads are placed on only one side of the shore.

a. deflection due to the wet concrete dead load (assumed 38 psf [1820 N/m²]), using the simple beam-deflection equation, is

$$\Delta_{CL} = .0092WL^4/EI$$

$$= \frac{0.0092(38)(5.5)^4}{29{,}500{,}000 \times 0.22} \times 1728$$

$$= 0.085 \text{ in. } (2.1 \text{ mm})$$

b. deflection due to 20-psf (960-N/m²) construction load

$$\Delta_{CL} = 0.85\left(\frac{20}{38}\right) = 0.045 \text{ in. } (1.1 \text{ mm})$$

c. deflection due to 150-lb (660-N) concentrated construction load

$$\Delta_{CL} = \frac{0.015PL^3}{EI}$$

$$= \frac{0.015(150)(5.5)^3}{29{,}500{,}000 \times .22}(1728)$$

$$= 0.10 \text{ in. } (2.5 \text{ mm})$$

d. compute the approximate ponding load of extra concrete due to the deflection occurring from the 150-lb (660-N) concentrated load. Assuming a parabolic deflec-

*Most metal deck systems experience the most critical loading during construction. The authors believe that this condition should always be investigated and that the loads given should be considered minimum requirements.

tion curve, the extra volume of concrete could be

$$\text{volume} = \tfrac{2}{3}\,\Delta C_L L b$$
$$= \frac{2}{3}\,\frac{0.10}{12}\,(5.5)(1)$$
$$= 0.031 \text{ ft}^3 \ (877 \text{ cm}^3)$$
$$wt = 150(0.031)$$
$$= 4.65 \text{ lb } (20.4 \text{ N})$$

or approximately 1 lb/ft² extra load. Since this is a small load when compared to the dead load of 38 psf (1820 N/m²) for this example, it will be neglected in further calculations. The authors wish to point out, however, that, in many cases, especially for single-span unshored conditions, the ponding load can be significant.

From the preceding calculation, it can be seen that the maximum deflection due to the wet concrete would be less than 0.10 in. (2.5 mm), which is considerably less than $L/180$ (0.37 in.) (9 mm) or $\frac{3}{4}$ in. (20 mm).

2. *Maximum positive-moment stresses.* For maximum position moment, only one span would be considered loaded as illustrated in Fig. 3.13.

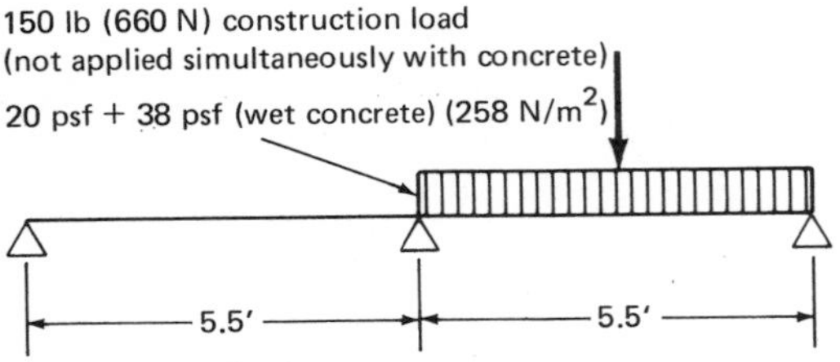

Fig. 3.13.

$$M_1^+ = \frac{49}{512}\,wl^2 = \frac{49}{512}\,(58)(5.5)^2 = 167.91 \text{ ft-lb}$$

(230 Nm) (due to concentrated load)

clearly uniform loading controls.

$$f = \frac{M}{S} = \frac{167.91 \times 12}{0.26} = 7750 \text{ psi } (53.4 \text{ MPa})$$
$$< 19{,}800 \text{ psi } (136.5 \text{ MPa}) \ (0.6F_y) \quad \text{OK}$$

3. *Maximum negative-moment stresses.*

$$M_1^- = \tfrac{1}{8}\,WL^2 = \tfrac{1}{8}\,(58)(5.5)^2 = 219.31 \text{ ft-lb}$$

(297 Nm) (due to uniform load)

$$M_2^- = \tfrac{3}{32}\,PL = \tfrac{3}{32}\,(150)5.5 = 77.34 \text{ ft-lb}$$

(104.9 Nm) (due to concentrated load)

$$f = \frac{M}{S} = \frac{219.31(12)}{0.28} = 9400 \text{ psi } (65 \text{ MPa})$$
$$< 19{,}800 \text{ psi } (136.5 \text{ MPa}) < 0.6F_y \quad \text{OK}$$

4. *Check ultimate moment.* Determine if the section is under- or over-reinforced:

$$\rho_b = \frac{0.85\beta_1 f_c'}{F_y}\,\frac{87{,}000(D - d_d)}{(87{,}000 + F_y)d}$$
$$\rho_b = \frac{0.85(0.85)3000}{33{,}000}\,\frac{87{,}000(4 - 1.5)}{(87{,}000 + 33{,}000)3.0}$$
$$= 0.040$$
$$\rho = \frac{A_s}{bd} = \frac{0.665}{12(3)}$$
$$= 0.018 < 0.04 \text{ (underreinforced)}$$
$$M_u = \frac{\phi A_s F_y}{12}\left(d - \frac{2}{2}\right)$$
$$a = \frac{A_s F_y}{0.85 f_c' b} = \frac{0.665(33{,}000)}{0.85(3000)(12)}$$
$$= 0.717 \text{ in. } (1.8 \text{ cm})$$
$$M_u = \frac{0.9(0.665)(33{,}000)}{12}\ 3.00 - \frac{0.717}{2}$$
$$= 4347 \text{ ft-lb } (5890 \text{ Nm})$$

Determine applied factored moment (Fig. 3.14).
Determine shore force, R (Fig. 3.15).

5. *Check shear.*

$$V_u = \phi\left[\frac{12d}{s}\left(\frac{m\rho d}{L'} + k\sqrt{f_c'}\right) + \gamma\,\frac{W_1 L}{2}\right]$$
$$M = \frac{881.73 + 473.73}{2}\,(4)$$
$$= 2710.92 \text{ ft-lb } (3.68 \text{ kN-m})$$

Determine L′ (see Fig. 3.16).

$$L' = \frac{M}{V} = \frac{2710.92}{744.27}$$
$$= 3.64 \text{ ft; i.e., } 43.71 \text{ in. } (111 \text{ cm})$$

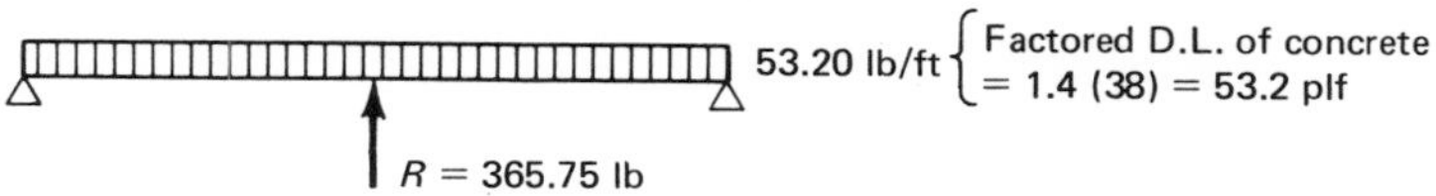

53.20 lb/ft {Factored D.L. of concrete = 1.4 (38) = 53.2 plf

R = 365.75 lb

$R = 53.2\ (1.25)\ (\frac{11}{2}) = 365.75$ lb

Fig. 3.14.

Loading condition after shore removal:

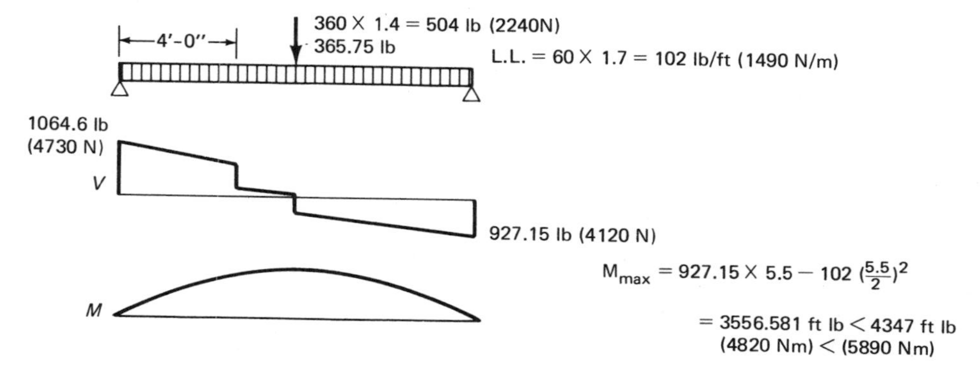

$M_{max} = 927.15 \times 5.5 - 102\ (\frac{5.5}{2})^2$

$= 3556.581$ ft lb < 4347 ft lb (4820 Nm) $<$ (5890 Nm)

∴ O.K. Maximum factored moment < Ultimate moment

Fig. 3.15.

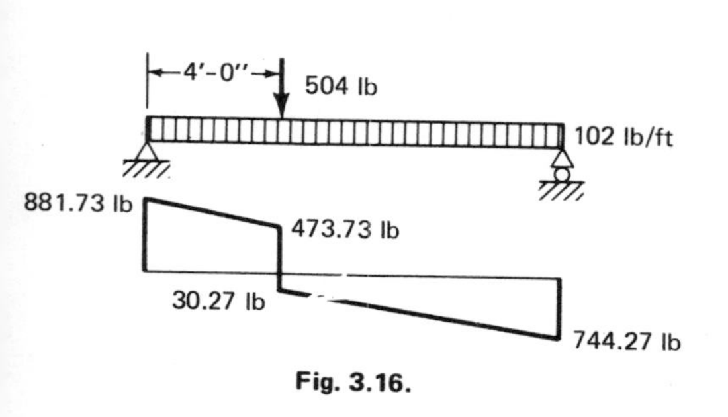

Fig. 3.16.

$W_1 = 38$ psf

$\gamma = 0.625$

$s = 1.0$

$\phi = 0.8$

$$V_u = 0.8 \left[\frac{12(3.00)}{1} \quad \frac{17{,}300 \times 0.018 \times 3.00}{45.72} \right.$$

$$\left. + 0.113\sqrt{3000} + \frac{0.625 \times 38 \times 11}{2} \right]$$

$= 871.14$ lb (3870 N) ≈ 881.73 lb (3920 N)

OK

6. *Check deflection of composite system.*

a. Determine I_c

$$E_c = 57{,}000\sqrt{f_c'} = 57{,}000\sqrt{3000}$$

$$= 3{,}122{,}000 \text{ psi } (21{,}500 \text{ MPa})$$

$$E_s = 29{,}500{,}000 \text{ psi } (203{,}400 \text{ MPa})$$

$n = 9.45$ USE $n = 9$

$$Y_{cc} = d\ \{2\rho n + (\rho n)^2]^{1/2} - \rho n\}$$

$$= 3.00\ \{[2(0.018)(9) + (0.018 \times 9)^2]^{1/2} - (0.018)(9)\}$$

$= 1.29$ in. (3.3 cm)

$$I_c = \frac{b}{3}(y_{cc})^3 + nA_s(y_{cs})^2 + nI_{sf}$$

$$Y_{cs} = d - y_{cc} = 3.00 - 1.25 = 1.71 \text{ in.}$$

$$I_c = \tfrac{12}{3}\ (1.29)^3 + 9(0.665)(1.71)^2 + 9(0.22) = 28.07 \text{ in.}^4\ (1168 \text{ cm}^4)$$

b. Determine I_u

$$y_{cc} = \frac{0.5bd^2 + nA_{sd} - nA_{sd} - (C_s - W_r)\dfrac{bd_d}{C_s}(D - 0.5d_d)}{bD + nA_s - \dfrac{b}{C_s}d_d\,(C_s - W_r)}$$

$$y_{cc} = \frac{0.5(12)(4)^2 + 9(0.665)(3.00) - (6 - 1.95)\dfrac{(12)(1.5)}{6}[4 - (0.5 \times 1.5)]}{12(4) + 9(0.665) - \dfrac{12}{6}(1.5)(6 - 1.95)}$$

$= 1.78$ in (4.5 cm)

$$I_u = \frac{bt_c3}{12} + bt_c(y_{cc} - 0.5t_c)^2 + nI_{sf} + nA_s y_{cs}^2 + \frac{W_{rb}d_d}{C_s}\left(\frac{d_d}{12}\right) + (D - y_{cc} - 0.5d_d)^2$$

$$= \frac{12(2.5)^3}{12} + 12(2.5)[1.78 - (0.5 \times 2.5)]^2 + 9(0.22) + 9(0.665)(1.22)^2 + \frac{1.95(12)(1.5)}{6}\left\{\frac{1.5}{12} + [4 - 1.78 - (0.5 \times 1.5)]^2\right\}$$

$= 48.31$ in.^{4} (2010 cm^4)

Moment of inertia for deflection, I

$$= \tfrac{1}{2}(I_c + I_u)$$
$$= \tfrac{1}{2}(28.07 + 48.31)$$
$$= 38.19 \text{ in.}^4 \ (1590 \text{ cm}^4)$$

Check using I_{gross}

$$I_g = \tfrac{1}{12}bh^3 = \tfrac{1}{12}(12)(4 - 1.0)^3$$
$$= 27 \text{ in.}^4 \ (1123 \text{ cm}^4)$$

c. Short-term deflection*

$$\Delta'_{ST} = \frac{5WL^4}{384EI} + \frac{\rho_a}{3EIL} + \frac{PL^3}{48EI}$$

$$= \left\{\frac{5(60)(11)^4}{384} + \frac{360(4)^2(7)^2}{3(11)} + \frac{365.75(11)^3}{48}\right\}\frac{1728}{3.122 \times 10^6 \times 38.19}$$

$$\Delta'_{ST} = 0.17 + 0.12 + 0.15$$
$$= 0.44 \text{ in. (11 mm)}$$

Since, in the last expression, 0.15 in. (3.8 mm) is the deflection that would occur immediately after shore removal, it would normally not have to be considered in the short-term deflection critieria.

d. Long-term deflection

$$\Delta_{LT} = \Delta'_{ST}[2 - 1.2(A'_s/A_s)]$$

Assuming no compression steel (mesh is negligible)

$$\Delta_{LT} = \Delta'_{ST}(2)$$

Assuming that one-half of the live load is sustained,

$$\Delta_{LT} = 2[0.17 + 0.12]$$
$$= 0.58 \text{ in. (14.7 mm)}$$

$$\Delta_T = 0.44 + .58 = 1.02 \text{ in. (25.9 mm)}$$

Deflections are excessive for the span. This would be expected since the span-to-depth ratio greatly exceeds the recommended value of 22 for a simple span.

The authors wish to point out the importance of the deflection criteria for all structural problems, particularly for structures consisting of composite systems. The majority of litigation and failure problems are the result of inadequate attention (on the part of designers) to stiffness criteria. Prior to the development of high-strength steels, ultimate strength design, and composite construction, designs based on lower allowable stress levels, in most cases, *automatically* provided stiffness characteristics, which prevented deflection-related problems such as ponding and floor vibrations. Today, designs

*The deflection obtained is slightly in error since the deflection determined from the second term was considered at the load point and not the beam centerline.

using higher strength steels and ultimate-strength behavior result in sections that are less stiff than designs of a few years ago, and thus stiffness-related problems are becoming increasingly more common. The authors urge the readers to carefully consider the stiffness aspects of designs in order to prevent such problems from occuring.

3.8 PRACTICAL CONSIDERATIONS

Several considerations are listed here. They are a direct result of the experience of the authors associated with the use of composite metal deck on various projects.

1. Consider the stiffness effects associated with the design. This includes deflection criteria during–as well as after–construction.
2. The critical stress and loading state for most composite deck systems occurs when the concrete is placed on the deck.
3. Extreme caution must be used by construction workers when working with metal decks of 22 gauge or less.
4. Take a micrometer to the job site to check deck thicknesses. (Paint and galvanizing do not add strength, but they do add thickness.)
5. Do not depend on composite action if vibrating loads are possible (forklift trucks).
6. Check to see if the deck embossments are present.
7. Carefully examine the trench header opening effects on the strength of metal deck systems.
8. Make sure that deck surfaces are free from oil and dirt prior to concrete placement. The elimination of surface bond can be catastrophic to the composite assumption.
9. Avoid using composite metal deck systems in applications where a corrosive environment exists. This includes areas where salt penetration through the concrete deck can cause deck corrosion.
10. Avoid prolonged exposure of the metal deck to the weather as heavy rusting can destroy the bond characteristics.
11. In deck calculations for long unshored span systems, the increased weight of the concrete due to "ponding" should be considered in the strength and deflection calculations.
12. In most applications, temperature and shrinkage reinforcement should be included in the concrete slab.

REFERENCES

ACI (1971), *Building Code Requirements for Reinforced Concrete (ACI- 318-71)*, American Concrete Institute, Detroit, Michigan (1976 Edition).

AISI (1968a), *Specification for the Design of Cold-Formed Steel Structural Members*, American Iron and Steel Institute, 1968 Edition.

AISI (1968b), *Illustrative Examples Based on the 1968 Edition of the Specification for the Design of Cold-Formed Steel Structural Members*, American Iron and Steel Institute.

Bryl, S. (1967), The Composite Effect of Profiled Steel Plate and Concrete in Deck Slabs, *Acier Stahl Steel*, Oct. 1967.

Eckberg, C. E., Jr., and Schuster, R. M. (1968), *Floor Systems with Composite Form-Reinforced Concrete Slabs*, Final Report, International Association for Bridge and Structural Engineering, 8th Congress, New York, Sept., pp. 385–394.

Fisher, J. W. (1970), Design of Composite Beams with Formed Metal Deck, *Engineering Journal*, American Institute of Steel Construction, V-1, 7, No. 3, July, pp. 88–96.

Friberg, B. F. (1954), Combined Form and Reinforcement for Concrete Slabs, *ACI Journal Proceedings*, 50, May, pp. 697–716.

INRYCO (1975), *INRYCO Lateral Diaphragm Data*, Manual 20-2, published by INRYCO, Inc., Melrose Park, Illinois.

Luttrell, L. D. and Division, J. H. (1971), Composite Slabs with Steel Deck Panels, *Proceedings of the Second Specialty Conference of Cold-Formed Steel Structures*, Department of Civil Engineering, University of Missouri, Rolla, Missouri.

Porter, M. L. (1968), Investigation of Light Gage Steel Forms as Reinforcement for Concrete Slabs, thesis presented to Iowa State University, at Ames, Iowa, in partial fulfillment of the requirements for the degree of Master of Science.

Porter, M. L. and Ekberg, C. E., Jr. (1971), Investigation of Cold-Formed Steel-Deck Reinforced Concrete Floor Slabs, *Proceedings of the First Specialty Conference on Cold-Formed Steel Structures*, University of Missouri-Rolla, Rolla, Missouri, Aug., pp. 179–185.

Porter, M. L. and Ekberg, C. E., Jr. (1972), *Summary of Full-Scale Laboratory Tests of Concrete Slabs*

Reinforced with Cold-Formed Steel Decking, Preliminary Report, International Association for Bridge and Structural Engineering, 9th Congress, Amsterdam, The Netherlands, May, pp. 173-183.

Porter, M. L. and Ekberg, C. E., Jr. (1976), Design Recommendations for Steel Deck Floor Slabs, *Journal of the Structural Division*, Vol. 102, No. ST11, Nov.

Robinson, H. (1969), Composite Beam Incorporating Cellular Steel Decking, *Journal of the Structural Division*, ASCE, Vol. 93, No. ST3, Proc., Paper 6447, March, pp. 355-380.

Sayed, G. A., Temple, M. C., and Madugula, M. K. S. (1974), Response of Composite Slabs to Dynamic Loads, *Canadian Journal of Civil Engineering*, Vol. 1, No. 1, pp. 62-70.

Schuster, R. M. (1972), *Composite Steel-Deck Reinforced Concrete Systems Failing in Shear-Bond*, Preliminary Report, International Association for Bridge and Structural Engineering, 9th Congress, Amsterdam, The Netherlands, May, pp. 185-191.

Schuster, R. M. (1976), Composite Steel-Deck Concrete Floor Systems, *Journal of the Structural Division*, ASCE, Vol. 102, No. ST5, May, pp. 899-916.

SDI (1973), *Tentative Recommendations for the Design of Steel Deck Diaphragms*, published by the Steel Deck Institute, 9836 West Roosevelt Road, Westchester, Illinois 60153.

4

Reinforced concrete composite flexural members

DAN E. BRANSON, Ph.D., P.E.
Professor of Civil Engineering
University of Iowa
Iowa City, Iowa

4.1 INTRODUCTION TO CHAPTERS 4 AND 5

In order to restrict the chapter lengths, the subjects of reinforced composite and prestressed composite flexural members are presented in separate chapters—4 and 5. The references, notation, and appendixes for both chapters can be found at the end of Chapter 5.

The composite concrete beam or girder usually consists of a cast-in-place reinforced concrete slab on a reinforced or prestressed concrete precast beam in which the two elements act together as a single unit under load. Typical sections of building and bridge composite flexural members of the types considered herein are shown in Figs. 4.1 through 4.3. The principles involved are, of course, the same for other types of sections, such as those shown in Fig. 4.4. Primarily because of the better quality control of precast units, it is usually economical to use higher concrete strengths for the precast elements than for the cast-in-place elements; this leads to the requirement of an elastic conversion of the cast-in-place unit when the elastic theory is used.

Reference is made in these chapters to the Joint ACI-ASCE Committee 333 on Composite Construction[1] (1960), ACI Code and Commentary (318-63)[2], ACI Code and Commentary*

*Unless otherwise noted, reference to the ACI Code will refer to the ACI Code and Commentary (318-71) plus supplements, and reference to AASHTO Specifications will refer to the AASHTO Highway Bridge Specifications, 1973, plus supplements.

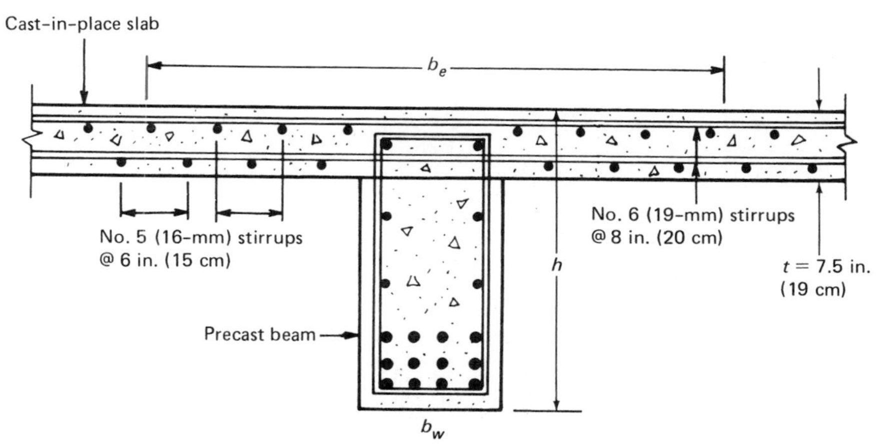

Fig. 4.1. Typical composite reinforced concrete bridge girder section.

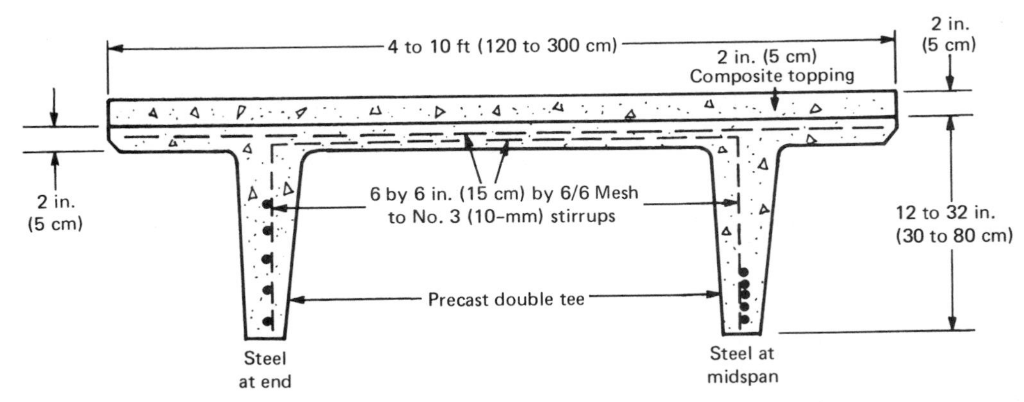

Fig. 4.2. Typical composite prestressed concrete building beam section.

(318-71)[3], PCI Handbook[4] (1971), British Standard Code of Practice and Manual[5] (1972), AASHTO Highway Bridge Specifications[6]* (1973), and Wang and Salmon[7] (1965). References 8 to 47 were also consulted.

4.1.1 Behavior of Composite Beam under Load

When a slab and beam member not connected at the interface or contact surface are subjected to load, the two elements will separate, as shown in Fig. 4.5a. In such a noncomposite case, each element supports a part of the total load in proportion to its respective stiffness or EI value. However, when horizontal shear is developed between the two elements, the resulting composite action enables the beam to support the load without separation, as shown in Fig. 4.5b. This composite behavior significantly increases both the flexural strength and stiffness of most members, as compared to the corresponding noncomposite members.

The transfer of shear at the interface can be accomplished by means of (1) bond between

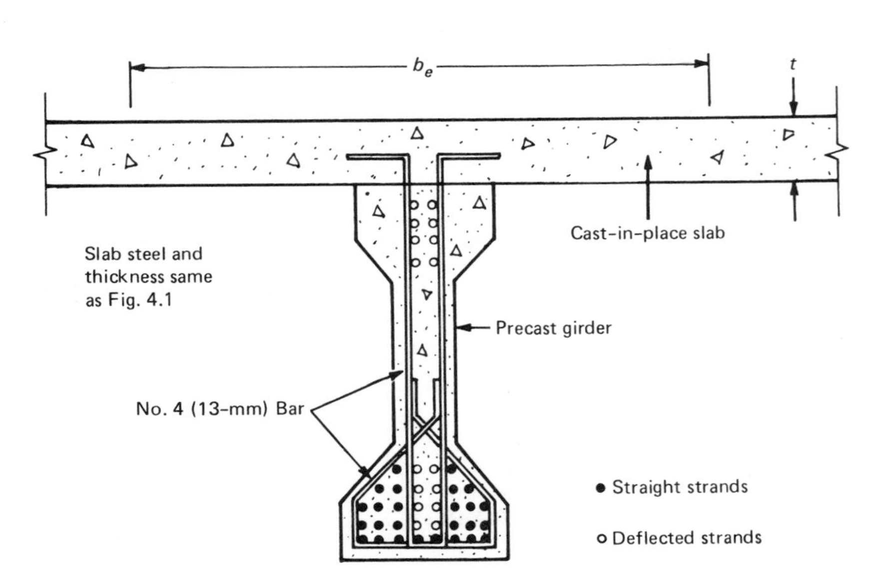

Fig. 4.3. Typical composite prestressed concrete bridge girder section.

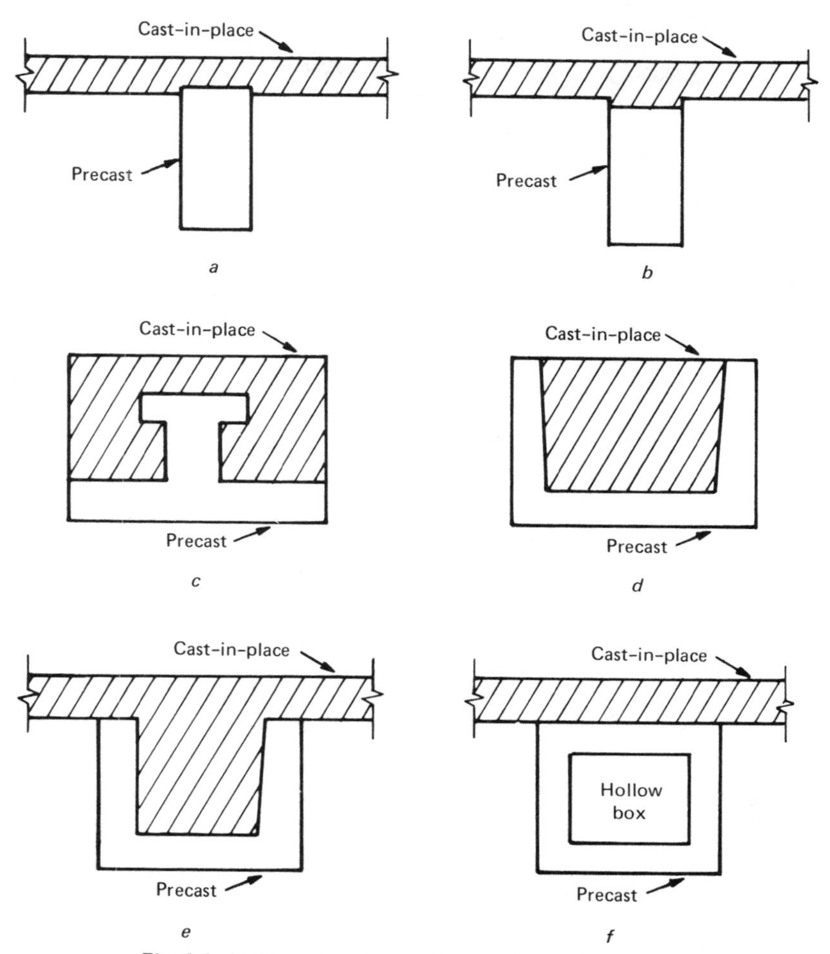

Fig. 4.4. Various other types of composite sections.

two elements (especially when the precast beam surface has been roughened), (2) bond plus the use of vertical ties to prevent separation, and (3) shear keys plus vertical ties. Such horizontal shear transfer in composite concrete beams is usually of little consequence at service-load levels, but becomes important in the overload range and at ultimate load. When vertical ties are used, it is usually convenient to use the same vertical steel for the direct shear requirements in the web of the beam and for horizontal shear.

For members with constant web widths, the direct shear or web shear capacity of composite members by either elastic or strength theories is essentially the same as for the precast unit by itself. However, for other types of sections that include cast-in-place concrete in the area where it can add to the shear strength of the member, such as in Fig. 4.4c–e, the shear capacity can be significantly greater for the composite member than for the precast member alone.

4.1.2 Sequence of Loading and the Use of Falsework or Shoring

The stress and deflection analysis of a composite beam will depend on whether or not falsework or shoring is used during construction. In the unshored case, the precast-beam dead load and all other loads (such as due to the slab, diaphragms, roofing, partitions, sidewalk, or railing, dead load, and construction loads) that are

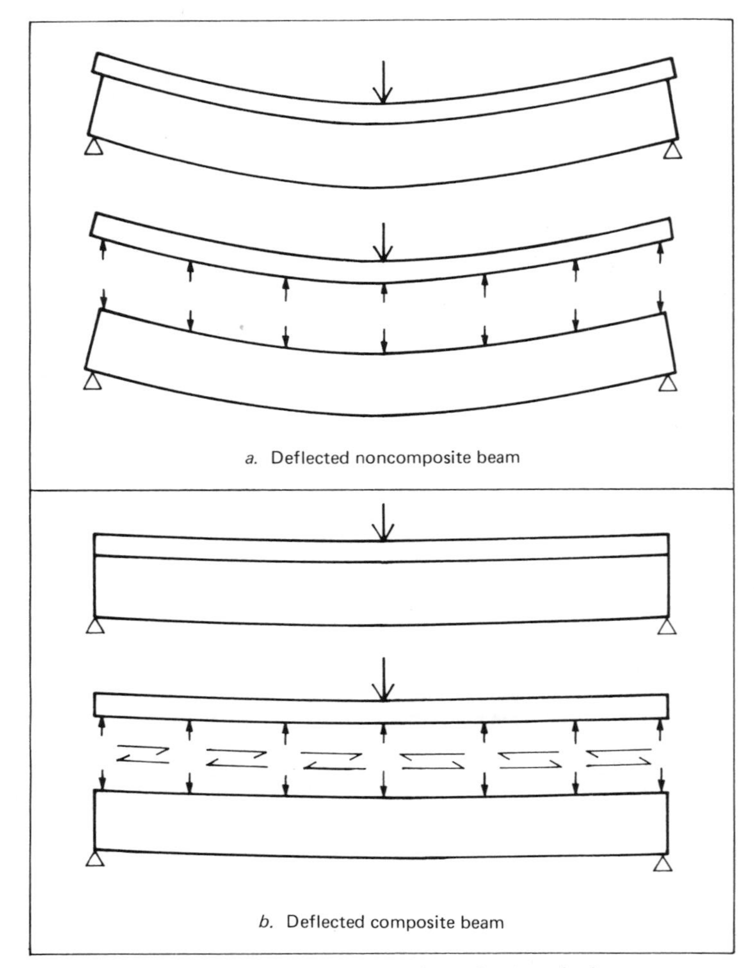

Fig. 4.5. Deflected noncomposite and composite beams.

applied before composite action is achieved* are carried by the precast beam; all loads applied after that (primarily live load) are carried by the composite beam.

In the shored case, the analysis depends on the mode of construction. The shoring may be used to support the precast beam under its own dead load plus the slab, etc., dead load, and then removed after composite action is achieved. In this case, the composite beam carries all dead load and all live load; that is, the total load. Shored reinforced composite beams will be assumed to be of this type. Or, the shoring may be used to support only the slab dead load, etc., but not the precast-beam dead load. In this case, the precast beam carries only its own dead load, and the composite beam carries the slab dead load, etc., plus live load; that is, all load except the precast-beam dead load. Prestressed precast beams are usually assumed to be of the latter type (and will be herein) when shoring is used (supporting their own load regardless of the method of shoring) because of the effect of camber growth. In all cases of unshored and shored construction, the live load is assumed to be carried by the composite beam.

In the preceding discussion, the shoring was assumed to be more or less continuous. For example, for one-point shoring at midspan of a simple beam (and similarly for more than one

*Usually assumed to be before the slab has attained 75% of the 28-day strength–about 1 week after slab casting for a moist cured, ordinary cement slab).

shore), the following analysis is required when the shore is used to support only the slab concrete and reinforcing steel (plus formwork, which is not included in the beam analysis): (1) the precast beam carries its own dead load as a simple beam, (2) the precast beam carries the slab dead load as a two-span continuous beam, and (3) the composite beam carries the shore reaction under the slab dead load but now applied as a concentrated load at midspan, plus the live load, as a simple beam. For most practical purposes, the author considers the use of three or more shores per span, including one at midspan, to have essentially the same effect as continuous shoring.

At ultimate load, the compression stress block in the positive-moment region (including simple beams, of course) is usually within the slab or flange, in which case the calculated ultimate moment is obtained in the same way as for a monolithic T-beam. For other cases, approximate results are usually satisfactory. In the negative-moment region, the ultimate moment is also obtained in the same way as for a monolithic beam. That is, the ultimate moment depends primarily on the strength of the concrete and steel materials and not on the stress history, including whether shoring was or was not used during construction. The direct shear capacity and the horizontal shear capacity are also essentially independent of the method of construction in most cases, although it should be remembered that horizontal shear at the interface of a composite member is produced only by that part of the total load carried by the composite member (producing stresses in the slab).

4.1.3 Effect of Continuity

In order to obtain continuity through the use of a continuous composite slab, longitudinal reinforcement is required in the cast-in-place slab, as shown in Fig. 4.6. Additional steel to provide a positive connection (to resist joint cracking due to positive restraint moments or just direct shrinkage and creep) between two prestressed precast beams and the joint is also necessary, as discussed in Section 5.2.1.8 and Ex. 5-2. A similar connection should also be used to resist joint cracking resulting from separation of the precast beams and the joint due to shrinkage in the case of reinforced precast beams, as shown in Ex. 4-2.

4.1.4 Differential Shrinkage and Creep

Differential shrinkage and creep between the precast-beam concrete and the slab concrete cast at different ages also produce stresses and deflections that are additive to those due to gravity loading in the case of simple spans, as computed in Appendix B. When continuity is obtained by means of reinforcement in the

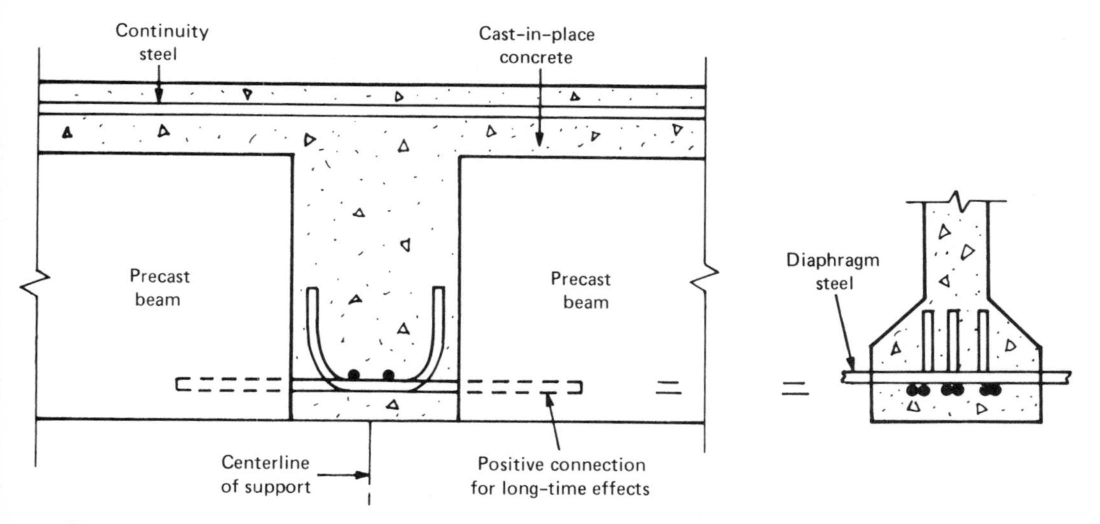

Fig. 4.6. Continuity steel and positive connection steel for typical continuous composite construction.

composite slab over the supports, moments and deflections due to differential shrinkage and creep also occur. These are summarized in Fig. B.3.

The effects of differential shrinkage and creep are usually not of great importance, but may be of some consequence where tensile stresses and/or deflections are critical. This is particularly true when the age of the precast-beam concrete at the time of slab casting is relatively large (i.e., more than a few weeks) and/or when the difference in the quality of the two concretes with regard to shrinkage and creep is significant. Differential shrinkage and creep cause a reduction in the cracking moment under load (effect neglected herein because of load sequence) but have a negligible effect on the ultimate strength of a composite member. The effect of differential shrinkage and creep, and the effect of shoring on this phenomenon, are analyzed in Appendix B and in some of the examples herein. The relative effects of shored and unshored construction and casting schedules of the two concretes for both reinforced and prestressed composite members are taken into account in the selection of the appropriate differential shrinkage and creep coefficient in Table B.1.

4.1.5 Effective Flange Width

The effect of lateral or transverse distribution of longitudinal stress in the flange of monolithic and composite T-beams is essentially the same, and hence the same simplified and conservative methods for determining the effective flange width are used for reinforced and prestressed composite T-beams and noncomposite or monolithic T-beams alike. For practical purposes, the concept of an effective flange width, b_e, is used to simplify the more complex lateral distribution effect, as shown in Fig. 4.7 for an infinitely wide flange. In this case, b_e is a function primarily of the type of loading and span length of the T-beam. In practical cases, the effects of the spacing of beams, the width of the beam web or stem, and the relative thickness of the slab with respect to the total beam depth are also important. These effects have been discussed in more detail by Wang and Salmon[14] and others.

The following procedures are used herein for composite T-beams. The effective flange width, b_e, shall not exceed the following (see Fig. 4.1 or 4.7 for the notation:

ACI Code

1. One-fourth of the span length of the beam
2. The distance center to center of beams
3. $16t + b_w$

AASHTO Specifications

1. Same as ACI Code
2. Same as ACI Code
3. $12t + b_w$

For isolated T-beams, and beams having a

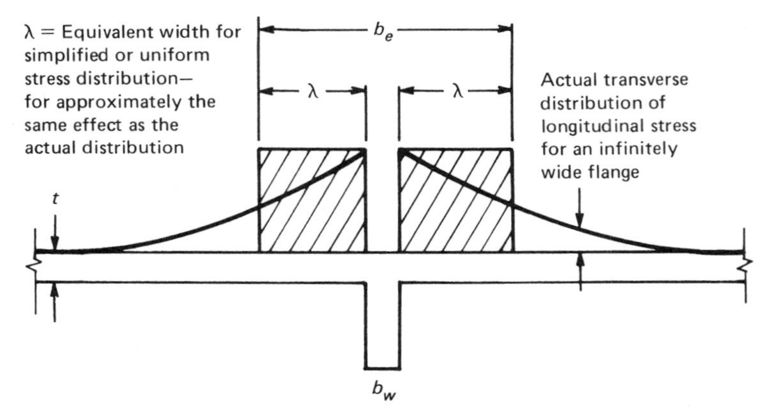

Fig. 4.7. An indication of the actual and simplified lateral or transverse distribution of longitudinal stress in a monolithic or composite T-beam, and the concept of the effective flange width, b_e, used in the flexural analysis of T-beams.

flange on one side only, see the ACI Code and AASHTO Specifications for appropriate effective flange widths.

The British Standard Code of Practice[5] specifies the following for the effective width of flanged beams: In the absence of any more accurate determination, the effective flange width for a T-beam should not exceed the lesser of the width of the web plus one-fifth of the distance between the points of zero moment, or the actual width of the flange. For a continuous beam, the distance between the points of zero moment may be taken as 0.7 times the effective span.

4.1.6 Advantages of Composite Construction

In addition to the economical utilization of precast units, the principal advantage of composite construction is that significantly stiffer and stronger beams can be obtained as compared to the same members without composite action. Also, longer spans and/or shallower sections may be used when composite action is ensured. In addition, the judicious use of shoring in critical stress and/or deflection situations can achieve improved results by using composite construction. When using precast construction in seismic zones, the composite slab or topping is normally required in order to serve as a diaphragm connecting the various units.

4.2 CAST-IN-PLACE SLAB ON REINFORCED CONCRETE PRECAST BEAM

The working stress and strength methods of analysis and design are considered in this section for the design of reinforced concrete composite beams for flexure, web shear, and horizontal shear. Both simple and continuous cast-in-place slabs or toppings using unshored and shored construction are included. As previously discussed, it will be assumed in the case of shored construction that the composite beam supports all loading.

A summary of additional design considerations and a section on deflection control by means of steel ratios, minimum thicknesses, and computed vs. allowable deflections are also presented. Both design and deflection examples are included for a simple-span composite beam and a three-span continuous composite beam. The effects of differential shrinkage and creep on both stresses and deflections are also summarized in Appendix B and included in the examples.

4.2.1 Working Stress Method

4.2.1.1 Section properties

Using the elastic theory and neglecting all concrete in tension, the cracked transformed section moment of inertia and other section properties for a composite reinforced T-beam are computed in Fig. 4.8 when the composite centroid is below the flange (and neglecting the stem area in compression), and in Fig. 4.9 when the composite centroid is within the flange or at the bottom of the flange. The 28-day concrete moduli of elasticity for both the precast beam and slab concrete are normally used in these calculations.

4.2.1.2 Flexure

Although the chapter in the 1971 ACI Code on composite concrete flexural members refers principally to the strength method, the alternate or working stress method is also permitted. The 1963 ACI Code referred specifically to the working stress method for composite beams as follows:

> The design of the composite reinforced con- concrete member shall be based on allowable stresses, working loads, and the accepted straight-line theory of flexure. The effects of creep, shrinkage, and temperature need not be considered except in unusual cases. The effects of shoring, or lack of shoring, on deflections and stresses shall be considered.

From the 1971 ACI Code for the working stress method, in general:

1. The modular ratio, $n = E_s/E_c$, may be taken as the nearest whole number (but not less than 6). Except in calculations for deflections, the value of n for lightweight concrete shall be assumed to be the same as for normal-weight concrete

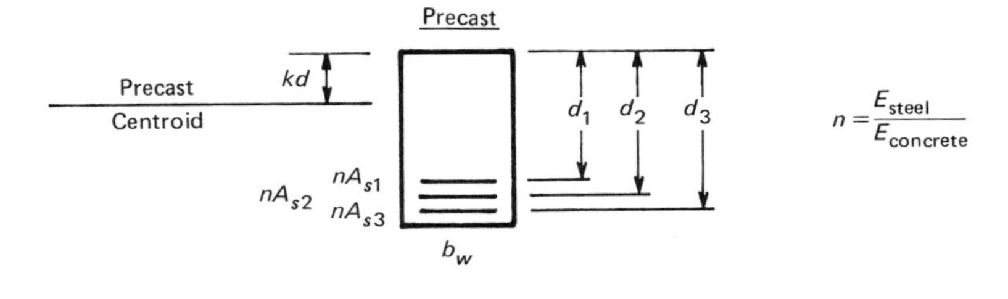

$$b_w(kd)^2/2 = (nA_{s1})(d_1 - kd) + (nA_{s2})(d_2 - kd) + (nA_{s3})(d_3 - kd),$$

Solve for kd.

$$S_t = I_{cr}/kd, \quad S_b = I_{cr}/(d_3 - kd)$$

where

$$I_{cr} = b_w(kd)^3/3 + nA_{s1}(d_1 - kd)^2 + nA_{s2}(d_2 - kd)^2 + nA_{s3}(d_3 - kd)^2$$

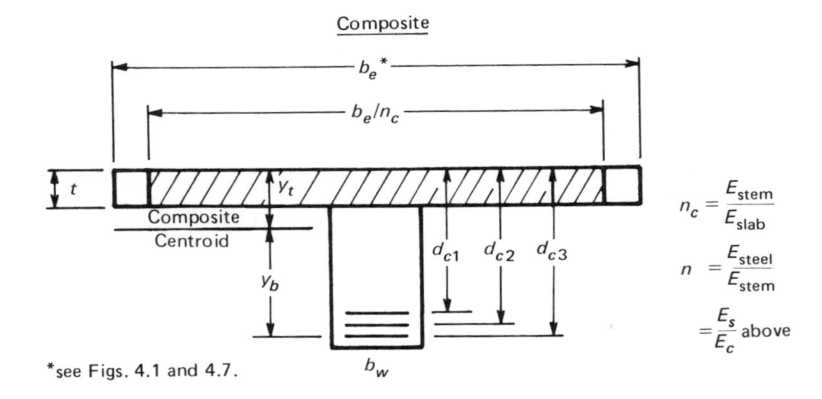

$$(b_e/n_c)(t)(y_t - 0.5t) = nA_{s1}(d_{c1} - y_t) + nA_{s2}(d_{c2} - y_t) + nA_{s3}(d_{c3} - y_t)$$

Solve for y_t.

$$y_b = d_{c3} - y_t. \quad S_t = I_{cr}/y_t, \quad S_b = I_{cr}/y_b$$

where

$$I_{cr} = (b_e/n_c)(t^3/12) + (b_e/n_c)(t)(y_t - 0.5t)^2 + nA_{s1}(d_{c1} - y_t)^2 + nA_{s2}(d_{c2} - y_t)^2 + nA_{s3}(d_{c3} - y_t)^2.$$

Also,

$$Q = (b_e/n_c)(t)(y_t - 0.5t)$$

for the shear stress at the interface.

Fig. 4.8. Precast and composite section properties, neglecting all concrete in tension and stem concrete in compression, when the centroidal axis is below the flange (see Fig. 4.1 for typical steel details).

of the same strength. Suggested values of n from the 1963 Code are

$f'_c =$	3000 psi (21 N/mm²)	4000 psi (28 N/mm²)	5000 psi (35 N/mm²)
$n =$	9	8	7

2. Allowable concrete compressive stress, $f_c = 0.45 f'_c$.
3. The allowable tensile stress in the reinforcement shall not be greater than 20,000 psi (138 N/mm²) for Grade 40 or Grade 50 steel, and 24,000 psi (165 N/mm²) for Grade 60 steel or for steels with yield strengths greater than 60,000 psi (414 N/mm²). For main reinforcement, $\frac{3}{8}$ in. (10 mm) or less in diameter, in one-way slabs of not more than 12-ft (3.7-m) span, the allowable stresses may be increased to 50% of the specified yield strength, but not to exceed 30,000 psi (207 N/mm²).
4. In doubly reinforced beams and slabs, an effective modular ratio of $2n$ shall be used to transform the compression reinforcement for stress computations. The allowable compressive stress in such reinforcement shall not be greater than the allowable tensile stress.

The AASHTO Specifications do not refer specifically to composite reinforced concrete

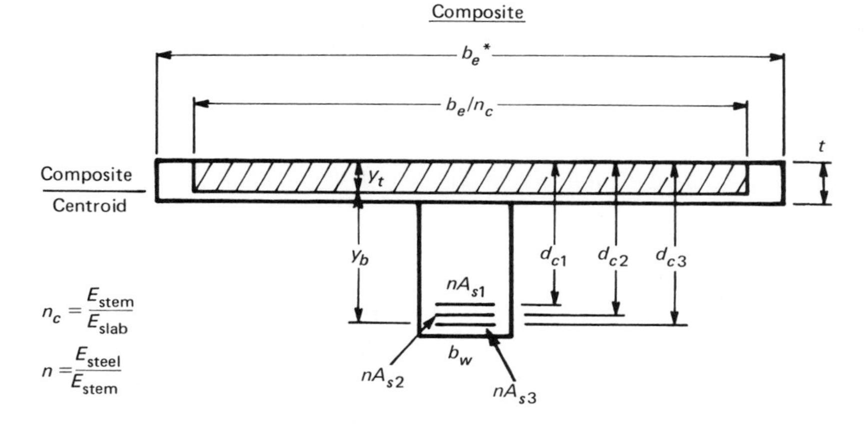

*see Figs. 4.1 and 4.7.

$$(b_e/n_c)(y_t)^2/2 = nA_{s1}(d_{c1} - y_t) + nA_{s2}(d_{c2} - y_t) + nA_{s3}(d_{c3} - y_t)$$

Solve for y_t.

$$y_b = d_{c3} - y_t. \quad S_t = I_{cr}/y_t, \quad S_b = I_{cr}/y_b$$

where

$$I_{cr} = (b_e/n_c)(y_t)^3/3 + nA_{s1}(d_{c1} - y_t)^2 + nA_{s2}(d_{c2} - y_t)^2 + nA_{s3}(d_{c3} - y_t)^2.$$

Also,

$$Q = (b_e/n_c)(y_t)^2/2$$

for the shear stress at the interface, since the shear stress is constant below the composite centroid.

Fig. 4.9. Composite section properties, neglecting all concrete in tension, when the centroidal axis is within the flange or at the bottom of the flange (see Fig. 4.1 for typical steel details).

beams. However, the working stress method for reinforced concrete in general is very similar to that of the ACI Code, as follows:

1. Values of $n = E_s/E_c$ are given for design as

f'_c =	2000 to 2400 psi (14 to 17 N/mm²)	2500 to 2900 psi (17 to 20 N/mm²)	3000 to 3900 psi (21 to 27 N/mm²)
n =	15	12	10

f'_c =	4000 to 4900 psi (28 to 34 N/mm²)	5000 psi or more (35 N/mm² or more)
n =	8	6

   and for deflection calculations, $n = 8$ for all concrete strengths.
2. Allowable concrete compressive stress, $f_c = 0.40 f'_c$.
3. Allowable steel tension and compression in flexural members = 20,000 psi (138 N/mm²) for Grade 40 steel and 24,000 psi (165 N/mm²) for Grade 60 steel.
4. For compression steel in flexural members, $2n$ may be used for the effective modular ratio.

The analysis of composite beams for unshored and shored construction was discussed in Section 4.1.2. For composite reinforced beams with shoring, it will be assumed herein that the shores are constructed so that the precast beam alone does not support any of the loading, including its own dead load. Thus, the recommendation of ACI-ASCE Committee 333[1] will be followed:

1. For unshored construction, the dead load of the precast beam and all other loads applied prior to the concrete slab attaining 75 percent of its specified 28-day strength should be assumed as carried by the precast beam alone. Live loads and dead loads applied after the concrete has attained 75 percent of its specified 28-day strength should be assumed as carried by the composite section.
2. For adequately shored construction all loads should be assumed as carried by the composite section.

Flexural stresses are computed for simple-span composite beams and the positive-moment region of continuous composite beams by Eqs. 4.1 and 4.2.

Unshored Construction

$$f = \frac{M_D}{S_p} + \frac{M_L}{S_c} \tag{4.1}$$

where

f = the flexural stress
M_D = moment due to precast beam plus slab dead load plus other loading applied before the slab concrete has attained 75% of its 28-day strength
M_L = moment due to live load plus other loading applied after the slab concrete has attained 75% of its 28-day strength
S_p = precast-beam section modulus
S_c = composite beam section modulus.

Shored Construction

$$f = \frac{M_D + M_L}{S_c} = \frac{M_{\text{Total Load}}}{S_c} \tag{4.2}$$

The effect of sustained loading must be included in deflection calculations but is normally not included in stress calculations.

Flexural stresses are computed in the negative-moment region for all loads carried by the continuous composite member in much the same way as for a monolithic T-beam. That is, the longitudinal steel in the slab carries the tension and the poured joint concrete as well as the web or stem of the precast beams carry the compression.

4.2.1.3 *Web shear*

Following a rational analysis consistent with Eqs. 4.1 and 4.2 for flexure, shear stresses are computed for simple-span composite beams and the positive-moment region of continuous composite beams by Eqs. 4.3 and 4.4.

Unshored Construction

$$v = \frac{V_D}{b_w d_p} + \frac{V_L}{b_w d_c} \tag{4.3}$$

where

v = the nominal shear stress
V_D = shear force due to precast beam

plus slab dead load plus other loading applied before the slab concrete has attained 75% of its 28-day strength

V_L = shear force due to live load plus other loading applied after the slab concrete has attained 75% of its 28-day strength

b_w = web width

d_p and d_c = distance from the extreme compression fiber of precast and composite sections, respectively, to the centroid of the bottom tension steel.

Shored Construction

$$v = \frac{V_D + V_L}{b_w d_c} = \frac{V_{\text{Total Load}}}{b_w d_c} \tag{4.4}$$

It should be recognized that, theoretically, the maximum shear stress in Eq. 4.3 occurs at the centroidal axis of the precast section in the first term and the composite section in the second term ($V/b_w d$ approximates $V/b_w jd$, which is theoretically equal to VQ/Ib_w for a rectangular section). However, only nominal web shear stresses are usually justifiable.

In the negative-moment region of continuous composite beams, Eqs. 4.3 and 4.4 also apply, with d_p in Eq. 4.3 still referring to the steel in the bottom of the precast beam, but now with d_c in both equations referring to the longitudinal steel in the slab.

According to the 1963 ACI Code, web reinforcement for the composite section shall be designed in the same manner as for an integral beam of the same shape. All stirrups so required shall be anchored into the cast-in-place slab, where their area may also be relied upon to provide some or all of the vertical tie steel required.

This seems to suggest that the web shear design for all loading, not just the part of the total load carried by the composite beam, should be based on the composite section. The author believes such an interpretation to be reasonable for both positive- and negative-moment regions, since the difference between a particular shear stress calculation for precast and composite sections by this method lies in the difference between d_p and d_c.

Accordingly, for most design purposes, Eq. 4.4 might be used to compute nominal web shear stresses for *simple and continuous* composite flexural members for *both unshored and shored construction*, where d_c = the distance from the top fiber of the composite section to the centroid of the bottom steel in *positive moment regions*, and d_c = the distance from the bottom fiber of the composite section to the centroid of the longitudinal slab steel in *negative moment regions*. A more conservative calculation could be made in the case of unshored construction by using d_p instead of d_c in Eq. 4.4 for both positive and negative moment regions.

When following the method of the AASHTO Specifications, Eqs. 4.3 and 4.4 may also be used with the addition of the parameter j (j = ratio of lever arm of resisting couple to depth d) in the denominator of each term.

Stirrups are designed by Eq. 4.5.

$$s = A_v f_s/(v - v_c)b_w \tag{4.5}$$

where

A_v = area of stirrups

s = spacing of stirrups

f_s = allowable steel stress

$v_c = 0.55(2)\sqrt{f_c'} = 1.1\sqrt{f_c'}$ (f_c' in psi by the ACI Code

$v_c = 0.03f_c'$ (maximum 90 psi = 0.62 N/mm^2) by the AASHTO Specifications.

For other limitations on the computed shear stress, stirrup area, and stirrup spacing, see the particular code or specification. Also see Ex. 4.1.

4.2.1.4 Horizontal shear

A rational analysis of horizontal shear stress at the interface between precast and cast-in-place elements would include only those loads that produce bending stresses in the slab. Thus, for unshored construction the precast beam plus slab dead load would be excluded. However, the allowable values for horizontal shear stress, in the 1963 ACI Code by the working stress method and in the 1971 ACI Code by the strength method, for example, are both based on the computed horizontal shear for the total load, regardless of whether shoring is used or not. Such a simplified and conservative approach (based on ultimate strength considerations) is justified by the uncertainty of the behavior. Consequently, the method of computing horizontal shear stress based on the total load will be followed herein for all cases.

The horizontal shear stress at the slab-beam interface is computed by Eq. 4.6:

$$v_h = \frac{VQ}{I_{cr} b_v} \tag{4.6}$$

where

- V = the shear force under the load-producing stress in the slab by a rational analysis—such as live load for unshored construction and total load for shored construction,—or V = the shear force under the total load for all cases by the 1963 ACI Code method, and using the allowable values given below
- Q = statical moment of the transformed area outside the contact surface about the centroidal axis of the composite section
- I_{cr} = moment of inertia of the cracked transformed composite section
- b_v = width of the area of contact between precast and cast-in-place concretes.

The allowable value of v_h computed in Eq. 4.6 is as follows:

1. a. When mechanical anchorages are not provided and the contact surface is rough and clean:

   allowable v_h = 40 psi (0.28 N/mm^2)

   b. When minimum steel tie requirements in (2) are followed and the contact surface is smooth (troweled, floated, or cast against a form):

   allowable v_h = 40 psi (0.28 N/mm^2)

   c. When minimum steel tie requirements in (2) are followed, and the contact surface is rough and clean:

   allowable v_h = 160 psi (1.10 N/mm^2)

   d. When additional vertical ties are used, the allowable bond stress on a rough surface may be increased at the rate of 75 psi (0.52 N/mm^2) for each additional area of steel ties equal to 1% of the contact area.

2. When mechanical anchorage in the form of vertical ties is provided, spacing of such ties shall not exceed four times the thickness of the slab nor 24 in. (61 cm). A minimum cross-sectional area of ties of 0.15% of the contact area shall be provided. It is preferable to provide all ties in the form of extended stirrups. Thus,

$$s_{max} = 4t \leqslant 24 \text{ in. (61 cm)} \tag{4.7}$$

$$s_{max} = A_s/0.0015 b_v \tag{4.8}$$

The above discussion of horizontal shear referred primarily to the positive-moment region in which the slab is acting in compression. The design for horizontal shear in the negative-moment region of a continuous composite beam is usually obtained by a nominal extension of the design in the positive-moment region, although a rational analysis involving the effective negative-moment section and forces may be required in certain cases. Also, one of the three strength methods described in Section 4.2.2.3 and illustrated in Ex. 4.2 will usually be simpler for the horizontal shear design of continuous beams.

See Section 4.2.3 for additional design considerations and Section 4.2.4 for deflection control in terms of steel ratios, minimum thicknesses, and deflections themselves.

Example 4.1

Design of simple-span composite reinforced concrete beam by the working stress method. Span = 35 ft (10.7 m), live load = 150 psf (7180 N/m^2), beam spacing c.c. = 8 ft (244 cm). See Fig. 4.10 for other design conditions and parameters, and Table 4.1 for the section properties.

Flexure

Slab design

$$\text{Bal. } k = \frac{f_c}{\frac{f_s}{n} + f_c} = \frac{1.35}{\frac{24}{9} + 1.35} = 0.336,$$

$$\text{Bal. } j = 1 - \text{Bal. } k/3$$

Simple-Span Composite Beam

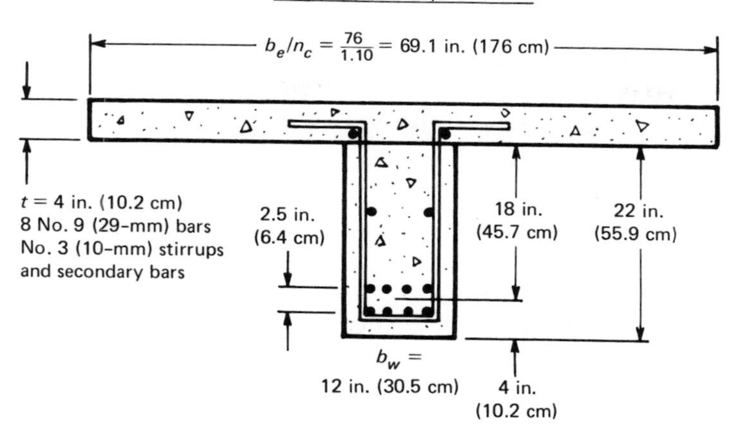

$h_p = 22$ in. (55.9 cm), $h_c = 26$ in. (66.0 cm)

$d_p = 18$ in. (45.7 cm), $d_c = 22$ in. (55.0 cm)

$A_s = 8 \text{ in}^2$ (51.6 cm^2)

$A_v = 0.22 \text{ in}^2$ (1.42 cm^2)

$\rho_p = A_s/bd = 8/(12)(18) = 0.0370$

$\rho_c = A_s/bd = 8/(69.1)(22) = 0.0053$

Beam span = 35 ft (10.7 m)

Beam spacing = 8 ft (244 cm) c.c.

All Concrete–Normal-Weight (See Table A.2, Section 4.2.1.2 for E_c, n)

Slab

$f'_c = 3000$ psi (21 N/mm^2)

$f_c = 0.45(3000) = 1350$ psi (9.3 N/mm^2)

$n = 9$

$E_c = 3.30 \times 10^6$ psi (22.8 kN/mm^2)

Precast Beam

$f'_c = 4000$ psi (28 N/mm^2)

$f_c = 0.45(4000) = 1800$ psi (12.4 N/mm^2)

$n = 8$

$E_c = 3.64 \times 10^6$ psi (25.1 kN/mm^2)

$n_c = 3.64/3.30 = 1.10$

$f_r = 495$ psi (3.42 N/mm^2) (Table A.1)

All Steel

$f_y = 60{,}000$ psi (414 N/mm^2)

$f_s = 24{,}000$ psi (165 N/mm^2)

From Section 4.1.3,

Max b_e = Beam Span/4 = (35)(12)/4

= 105 in. (267 cm)

= Beam Spacing = (8)(12)

= 96 in. (244 cm)

$= 16t + b_w = (16)(4) + 12$

= 76 in. (193 cm) USE

Fig. 4.10. Material properties and design conditions for Ex. 4.1.

$= 1 - 0.112 = 0.888$

Bal. ρ = Bal. $k(f_c)/2f_s = (0.336)(1.35)/(2)(24)$

$= 0.0095$ (Eq. 4.24)

Bal. $R = (f_c/2)$(Bal. j)(Bal. k)

$= (1350/2)(0.888)(0.336)$

$= 201$ psi (1.39 N/mm^2)

From Table 4.2,

Min. $t = \ell/28 = (8)(12)/28$

$= 3.4$ in. (8.6 cm).

Trying $t = 4$ in. (10.2 cm),

$w_D = 150/3 = 50$ psf (2390 N/m^2),

$w_L/w_D = 150/50 = 3.0.$

Since the live load to dead load ratio does not exceed 3, the ACI Code moment coefficients of $-\frac{1}{10}$ and $+\frac{1}{14}$ may be used with the clear span = $\ell_n = 8 - 1 = 7$ ft (2.1 m). As a practical matter, use for both positive and negative moments—$\frac{1}{10}$,

$$\pm M_{D+L} = 0.10\, w_{D+L}(\ell_n)^2 = 0.10(200)(7)^2$$
$$= 980 \text{ ft-lb } (1330 \text{ N-m}).$$

$$\text{Req. } d = \sqrt{\frac{M_{D+L}}{Rb}} = \sqrt{\frac{(980)(12)}{(201)(12)}}$$
$$= 2.21 \text{ in. } (5.57 \text{ cm}).$$

Assuming No. 3 (10-mm) bars,

$$\text{Req. } t = 2.21 + 0.75 \text{ (cover)} + 0.19 \text{ (radius)}$$
$$= 3.15 \text{ in. } (8.0 \text{ cm}) \quad \text{(Section 4.2.3.6)}.$$

Could use $t = 3.5$ in. (8.9 cm), but use $t = 4.0$ in. (10.2 cm). Actual $d = 4 - 0.75 - 0.19 = 3.06$ in. (7.8 cm).

$$\pm \text{ Req. } A_s = \frac{M_{D+L}}{f_s jd} = \frac{(980)(12)}{(24{,}000)(0.888)(3.06)}$$
$$= 0.180 \text{ in.}^2\text{/ft width } (3.8 \text{ cm}^2\text{/m})$$

$$\pm S_p. = \frac{A_s \text{ of 1 bar}}{\text{Req. } A_s/\text{ft}}(12) = \frac{0.11}{0.180}(12)$$
$$= 7.3 \text{ in. } (18.5 \text{ cm}).$$

$$\text{Maximum ACI spacing} = 3t = 3(4)$$
$$= 12 \text{ in. } (30.5 \text{ cm}) \quad \text{OK}$$
(Section 4.2.3.6).

Use No. 3 (10-mm) bars at 7 in. (18 cm) c.c. in both top and bottom of slab.

Actual $\rho = A_s/bd = 0.11/(7)(3.06)$

Table 4.1. Section properties computed for Ex. 4.1.

The cracked-section properties are computed from Fig. 4.8.
For precast beam,

$$I_g = b_w h^3/12 = (12)(22)^3/12 = 10.648 \text{ in.}^4 \ (443{,}200 \text{ cm}^4)$$

$$b_w(kd)^2/2 = nA_s(d - kd),\ 12(kd)^2/2 = (8)(8)(18 - kd),\ kd = 9.51 \text{ in. } (24.2 \text{ cm})$$

$$I_{cr} = b_w(kd)^3/3 + nA_{s1}(d_1 - kd)^2 + nA_{s2}(d_2 - kd)^2 = (12)(9.51)^3/3 + (8)(4)(18 - 1.25 - 9.51)^2$$
$$+ (8)(4)(18 + 1.25 - 9.51)^2 = 3440 + 1677 + 3036 = 8153 \text{ in.}^4 \ (339{,}300 \text{ cm}^4)$$

$$S_t = I_{cr}/kd = 8153/9.51 = 857 \text{ in.}^3 \ (14{,}100 \text{ cm}^3)$$

$$S_b = I_{cr}/(d_2 - kd) = 8153/(18 + 1.25 - 9.51) = 837 \text{ in.}^3 \ (13{,}700 \text{ cm}^3)$$

For composite beam,

$$(y_b)_g = \frac{(12 \times 22)(11) + (69.1 \times 4)(24)}{(12 \times 22) + (69.1 \times 4)} = \frac{(264)(11) + (276.4)(24)}{264 + 276.4} = 17.65 \text{ in. } (44.8 \text{ cm})$$

$$I_g = 10{,}648 + 264(17.65 - 11)^2 + (69.1)(4)^3/12 + (276.4)(24 - 17.65)^2 = 33{,}840 \text{ in.}^4 \ (1{,}408{,}400 \text{ cm}^4)$$

$$(b_e/n_c)(t)(y_t - 0.5t) = nA_s(d_c - y_t), (69.1)(4)(y_t - 2) = (8)(8)(22 - y_t),$$

$$y_t = 5.76 \text{ in. } (14.6 \text{ cm}) > 4 \text{ in.} \quad \text{OK}$$

$$I_{cr} = (b_e/n_c)(t^3)/12 + (b_e/n_c)(t)(y_t - 0.5t)^2 + nA_{s1}(d_{c1} - y_t)^2 + nA_{s2}(d_{c2} - y_t)^2 = (69.1)(4)^3/12$$
$$+ (69.1)(4)(5.76 - 2)^2 + (8)(4)(22 - 1.25 - 5.76)^2 + (8)(4)(22 + 1.25 - 5.76)^2$$
$$= 21{,}260 \text{ in.}^4 \ (884{,}800 \text{ cm}^4)$$

$$S_t = I_{cr}/y_t = 21{,}260/5.76 = 3691 \text{ in.}^3 \ (60{,}500 \text{ cm}^3)$$

$$S_b = I_{cr}/(d_{c2} - y_t) = 21{,}260/(22 + 1.25 - 5.76) = 1216 \text{ in.}^3 \ (19{,}900 \text{ cm}^3)$$

Also,

$$Q = (b_e/n_c)(t)(y_t - 0.5t) = (69.1)(4)(5.76 - 2) = 1039 \text{ in.}^3 \ (17{,}000 \text{ cm}^3)$$

$= 0.0051 < \rho_b = 0.0095$ OK (4.2.3.1)

> 0.0018 OK

(Section 4.2.3.1)

$b(kd)^2/2 = nA_s(d - kd)$

$7(kd)^2/2 = 9(0.11)(3.06 - kd),$

$kd = 0.80$ in. (2.03 cm)

$jd = d - kd/3 = 3.06 - 0.80/3$

$= 2.79$ in. (7.08 cm)

$M = (f_c/2)bkdjd$

$= (1350/2)(12)(0.80)(2.79)/12$

$= 1507$ ft-lb (2043 N-m)

> 980 ft-lb (1330 N-m) OK

$M = A_s f_s jd$

$= (0.11)(12/7)(24{,}000)(2.79)/12$

$= 1052$ ft-lb (1427 N-m)

> 980 ft-lb (1330 N-m) OK

Provide shrinkage and temperature reinforcement perpendicular to main reinforcement:

Req. $A_s = (0.0018)(12)(3.06) = 0.066$ in.^{2}

Use No. 3 (10-mm) bars at 18 in. (45-cm),

$A_s = 0.073$ in.^{2}/ft width (1.6 cm^2/m)

(Section 4.2.3.1)

Checking ACI + $\rho_{min} = 200/f_y$ (need not apply to slabs of uniform thickness) = 200/60,000 = 0.0033 < 0.0051 OK. (Eq. 4.28)

Checking ACI crack-control parameter in Section 4.2.3.3 (Eq. 4.29),

f_s = computed value or $0.6 f_y$

= 36 ksi (248 N/mm^2)

$d_c = 0.75 + 0.19 = 0.94$ in. (2.39 cm)

$A = (2)(0.94)(7)$

$= 13.16$ in.^{2}/bar (84.9 cm^2/bar)

$z = f_s \sqrt[3]{d_c A} = 36\sqrt[3]{(0.94(13.16)}$

= 83 kips/in. (145 kN/cm)

< 175 kips/in. (306 kN/cm) for interior exposure and < 145 kips/in. (254 kN/cm) for exterior exposure. OK

To show that shear is usually small in such a slab, neglecting the distance d from supports:

$V = 1.15\, w\ell_n/2 = (1.15)(200)(7)/2$

= 805 lb (3580 N)

$v = V/bd = 805/(12)(3.06)$

= 22 psi (0.15 N/mm^2) $< 1.1\sqrt{f_c'}$

= 60 psi (0.41 N/mm^2) OK

Beam design

For the composite beam from Table 4.2, estimate $h = \ell/16 = (35)(12)/16 = 26.3$ in. (66.8 cm). Or, based on deflection considerations from Table 4.3, estimate h to range from $\ell/18 = 23.3$ in. (59.2 cm) to $\ell/10 = 42.0$ in. (107 cm). Try $h_c = 26$ in. (66.0 cm), and check deflections in Section 4.2.4.3.

Initially assume unshored construction. For slab,

$w_s = (8)(4/12)(150) = 400$ lb/ft (5840 N/m)

$M_s = w_s\ell^2/8 = (400)(35)^2/8$

$= 61{,}250$ ft-lb (83.1 kN-m)

For precast beam 12 × 22 in. (30.5 × 55.9 cm),

$w_p = (12 \times 22)(150)/144$

$= 275$ lb/ft (4010 N/m)

$M_p = w_p\ell^2/8 = (275)(35)^2/8$

$= 42{,}110$ ft-lb (57.1 kN-m)

$M_D = M_s + M_p = 61{,}250 + 42{,}110$

$= 103{,}360$ ft-lb (140.2 kN-m)

For precast beam (from Fig. 4.10), $f_c = 1800$ psi (12.4 N/mm^2), $f_s = 24{,}000$ psi (165 N/mm^2):

Bal. $j_p = 0.860$

Bal. $\rho = 0.0188$

Bal. $R = 324$ psi (2.23 N/mm^2).

Calc. $R_p = M_D/bd^2 = (103{,}360)(12)/(12)(18)^2$

= 319 psi (2.20 N/mm^2)

< 324 psi (2.23 N/mm^2)

OK for concrete stress in the unshored case.

From Fig. 4.10, $\rho_p = 0.0370 > 0.0188$ as expected, but $\rho_c = 0.0053 < 0.0188$ OK.

For composite beam, assume $j_c = 0.90$

$$M_L = (150)(8)(35)^2/8$$
$$= 183{,}800 \text{ ft-lb } (249 \text{ kN-m})$$

$$\text{Approx. } A_s = \frac{M_D}{f_s j_p d_p} + \frac{M_L}{f_s j_c d_c}$$
$$= \frac{(103{,}360)(12)}{(24{,}000)(0.860)(18)} + \frac{(183{,}800)(12)}{(24{,}000)(0.90)(22)}$$
$$= 3.34 + 4.64$$
$$= 7.98 \text{ in.}^2 \ (51.5 \text{ cm}^2)$$

Use eight No. 9 (29-mm) bars, $A_s = 8.00$ in.^{2} (51.6 cm^2)

ACI Code Req. beam width = 7(1.128)

+ 1 (No. 3 or No. 4 stirrups) + 3 (cover)

= 11.90 in. < 12 in. (30.5 cm) OK

(Section 4.2.3.6)

ACI Code Req. $(h - d) = 1.5$ (cover)

+ 0.5 (No. 3 or No. 4 stirrups) + 1.13 + 0.4

= 3.63 in. < 4 in. (10.2 cm) OK

(Section 4.2.3.6)

As outlined in Section 4.2.3.6, cover requirements in certain cases can be less than these for precast members.

The section properties are computed in Table 4-1.

For precast beam,

$$f_c = f_t = M_D/S_t = (103{,}360)(12)/857 \quad \text{(Eq. 4.1)}$$
$$= 1447 \text{ psi } (9.98 \text{ N/mm}^2) < 0.45\,(4000)$$
$$= 1800 \text{ psi } (12.41 \text{ N/mm}^2) \quad \text{OK}$$

$$f_s = nf_b = nM_D/S_b = (8)(103{,}360)(12)/837 \quad \text{(Eq. 4.1)}$$
$$= 11{,}850 \text{ psi } (81.7 \text{ N/mm}^2)$$
$$< 24{,}000 \text{ psi } (165 \text{ N/mm}^2) \quad \text{OK}$$

Assuming a one-third temporary construction live-load moment = $M_{CL} = M_L/3 = 183{,}800/3 =$ 61,270 ft-lb (83.1 kN-m)

For $M = M_D + M_{CL} = 103{,}360 + 61{,}270 =$ 164,630 ft-lb (223 kN-m) $f_c = (164.6/103.4)$ (1447) = 2300 psi (15.9 N/mm^2) > 1800 psi, but a temporary overstress of 28% should be acceptable.

$$f_s = (164.6/103.4)(11{,}850)$$
$$= 18{,}860 \text{ psi } (130 \text{ N/mm}^2) < 24{,}000 \text{ psi} \quad \text{OK}$$

For composite beam,

$$f_c = f_t/n_c = M_L/n_c S_t$$
$$= (183{,}800)(12)/(1.10)(3691) \quad \text{(Eq. 4.1)}$$
$$= 543 \text{ psi } (3.74 \text{ N/mm}^2) < 0.45(3000)$$
$$= 1350 \text{ psi } (9.31 \text{ N/mm}^2) \quad \text{OK}$$

f_c at top of precast beam

$$= 1447 + 543 n_c(y_t - t)/y_t$$
$$= 1447 + (543)(1.10)(1.76/5.76)$$
$$= 1630 \text{ psi } (11.2 \text{ N/mm}^2)$$
$$< 0.45(4000)$$
$$= 1800 \text{ psi } (12.41 \text{ N/mm}^2) \quad \text{OK}$$

$$f_s = nf_b = 11{,}850 \text{ (above)} + nM_L/S_b \quad \text{(Eq. 4.1)}$$
$$= 11{,}850 + (8)(183{,}800)(12)/1216$$
$$= 26{,}360 \text{ psi } (182 \text{ N/mm}^2)$$
$$> 24{,}000 \text{ psi } (165 \text{ N/mm}^2) \text{ or a 10\% overstress.}$$

The option might be to allow the overstress, to redesign, to check the strength design method, or to check the use of shored construction.

For shored construction,

$$f_c = f_t/n = (M_D + M_L)/n_c S_t \quad \text{(Eq. 4.2)}$$
$$= (103{,}360 + 183{,}800)(12)/(1.10)(3691)$$
$$= 849 \text{ psi } (5.85 \text{ N/mm}^2) < 0.45\,(3000)$$
$$= 1350 \text{ psi} \quad \text{OK}$$

$$f_s = nf_b = n(M_D + M_L)/S_b \quad \text{(Eq. 4.2)}$$
$$= (8)(287{,}160)(12)/1216$$

$= 22{,}670$ psi (156 N/mm²)

$< 24{,}000$ psi OK.

The flexural design is now satisfactory using shored construction.

Checking ACI + ρ_{min}

$= 200/f_y = 200/60{,}000 = 0.0033 < A_s/b_w d$

$= 8.00/(12)(22) = 0.0303$ OK

(Eq. 4.28)

Checking the ACI crack-control parameter in Section 4.2.3.3 (Eq. 4.29),

f_s = computed value or $0.6 f_y$

= 36 ksi (248 N/mm²)

$d_c = 4 - 1.25 = 2.75$ in. (7.0 cm)

$A = (2)(4)(12)/8$

= 12 in.²/bar (77.4 cm²/bar)

$z = f_s \sqrt[3]{d_c A} = 36 \sqrt[3]{(2.75)(12)}$

= 116 kips/in. (203 kN/cm)

$<$ 175 kips/in. (306 kN/cm)

for interior exposure and $<$ 145 kips/in. (254 kN/cm) for exterior exposure. OK

The stresses due to differential shrinkage and creep are checked as an illustration, although this is normally not required for reinforced concrete (cracked) sections, since this effect occurs gradually over time.

From Fig. 4.10,

$E_1 = (E_c)_{28d} = 3.30 \times 10^6$ psi (22.8 kN/mm²)

(Eq. B.9)

$E_2 = (E_c)_{28d} = 3.64 \times 10^6$ psi (25.1 kN/mm²)

(Eq. B.10)

For the differential shrinkage and creep coefficient—$(\text{C.F.})_H = 0.70$ from Table A.5 assuming 70% humidity, and $(\text{C.F.})_T = 1.17$ for 4-in. (10.2-cm) slab thickness from Table A.6. From Table B.1, $D_u = 395 \times 10^{-6}$ in./in. (mm/mm) for the shored case, and $D_u = 355 \times 10^{-6}$ in./in. (mm/mm) for the unshored case, before the correction factors are applied. Using the larger of the two (shored case) and applying the correction factors,

$$D_u = (0.70)(1.17)(395 \times 10^{-6})$$

$$= 324 \times 10^{-6} \text{ in./in. (mm/mm)}$$

From Figs. 4.8, 4.10 and B.1, and Table 4.1,

$y_{cs} = y_t - t/2 = 5.76 - 2 = 3.76$ in. (9.55 cm)

$y_{ct} = y_t = 5.76$ in. (14.6 cm)

$y_{ci} = y_t - t = 5.76 - 4 = 1.76$ in. (4.47 cm)

$y_{cb} = y_b = d_{c3} - y_t = 22 + 1.25 - 5.76$

= 17.49 in. (44.4 cm)

$A_1 = b_e t = (78)(4) = 312$ in.² (2013 cm²)

$A_c = (b_e/n_c)(t) + nA_s = (69.1)(4) + (8)(8)$

= 340 in.² (2194 cm²)

$I_c = I_{cr} = 21{,}260$ in.⁴ (884,800 cm⁴).

By the elastic theory, the differential shrinkage and creep force is calculated as

$$Q = D_u A_1 E_1 = (324)(312)(3.30)$$

(Eq. B.1)

= 333,600 lb (1,484,000 N)

However, using Eq. B.11 for creep relaxation effects,

$$Q_{DS} = Q/2 = 333{,}600/2$$

$$= 166{,}800 \text{ lb } (742{,}000 \text{ N})$$

(Eq. B.11)

Computing the concrete stresses due to differential shrinkage and creep,

$$\sigma_{1t} = \frac{Q_{DS}}{A_1} + \left(-\frac{Q_{DS}}{A_c} - \frac{Q_{DS} y_{cs} y_{ct}}{I_c}\right) \frac{E_1}{E_2}$$

(Eq. B.2)

$$= \frac{166{,}800}{312} + \left(-\frac{166{,}800}{335} - \frac{166{,}800 \times 3.76 \times 5.76}{21{,}260}\right) \frac{3.30}{3.64}$$

$$= 535 + (-498 - 170)(0.907)$$

$$= -71 \text{ psi } (0.49 \text{ N/mm}^2) \text{ (compression)}$$

$$\sigma_{1b} = \frac{Q_{DS}}{A_1} + \left(-\frac{Q_{DS}}{A_c} - \frac{Q_{DS}y_{cs}y_{ci}}{I_c}\right)\frac{E_1}{E_2}$$

(Eq. B.3)

$= 535 + [-498 - (170)(1.76/5.76)](0.907)$

$= 535 + [-498 - 52](0.907)$

$= 36$ psi (0.25 N/mm^2) (tension)

$$\sigma_{2t} = -\frac{Q_{DS}}{A_c} - \frac{Q_{DS}y_{cs}y_{ci}}{I_c}$$ (Eq. B.4)

$= -498 - 52$

$= -550$ psi (3.79 N/mm^2) (compression)

For unshored construction, $\sigma_{2t} = (-550)(355/395) = -494$ psi (3.41 N/mm^2).

Computing the steel stress due to differential shrinkage and creep,

$$\sigma_{2b} = n\left(-\frac{Q_{DS}}{A_c} + \frac{Q_{DS}y_{cs}y_{cb}}{I_c}\right)$$

$n \times$ (Eq. B.5)

$= (8)[-498 + (170)(17.49/5.76)]$

$= (8)(-498 + 516) = (8)(18)$

$= 144$ psi (0.99 N/mm^2) (tension)

These computed stresses due to differential shrinkage and creep are very small, except for the concrete stress at the top of the precast beam, σ_{2t}. The stress at the top of the precast beam under load would be small in the shored case. However, in the unshored case, the combined stress at the top of the precast beam is $\sigma_{2t} = 1630$ (previous calculation) $+ 494 = 2124$ psi (14.6 N/mm^2), compression; this is greater than the allowable stress of $0.45(4000) = 1800$ psi (12.41 N/mm^2). This result might further indicate the need to use shored construction in this design.

Deflections are checked in Section 4.2.4.4.

Web shear

For either unshored or shored construction, use Eq. 4.4 as previously discussed. Using the shear at the end of the span (neglecting distance d out),

$V_D = (w_p + w_s)\ell/2 = (275 + 400)(35)/2$

$= 11{,}810$ lb ($52{,}500$ N)

$V_L = w_L \ell/2 = (150)(8)(35)/2$

$= 21{,}000$ lb ($93{,}400$ N)

$V_D + V_L = 11{,}810 + 21{,}000$

$= 32{,}810$ lb ($145{,}900$ N)

$$v = \frac{V_D + V_L}{b_w d_c} = \frac{32{,}810}{(12)(22)}$$

$= 124$ psi (0.85 N/mm^2) (Eq. 4.4)

$< 5\sqrt{f_c'} = 5\sqrt{4000}$

$= 316$ psi upper limit OK

(1963 ACI Code)

$< 3\sqrt{f_c'} = 3\sqrt{4000} = 190$ psi for $s_{max} = d/2$ (1963 ACI Code)

$v_c = 1.1\sqrt{f_c'} = 1.1\sqrt{4000}$

$= 70$ psi (0.48 N/mm^2)

For No. 3 (10-mm) stirrups, $A_v = 0.22$ in.^{2} (1.42 cm^2)

$s = A_v f_s/(v - v_c)b_w$ (Eq. 4.5)

$= (0.22)(24{,}000)/(124 - 70)(12)$

$= 8.1$ in. (20.6 cm)

For No. 4 (13-mm) stirrups, $A_v = 0.40$ in.^{2} (2.58 cm^2)

$s = (0.40/0.22)(8.1) = 14.7$ in. (37.3 cm)

$s_{max} = d/2 = 18/2 = 9$ in. (22.9 cm) or $22/2$

$= 11$ in. (27.9 cm)

Could use No. 3 (10-mm) stirrups at 8 in. or No. 4 (13-mm) stirrups, e.g., at 10 or 11 in. Use No. 3 (10-mm) stirrups and $s = 8$ in. (20 cm) for web shear, e.g., for the end quarter-spans and $s = 11$ in. (28 cm) for the middle half-span.

Horizontal shear

For either unshored or shored construction, use Eq. 4.6 as previously discussed.

$$v_h = \frac{V_{\text{Total Load}} Q}{I_{cr} b_v} = \frac{(32{,}810)(1039)}{(21{,}260)(12)} \qquad \text{(Eq. 4.6)}$$

$= 134$ psi (0.92 N/mm^2) > 40 psi and <160 psi.

Hence, it is required that the following minimum tie design be used plus the contact surface to be rough and clean:

$$s_{max} = 4t = (4)(4) = 16 \text{ in. } (40.6 \text{ cm}) \qquad \text{(Eq. 4.7)}$$

$$s_{max} = A_v/(0.0015)b_v = 0.22/(0.0015)(12) \qquad \text{(Eq. 4.8)}$$

$$= 12.2 \text{ in. } (31.0 \text{ cm})$$

Hence, use the web stirrup design of No. 3 (10-mm) stirrups extended into the slab.

4.2.2 Strength Method

The ACI Code and Commentary suggest that tests to destruction indicate no difference in the strength of unshored and shored members, and hence no distinction need be made in their design. That is, the strength of the composite beam depends primarily on the strength of its constituent materials, and not on the stress history, including whether or not shoring was used. The same might also be said of the effects of differential shrinkage and creep, although the force resulting from differential shrinkage and creep is normally not included with the factored gravity loads that produce the stresses up to failure. However, for practical purposes, it will be assumed that the effects of differential shrinkage and creep can be neglected in the strength design of reinforced concrete composite members, as per the British Standard Code of Practice.[5]

4.2.2.1 Flexure

Using the ACI Code approach for the moment-strength design of unshored and shored composite beams in both positive and negative moment regions under gravity loading,

$$\text{Design } M_u = 1.4M_D + 1.7M_L \qquad (4.9)$$

$$\rho_{max} = 0.75\rho_b \qquad (4.10)$$

where ρ_b is given by Eqs. 4.25 through 4.27 for rectangular beams and T-beams.

The usable ultimate moment, M_u, is computed by the equations in Figs. 4.11 and 4.12 for underreinforced sections in positive and negative moment regions.

The following limitations were specified in the 1963 ACI Code:

1. For beams designed on the basis of ultimate strength and built without shores, Eq. 4.11 should be satisfied.

   $$d_c/d_p < 1.15 + 0.24M_L/M_D \qquad (4.11)$$

   where d_c and d_p refer to the effective depths of the composite and precast sections, respectively.
2. When the specified yield point of the tension reinforcement exceeds 40,000 psi (276 N/mm^2), beams designed on the basis of ultimate strength should always be built with shores unless provisions are made to prevent excessive tensile cracking.

The same ACI Code further stated, with regard to construction loads, that the nonprestressed precast element shall be investigated separately to assure that the loads applied before the cast-in-place concrete has attained 75% of its specified 28-day strength do not cause moments in excess of 60% of the ultimate capacity of the precast section. The author suggests that these rules might still be used as a guide to indicate the possible need for shores and limitations due to construction loads.

The 1971 ACI Code[3] states:

> When used, shoring shall not be removed until the supported elements have developed the design properties required to support all loads and limit deflections and cracking at the time of shoring removal.

The following approach by Wang and Salmon[7] is used to indicate the development of Eq. 4.11: Letting the steel stress for the limiting case of no shoring $= 0.70f_y$,

$$\frac{M_D}{A_s j_p d_p} + \frac{M_L}{A_s j_c d_c} = 0.70f_y \qquad \text{(a)}$$

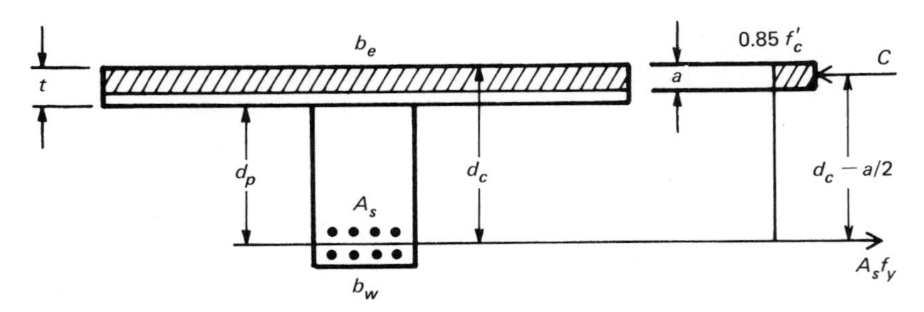

a. $a < t$ (Usual case)

$$C = 0.85 f'_c b_e a = A_s f_y$$

or

$$a = A_s f_y / (0.85)(\text{Slab } f'_c) b_e$$

and

$$M_u/\phi = A_s f_y (d_c - a/2), \quad \phi = 0.90 \text{ for bending}$$

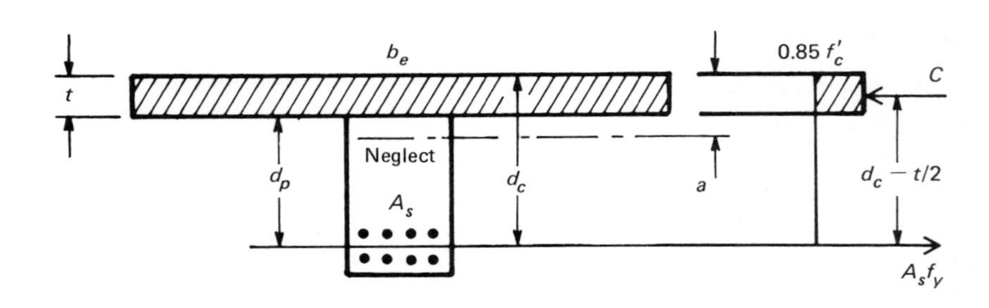

b. $a > t$

$$M_u/\varphi = A_s f_y (d_c - t/2), \varphi = 0.90 \text{ for bending}$$

Fig. 4.11. Ultimate bending moment of underreinforced composite beams (same as monolithic T-beams) for two cases of concrete stress blocks in positive moment regions.

For the ultimate moment by the 1963 ACI Code and Fig. 4.11a,

$$M_u = 1.5M_D + 1.8M_L = 0.90 A_s f_y (d_c - a/2) \tag{b}$$

or

$$A_s = \frac{1.5M_D + 1.8M_L}{0.90 f_y (d_c - a/2)} \tag{c}$$

Substituting (c) into (a),

$$\frac{M_D}{j_p d_p} + \frac{M_L}{j_c d_c} = \frac{0.70 f_y}{0.90 f_y} \left(\frac{1.5M_D + 1.8M_L}{d_c - a/2} \right) \tag{d}$$

Multiplying (d) by $d_p d_c / M_D$,

$$\frac{d_c}{j_p} + \frac{d_p}{j_c} \frac{M_L}{M_D} = \left(1.17 + 1.40 \frac{M_L}{M_D} \right) \frac{d_p d_c}{d_c - a/2} \tag{e}$$

or

$$d_c = \left[1.17 \frac{j_p d_c}{d_c - a/2} + \frac{M_L}{M_D} \left(- \frac{j_p}{j_c} + 1.40 \frac{j_p d_c}{d_c - a/2} \right) \right] d_p \tag{f}$$

Assuming $j_p d_c / (d_c - a/2) = 0.91, j_p/j_c = 0.94$,

$$\frac{d_c}{d_p} = (1.17 \times 0.91) + [(1.40)(0.91) - 0.94](M_L/M_D)$$

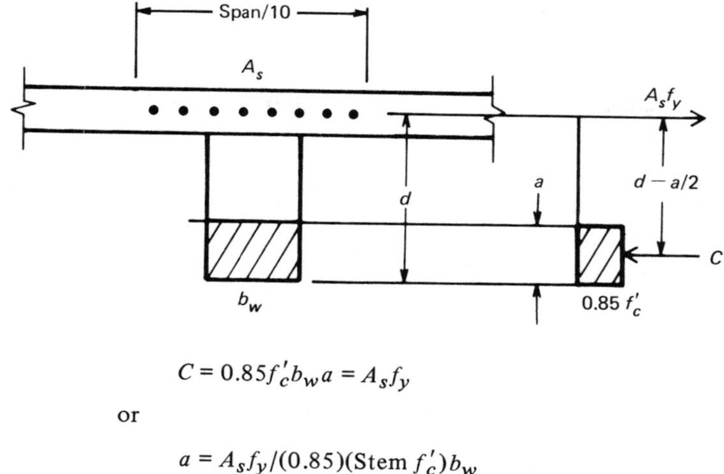

$$C = 0.85 f'_c b_w a = A_s f_y$$

or

$$a = A_s f_y / (0.85)(\text{Stem } f'_c) b_w$$

and

$$M_u/\phi = A_s f_y (d - a/2), \quad \phi = 0.90 \text{ for bending}$$

Fig. 4.12. Ultimate bending moment for underreinforced case in negative moment regions.

$$= 1.06 + (1.27 - 0.94)(M_L/M_D)$$

$$= 1.06 + 0.33 M_L/M_D \qquad \text{(g)}$$

which is the same as Eq. 4.11 when $M_L/M_D = 1$, and similar to the same equation for other values of M_L/M_D.

The AASHTO Specifications do not refer specifically to composite reinforced concrete beams. However, for load-factor design, the ultimate bending moment for T-beams is computed by the usual equations for such members. In the AASHTO method, different load factors are used than in Eq. 4.9, and $\rho_{max} = 0.50\rho_b$, instead of $0.75\rho_b$ in Eq. 4.10.

4.2.2.2 Web Shear

When the entire composite member is assumed to resist the vertical shear, the web shear design is determined the same as for a monolithically cast member of the same cross-sectional shape, according to the ACI Code, as follows:

$$v_u = \frac{V_u}{\phi b_w d} \qquad (4.12)$$

where

$V_u = 1.4V_D + 1.7V_L$ for both unshored and shored members under gravity loading, etc.

$d = d_c$ as shown in Fig. 4.11

$\phi = 0.85$ for shear.

In the case of a continuous slab, Eq. 4.12 may also be used, although the effective depth, d, might logically refer to the negative-moment region section, as shown in Fig. 4.12.

Using the ACI Code method, stirrups are designed by Eq. 4.13.

$$s = A_v f_y / (v_u - v_c) b_w \qquad (4.13)$$

where

s = spacing of stirrups

A_v = area of stirrups

v_c may be taken as $2\sqrt{f'_c}$ for normal-weight concrete. The value of $(v_u - v_c)$ shall not exceed $8\sqrt{f'_c}$. Also,

$$s_{max} = A_v f_y / 50 b_w \qquad (4.14)$$

$$s_{max} = d/2 \text{ when } (v_u - v_c) \leqslant 4\sqrt{f'_c} \qquad (4.15)$$

$$s_{max} = d/4 \text{ when } (v_u - v_c) > 4\sqrt{f'_c} \qquad (4.16)$$

$$s_{max} = 24 \text{ in. } (61 \text{ cm}) \qquad (4.17)$$

Stirrups are not required when v_u in Eq. 4.12 is less than $v_c/2$.

When f_{ct} is not specified, all values of $\sqrt{f'_c}$

affecting v_c shall be multiplied by 0.75 for all-lightweight concrete and 0.85 for sand-lightweight concrete. The web reinforcement should be fully anchored into the components. Extended and anchored web reinforcement may be included as ties for horizontal shear.

The web shear design of reinforced concrete members by the AASHTO Specifications for load-factor design used the same approach as the procedure above.

4.2.2.3 Horizontal Shear

According to the ACI Code, full transfer of horizontal shear forces may be assumed when all of the following are satisfied: (1) the contact surfaces are clean and intentionally roughened, (2) minimum ties are provided as described below, (3) web members are designed to resist the entire vertical shear, and (4) all stirrups are fully anchored into all intersecting components. Otherwise, horizontal shear shall be fully investigated.

The design horizontal shear stress is computed by Eq. 4.18. The elastic counterpart of this equation is derived in Fig. 4.13 from the basic shear-stress equation for the case in which the centroidal axis is assumed at the bottom of the slab. The result is an approximation of this and other cases:

$$v_{dh} = \frac{V_u}{\phi b_v d_c} \tag{4.18}$$

where $\phi = 0.85$, and b_v is the width of the cross section being investigated for horizontal shear. In most cases, Eq. 4.18 is the same as Eq. 4.12.

Alternatively, the actual compressive or tensile force in any segment may be computed, and provisions made to transfer the force as horizontal shear to the supporting element. In this case, the ϕ factor for shear shall be used with the compressive or tensile force.

The design shear force may be transferred at contact surfaces using the permissible horizontal shear stresses v_h stated in the following:

1. When ties are not provided, but the contact surfaces are clean and intentionally roughened,

   permissible $v_h = 80$ psi ($0.55\ \text{N/mm}^2$)

2. When the minimum tie requirements are provided and the contact surfaces are clean but not intentionally roughened,

   permissible $v_h = 80$ psi ($0.55\ \text{N/mm}^2$)

3. When the minimum tie requirements are provided and the contact surfaces are clean and intentionally roughened,

   permissible $v_h = 350$ psi ($2.41\ \text{N/mm}^2$)

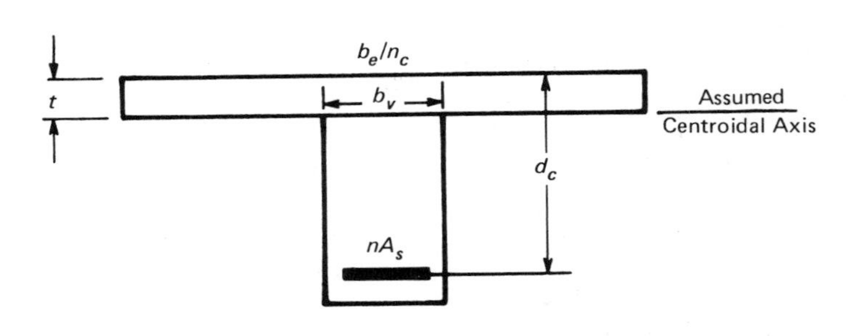

$$Q = (b_e/n_c)(t)^2/2 = nA_s(d_c - t)$$

$$I = (b_e/n_c)(t)^3/3 + nA_s(d_c - t)^2$$

$$I/Q = (2/3)t + (d_c - t) = d_c - t/3$$

$$v = \frac{VQ}{Ib_v} = \frac{V}{(d_c - t/3)b_v} \simeq \frac{V}{d_c b_v}$$

Fig. 4.13. Derivation of approximate horizontal shear stress at the stem-flange interface when the centroidal axis is assumed to be located at the interface and the section is fully cracked in tension.

4. When v_{dh} in Eq. 4.18 exceeds 350 psi, design for horizontal shear shall be made by the shear-friction method.

Intentional roughness

Intentional roughness may be assumed only when the interface is roughened with a full amplitude of approximately $\frac{1}{4}$ in. (0.64 cm).

Minimum tie requirements

The minimum tie requirements are given by Eqs. 4.14, 4.17, and 4.19.

$$s_{max} = 4t \qquad (4.19)$$

Shear friction method applied to tie design for horizontal shear

Using the ACI Code method, the required area of horizontal shear reinforcement for half the span is computed by Eq. 4.20, provided that the precast-beam contact surface is intentionally roughened. Such reinforcement must be well distributed (perhaps uniformly spaced).

$$A_{vf} = \frac{V_u}{\phi f_y \mu} \qquad (4.20)$$

where

A_{vf} = the required tie area for half the span
$\phi = 0.85$
the coefficient of friction $\mu = 1.0$
V_u = the ultimate horizontal shear force at the interface

and f_y shall not exceed 60,000 psi (414 N/mm^2). Equation 4.20 may be applied when v_{dh} in Eq. 4.18 does not exceed $0.2f'_c$ nor 800 psi (5.52 N/mm^2).

In Eq. 4.20, V_u may be computed as the volume of the shear stress block or diagram for half the span. For a triangular shear diagram due to uniform loading, for example, V_u is given by Eq. 4.21.

$$\frac{V_u}{\phi} = (1/2)(v_{dh} \text{ in Eq. 4.18 at the end of the span})(b_v)(\ell/2) \qquad (4.21)$$

Alternatively, V_u in Eq. 4.20 may be determined as the smaller of the ultimate horizontal forces that could be developed at the interface for half the span (C or T) by Eqs. 4.22 and 4.23 in positive-moment regions.

$$\frac{V_u}{\phi} = C = 0.85(\text{Slab } f'_c)b_e t \qquad (4.22)$$

or

$$\frac{V_u}{\phi} = T = A_s f_y \qquad (4.23)$$

where C and T are the ultimate longitudinal compressive and tensile forces in the slab and tension reinforcement. For values in the negative-moment regions, Eq. 4.23 applies, but Eq. 4.22 would, of course, be replaced by $C = 0.85(\text{Stem } f'_c)b_w a$, as shown in Fig. 4.12.

The ϕ factor in these equations has been used in such a way that the factor is included in the results only when applied loading (with the load factors to yield the total safety factor) is involved—in the calculation of v_{dh} in Eq. 4.18 for use in Eq. 4.21. Otherwise, the ϕ factor is cancelled out.

Based on the assumptions previously discussed for the strength design of composite beams (primarily in not distinguishing between shored and unshored construction), the design procedures for web shear and horizontal shear may be applied to both simple and continuous beams using either unshored or shored construction. In the case of the continuous cast-in-place slab, the effective depth d for the negative region composite section might logically be used, as shown in Fig. 4.12.

See Section 4.2.3 for additional design considerations and Section 4.2.4 for deflection control in terms of steel ratios, minimum thicknesses, and deflections themselves.

4.2.2.4 Example 4.2

Design of three-span continuous composite reinforced concrete beam by the strength method. Clear spans = 30 ft (9.1 m), live load = 100 psf (4790 N/m^2), beam spacing c.c. = 10 ft (305 cm). See Fig. 4.14 for other design conditions and parameters.

Flexure

Slab design

For $f'_c = 3000$ psi and $f_y = 40{,}000$ psi: $\rho_b = 0.0278$ (Eq. 4.25). For reasonable deflection

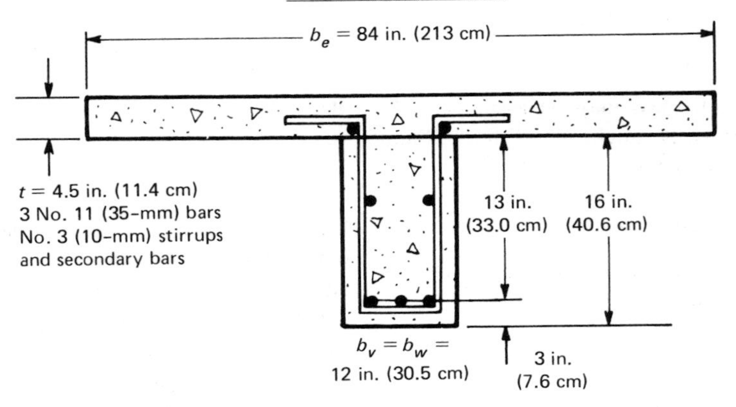

$$h_p = 16 \text{ in. } (40.6 \text{ cm}), \quad h_c = 20.5 \text{ in. } (52.1 \text{ cm})$$

$$d_p = 13 \text{ in. } (33.0 \text{ cm}), \quad d_c = 17.5 \text{ in. } (44.5 \text{ cm})$$

$$A_s = 4.68 \text{ in.}^2 \ (30.2 \text{ cm}^2)$$

$$A_v = 0.22 \text{ in.}^2 \ (1.42 \text{ cm}^2)$$

$$\rho_p = 4.68/(12)(13) = 0.0300$$

$$\rho_c = 4.68/(84)(17.5) = 0.00318$$

Beam Clear Span = 30 ft (9.1 m)

Beam Spacing = 10 ft (305 cm) c.c.

Fig. 4.14. Material properties and design conditions for Ex. 4.2.

control, from Section 4.2.4.1: try $\rho_{max} = 0.18f'_c/f_y = (0.18)(3)/40 = 0.014$ (checked even though $f_y = 40$ ksi), or ρ_{max} = 25% to 35% ρ_b for rectangular beams = (e.g., (0.30)(0.0278) = 0.0083.

$$\omega = \rho f_y/f'_c = (0.0083)(40)/3 = 0.111$$

$$R_u = \omega f'_c(1 - 0.59\omega)$$

$$= (0.111)(3000)[1 - (0.59)(0.111)]$$

$$= 311 \text{ psi } (2.14 \text{ N/mm}^2)$$

From Table 4.2, Min. $t = \ell/28 = (10)(12)/28 =$ 4.3 in. (10.9 cm). Trying t = 4.5 in. (11.4 cm), $w_D = 150(4.5/12) = 56.3$ psf (2700 N/m²), $w_L/w_D = 100/56.3 = 1.8 < 3.0$. Since the live-load to dead-load ratio does not exceed 3, the ACI Code moment coefficients of $-1/10$ and $+1/14$ may be used with the clear span = ℓ_n = 10 − 1 = 9 ft (2.7 m). As a practical matter, use for both positive and negative moments: 1/10,

$$\pm M_D = 0.10 w_D(\ell_n)^2 = (0.10)(56.3)(9)^2$$

$$= 456 \text{ ft-lb } (618 \text{ N-m})$$

$$\pm M_L = 0.10 w_L(\ell_n)^2 = (0.10)(100)(9)^2$$

$$= 810 \text{ ft-lb } (1100 \text{ N-m})$$

$$\pm M_u = 1.4M_D + 1.7M_L = 1.4(456) + 1.7(810)$$

$$= 2015 \text{ ft-lb } (2730 \text{ N-m})$$

$$\text{Req. } d = \sqrt{\frac{M_u}{\phi R_u b}} = \sqrt{\frac{(2015)(12)}{(0.90)(311)(12)}}$$

$$= 2.68 \text{ in. } (6.81 \text{ cm})$$

Assuming No. 4 (13-mm) bars,

$$\text{Req. } t = 2.68 + 0.75 \text{ (cover)} + 0.25 \text{ (radius)}$$

Negative Moment Section

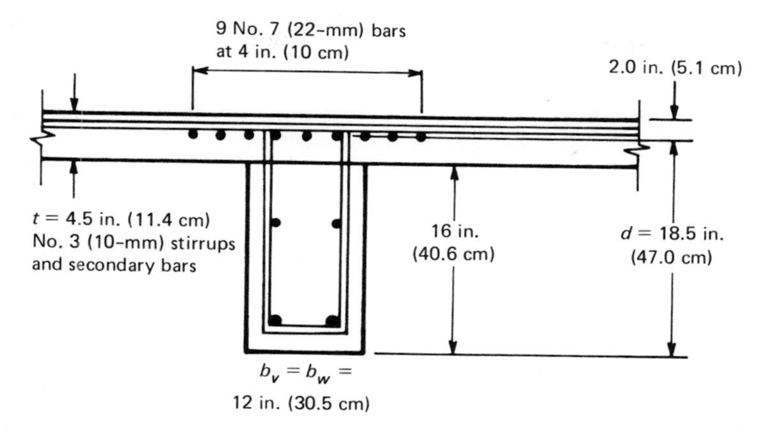

$A_s = 5.40 \text{ in.}^2$ (34.8 cm^2), $\rho = 5.40/(12)(18.5) = 0.0243$

Material Properties

Slab $f'_c = 3000$ psi (21 N/mm²),

Normal-Weight Concrete

Precast Beam $f'_c = 4000$ psi (28 N/mm²),

Sand-Lightweight Concrete

($w = 120$ pcf $= 1920$ kg/m³)

All Steel $f_y = 40{,}000$ psi (276 N/mm²)

Effective Flange Width (From Section 4.1.3)

Max b_e = Beam Span/4 = (30)(12)/4

= 90 in. (229 cm)

= Beam Spacing = (10)(12)

= 120 in. (305 cm)

$= 16t + b_w = (16)(4.5) + 12$

= 84 in. (213 cm) USE

Fig. 4.14 (cont.)

= 3.68 in. (9.3 cm) (Section 4.2.3.6)

Could use $t = 4.0$ in. (10.2 cm), but use $t =$ 4.5 in. (11.4 cm) based on deflection considerations in Tables 4.2 and 4.3.

Actual $d = 4.5 - 0.75 - 0.25 = 3.50$ in. (8.9 cm).

Trying $a = 0.07d = (0.07)(3.50) = 0.25$ in. (0.64 cm),

$$\pm \text{Req. } A_s = \frac{M_u}{\phi f_y(d - a/2)}$$

$$= \frac{(2015)(12)}{(0.90)(40{,}000)(3.50 - 0.125)}$$

$= 0.20 \text{ in.}^2/\text{ft}$ width ($4.2 \text{ cm}^2/\text{m}$)

Checking,

$$a = \frac{A_s f_y}{0.85 f'_c b} = \frac{(0.20)(40)}{(0.85)(3)(12)}$$

$$= 0.26 \text{ in. OK}$$

$$\pm \text{ Req. Sp.} = \frac{A_s \text{ of 1 bar}}{\text{Req. } A_s/\text{ft}}(12) = \frac{0.20}{0.20}(12)$$

$$= 12 \text{ in. } (30.5 \text{ cm})$$

Maximum ACI spacing = $3t$

$$= 3(4.5)$$

$$= 13.5 \text{ in. } (34.3 \text{ cm}) \text{ OK}$$

(Section 4.2.3.6)

Use No. 4 (13-mm) bars at 12 in. (30 cm) c.c. in both top and bottom of slab.

Actual $\rho = A_s/bd = 0.20/(12)(3.50) = 0.00476 < 0.0083$ above for deflections; $< 0.75\rho_b = (0.75)(0.0278) = 0.0209$ for ACI Code strength design and Section 4.2.3.1 herein; > 0.0020 minimum ratio based on shrinkage and temperature reinforcement (Sections 4.2.3.1, 4.2.3.2). OK

Provide shrinkage and temperature reinforcement perpendicular to main reinforcement: Req. $A_s = (0.0020)(12)(3.50) = 0.084$ in.^{2}/ft (1.8 cm^2/m).

Use No. 4 (13-mm) bars at 18 in. (45 cm), $A_s = 0.20/1.5 = 0.133$ in.^{2}/ft (2.8 cm^2/m) (Section 4.2.3.1). Checking the calculations, $\omega = \rho f_y/f'_c = (0.00476)(40)/3 = 0.0635$,

$$R_u = \omega f'_c(1 - 0.59\omega)$$

$$= (0.0635)(3000)[1 - (0.59)(0.0635)]$$

$$= 183 \text{ psi } (1.26 \text{ N/mm}^2)$$

$$M_u = \phi R_u b d^2 = (0.90)(183)(12)(3.50)^2/12$$

$$= 2018 \text{ ft-lb } (2740 \text{ N-m})$$

$$> 2015 \text{ ft-lb } (2730 \text{ N-m}) \text{ OK}$$

Alternatively,

$$a = A_s f_y / 0.85 f'_c b = (0.20)(40)/(0.85)(3)(12)$$

$$= 0.261 \text{ in. } (0.66 \text{ cm})$$

$$M_u = \phi A_s f_y (d - a/2)$$

$$= (0.90)(0.20)(40{,}000)(3.50 - 0.261/2)/12$$

$$= 2021 \text{ ft-lb } (2740 \text{ N-m}) > 2015 \text{ ft-lb OK}$$

Checking ACI + $\rho_{min} = 200/f_y$ (need not apply to slabs of uniform thickness) = 200/40,000 = 0.0050 $\simeq$ 0.00476 OK (Eq. 4.28)

Checking ACI crack-control parameter in Section 4.2.3.3 (Eq. 4.29) (need not apply in this case since f_y = 40 ksi = 276 N/mm^2),

$$f_s = \text{computed value or } 0.6 f_y$$

$$= 24 \text{ ksi } (165 \text{ N/mm}^2)$$

$$d_c = 0.75 + 0.25 = 1.00 \text{ in. } (2.54 \text{ cm})$$

$$A = (2)(1.00)(12)$$

$$= 24.00 \text{ in.}^2/\text{bar } (155 \text{ cm}^2/\text{bar})$$

$$z = f_s \sqrt[3]{d_c A} = 24 \sqrt[3]{(1.00)(24.00)}$$

$$= 69 \text{ kips/in. } (121 \text{ kN/cm})$$

$$< 175 \text{ kips/in. } (306 \text{ kN/cm})$$

for interior exposure

and

$$< 145 \text{ kips/in. } (254 \text{ kN/cm})$$

for exterior exposure. OK

To show that shear is usually small in such cases, neglecting the distance d from supports:

$$w_u = 1.4 w_D + 1.7 w_L = 1.4(56.3) + 1.7(100)$$

$$= 249 \text{ lb/ft } (3630 \text{ N/m})$$

$$V_u = 1.15 w_u \ell_n / 2 = (1.15)(249)(9)/2$$

$$= 1290 \text{ lb } (5740 \text{ N})$$

$$v_u = V_u / \phi b d = 1290/(0.85)(12)(3.50)$$

$$= 36 \text{ psi } (0.22 \text{ N/mm}) < 2\sqrt{f'_c} =$$

$$110 \text{ psi } (0.76 \text{ N/mm})$$

Beam design

For composite beam from Table 4.2, estimate $h = (\ell/18.5)$ (0.80 for f_y = 40 ksi or 276 N/mm^2) (1.09 for precast w = 120 pcf or 1920 kg/m^3, since the lightweight concrete is in compression and thus subject to creep in the negative moment region) = (30)(12)(0.80)(1.09)/18.5 = 17.0 in. (43.2 cm).

Or from Table 4.3, estimate $h = (\ell/13)$ (0.80)(1.09) = (30)(12)(0.80)(1.09)/13 = 24.0 in. (61.0 cm).

For such a continuous composite beam, an

economical depth that is satisfactory for deflections will usually result in a relatively high steel percentage (as greater than about 1.5%) in the negative moment region and relatively low steel ratio (as less than about 0.5% with the large flange or b) in the positive moment region.

Try $h = 20.5$ in. (52.1 cm), precast beam 12 × 16 in. (30.5 × 40.6 cm), Pos. $d_p = 13$ in. (33.0 cm), Pos. $d_c = 17.5$ in. (44.5 cm), Neg. $d_c = 18.5$ in. (47.0 cm). These dimensions are shown in Fig. 4.14.

For slab,

$$w_s = (10)(4.5/12)(150) = 563 \text{ lb/ft } (8210 \text{ N/m})$$

For precast beam,

$$w_p = (12 \times 16)(125)/144 = 167 \text{ lb/ft } (2440 \text{ N/m})$$

$$w_D = w_s + w_p = 563 + 167 = 730 \text{ lb/ft } (10{,}700 \text{ N/m})$$

$$w_L/w_D = 1000/730 < 3.0.$$

Since the live-load to dead-load ratio does not exceed 3, the ACI Code moment coefficients of -1/10 and +1/11 may be used.

$$d_c/d_p < 1.15 + 0.24 M_L/M_D \qquad \text{(Eq. 4.11)}$$

$$17.5/13 < 1.15 + 0.24(1000/730)$$

$1.35 < 1.48$ OK to use without shores according to this criterion.

Negative moment region

$$M_D = w_D \ell_n^2/10 = (730)(30)^2/10 = 65{,}700 \text{ ft-lb } (89.1 \text{ kN-m})$$

$$M_L = w_L \ell_n^2/10 = (10)(100)(30)^2/10 = 90{,}000 \text{ ft-lb } (122.0 \text{ kN-m})$$

$$M_u = 1.4 M_D + 1.7 M_L = 1.4(65{,}700) + 1.7(90{,}000) = 91{,}980 + 153{,}000 = 244{,}980 \text{ ft-lb } (332 \text{ kN-m}) \qquad \text{(Eq. 4.9)}$$

From Fig. 4.12, assume $a = 0.25d = (0.25)(18.5) = 4.62$ in. (11.7 cm), and neglect the bars in compression (which contribute a small amount to the moment resistance).

$$\text{Req. } A_s = \frac{M_u}{\phi f_y(d - a/2)} = \frac{(244{,}980)(12)}{(0.90)(40{,}000)(18.5 - 4.62/2)} = 5.04 \text{ in.}^2 \ (32.5 \text{ cm}^2)$$

Checking,

$$a = \frac{A_s f_y}{0.85(\text{stem } f_c')b_w} = \frac{(5.04)(40)}{(0.85)(4)(12)} = 4.94 \text{ in. } (12.5 \text{ cm})$$

$$\text{Revised } A_s = \frac{(244{,}980)(12)}{(0.90)(40{,}000)(18.5 - 4.94/2)} = 5.09 \text{ in.}^2 \ (32.8 \text{ cm}^2)$$

Use nine No. 7 (22-mm) bars,

$$A_s = (9)(0.60) = 5.40 \text{ in.}^2 \ (34.8 \text{ cm}^2)$$

$$\rho = A_s/b_w d = 5.40/(12)(18.5) = 0.0243.$$

For $f_c' = 4000$ psi (27.6 N/mm²) and $f_y = 40$ ksi (276 N/mm²), $\rho_b = 0.0371$ in Eq. 4.25.

Hence,

$$0.75\rho_b = (0.75)(0.0371) = 0.0278 > 0.0243$$

OK for strength design (Eq. 4.10)

Effective width for tensile steel (Fig. 4.12) = span/10 (Section 4.2.3.3) = (30)(12)/10 = 36 in. (91 cm). Spacing = 36/9 = 4.0 in. (10.2 cm).

Use nine No. 7 (22-mm) bars spaced at 4 in. (10 cm) c.c. Use previously designed slab shrinkage and temperature steel outside the 36-in. width. (See Fig. 4.14.)

Checking the calculations,

$$\omega = \rho f_y/f_c' = 0.0243(40)/4 = 0.243,$$

$$R_u = \omega f_c'(1 - 0.59\omega) = (0.243)(4000)[1 - (0.59)(0.243)] = 833 \text{ psi } (5.74 \text{ N/mm}^2)$$

$$M_u = \phi R_u b_w d^2 = (0.90)(833)(12)(18.5)^2/12$$

$= 256{,}600$ ft-lb (348 kN-m)

$> 244{,}980$ ft-lb (332 kN-m) OK

Alternatively,

$a = A_s f_y / 0.85 f'_c b_w = (5.40)(40)/(0.85)(4)(12)$

$= 5.29$ in. (13.4 cm)

$M_u = \phi A_s f_y (d - a/2)$

$= (0.90)(5.40)(40{,}000)(18.5 - 5.29/2)/12$

$= 256{,}900$ ft-lb (348 kN-m)

$> 244{,}980$ ft-lb OK

Checking ACI crack-control parameter in Section 4.2.3.3 (Eq. 4.29) (need not apply in this case since $f_y = 40$ ksi $= 276$ N/mm²),

$f_s =$ computed value or $0.6 f_y$

$= 24$ ksi (165 N/mm²)

$d_c = 2.00$ in. (5.08 cm)

$A = 2(2.00)(3.50)$

$= 14.0$ in.²/bar (90 cm²/bar)

$z = f_s \sqrt[3]{d_c A} = 24 \sqrt[3]{(2.00)(14.0)}$

$= 73$ kips/in. (128 kN/cm)

< 175 kips/in. (306 kN/cm)

for interior exposure OK

< 145 kips/in. (254 kN/cm)

for exterior exposure OK

The author recommends the following nominal design for the bottom shrinkage connection steel between the precast reinforced beams and the joint concrete, as shown in Fig. 4.6:

$A_s/4 = 4.68/4 = 1.17$ in.² (7.6 cm²)

Use two No. 7 (22-mm) bars, $A_s = 1.20$ in.² (7.7 cm²).

Positive moment region

$M_D = w_D \ell_n^2 / 11 = (730)(30)^2/11$

$= 59{,}730$ ft-lb (81.0 kN-m)

$M_L = w_L \ell_n^2 / 11 = (10)(100)(30)^2/11$

$= 81{,}820$ ft-lb (111.0 kN-m)

$M_u = 1.4 M_D + 1.7 M_L$

$= 1.4(59{,}730) + 1.7(81{,}820)$

$= 83{,}620 + 139{,}090$

$= 222{,}700$ ft-lb (302 kN-m) (Eq. 4.9)

From Fig. 4.11a, assume $a = 1$ in. (2.54 cm),

$$\text{Req. } A_s = \frac{M_u}{\phi f_y (d - a/2)}$$

$$= \frac{(222{,}700)(12)}{(0.90)(40{,}000)(17.5 - 0.5)}$$

$= 4.37$ in.² (28.2 cm²)

Checking,

$$a = \frac{A_s f_y}{0.85(\text{Slab } f'_c) b_e} = \frac{(4.37)(40)}{(0.85)(3)(84)}$$

$= 0.82$ in. (2.08 cm)

$$\text{Revised } A_s = \frac{(222{,}700)(12)}{(0.90)(40{,}000)(17.5 - 0.82/2)}$$

$= 4.34$ in.² (28.0 cm²)

Use three No. 11 (35-mm) bars,

$A_s = 3(1.56) = 4.68$ in.² (30.2 cm²)

ACI Code Req. beam width $= 5(1.41)$

$+ 1$ (No. 3 or No. 4 stirrups) $+ 3$ (cover)

$= 11.05$ in.

< 12 in. (30.5 cm) OK (Section 4.2.3.6)

ACI Code Req. $(h - d) = 1.5$ (cover)

$+ 0.5$ (No. 3 or No. 4 stirrups) $+ 0.71$ (radius)

$= 2.71$ in.

< 3.0 in. (7.6 cm) OK (Section 4.2.3.6)

As outlined in Section 4.2.3.6, cover requirements in certain cases can be less than these for precast members.

$\rho = A_s / b_e d = 4.68/(84)(17.5) = 0.00318.$

For

$f'_c = 3000$ psi (20.7 N/mm²)

and

$f_y = 40$ ksi (276 N/mm²),

$\rho_b = 0.0278$ in Eq. 4.25.

Hence,

$$0.75\rho_b = (0.75)(0.0278)$$
$$= 0.0209 > 0.00318$$

OK for strength design (Eq. 4.10).

Checking the calculations,

$$\omega = \rho f_y/f'_c = 0.00318(40)/3 = 0.0424,$$

$$R_u = \omega f'_c(1 - 0.59\omega)$$
$$= (0.0424)(3000)[1 - (0.59)(0.0424)]$$
$$= 124 \text{ psi } (0.85 \text{ N/mm}^2)$$

$$M_u = \phi R_u b_e d^2 = (0.90)(124)(84)(17.5)^2/12$$
$$= 239{,}200 \text{ ft-lb } (324 \text{ kN-m})$$
$$> 222{,}700 \text{ ft-lb } (302 \text{ kN-m}) \quad \text{OK}$$

Alternatively,

$$a = A_s f_y/0.85 f'_c b_e = (4.68)(40)/(0.85)(3)(84)$$
$$= 0.874 \text{ in. } (2.22 \text{ cm})$$

$$M_u = \phi A_s f_y(d - a/2)$$
$$= (0.90)(4.68)(40{,}000)(17.5 - 0.874/2)/12$$
$$= 239{,}600 \text{ ft-lb } (325 \text{ kN-m})$$
$$> 222{,}700 \text{ ft-lb} \quad \text{OK}$$

Checking precast beam for the case of unshored construction (Section 4.2.2.1),

$$a = A_s f_y/0.85 f'_c b_w = (4.68)(40)/(0.85)(4)(12)$$
$$= 4.59 \text{ in. } (11.7 \text{ cm})$$

$$M_u = \phi A_s f_y(d - a/2)$$
$$= (0.90)(4.68)(40{,}000)(13 - 4.59/2)/12$$
$$= 150{,}300 \text{ ft-lb } (204 \text{ kN-m})$$

Alternatively,

$$\rho = A_s/b_w d = 4.68/(12)(13) = 0.0300$$

$$\omega = \rho f_y/f'_c = (0.0300)(40)/4 = 0.300$$

$$R_u = \omega f'_c(1 - 0.59\omega)$$
$$= (0.300)(4000)[1 - (0.59)(0.300)]$$
$$= 988 \text{ psi } (6.81 \text{ N/mm}^2)$$

$$M_u = \phi R_u b_w d^2 = (0.90)(988)(12)(13)^2/12$$
$$= 150{,}300 \text{ ft-lb}$$

$$(0.60)(150{,}300)$$
$$= 90{,}200 \text{ ft-lb } (122.3 \text{ kN-m}) > 1.4 M_D$$
$$= 83{,}620 \text{ ft-lb } (113.4 \text{ kN-m}) \quad \text{OK}$$

Hence, the member may be used without shoring according to this criterion.

Checking ACI + $\rho_{min} = 200/f_y = 200/40{,}000$

$$= 0.0050 \quad \text{(Eq. 4.28)}$$
$$< A_s/b_w d = 4.68/(12)(17.5) = 0.0223 \quad \text{OK}$$

Checking ACI crack-control parameter in Section 4.2.3.3 (Eq. 4.29) (need not apply in this case since $f_y = 40$ ksi (276 N/mm²)),

$$f_s = \text{computed value or } 0.6 f_y$$
$$= 24 \text{ ksi } (165 \text{ N/mm}^2)$$

$$d_c = 3.00 \text{ in. } (7.62 \text{ cm})$$

$$A = 2(3.00)(12)/3$$
$$= 24.0 \text{ in.}^2/\text{bar } (155 \text{ cm}^2/\text{bar})$$

$$z = f_s \sqrt[3]{d_c A} = 24 \sqrt[3]{(3.00)(24.0)}$$
$$= 100 \text{ kips/in. } (175 \text{ kN/cm})$$

< 175 kips/in. (306 kN/cm)

for interior exposure OK

< 145 kips/in. (254 kN/cm)

for exterior exposure OK

The beam may be used with or without shores insofar as the flexural strength criteria (including Eq. 4.11) are concerned. Deflections are checked in Section 4.2.4.5.

Web shear

As indicated in Section 4.2.2.2, use Eq. 4.12 for unshored or shored construction, with the negative region d in the case of continuous composite beams.

$$V_D = 1.15 w_D \ell_n/2 = (1.15)(730)(30)/2$$
$$= 12{,}590 \text{ lb } (56{,}000 \text{ N})$$

$$V_L = 1.15 w_L \ell_n/2 = (1.15)(1000)(30)/2$$
$$= 17{,}250 \text{ lb } (76{,}700 \text{ N})$$

$$V_u = 1.4 V_D + 1.7 V_L$$
$$= 1.4(12{,}590) + 1.7(17{,}250)$$

$= 17{,}630 + 29{,}330$

$= 46{,}960$ lb (208,900 N)

$$v_u = \frac{V_u}{\phi b_w d} = \frac{46{,}960}{(0.85)(12)(18.5)}$$

$= 249$ psi (1.72 N/mm²) (Eq. 4.12)

$v_c = (0.85$ for sand-lightweight concrete$)2\sqrt{f'_c}$

$= (0.85)(2)\sqrt{4000} = 108$ psi (0.74 N/mm²)

$v_u - v_c = 249 - 108 = 141$ psi (0.97 N/mm²)

$< (0.85)(8)\sqrt{4000}$

$= 430$ psi (2.96 N/mm²) upper limit OK

$< (0.85)(4)\sqrt{4000}$

$= 215$ psi (1.48 N/mm²) for $s_{max} = d/2$

(Eq. 4.15)

For No. 3 (10-mm) stirrups, $A_v = 0.22$ in.² (1.42 cm²),

$s = A_v f_y/(v_u - v_c)b_w$ (Eq. 4.13)

$= (0.22)(40{,}000)/(141)(12)$

$= 5.20$ in. (13.2 cm)

$s_{max} = A_v f_y / 50 b_w$ (Eq. 4.14)

$= (0.22)(40{,}000)/(50)(12)$

$= 14.7$ in. (37.3 cm)

For No. 4 (13-mm) stirrups, $A_v = 0.40$ in.² (2.58 cm²),

$s = (0.40)(40{,}000)/(141)(12)$

$= 9.46$ in. (24.0 cm) (Eq. 4.13)

$s_{max} = (0.40)(40{,}000)/(50)(12)$

$= 26.7$ in. (67.8 cm) (Eq. 4.14)

For No. 3 or No. 4 stirrups,

$s_{max} = d/2 = 18.5/2 = 9.25$ in. (23.5 cm)

(Eq. 4.15)

$s_{max} = 24$ in. (61 cm) (Eq. 4.17)

Could use for web shear No. 3 (10-mm) stirrups at 5 in. (13 cm) c.c. for the end quarter-span, for instance, and at 9 in. (23 cm) c.c. for the middle half-span. Or, could use for web shear No. 4 (13-mm) stirrups at 9 in. (23 cm) c.c. for the entire length of beam.

Horizontal shear

Using Eq. 4.18 for unshored or shored construction, with the negative region d_c,

$$v_{dh} = \frac{V_u}{\phi b_v d_c} = \frac{46{,}960}{(0.85)(12)(18.5)}$$

$= 249$ psi (1.72 N/mm²)

(Eq. 4.18)

>80 psi (0.55 N/mm²) and

<350 psi (2.41 N/mm²)

Hence, it is required that the following minimum tie design be used, plus the contact surface to be clean and intentionally roughened:

$s_{max} = 4t = (4)(4.5) = 18$ in. (45.7 cm)

(Eq. 4.19)

$s_{max} = A_v f_y / 50 b_w$

(see web shear calculations) (Eq. 4.14)

$s_{max} = 24$ in. (61 cm) (Eq. 4.17)

Hence, use either of the web stirrup designs extended into the slab.

The following two examples illustrate the shear-friction method for the same composite beam design:

1. Using Eqs. 4.20 and 4.21, and assuming $v_{dh} = 400$ psi (2.76 N/mm²) instead of 249 psi (1.72 N/mm²) previously computed. Check

   $v_{dh} \leqslant 0.2 f'_c = 0.2(4000)$

   $= 800$ psi (5.52 N/mm²) OK

   $V_u = (1/2)\phi v_{dh} b_v \ell/2$ (Eq. 4.21)

   $= (0.85)(400)(12)(30)(12)/4$

   $= 367{,}200$ lb (1,633,000 N)

   $$A_{vf} = \frac{V_u}{\phi f_y \mu} = \frac{367{,}200}{(0.85)(40{,}000)(1.0)}$$

   (Eq. 4.20)

   $= 10.8$ in.² (70 cm²)

   for half the span

   Using No. 4 (13-mm) stirrups, No. = $10.8/0.40 = 27.0$ stirrups per half-span. Req. spacing = $(15)(12)/27.0 = 6.7$ in.

(17.0 cm). Use No. 4 (13-mm) stirrups at 6.5 in. (17 cm) c.c. for the entire beam length.

2. Using Eqs. 4.20, 4.22, and 4.23,
In the positive-moment region:

$$V_u = \phi(0.85)(\text{Slab } f_c')b_e t \quad \text{(Eq. 4.22)}$$
$$= (0.85)(0.85)(3000)(84)(4.5)$$
$$= 819{,}300 \text{ lb } (3{,}644{,}000 \text{ N})$$

or

$$V_u = \phi A_s f_y \quad \text{(Eq. 4.23)}$$
$$= (0.85)(4.68)(40{,}000)$$
$$= 159{,}100 \text{ lb } (707{,}700 \text{ N})$$

Hence, V_u = 159,100 lb (707,700 N).
In the negative-moment region:

$$V_u = \phi(0.85)(\text{Stem } f_c')b_w a$$
$$= (0.85)(0.85)(4000)(12)(5.29 \text{ from flexural calculations})$$
$$= 183{,}500 \text{ lb } (816{,}000 \text{ N})$$

or

$$V_u = \phi A_s f_y = (0.85)(5.40)(40{,}000)$$
$$= 183{,}600 \text{ lb } (816{,}700 \text{ N})$$

Hence, V_u = 183,500 lb (816,000 N). Use

$$A_{vf} = \frac{V_u}{\phi f_y \mu} = \frac{183{,}500}{(0.85)(40{,}000)(1.0)} \quad \text{(Eq. 4.20)}$$
$$= 5.40 \text{ in.}^2 \ (34.8 \text{ cm}^2) \text{ for half the span}$$

Using No. 3 (10-mm) stirrups, No. = 5.40/0.22 = 24.5 stirrups per half-span. Req. spacing = (15)(12)/24.5 = 7.3 in. (18.5 cm). Use No. 3 (10-mm) stirrups at 7 in. (18 cm) for the entire beam length.

4.2.3 Additional Design Considerations

A brief summary of the principal design considerations required for reinforced concrete composite beams, in addition to those already described, is given in this section.

4.2.3.1 Steel Ratios for Working Stress and Strength Design Methods

In the stress method of design, the tension steel ratio, $\rho = A_s/bd$, is usually kept approximately equal to, or less than, the balanced ratio given by Eq. 4.24.

$$\text{Working stress bal. } \rho = kf_c/2f_s \quad (4.24)$$

where f_c and f_s are the allowable concrete and steel stresses, respectively.

In the strength design method, the ACI Code requires that the tension-steel ratio, $\rho = A_s/bd$, not exceed $0.75\rho_b$ (Eq. 4.10), where ρ_b is given by Eqs. 4.25 through 4.27—derived in ACI Commentary.

For rectangular beams with tension steel only,

$$\rho_b = \overline{\rho}_b = \frac{0.85\beta_1 f_c'}{f_y}\left(\frac{87{,}000}{87{,}000 + f_y}\right) \quad (4.25)$$

For rectangular beams with compression steel,

$$\rho_b = \overline{\rho}_b + \rho' \frac{f_s'}{f_y} \quad (4.26)$$

For T-beams (neutral axis below flange) without compression steel,

$$\rho_b = \frac{b_w}{b}(\overline{\rho}_b + \rho_f) \quad (4.27)$$

where

β_1 = 0.85 for strengths up to 4000 psi, and reduced continuously at a rate of 0.05 for each 1000 psi of strength in excess of 4000 psi, with a minimum of 0.65 (27.6 N/mm²)

$\rho' = A_s'/bd$

f_s' = stress in compression steel

$$= 87{,}000\left(1 - \frac{d'}{d}\,\frac{87{,}000 + f_y}{87{,}000}\right) \leq f_y$$

$\rho_f = A_{sf}/b_w d$, $A_{sf} = 0.85 f_c'(b - b_w)t/f_y$.

Minimum shrinkage and temperature reinforcement by the ACI Code is 0.0020 for slabs with Grade 40 or 50 deformed bars and 0.0018 for slabs with Grade 60 deformed bars or welded wire fabric, deformed or plain. The maximum spacing of such steel is 5 times the slab thickness or 18 in. (45.7 cm). The AASHTO requirement for shrinkage and temperature reinforcement is not less than $\frac{1}{8}$ in.^{2} (0.81 cm^2) of reinforcement per foot (as No. 4 bars at 18 in. = 0.133 in.^{2}/ft = 2.8 cm^2/m) shall be placed in each direction of all concrete surfaces. The maximum spacing of bars is also 18 in. (45 cm).

4.2.3.2 Minimum Positive Flexural Reinforcement

Where positive reinforcement is required by analysis, the ACI Code provides that

$$\rho_{min} = 200/f_y \tag{4.28}$$

unless the area of steel provided at every section, positive or negative, is at least one-third greater than that required by analysis. In T-beams, where the stem is in tension, the ratio, ρ, shall be computed for this purpose using the width of the stem. This provision does not apply to slabs of uniform thickness. Equation 4.28 may apply in cases where a beam is considerably deeper than that required for strength, and is derived[19] from $M_{plain} = f_r I_g / y_{ten} = M_{reinforced} = \phi A_s f_y (d - a/2)$.

In structural slabs of uniform thickness, the minimum amount of reinforcement in the direction of the span shall not be less than that required for shrinkage and temperature.

4.2.3.3 Distribution of Flexural Reinforcement in Tension for Beams and One-Way Slabs

When f_y exceeds 40,000 psi (276 N/mm^2), the ACI Code provides for an adequate distribution of flexural tensile reinforcement in beams and one-way slabs by means of the crack control parameter, z, as follows:

$$z = f_s \sqrt[3]{d_c A} \tag{4.29}$$

where

- f_s = service-load stress or $0.6f_y$, in ksi
- d_c = concrete cover measured from the extreme tension fiber to the center of the bar located closest thereto
- A = effective area of concrete per bar = effective tension area of concrete surrounding the main tension reinforcing bars and having the same centroid as the reinforcement, divided by the number of bars, in in.^{2}

When the main reinforcement consists of several bar sizes, the number of bars shall be computed as the total steel area divided by the area of the largest bar used. The quantity z in Eq. 4.29 shall not exceed 175 kips/in. (306 kN/cm) for interior exposure and 145 kips/in. (254 kN/cm) for exterior exposure. Equation 4.29 is based on a similar expression developed by Gergely and Lutz.[20]

In addition, where flanges are in tension, the ACI Code provides that a part of the main tension reinforcement shall be distributed over the effective flange width or a width equal to one-tenth of the span, whichever is smaller. If the effective flange width exceeds one-tenth of the span, some longitudinal reinforcement shall be provided in the outer portions of the flange (see Ex. 4-2).

If the depth of the web exceeds 3 ft (0.90 m), longitudinal reinforcement having a total area at least equal to 10% of the main tension steel area shall be placed near the faces of the web and distributed in the zone of flexural tension with a spacing not more than 12 in. (30 cm) or the width of the web, whichever is less.

4.2.3.4 Redistribution of Moments in Continuous Nonprestressed Flexural Members Designed by the Strength Method

Negative moments calculated by the elastic theory may be increased or decreased (with the positive moment adjusted accordingly in order to maintain the correct statical moments per span) by not more than the percentage in Eq. 4.30.

$$\text{Percentage} = 20\left(1 - \frac{\rho - \rho'}{\rho_b}\right) \tag{4.30}$$

This equation applies when ρ or $\rho - \rho'$ is equal to or less than $0.50\rho_b$, where ρ_b is given by Eq. 4.25. In this application, Eqs. 4.25 and 4.30 apparently are intended to apply to both rectangular and T-beams in the ACI Code. This provision does not apply when approximate moment coefficients are used. It also does not apply when the stress design method is used.

Such redistribution of moments might be applied to composite reinforced beams in the case of live-load moments for unshored construction and total load moments for shored construction. The moment-redistribution procedure is illustrated, for example, by the Portland Cement Association[19] (1972).

4.2.3.5 Development of Reinforcement

The usual bond and development length requirements should, of course, be followed for all principal reinforcement in composite members, including the vertical ties.

4.2.3.6 Spacing and Cover of Reinforcement

ACI Code. The clear distance between parallel bars in a layer shall be not less than the nominal diameter of the bars, nor less than 1 in. (2.5 cm). Where parallel reinforcement is placed in two or more layers, the bars in the upper layers shall be placed directly above those in the bottom layer with the clear distance between layers not less than 1 in. (2.5 cm). Limitations are also given for the maximum size aggregate. In slabs other than concrete joist construction, the principal reinforcement shall be spaced not farther apart than three times the slab thickness, nor more than 18 in. (45 cm).

The following minimum concrete cover shall be provided for reinforcing bars.

Cast-in-place concrete (nonprestressed)	*Cover*
Exposed to earth or weather,	
No. 6–18 bars	2 in. (5.1 cm)
No. 5 bar, $\frac{5}{8}$-in. wire, and smaller	$\frac{3}{2}$ in. (3.8 cm)
Not exposed to weather or in contact with the ground;	
Slabs, joists, No. 14 and No. 18 bars	$\frac{3}{2}$ in. (3.8 cm)
Slabs, joists, No. 11 bar and smaller	$\frac{3}{4}$ in. (1.9 cm)
Beams, girders (principal steel, stirrups, etc.)	$\frac{3}{2}$ in. (3.8 cm)

Precast concrete (manufactured under plant control conditions)	*Cover*
Exposed to earth or weather,	
No. 14 and No. 18 bars	2 in. (5.1 cm)
No. 6–11 bars	$\frac{3}{2}$ in. (3.8 cm)
No. 5 bar, $\frac{5}{8}$-in. wire, and smaller	$\frac{5}{4}$ in. (3.2 cm)
Not exposed to weather or in contact with the ground;	
Slabs, joists, No. 14 and No. 18 bars	$\frac{5}{4}$ in. (3.2 cm)
Slabs, joists, No. 11 bar and smaller	$\frac{5}{8}$ in. (1.6 cm)
Beams, girders (principal steel)	bar or wire diameter, but not less than $\frac{5}{8}$ in. (1.6 cm) and need not exceed $\frac{3}{2}$ in. (3.8 cm)
Beams, girders (stirrups, etc.)	$\frac{3}{8}$ in. (1.0 cm)

The approximate metric equivalent bar sizes are:

No.	*mm*	*No.*	*mm*	*No.*	*mm*
3	10	7	22	11	35
4	13	8	25	14	44
5	16	9	29	18	57
6	19	10	32		

AASHTO Highway Bridge Specifications. The clear distance between parallel bars shall not be more than 18 in. (45 cm) adjacent to concrete surfaces nor less than 1.5 times the nominal diameter of the bars, 1.5 times the maximum size of coarse aggregate, nor 1.5 in. (3.8 cm). The maximum spacing of bars carrying stress in a slab shall be 1.5 times the thickness of the slab.

The minimum covering measured from the surface of the concrete to the face of any reinforcing bar, shall be not less than 2 in. (5.1 cm), except as follows:

Top of slab	$\frac{3}{2}$ in. (3.8 cm)
Bottom of slab	1 in. (2.5 cm)

Stirrups in T-beams	$\frac{3}{2}$ in. (3.8 cm)
Stirrups at outside faces of box girders	$\frac{3}{2}$ in. (3.8 cm)
Stirrups at inside faces of box girders	1 in. (2.5 cm)

4.2.4 Deflection Control—Steel Ratios, Minimum Thicknesses, Deflections

4.2.4.1 Steel ratios

One way to minimize deflection problems is to use a relatively small tension steel ratio in the flexural design of reinforced members, since this will result in a relatively deep beam because the required internal moment arm in design will be relatively large. For example, the 1963 ACI Code provided in the strength design method that deflections under service loads must be checked when ρ for singly reinforced beams, $(\rho - \rho')$ for doubly reinforced beams, and $(\rho_w - \rho_f)$ for T-beams, exceeds $0.18 f_c'/f_y$, or whenever f_y exceeds 40 ksi (276 N/mm^2). This was simply a "red flag" to check deflections at or above the steel ratio of $0.18 f_c'/f_y$, which is close to the balanced steel ratio by the elastic theory and less than one-half the balanced steel ratio by the strength theory, since $\rho_b = (0.50, 0.45, 0.43$, and $0.39) f_c'/f_y$ in Eq. 4.25 for $f_y = 40$, 50, 60, 75 ksi (276, 345, 414, 517 N/mm^2), respectively, and $\beta_1 = 0.85$ for $f_c' = 4{,}000$ psi (27.6 N/mm^2). In addition to the previously defined steel ratios, $\rho_w = A_s/b_w d$.

More recently, ACI Committee 435[21] suggested that normal-weight concrete nonprestressed one-way structural members will normally be of sufficient size so that deflections will be within acceptable limits when the tension steel reinforcement used in the positive-moment zone does not exceed the following percentages of that in the balanced condition: For members not supporting or not attached to nonstructural elements likely to be damaged by large deflections: $\rho \leqq 35\%$ (30%) for rectangular and $\rho_w \leqq 40\%$ (35%) for T- or box beams, of the balanced ratio; and for members supporting or attached to nonstructural elements likely to be damaged by large deflections: $\rho \leqq 25\%$ (20%) for rectangular and $\rho_w \leqq 30\%$ (25%) for T- or box beams, of the balanced ratio. The values in parentheses refer to lightweight concrete members. For composite beams, the critical case is usually the composite T-beam.

As shown in Section 4.2.4.3, time-dependent deflections can be reduced by means of compression steel in a reinforced concrete flexural member, in accordance with the compression steel ratio $\rho' = A_s'/bd$; or by A_s'/A_s.

4.2.4.2 Minimum thicknesses

The ACI Code states that the minimum thicknesses in Table 4.2 shall apply for one-way construction unless the computation of deflection indicates that lesser thicknesses may be used without adverse effects. However, Table 4.2 applies only to members *not* supporting or attached to partitions or other construction likely to be damaged by large deflections. The table applies directly to normal-weight concrete and $f_y = 60$ ksi (414 N/mm^2), with modification factors in the footnotes for lightweight concrete and other steel grades.

In the case of shored composite members, the minimum thicknesses in Table 4.2 apply as for monolithic T-beams. In the case of unshored construction, if the thickness of a nonprestressed precast member meets the requirements of Table 4.2, deflections need not be computed, according to the ACI Code. The code also states that, if the thickness of an unshored nonprestressed composite member meets the requirements of Table 4.2, deflections occurring after the member becomes composite need not be calculated, but the long-time deflection of the precast member should be investigated for the magnitude and duration of load prior to the beginning of effective composite action.

Values in Table 4.2 have been modified by ACI Committee 435[21] and extended in Table 4.3 to include members that *are* supporting or attached to nonstructural elements likely to be damaged by large deflections. The Committee stated that deflections under uniform loads commonly encountered in buildings will normally be satisfactory when the minimum thicknesses specified in Table 4.3 for one-way members are met or exceeded.

It should be noted that the preceding methods that use minimum thicknesses as an approximate means of controlling deflections do not include the effects of load level, compression steel, environmental conditions, etc. As in the British Standard Code of Practice[5], such effects can be approximated for application with basic minimum thicknesses or span-depth ratios. However, as expressly stated in the British Code Manual, the basic span-depth ratios are still

Table 4.2. Minimum thickness of beams or one-way slabs unless deflections are computed[a] (ACI Code 381-71).[3]

Member	Minimum Thickness: Simply supported	One end continuous	Both ends continuous	Cantilever
	Members not supporting or attached to partitions or other construction likely to be damaged by large deflections			
Solid one-way slabs	$\ell/20$	$\ell/24$	$\ell/28$	$\ell/10$
Beams or ribbed one-way slabs	$\ell/16$	$\ell/18.5$	$\ell/21$	$\ell/8$

[a]The values given in this table shall be used directly for nonprestressed reinforced concrete members made with normal-weight concrete (w = 145 pcf) and Grade 60 reinforcement. For other conditions, the values shall be modified as follows:

1. For structural lightweight concrete having unit weights in the range 90 to 120 pcf, the values in the table shall be multiplied by $1.65 - 0.005\,w$ but not less than 1.09, where w is the unit weight in pcf.
2. For nonprestressed reinforcement having yield strengths other than 60,000 psi, the values in the table shall be multiplied by $0.4 + f_y/100{,}000$.

Examples of these correction factors are as follows:

f_y (ksi) =	40	50	60	75	w =	90	100	110	120
Material Corr. Factor =	0.80	0.90	1.00	1.15		1.20	1.15	1.10	1.09

Table 4.3. Minimum thickness of beams or one-way slabs used in roof and floor construction unless deflections are computed[a] (ACI Committee 435).[21]

	Members not supporting or not attached to nonstructural elements likely to be damaged by large deflections				Members supporting or attached to nonstructural elements likely to be damaged by large deflections			
	Simply supported	One end continuous	Both ends continuous	Cantilever	Simply supported	One end continuous	Both ends continuous	Cantilever
Roof slab[b]	$\ell/22$	$\ell/28$	$\ell/35$	$\ell/9$	$\ell/14$	$\ell/18$	$\ell/22$	$\ell/5.5$
Floor slab, and roof beam[b] or ribbed roof slab[b]	$\ell/18$	$\ell/23$	$\ell/28$	$\ell/7$	$\ell/12$	$\ell/15$	$\ell/19$	$\ell/5$
Floor beam or ribbed floor slab	$\ell/14$	$\ell/18$	$\ell/21$	$\ell/5.5$	$\ell/10$	$\ell/13$	$\ell/16$	$\ell/4$

[a]See footnote a, Table 4.2.
[b]Refers to roofs subjected to normal snow or construction live loads only, and with minimal ponding problems.

based on experience or judgment for typical cases. And, of course, the same is true in the case of allowable deflections to be satisfied when computing deflections.

4.2.4.3 *Deflections*

The equations for computing deflections of composite beams are based primarily on the work of Branson and colleagues (1963-1977)[17, 22-29] and include the use of the I_e method for computing initial cracked deflections—which is used in the ACI Code and the AASHTO Specifications for load-factor design. The general procedure for computing initial and time-dependent deflections by this approach is given in Appendix C.

The ultimate or final deflection of reinforced concrete composite beams using unshored or shored construction is computed by Eqs. 4.31 through 4.34. For more information on the calculation of deflections of composite beams at any time (as opposed to ultimate values), see Branson[17] and Refs. 22 through 29. With the continuous material parameter equations in Appendix A, the procedure readily lends itself to computer programming in which the material and time factors are read-in for a particular problem. Results by this approach for composite reinforced beams have been favorably compared with experimental data by Kripanarayanan and Branson.[29]

These equations could be greatly shortened by combining terms, but are kept in the form of separate terms to show the individual effects due to cracking, creep, shrinkage, etc., that take place both before and after slab casting. Subscripts 1 and 2 are used to refer to the slab (or effect of the slab such as under slab dead load) and precast beam, respectively, as shown in Fig. B.1.

Since the edge of the support is the fixed point for determining deflections, which can occur only in the clear part of the span, the author recommends using the clear span for all deflection calculations.[17]

General method for unshored composite beams

$$\begin{aligned}\Delta_u = {} & \overbrace{(\Delta_i)_2}^{(1)} + \overbrace{(\alpha_s)_{cp} k_r C_{u2} (\Delta_i)_2}^{(2)} \\ & + \overbrace{[1 - (\alpha_s)_{cp}] k_r C_{u2} (\Delta_i)_2 \frac{I_2}{I_c}}^{(3)} \\ & + \overbrace{(\alpha_s)_{sh} \Delta_{sh}}^{(4)} + \overbrace{[1 - (\alpha_s)_{sh}] \Delta_{sh} \frac{I_2}{I_c}}^{(5)} \\ & + \overbrace{(\Delta_i)_1}^{(6)} + \overbrace{k_r C_{u1} (\Delta_i)_1 \frac{I_2}{I_c}}^{(7)} + \overbrace{\Delta_{DS}}^{(8)} \\ & + \overbrace{\Delta_L}^{(9)} + \overbrace{(\Delta_{cp})_L}^{(10)} \end{aligned} \tag{4.31}$$

Term 1 is the *initial dead-load deflection of the precast beam* using Eq. C.4 for the deflection: $(\Delta_i)_2 = K(5/48)M_2 \ell^2/E_{ci}I_2$, Eq. C.5 for K, and Eq. C.1 for $I_2 = I_a = I_e$ at midspan (unless Eqs. C.2 and C.3 are used to obtain I_a). M_2 = midspan moment due to precast-beam dead load. Assuming reinforced precast beams are not subjected to dead-load moment in the storage yard, $E_{ci} = E_c$ (e.g., 28 days) will be used. For computing I_2 in Eq. C.1, M_a refers to the precast beam dead load, and M_{cr}, I_g, and I_{cr}, to the precast beam in this term.

Term 2 is the *dead-load creep deflection of the precast beam up to the time of slab casting*, using Eq. C.12. As defined in Table 4.4, $(\alpha_s)_{cp}$ is the ratio of the creep coefficient (defined as ratio of creep strain to initial strain) up to the time of slab casting to the ultimate creep coefficient. The reduction factor k_r is given by Eq. C.8 and refers to the compression steel in the precast beam at midspan, unless Eqs. C.10 and C.11 are used to obtain $(k_r)_a$. The ultimate creep coefficient C_{u2}, based on the concrete age when the precast-beam dead-load moment is applied, is given in Appendix A (see Table A.8 plus the size correction factor for the precast beam in Table A.6 or A.7, for example). For average conditions, $C_{u2} = 1.6$ may normally be used.

Term 3 is the *creep deflection of the com-*

Table 4.4. Representative values of $(\alpha_s)_{cp}$ in Eq. 4.31 for unshored beams and γ_s in Eq. 4.32 for shored beams.[a]

t = Time in Days Between Precast-Beam Dead-Load Application and Slab Casting for $(\alpha_s)_{cp}$, and Concrete Age for γ_s	$(\alpha_s)_{cp}$ [b]	Moist-Cured γ_s [c]	Steam-Cured γ_s [c]
2	0.13	–[d]	0.98
4	0.19	–	0.95
7	0.24	*0.83*	0.90
10	0.28	*0.77*	0.86
20	0.38	*0.61*	0.74
28	0.42	*0.52*	0.67
60	0.54	*0.33*	0.48
90	0.60	*0.25*	0.38
120	0.64	*0.20*	0.32

[a]The italicized values refer to moist-cured slab concrete (usual case) for γ_s. The values in the table apply to different-weight concrete.

[b]$(\alpha_s)_{cp}$ is the ratio of the precast-beam creep that takes place up to the time of slab casting to the total creep. From Eq. A.10 for both moist- and steam-cured concrete: $(\alpha_s)_{cp} = C_t/C_u = t^{0.60}/(10 + t^{0.60})$, where t = time in days between precast-beam dead-load application and slab casting. See Terms 2 and 3 of Eq. 4.31.

[c]γ_s is the ratio of the differential shrinkage and creep that takes place after the shores are removed to the ultimate value, where t is the age of the slab concrete when the shores are removed. Using Eq. A.11 for moist-cured:

$$\gamma_s = \left(1 - \frac{t - 7}{35 + t - 7}\right)/1.2$$

in which the 1.2 factor was used in Appendix A to convert from 7-day to 1-day shrinkage. Using Eq. A.12 for steam-cured:

$$\gamma_s = \left(1 - \frac{t - 1}{55 + t - 1}\right)$$

See Term 4 of Eq. 4.32.

[d]Not applicable.

posite beam following slab casting due to the precast-beam dead load. The ratio I_2/I_c, modifies the initial stress (strain) and accounts for the effect of the composite section in restraining additional creep curvature (strain) after the composite section becomes effective. An accurate evaluation of this effect is complicated by the fact that the strain distribution in the precast beam has been sharply altered at the time of slab casting. Since the composite section *per se* is not supporting any part of the slab and precast beam dead load following slab casting in this unshored case, there appears to be no simple way to evaluate such an effective composite section I. Under the slab and precast beam loading, the precast beam will normally be rather severely cracked. But the neutral axis under this loading is usually not close to the fully cracked composite section centroid. Hence, in estimating the deflection due to creep of the precast beam in Terms 3 and 7; and due to shrinkage of the precast beam in Term 5; it is not clear whether the ratio of the gross section Is or the cracked section Is would yield the better solution. As a simple approximation, it is recommended that an average of the gross and cracked section ratios at midspan be used in this term, or $I_2/I_c = [(I_2/I_c)_g + (I_2/I_c)_{cr}]/2$. For computing I_c, the 28-day values of E_c are recommended. The other parameters in Term 3 are the same as in Term 2.

Term 4 is the *deflection due to shrinkage warping of the precast beam up to the time of slab casting*, using Eq. C.14 for the total deflection—$\Delta_{sh} = K_{sh}(\phi_{sh})_u \ell^2$, Eqs. C.16 through C.21 for K_{sh}, and Eq. C.15 for the curvature—

$$(\phi_{sh})_u = A_{sh}(\epsilon_{sh})_u/h_p$$

where A_{sh} is obtained in Fig. C.2 or Table C.1 based on the precast beam steel percentages p and p'. As defined in Table 4.5, $(\alpha_s)_{sh}$ is the ratio of the precast beam shrinkage that takes place between the time of dead-load application and slab casting to the ultimate shrinkage. This assumes that shrinkage warping begins at the time of application of the precast-beam dead-load moment. The ultimate shrinkage $(\epsilon_{sh})_u$, based on the concrete age when the precast beam dead-load moment is applied, is given in Appendix A (see Table A.8 plus the size correction factor for the precast beam in Table A.6 or A.7, for example). For average conditions, $(\epsilon_{sh})_u = 400 \times 10^{-6}$ in./in. (mm/mm) may normally be used.

Term 5 is the *shrinkage deflection of the composite beam following slab casting due to the shrinkage of the precast-beam concrete*, but not including the effect of differential shrinkage and creep, which is given by Term 8. The ratio, I_2/I_c, is taken to be the same as in Term 3 (see discussion of Term 3), and accounts for the effect of the composite section in restraining the shrinkage warping of the precast beam. The other parameters are the same as in Term 4.

Term 6 is the *initial deflection of the precast beam under slab dead load.* As discussed in Appendix C and Fig. C.1 for incremental deflections,

$$(\Delta_i)_1 = [(\Delta_i)_1 + (\Delta_i)_2] - (\Delta_i)_2$$

where

$$[(\Delta_i)_1 + (\Delta_i)_2] = K(5/48)(M_1 + M_2)\ell^2/E_c I_{2s}$$

by Eq. C.4

and $(\Delta_i)_2$ is the same as in Term 1. See Eq. C.5 for K, and Eq. C.1 for $I_{2s} = I_a = I_e$ at midspan (unless Eqs. C.2 and C.3 are used to obtain I_a). M_1 = midspan moment due to slab dead load. E_c at age 28 days will be used. For computing I_{2s} in Eq. C.1, M_a refers to the precast beam plus slab dead load and M_{cr}, I_g, and I_{cr} to the precast beam. When diaphragms, sidewalks, partitions, ceilings, roofing, etc. are cast at the same time as the slab or soon thereafter, their dead load is also included in M_1 and M_a.

Term 7 is the *creep deflection of the composite beam due to slab dead load (plus diaphragm, sidewalk, etc., dead load when cast at the same time as the slab)*. The ratio, I_2/I_c, is taken to be the same as in Term 3 (see discussion of Term 3), and k_r is the same as in Term 2. The ultimate creep coefficient C_{u1}, based on the age of the precast-beam concrete when the slab is cast, is given in Appendix A (see Table A.8 plus the size correction factor for the precast beam in Table A.6 or A.7, for example).

Term 8 is the *deflection due to differential shrinkage and creep*, as presented in Appendix B. See Figs. B.1 and B.2 for the notation and description of the differential shrinkage and

Table 4.5. Representative values of $(\alpha_s)_{sh}$ for unshored beams.[a]

	t_s = Age of Precast-Beam Concrete When the Slab Is Cast in Days						
t_p[b]	7	10	20	28	60	90	120
4	*0.05*, –	*0.09*, –	*0.21*, –	*0.28*, –	*0.47*, –	*0.57*, –	*0.63*, –
7	–	*0.04*, 0.08	*0.16*, 0.27	*0.23*, 0.38	*0.42*, 0.60	*0.52*, 0.70	*0.59*, 0.76
10	–	–	*0.12*, 0.19	*0.19*, 0.30	*0.38*, 0.52	*0.48*, 0.62	*0.54*, 0.69
20	–	–	–	*0.07*, 0.10	*0.26*, 0.33	*0.36*, 0.43	*0.43*, 0.49
28	–	–	–	–	*0.19*, 0.23	*0.29*, 0.33	*0.36*, 0.39
60	–	–	–	–	–	*0.10*, 0.10	*0.17*, 0.16
90	–	–	–	–	–	–	*0.07*, 0.06

[a]The italicized values refer to steam-cured precast-beam concrete (usual case), and the other values to moist-cured precast-beam concrete. The values in the table apply to different-weight concrete. The term $(\alpha_s)_{sh}$ is the ratio of the precast-beam shrinkage that takes place between the time of dead-load application and slab casting to the ultimate shrinkage. Using Eq. A.12 for steam-cured: $(\alpha_s)_{sh} = [(t_s - 1)/(55 + t_s - 1)] - [(t_p - 1)/(55 + t_p - 1)]$. Using Eq. A.11 for moist-cured: $(\alpha_s)_{sh} = [(t_s - 7)/(35 + t_s - 7)] - [(t_p - 7)/(35 + t_p - 7)]$. See Terms 4 and 5 of Eq. 4.31.

[b]The term t_p = age of precast-beam concrete when the precast-beam dead-load moment is applied in days.

creep force Q (theoretically computed for elastic assumption in Eq. B.1 and modified in Eq. B.11), Eqs. B.6 through B.12 for the various parameters and the deflection, Fig. B.3 for K_{DS}, and Table B.1 for D_u. For a slab thickness other than 6 in. (15 cm), the size correction factor for D_u is obtained in Table A.6. For humidity other than 70%, the humidity correction factor for D_u is obtained in Table A.5. The gross composite section properties for y_{cs} and I_c are recommended in this term for computing deflections.

Term 9 is the *initial or short-time live load (plus all other loads applied to the composite beam and not included in Term 6)* deflection of the composite beam using Eq. C.4 for the deflection, $\Delta_L = K(5/48)M_L \ell^2/E_c(I_c)_{D+L}$; Eq. C.5 or C.6 for K, and Eq. C.1 for $(I_c)_{D+L} = I_a = I_e$ at midspan (unless Eqs. C.2 and C.3 are used to obtain I_a). M_L = midspan moment due to live load plus any additional load applied to the composite beam such as due to diaphragms, sidewalks, partitions, ceilings, roofing, etc., which is placed after the composite section is effective. E_c (of precast-beam concrete, since the slab is transformed into equivalent precast-beam concrete) at age 28 days will be used. For computing $(I_c)_{D+L}$ in Eq. C.1, M_a refers to the total load (including precast beam and slab dead load, all other dead load, and live load), and M_{cr}, I_g, and I_{cr}, refer to the composite section. This solution for Δ_L is an approximation which does not provide for the incremental deflection as the composite beam deflection under total load minus the fictitious composite beam deflection under precast beam plus slab dead load, which would also be an approximation. The procedure recommended was deemed to be adequate in most cases in Branson.[17]

Term 10 is the *partial live load (also including other sustained loads in Term 9) creep deflection for any permanent or sustained live load applied to the composite beam.*

$$(\Delta_{cp})_L = (M_P/M_L)k_{rs}C_{up}\Delta_L$$

where M_P is the moment due to the permanent load supported by the composite beam; M_L is the same as in Term 9; k_{rs} is given by Eq. C.8 and refers to any slab compression steel that may be taken into account at midspan (unless Eqs. C.10 and C.11 are used to obtain $(k_{rs})_a$); C_{up} is the ultimate creep coefficient based on the slab concrete age when the permanent load is applied to the composite beam (see Table A.8 plus the size correction factor for the slab in Table A.6 or A.7, for example; and Δ_L is obtained in Term 9.

It can be seen by inspection that Eq. 4.31 for composite beams reduces to the basic equation for noncomposite reinforced beams of

$$\Delta_u = (\Delta_i)_D + k_r C_u (\Delta_i)_D + \Delta_{sh} + \Delta_L + (\Delta_{cp})_L$$

when $I_2/I_c = 1$, subscript $2 = D$, $C_{u2} = C_u$, and Terms 6, 7, and 8—which refer to the composite beam—are deleted. Of course, the last two terms in this case also apply to the noncomposite beam.

General method for shored composite beams, assuming the composite beam supports all dead and live load

$$\Delta_u = \overbrace{(\Delta_i)_{1+2}}^{(1)} + \overbrace{(\Delta_{cp})_{1+2}}^{(2)} + \overbrace{\Delta_{sh}\frac{I_2}{I_c}}^{(3)} + \overbrace{\Delta_{DS}}^{(4)} + \overbrace{\Delta_L}^{(5)} + \overbrace{(\Delta_{cp})_L}^{(6)} \quad (4.32)$$

Term 1 is the *initial deflection of the composite beam due to slab plus precast-beam dead load (plus diaphragm, sidewalk, partition, ceiling, roofing, etc., dead load that is cast before the shores are removed)* using Eq. C.4 for the deflection—$(\Delta_i)_{1+2} = K(5/48)(M_1 + M_2)\ell^2/E_c(I_c)_{1+2}$, Eq. C.6 for K, and Eq. C.1 for $(I_c)_{1+2} = I_a = I_e$ at midspan (unless Eqs. C.2 and C.3 are used to obtain I_a). $M_1 + M_2$ = midspan moment due to total dead load applied before the shores are removed, which is at least the slab plus precast-beam dead load. E_c (of precast-beam concrete, since the slab is transformed into equivalent precast-beam concrete) at age 28 days will be used. For computing $(I_c)_{1+2}$ in Eq. C.1, M_a refers to the moment $M_1 + M_2$, and M_{cr}, I_g, and I_{cr}, to the composite section. This term corresponds to Terms 1 and 6 of Eq. 4.31.

Term 2 is the *creep deflection of the composite beam due to the dead load* in Term 1 using Eq. C.12—$(\Delta_{cp})_{1+2} = k_{rs}C_{us}(\Delta_i)_{1+2}$. The reduction factor k_{rs} is given by Eq. C.8 and refers to any slab compression steel that may be taken into account at midspan—unless Eqs. C.10 and C.11 are used to obtain $(k_{rs})_a$. The ultimate slab creep coefficient C_{us}, based on the slab concrete age when the shores are removed, is given in Appendix A (see Table A.8 plus the size correction factor for the slab in Table A.6 or A.7 for example). This term corresponds to Terms 2, 3 and 7 of Eq. 4.31.

Term 3 is the *shrinkage deflection of the composite beam following the time the shores are removed due to the shrinkage of the precast-beam concrete*, but not including the effect of differential shrinkage and creep given by Term 4. Using Eq. C.14,

$$\Delta_{sh} = K_{sh}(\phi_{sh})_u \ell^2$$

where K_{sh} is obtained from Eqs. C.16 through C.21, $(\phi_{sh})_u = A_{sh}(\epsilon_{sh})_u/h_p$ from Eq. C.15, and A_{sh} is obtained in Fig. C.2 or Table C.1 based on the precast beam steel percentages, p and p'. The ultimate shrinkage strain $(\epsilon_{sh})_u$, based on the age of the precast beam concrete when the shores are removed, is given in Appendix A (for example, see Table A.8 plus the size correction factor for the precast beam in Table A.6 or A.7). The ratio, I_2/I_c, accounts for the effect of the composite section in restraining the shrinkage warping of the precast beam after the shores are removed. Since both sections are normally partially cracked at this stage, as a simple approximation it is recommended that an average of the gross and cracked section ratios at midspan be used in this term (the same as in Term 3 of Eq. 4.31), or $I_2/I_c = [(I_2/I_c)_g + (I_2/I_c)_{cr}]/2$. There is some question in this equation about Term 4 of Eq. 4.31, which is the shrinkage warping up to the time of slab casting, since the precast beam is now shored or supported. It may be reasonable to exclude this effect completely. Term 3 of Eq. 4.32 replaces Terms 4 and 5 of Eq. 4.31.

Term 4 is the *deflection due to differential shrinkage and creep*, as presented in Appendix B. This term is the same as Term 8 of Eq. 4.31, except that the differential shrinkage and creep coefficient D_u in Table A.1 takes into account the different creep effects under gravity loads for shored construction. In addition, this term should be adjusted to reflect only the differential shrinkage and creep effect that takes place after the shores are removed, by using a reduced coefficient $D_u = \gamma_s$ (D_u from Table B.1). As defined in Table 4.4, γ_s is the ratio of the differential shrinkage and creep that takes place after the shores are removed to the ultimate value. Also, see Term 8 of Eq. 4.31 for possible size and humidity correction factors for D_u.

Term 5 is the *initial or short-time live-load deflection of the composite beam.* This term is the same as would be calculated for the incremental live-load deflection of a monolithic beam. It should not include any dead load due to diaphragms, partitions, sidewalks, ceilings, roofing, etc., which was placed before the shores were removed and hence was included in Term 1.

Term 6 is the *partial live-load creep deflection for any permanent or sustained live load plus other sustained loads placed after the shores were removed.* This term is the same as Term 10 of Eq. 4.31, except that it should not include any dead load that was included in Term 2.

The parameters, $(\alpha_s)_{cp}$, $(\alpha_s)_{sh}$, and γ_s, are defined and representative values tabulated in Tables 4.4 and 4.5.

Equation 4.31 for unshored construction reduces to the simplified Eqs. 4.31a and 4.31b when the following representative parameters are used: For steam-cured precast-beam dead load applied at age 28 days, moist-cured slab cast at precast-beam age of 2 months, H = 70% humidity, and neglecting the size correction factors in Table A.6 or A.7—C_{u2} = 1.56 for steam-cured loading age 28 days (Table A.8), C_{u1} = 1.43 for steam-cured loading age 2 months (Table A.8), and $(\epsilon_{sh})_u = 362 \times 10^{-6}$ in./in. (mm/mm) for steam-cured precast beam shrinkage following age 28 days (Table A.8). Also, $(\alpha_s)_{cp}$ = 0.42 (Table 4.4) and $(\alpha_s)_{sh}$ = 0.19 (Table 4.5).

Simplified equations for unshored composite beams, as by Eq. 4.31

$$\Delta_u = \overbrace{(\Delta_i)_2}^{(1)} + \overbrace{(0.42)(1.56k_r)(\Delta_i)_2}^{(2)}$$

$$+ \overbrace{(0.58)(1.56k_r)(\Delta_i)_2 \frac{I_2}{I_c}}^{(3)}$$

$$+ \overbrace{(0.19)\Delta_{sh}}^{(4)} + \overbrace{(0.81)\Delta_{sh} \frac{I_2}{I_c}}^{(5)}$$

$$+ \overbrace{(\Delta_i)_1}^{(6)} + \overbrace{(1.43k_r)(\Delta_i)_1 \frac{I_2}{I_c}}^{(7)}$$

$$+ \overbrace{\Delta_{DS}}^{(8)} + \overbrace{\Delta_L}^{(9)} + \overbrace{(\Delta_{cp})_L}^{(10)}$$

$$= \overbrace{\left(1.00 + 0.66k_r + 0.90k_r \frac{I_2}{I_c}\right)(\Delta_i)_2}^{(1+2+3)}$$

$$+ \overbrace{\left(0.19 + 0.81 \frac{I_2}{I_c}\right)(362 \times 10^{-6}) K_{sh}A_{sh}\ell^2/h_p}^{(4+5)}$$

$$+ \overbrace{\left(1.00 + 1.43k_r \frac{I_2}{I_c}\right)(\Delta_i)_1}^{(6+7)}$$

$$+ \overbrace{\Delta_{DS}}^{(8)} + \overbrace{\Delta_L}^{(9)} + \overbrace{(\Delta_{cp})_L}^{(10)} \qquad (4.31a)$$

When $k_r = 0.85$ (no compression steel in precast beam), Eq. 4.31a reduces to Eq. 4.31b.

$$\Delta_u = \overbrace{\left(1.56 + 0.77 \frac{I_2}{I_c}\right)(\Delta_i)_2}^{(1+2+3)}$$

$$+ \overbrace{\left(0.19 + 0.81 \frac{I_2}{I_c}\right)(362 \times 10^{-6}) K_{sh}A_{sh}\ell^2/h_p}^{(4+5)}$$

$$+ \overbrace{\left(1.00 + 1.22 \frac{I_2}{I_c}\right)(\Delta_i)_1}^{(6+7)}$$

$$+ \overbrace{\Delta_{DS}}^{(8)} + \overbrace{\Delta_L}^{(9)} + \overbrace{(\Delta_{cp})_L}^{(10)} \qquad (4.31b)$$

Equation 4.32 for shored construction reduces to the simplified Eq. 4.32a when the following representative parameters are used: For shores removed when the moist-cured slab age is 10 days and the steam-cured precast beam age is 2 months, $H = 70\%$ humidity, and neglecting the size correction factors in Table A.6 or A.7, Slab $C_{us} = 1.79$ for moist-cured slab loading age 10 days (Table A.8), and precast beam $(\epsilon_{sh})_u = 261 \times 10^{-6}$ in./in. (mm/mm) for steam-cured precast beam shrinkage following age 2 months (Table A.8). Also, $k_{rs} = 0.85$ when neglecting the effect of any slab compression steel on creep in Term 2.

Simplified equation for shored composite beams, as by Eq. 4.32

$$\Delta_u = \overbrace{(2.52)(\Delta_i)_{1+2}}^{(1+2)}$$

$$+ \overbrace{(261 \times 10^{-6})(K_{sh}A_{sh}\ell^2/h_p) \frac{I_2}{I_c}}^{(3)}$$

$$+ \overbrace{\Delta_{DS}}^{(4)} + \overbrace{\Delta_L}^{(5)} + \overbrace{(\Delta_{cp})_L}^{(6)} \qquad (4.32a)$$

Rough approximate method for unshored composite beams

$$\Delta_u = \Delta_D + (k_r T)\Delta_D(I_2/I_c) + \Delta_L + (\Delta_{cp})_L \qquad (4.33)$$

where

$\Delta_D = K(5/48)M_D\ell^2/E_cI_{2s}$ in Eq. C.4
K is given by Eq. C.5

M_D = slab plus precast beam (plus diaphragm, sidewalk, etc., when cast at the same time as the slab) dead-load moment at midspan

$E_c = (E_c)_{28d}$ of the precast-beam concrete

$I_{2s} = I_e$ for the precast beam at midspan in Eq. C.1
$(k_r T) = T/(1 + 50\rho')$
ρ' = the precast-beam compression steel ratio at midspan in Eq. C.8
T = 3.5 for the combined effects of precast-beam creep, precast-beam shrinkage, and differential shrinkage and creep of the composite beam.

For other values of T, see Tables A.9 and A.10. Note, however, that the values in these tables refer to noncomposite beams and do not include the effect of differential shrinkage and creep. For computing I_{2s} in Eq. C.1, $M_a = M_D$, and M_{cr}, I_g, and I_{cr}, refer to the precast beam. As discussed in Term 3 of Eq. 4.31, use $I_2/I_c = [(I_2/I_c)_g + (I_2/I_c)_{cr}]/2$. The terms Δ_L and $(\Delta_{cp})_L$ are the same as Terms 9 and 10 of Eq. 4.31, respectively.

Rough approximate method for shored composite beams, assuming the composite beam supports all dead and live load

$$\Delta_u = \Delta_D + (k_{rs}T)\Delta_D + \Delta_L + (\Delta_{cp})_L \tag{4.34}$$

where

$k_{rs}T = T/(1 + 50\rho')$
ρ' = the compression steel ratio for the slab steel and composite section at midspan in Eq. C.8
T = 3.5 for the combined effects of slab creep, precast-beam shrinkage, and differential shrinkage and creep of the composite beam (all following the removal of shores)

and Δ_D is the same as Term 1 of Eq. 4.32, Δ_L is the same as Term 5 of Eq. 4.32, and $(\Delta_{cp})_L$ is the same as Term 6 of Eq. 4.32. For other values of T, see Tables A.9 and A.10, except that the values in these tables refer to noncomposite beams and do not include the effect of differential shrinkage and creep.

Allowable deflections

The limiting deflections in the ACI Code are shown in Table 4.6. For more detailed information on allowable deflections, see the ACI Deflection Committee 435 Report.[30]

4.2.4.4 Example 4.3

Deflection of simple-span composite beam designed in Ex. 4.1. Assume that shored construction is used (as suggested in Ex. 4.1), and compute deflections by Eqs. 4.32, 4.32a, and 4.34.

See Ex. 4.1 and Fig. 4.10 for the material properties and design conditions, and Table 4.1 for the section properties used in design. See Table 4.7 for a summary of parameters and conditions, and Table 4.8 for the section properties, used in the deflection calculations. The deflection results are summarized and compared with allowable values in Table 4.9.

Solution by the General Eq. 4.32

Term 1. *Initial deflection of the composite beam due to slab plus precast-beam dead load:*

$$(\Delta_i)_{1+2} = \frac{K(5/48)(M_1 + M_2)\ell^2}{E_c(I_c)_{1+2}}$$

$$= \frac{(1)(5/48)(61{,}250 + 42{,}110)(12)(35)^2(12)^2}{(3.64 \times 10^6)(26{,}900)}$$

$$= 0.233 \text{ in. } (5.9 \text{ mm})$$

Term 2. *Creep deflection of the composite beam due to slab plus precast-beam load:*

$$(\Delta_{cp})_{1+2} = k_{rs}C_{us}(\Delta_i)_{1+2} = (0.85)(1.99)(0.233)$$

$$= 0.39 \text{ in. } (9.9 \text{ mm})$$

Term 3. *Shrinkage deflection of the composite beam following the time the shores are removed due to shrinkage of the precast-beam concrete:*

For $p = 3.70\%$, $p' = 0$–$A_{sh} = 1.00$ (Fig. C.2 or Table C.1

$$(\phi_{sh})_u = A_{sh}(\epsilon_{sh})_u/h_p = (1.00)(214 \times 10^{-6})/22$$

$$= 9.73 \times 10^{-6} \text{ 1/in. } (3.83 \times 10^{-6} \text{ 1/cm})$$

$$I_2/I_c = [(I_2/I_c)_g + (I_2/I_c)_{cr}]/2$$

$$= (0.315 + 0.383)/2 = 0.349$$

$$\Delta_{sh}\frac{I_2}{I_c} = K_{sh}(\phi_{sh})_u\ell^2\frac{I_2}{I_c}$$

$$= (1/8)(9.73 \times 10^{-6})(35)^2(12)^2(0.349)$$

$$= 0.07 \text{ in. } (1.8 \text{ mm})$$

Table 4.6. ACI Code (318-71) allowable computed deflections.[3]

<table>
<tr><th>Type of Member</th><th>Deflection to Be Considered</th><th>Deflection Limitation</th></tr>
<tr><td>Flat roofs not supporting or attached to nonstructural elements likely to be damaged by large deflections</td><td>Immediate deflection due to the live load</td><td>$\ell/180$[a]</td></tr>
<tr><td>Floors not supporting or attached to nonstructural elements likely to be damaged by large deflections</td><td>Immediate deflection due to the live load</td><td>$\ell/360$</td></tr>
<tr><td>Roof or floor construction supporting or attached to nonstructural elements likely to be damaged by large deflections.</td><td rowspan="2">That part of the total deflection which occurs after attachment of nonstructural elements, the sum of the long-time deflection due to all sustained loads, and the immediate deflection due to any additional live load[c]</td><td>$\ell/480$[b]</td></tr>
<tr><td>Roof or floor construction supporting or attached to non-structural elements not likely to be damaged by the large deflections.</td><td>$\ell/240$[d]</td></tr>
</table>

[a]This limit is not intended to safeguard against ponding. Ponding should be checked by suitable calculations of deflection, including the added deflections due to ponded water, and considering long-time effects of all sustained loads, camber, construction tolerances, and reliability of provisions for drainage.
[b]This limit may be exceeded if adequate measures are taken to prevent damage to supported or attached elements.
[c]The long-time deflection shall be determined in accordance with Section 9.5.2.3 or 9.5.4.2 of the Code, but may be reduced by the amount of deflection that occurs before attachment of the nonstructural elements. This amount shall be determined on the basis of accepted engineering data relating to the time-deflection characteristics of members similar to those being considered.
[d]But not greater than the tolerance provided for the nonstructural elements. This limit may be exceeded if camber is provided so that the total deflection minus the camber does not exceed the limitation.

The precast-beam shrinkage-warping deflection appears to be typically quite small in the case of shored construction.

Term 4. *Deflection due to differential shrinkage and creep:*

For slab age of 10 days when shores removed, $\gamma_s = 0.77$ (Table 4.4).

From Ex. 4.1 differential shrinkage and creep stress calculations,

$$Q_{DS} = 166{,}800 \text{ lb } (742{,}000 \text{ N}).$$

Use $Q = \gamma_s Q_{DS} = (0.77)(166{,}800) = \underline{128{,}400 \text{ lb}}$ (571,100 N).

Although the cracked-section values were used in the stress calculations, the gross-section values for y_{cs} and I_c will be used for computing differential shrinkage and creep deflections, since the composite beam is only partially cracked in the early and dominant stages of differential shrinkage.

$$y_{cs} = h_c - t/2 - y_b = 26 - 4/2 - 17.65$$

$$= 6.35 \text{ in. } (16.1 \text{ cm})$$

$$I_c = (I_c)_g$$

$$= 33{,}840 \text{ in.}^4 \; (1{,}408{,}400 \text{ cm}^4) \qquad \text{(Table 4.8)}$$

$$\Delta = \frac{K_{DS} Q y_{cs} \ell^2}{E_c I_c}$$

$$= \frac{(1/8)(128{,}400)(6.35)(35)^2(12)^2}{(3.64 \times 10^6)(33{,}840)} \qquad \text{(Eq. B.6 or B.7)}$$

$$= 0.146 \text{ in. } (3.7 \text{ mm})$$

$$\Delta_{DS} = (4/3)\Delta = (4/3)(0.146)$$

$$= 0.19 \text{ in. } (4.8 \text{ mm}) \qquad \text{(Eq. B.12)}$$

Table 4.7. Summary of parameters and conditions for Ex. 4.3.

From Ex. 4.1 and Fig. 4.10

Span = 35 ft (10.7 m), Precast Beam f_r = 496 psi (3.42 N/mm^2),
Precast Beam $E_c = (E_c)_{28d} = 3.64 \times 10^6$ psi (25.1 kN/mm^2)

$$M_1 = M_s = 61{,}250 \text{ ft-lb } (83.1 \text{ kN-m})$$

$$M_2 = M_p = 42{,}100 \text{ ft-lb } (57.1 \text{ kN-m}),$$

$$M_L = 183{,}800 \text{ ft-lb } (249 \text{ kN-m})$$

Other Parameters and Conditions

For simple beam, $K = 1$ (Eq. C.5), $K_{sh} = K_{DS} = \frac{1}{8}$ (Eq. C.17, Fig. B.3). Neglecting any slab compression steel effect at midspan, $k_{rs} = 0.85$ (Eq. C.8).

Average Relative Humidity = 70%, Precast-Beam Steam-Cured and Slab Moist-Cured.
For 4-in. (10.2-cm) Slab, Size Creep $(\text{C.F.})_T = 1.11$ (Table A.6).
For 12-in. Precast Beam, Size Shrinkage $(\text{C.F.})_T = 0.82$ (Table A.6).
Alternatively, $v/s = (12 \times 22)/(24 + 44) = 3.9$ in., and Size Shrinkage $(\text{C.F.})_T = 0.79$ (Table A.7). Use Size Shrinkage $(\text{C.F.})_T = 0.82$.
For shored construction with the shores removed when the slab age is 10 days and the precast-beam age is 2 months. From Table A.8:
Basic Slab $C_{us} = 1.79$, Basic Precast Beam $(\epsilon_{sh})_u = 261 \times 10^{-6}$ in./in. (mm/mm)
Use Slab $C_{us} = (1.11)(1.79) = \underline{1.99}$
Use Precast Beam $(\epsilon_{sh})_u = (0.82)(261 \times 10^{-6}) = \underline{214 \times 10^{-6}}$ in./in. (mm/mm)

Assume 30% of the live load is a sustained load applied at a slab age of 2 months.
Basic Slab $C_{up} = 1.45$, Use Slab $C_{up} = (1.11)(1.45) = \underline{1.61}$

Term 5. *Initial or short-time live-load deflection of the composite beam:*

$$\Delta_L = \frac{K(5/48)(M_1 + M_2 + M_L)\ell^2}{E_c(I_c)_{D+L}} - (\Delta_i)_{1+2}$$

$$= \frac{(1)(5/48)(287{,}150)(12)(35)^2(12)^2}{(3.64 \times 10^6)(21{,}520)} - 0.233$$

$$= 0.575 \text{ in. } (14.6 \text{ mm})$$

Term 6. *Partial live-load creep deflection, assuming 30% of the live load to be a sustained or permanent load:*

$$(\Delta_{cp})_L = (M_P/M_L)k_{rs}C_{up}\Delta_L$$

$$= (0.30)(0.85)(1.61)(0.575)$$

$$= 0.24 \text{ in. } (6.1 \text{ mm})$$

Total ultimate deflection by Eq. 4.32, which is also summarized in Table 4.9:

$$\Delta_u = \overset{(1)}{0.23} + \overset{(2)}{0.39} + \overset{(3)}{0.07} + \overset{(4)}{0.19} + \overset{(5)}{0.58} + \overset{(6)}{0.24}$$

$$= \underline{1.70 \text{ in. } (43.2 \text{ mm})}$$

Solution by the Simplified Eq. 4.32a

$$\Delta_u = (2.52)(\Delta_i)_{1+2} + (261 \times 10^{-6})(K_{sh}A_{sh}\ell^2/h_p)\frac{I_2}{I_c} + \Delta_{DS} + \Delta_L + (\Delta_{cp})_L$$

$$= (2.52)(0.233) + (261 \times 10^{-6})(1/8)(1.00)(35)^2(12)^2(0.349)/22 + 0.19 + 0.58 + 0.24 = \underline{1.69 \text{ in. } (42.9 \text{ mm})} \quad \text{(Eq. 4.32a)}$$

versus 1.70 in. (43.2 mm) by the more general Eq. 4.32. The difference in the two solutions in this example was due to the nonusage of creep and shrinkage size correction factors in the simplified Eq. 4.32a, which had an offsetting effect between the creep and shrinkage factors used (see values preceding Eq. 4.32a and in Table 4.7).

Solution by the rough approximate method (Eq. 4.34)

Table 4.8. Section properties for Ex. 4.3.

From Ex. 4.1, Fig. 4.10, and Table 4.1

Precast Beam I_g = 10,648 in.^{4} (443,200 cm^4)

I_{cr} = 8153 in.^{4} (339,300 cm^4), I_g/I_{cr} = 1.31

Composite Beam I_g = 33,840 in.^{4} (1,408, 400 cm^4)

$y_{ten} = (y_b)_g$ = 17.65 in. (44.8 cm)

I_{cr} = 21,260 in.^{4} (884,800 cm^4), I_g/I_{cr} = 1.59

$(I_2/I_c)_g$ = 10,648/33,840 = 0.315, $(I_2/I_c)_{cr}$ = 8153/21,260 = 0.383

For precast-beam shrinkage calculations,

Pos. $\rho_p = 0.0370$, $p_p = 3.70$, $\rho'_p = p'_p = 0$, h_p = 22 in. (55.0 cm)

Effective Section Properties for the Composite Beam Section

$$M_{cr} = f_r I_g / y_{ten} = (495)(33{,}840)/17.65 = 949{,}000 \text{ in-lb}$$
$$= 79{,}080 \text{ ft-lb } (107.2 \text{ kN-m})$$
$$M_1 + M_2 = 61{,}250 + 42{,}100 = 103{,}350 \text{ ft-lb } (140.1 \text{ kN-m})$$
$$(M_1 + M_2)/M_{cr} = 103{,}350/79{,}080 = 1.307,$$
$$M_{cr}/(M_1 + M_2) = 79{,}080/103{,}350 = 0.765, \quad 0.765^3 = 0.448,$$
$$(I_c)_{1+2} = (M_{cr}/M_a)^3 I_g + [1 - (M_{cr}/M_a)^3] I_{cr} \quad \text{(Eq. C.1)}$$
$$= (0.448)(33{,}840) + (0.552)(21{,}260)$$
$$= 15{,}160 + 11{,}740 = \underline{26{,}900 \text{ in.}^4 \ (1{,}119{,}600 \text{ cm}^4)}$$
$$M_1 + M_2 + M_L = 61{,}250 + 42{,}100 + 183{,}800 = 287{,}150 \text{ ft-lb } (389 \text{ kN-m})$$
$$(M_1 + M_2 + M_L)/M_{cr} = 287{,}150/79{,}080 = 3.631,$$
$$M_{cr}/(M_1 + M_2 + M_L) = 79{,}080/287{,}150 = 0.275, \quad 0.275^3 = 0.021,$$
$$(I_c)_{D+L} = (M_{cr}/M_a)^3 I_g + [1 - (M_{cr}/M_a)^3] I_{cr} \quad \text{(Eq. C.1)}$$
$$= (0.021)(33{,}840) + (0.979)(21{,}260)$$
$$= 711 + 20{,}814 = \underline{21{,}520 \text{ in.}^4 \ (895{,}700 \text{ cm}^4)}$$

Δ_D = 0.233 in. (5.9 mm)

(Same as Term 1, Eq. 4.32)

$(k_{rs}T)\Delta_D$ = (3.5)(0.233) = 0.82 in. (20.8 mm)

Δ_L = 0.58 in. (14.6 mm)

(Same as Term 5, Eq. 4.32)

$(\Delta_{cp})_L$ = 0.24 in. (6.1 mm)

(Same as Term 6, Eq. 4.32)

Total ultimate deflection by Eq. 4.34:

$$\Delta_u = \Delta_D + (k_{rs}T)\Delta_D + \Delta_L + (\Delta_{cp})_L \quad \text{(Eq. 4.34)}$$
$$= 0.23 + 0.82 + 0.58 + 0.24$$
$$= \underline{1.87 \text{ in. } (47.5 \text{ mm})}$$

vs. 1.70 in. (43.2 mm) by the more general Eq. 4.32, and 1.69 in. (42.9 mm) by the simplified Eq. 4.32a.

4.2.4.5 Example 4.4. Deflection of Three-Span Continuous Composite Beam Designed in Ex. 4.2.

Assume that unshored construction is used (either unshored or shored construction was found to be permissible in the design), and

Table 4.9. Term-by-term deflections in Eq. 4.32 for the shored composite beam in Ex. 4.3, and comparison with the ACI Code (318-71) allowable deflections from Table 4.6.

Initial Slab + Precast-Beam Dead-Load Deflection (1)		Creep Deflection Due to Slab + Precast-Beam Dead Load (2)		Composite Beam Deflection Due to Precast-Beam Shrinkage (3)		Differential Shrinkage and Creep Deflection (4)		Initial Live-Load Deflection (5)		Live-Load Creep Deflection (6)		Total	
in.	mm	in.	mm	in.	mm	in.	mm	in.	mm	in.	mm	in.	mm
0.23	5.9	0.39	9.9	0.07	1.8	0.19	4.8	0.58	14.6	0.24	6.1	1.70	43.2

$$\Delta_L = 0.58 \text{ in. } (14.6 \text{ mm})$$

Assuming nonstructural elements are installed after the shores are removed (after slab age of 10 days), $\Delta_t + \Delta_L$ will be taken as $\Delta_u - (\Delta_i)_{1+2} = 1.70 - 0.23 = 1.47$ in. (37.3 mm). Hence,

$$0.58 \text{ in. } (15 \text{ mm}) < \ell/180 = (35)(12)/180 = 2.33 \text{ in. } (59 \text{ mm}) \quad \text{OK}$$

$$0.58 \text{ in. } (15 \text{ mm}) < \ell/360 = 1.17 \text{ in. } (30 \text{ mm}) \quad \text{OK}$$

$$1.47 \text{ in. } (37 \text{ mm}) > \ell/480 = 0.88 \text{ in. } (22 \text{ mm}) \quad \text{NG}$$

$$1.47 \text{ in. } (37 \text{ mm}) > \ell/240 = 1.75 \text{ in. } (44 \text{ mm}) \quad \text{OK}$$

Thus, the beam meets three of the four deflection criteria in Table 4.6, but does not meet the requirement for roof or floor construction supporting or attached to nonstructural elements likely to be damaged by large deflections, unless Footnotes b, d, or both are followed. This $\ell/480$ limitation is typically a rather stringent criterion.

Table 4.10. Summary of parameters and conditions for Ex. 4.4.

From Ex. 4.2 and Fig. 4.14

Clear Spans = 30 ft (9.1 m)

$$w_s = 563 \text{ lb/ft } (8214 \text{ N/m}), \quad w_p = 167 \text{ lb/ft } (2440 \text{ N/m})$$

$$\text{Pos. } M_L = w_L\ell^2/11 = 81{,}820 \text{ ft-lb } (111.0 \text{ kN-m})$$

$$\text{Neg. } M_L = w_L\ell^2/10 = 90{,}000 \text{ ft-lb } (122.0 \text{ kN-m})$$

For precast-beam shrinkage calculations,

$$\text{Pos. } \rho_p = 0.0300, \quad p_p = 3.00, \quad \rho' = p' = 0$$

Positive Moments for Slab and Precast-Beam Dead Load

$$M_1 = M_s = w_s\ell^2/8 = (563)(30)^2/8 = 63{,}340 \text{ ft-lb } (85.9 \text{ kN-m})$$

$$M_2 = M_p = w_p\ell^2/8 = (167)(30)^2/8 = 18{,}790 \text{ ft-lb } (25.5 \text{ kN-m})$$

Other Parameters and Conditions

For simple precast beam (for all dead load): $K = 1$ (Eq. C.5), and $K_{sh} = \frac{1}{8}$ (Eq. C.17).

For continuous beam end span (for live-load and differential shrinkage and creep): $K = 1.20 - 0.20M_o/M_m = 1.20 - 0.20(11/8) = 0.925$ (Eq. C.5), and $K_{DS} = 0.0528$ (Fig. B.3).

The beam will have no negative-moment creep or k_r effect, since short-time live-load is the only loading applied to the continuous beam. Hence, Eq. C.10 for $(k_r)_a$ is not applicable. Neglecting any precast-beam compression steel effect at midspan,

$$k_r = 0.85 \text{ (Eq. C.8)}$$

Average Relative Humidity = 60%, Precast-Beam Steam-Cured and Slab Moist-Cured compute the end-span deflections by Eqs. 4.31, 4.31b, and 4.33.

See Ex. 4.2 and Fig. 4.14 for the material and design conditions. See Table 4.10 for a summary of parameters and conditions, Table 4.11 for the material properties, and Table 4.12 for the section properties (including the effective moments of inertia) used in the deflection calculations. The deflection results are summarized and compared with allowable values in Table 4.13.

Solution by the General Eq. 4.31

Term 1. *Initial dead-load deflection of the precast beam:*

$$(\Delta_i)_2 = \frac{K(5/48)M_2\ell^2}{E_c I_2}$$

$$= \frac{(1)(5/48)(18{,}790)(12)(30)^2(12)^2}{(2.74 \times 10^6)(4096)}$$

$$= 0.271 \text{ in. } (6.9 \text{ mm})$$

Table 4.11. Material properties for Ex. 4.4.

Modulus of Rupture and Modulus of Elasticity

For normal-weight slab and $f'_c = 3000$ psi (21 N/mm^2):

$$f_r = 429 \text{ psi } (2.96 \text{ N/mm}^2) \quad \text{(Table A.1)}$$

$$E_c = 3.30 \times 10^6 \text{ psi } (22.8 \text{ kN/mm}^2) \quad \text{(Table A.2)}$$

For sand-lightweight precast beam and $f'_c = 4000$ psi (28 N/mm^2):

$$f_r = 450 \text{ psi } (3.10 \text{ N/mm}^2) \quad \text{(Table A.1)}$$

$$E_c = 2.74 \times 10^6 \text{ psi } (18.9 \text{ kN/mm}^2) \quad \text{(Table A.2)}$$

$$n = 29/2.74 = 10.6$$

$$n_c = 2.74/3.30 = 0.83$$

$$b_e/n_c = 84/0.83 = 101 \text{ in. } (257 \text{ cm}).$$

The n and n_c values are used to transform both slab concrete and steel to equivalent precast-beam concrete in both Positive and Negative regions.

Time-Dependent Properties

For 12-in. (30.5-cm) Precast Beam, Size Creep $(\text{C.F.})_T = 0.90$ and Size Shrinkage $(\text{C.F.})_T = 0.82$ (Table A.6). Alternatively, $v/s = (12 \times 16)/(24 + 32) = 3.4$ in. (8.6 cm), Size Creep $(\text{C.F.})_T = 0.85$, and Size Shrinkage $(\text{C.F.})_T = 0.83$ (Table A.7).

Use Size Creep $(\text{C.F.})_T = \underline{0.90}$ and Size Shrinkage $(\text{C.F.})_T = \underline{0.82}$.

For unshored construction with steam-cured precast-beam dead load applied at age 28 days and moist-cured slab cast at precast-beam age of 2 months (and $H = 60\%$):

Basic Precast Beam $C_{u2} = 1.70$ for loading at age 28 days (Table A.8).
Use Precast Beam $C_{u2} = (0.90)(1.70) = \underline{1.53}$ for loading at age 28 days.
Basic Precast Beam $C_{u1} = 1.55$ for loading at age 2 months (Table A.8).
Use Precast Beam $C_{u1} = (0.90)(1.55) = \underline{1.40}$ for loading at age 2 months
Basic Precast Beam $(\epsilon_{sh})_u = 414 \times 10^{-6}$ in./in. (mm/mm) for shrinkage following age 28 days (Table A.8)
Use Precast Beam $(\epsilon_{sh})_u = (0.82)(414 \times 10^{-6}) = \underline{339 \times 10^{-6} \text{ in./in. (mm/mm)}}$ for shrinkage following age 28 days.

From Table B.1 for 6-in. slab and 70% humidity,

$$D_u = 355 \times 10^{-6} \text{ in./in. (mm/mm)}$$

For 4.5-in. (11.4-cm) slab, Size Shrinkage $(\text{C.F.})_T = 1.13$ (Table A.6). For humidity, $(\text{C.F.})_H = 0.70$ at 70% and 0.80 at 60% (Table A.5). Hence, use for the differential shrinkage and creep coefficient–

$$\underline{D_u} = (1.13)(0.80/0.70)(355 \times 10^{-6}) = \underline{458 \times 10^{-6} \text{ in./in. (mm/mm)}}$$

$$(\alpha_s)_{cp} = 0.42 \quad \text{(Table 4.4)}, \qquad (\alpha_s)_{sh} = 0.19 \quad \text{(Table 4.5)}$$

Table 4.12. Section properties for Ex. 4.4.

The gross and cracked section properties are computed from Figs. 4.8, 4.9, and 4.14.

Precast-Beam Section Properties

$$I_g = b_w h_p^3/12 = (12)(16)^3/12 = 4096 \text{ in.}^4 \text{ } (170{,}500 \text{ cm}^4)$$

$$b_w(kd)^2/2 = nA_s(d - kd)$$

$$12(kd)^2/2 = (11.0)(4.68)(13.0 - kd)$$

$$kd = 7.11 \text{ in. } (18.1 \text{ cm})$$

$$I_{cr} = b_w(kd)^3/3 + nA_s(d - kd)^2 = (12)(7.11)^3/3 + (11.0)(4.68)(13.0 - 7.11)^2$$

$$= 3224 \text{ in.}^4 \text{ } (134{,}200 \text{ cm}^4)$$

$$I_g/I_{cr} = 4096/3224 = 1.27$$

Composite Beam Positive Region Section Properties

$$(y_b)_g = \frac{(12 \times 16)(8) + (101 \times 4.5)(18.25)}{(12 \times 16) + (101)(4.5)} = \frac{(192)(8) + (455)(18.25)}{192 + 455}$$

$$= 15.21 \text{ in. } (38.6 \text{ cm})$$

$$I_g = 4096 + (192)(15.21 - 8)^2 + (101)(4.5)^3/12$$

$$+ (455)(18.25 - 15.21)^2 = 19{,}050 \text{ in.}^4 \text{ } (792{,}900 \text{ cm}^4)$$

$$(I_2/I_c)_g = 4096/19{,}050 = 0.215$$

$$(b_e/n_c)(y_t)^2/2 = nA_s(d_c - y_t)$$

$$(101)(y_t)^2/2 = (10.6)(4.68)(17.5 - y_t)$$

$$y_t = 3.68 \text{ in. } (9.3 \text{ cm}) < 4.5 \text{ in. } (114 \text{ cm})$$

$$I_{cr} = (b_e/n_c)(y_t)^3/3 + nA_s(d_c - y_t)^2$$

$$= (101)(3.68)^3/3 + (10.6)(4.68)(17.5 - 3.68)^2 = 11{,}150 \text{ in.}^4 \text{ } (464{,}100 \text{ cm}^4)$$

$$I_g/I_{cr} = 19{,}050/11{,}150 = 1.71$$

$$(I_2/I_c)_{cr} = 3224/11{,}150 = 0.289$$

Composite Beam Negative-Region Section Properties

Gross-section properties are the same as the positive-region values.

$$b_w(kd)^2/2 = nA_s(d - kd)$$

$$12(kd)^2/2 = (10.6)(5.40)(18.5 - kd)$$

$$kd = 9.35 \text{ in. } (23.8 \text{ cm})$$

$$I_{cr} = b_w(kd)^3/3 + nA_s(d - kd)^2 = (12)(9.35)^3/3 + (10.6)(5.40)(18.5 - 9.35)^2$$

$$= 8060 \text{ in.}^4 \text{ } (335{,}500 \text{ cm}^4)$$

$$I_g/I_{cr} = 19{,}050/8060 = 2.36$$

Precast-Beam Effective Moment of Inertia for Positive Region

$$M_{cr} = f_r I_g/y_{ten} = (450)(4096)/8 = 230{,}400 \text{ in.-lb} = 19{,}200 \text{ ft-lb } (26.0 \text{ kN-m})$$

$$M_{cr}/M_2 = 19{,}200/18{,}790 > 1$$

Hence $I_2 = I_g = \underline{4096 \text{ in.}^4 \text{ } (170{,}500 \text{ cm}^4)}$

$$M_{cr}/(M_1 + M_2) = 19{,}200/(63{,}330 + 18{,}790) = 0.234$$

$$0.234^3 = 0.013$$

Table 4.12. (cont.)

$$I_{2s} = (M_{cr}/M_a)^3 I_g + [1 - (M_{cr}/M_a)^3] I_{cr} \quad \text{(Eq. C.1)}$$

$$= (0.013)(4096) + (0.987)(3224) = \underline{3235 \text{ in.}^4 \ (134{,}600 \text{ cm}^4)}$$

Composite Beam Effective Moment of Inertia for Positive Region

$$M_{cr} = f_r I_g / y_{\text{ten}} = (450)(19{,}050)/15.21 = 563{,}600 \text{ in-lb}$$

$$= 46{,}970 \text{ ft-lb } (63.7 \text{ kN-m})$$

$$M_{cr}/(M_1 + M_2 + M_L) = 46{,}970/(63{,}340 + 18{,}790 + 81{,}820) = 0.286$$

$$0.286^3 = 0.023$$

$$(I_c)_{D+L} = (M_{cr}/M_a)^3 I_g + [1 - (M_{cr}/M_a)^3] I_{cr} \quad \text{(Eq. C.1)}$$

$$= (0.023)(19{,}050) + (0.977)(11{,}150) = \underline{11{,}330 \text{ in.}^4 \ (471{,}600 \text{ cm}^4)}$$

Composite Beam Effective Moment of Inertia for Negative Region

$$M_{cr} = f_r I_g / y_{\text{ten}} = (429)(19{,}050)/(20.5 - 15.21) = 1{,}545{,}000 \text{ in.-lb}$$

$$= 128{,}800 \text{ ft-lb } (174.7 \text{ kN-m})$$

Hence, use $I_e = (I_c)_{D+L} = (I_c)_L = I_g = \underline{19{,}050 \text{ in.}^4 \ (792{,}900 \text{ cm}^4)}$

Composite Beam Average Effective Moment of Inertia

$$(I_c)_{D+L} = I_a = 0.85 I_m + 0.15(I_{\text{cont. end}}) \quad \text{(Eq. C.2)}$$

$$= (0.85)(11{,}330) + (0.15)(19{,}050) = \underline{12{,}490 \text{ in.}^4 \ (519{,}800 \text{ cm}^4)}$$

as compared to the midspan value of $(I_c)_{D+L} = 11{,}330$ in.^{4}.

Term 2. *Dead-load creep deflection of the precast beam up to the time of slab casting:*

$$(\alpha_s)_{cp} k_r C_{u2} (\Delta_i)_2 = (0.42)(0.85)(1.53)(0.271)$$

$$= 0.15 \text{ in. } (3.8 \text{ mm})$$

Term 3. *Creep deflection of the composite beam following slab casting due to the precast-beam dead load:*

Using

$$I_2/I_c = [(I_2/I_c)_g + (I_2/I_c)_{cr}]/2$$

$$= (0.215 + 0.289)/2$$

$$= 0.252 \text{ (Table 4.12)}$$

$$[1 - (\alpha_s)_{cp}] k_r C_{u2} (\Delta_i)_2 \frac{I_2}{I_c}$$

$$= (0.58)(0.85)(1.53)(0.271)(0.252)$$

$$= 0.05 \text{ in. } (1.3 \text{ mm})$$

Term 4. *Deflection due to shrinkage warping of the precast beam up to the time of slab casting:*

For $p = 3.00\%$, $p' = 0$—$A_{sh} = 1.00$ (Fig. C.2 or Table C.1)

$$(\phi_{sh})_u = A_{sh} (\epsilon_{sh})_u / h_p$$

$$= (1.00)(339 \times 10^{-6})/16$$

$$= 21.19 \times 10^{-6} \text{ 1/in. } (8.34 \times 10^{-6} \text{ 1/cm})$$

$$\Delta_{sh} = K_{sh} (\phi_{sh})_u \ell^2$$

$$= (1/8)(21.19 \times 10^{-6})(30)^2(12)^2$$

$$= 0.343 \text{ in. } (8.7 \text{ mm})$$

$$(\alpha_s)_{sh} \Delta_{sh} = (0.19)(0.343) = 0.07 \text{ in. } (1.8 \text{ mm})$$

Term 5. *Shrinkage deflection of the composite beam following slab casting due to the shrinkage of the precast-beam concrete:*

$$[1 - (\alpha_s)_{sh}] \Delta_{sh} \frac{I_2}{I_c} = (0.81)(0.343)(0.252)$$

$$= 0.07 \text{ in. } (1.8 \text{ mm})$$

Term 6. *Initial deflection of the precast beam under slab dead load:*

$$(\Delta_i)_1$$

$$= [(\Delta_i)_1 + (\Delta_i)_2] - (\Delta_i)_2$$

$$= \frac{K(5/48)(M_1 + M_2)\ell^2}{E_c I_{2s}} - (\Delta_i)_2$$

Table 4.13. Term-by-term deflections in Eq. 4.31 for the unshored composite beam in Ex. 4.4, and comparison with the ACI Code (318-71) allowable deflections from Table 4.6.[a]

Initial Dead-Load Deflection of Precast Beam (1)	Beam Dead-Load Creep Deflection Up To Time Slab Cast (2)	Beam Dead-Load Creep Deflection After Slab Cast (3)	Shrinkage Deflection Up To Time Slab Cast (4)	Shrinkage Deflection After Slab Cast (5)	Initial Deflection Due To Slab Dead-Load (6)	Creep Deflection Due To Slab Dead-Load (7)	Differential Shrinkage and Creep Deflection (8)	Live-Load Deflection (9)	Total[b]
0.27	*0.15*	*0.05*	*0.07*	*0.07*	*1.23*	*0.37*	*0.15*	*0.36*	*2.72*
6.9	3.8	1.3	1.8	1.8	31.2	9.4	3.8	9.1	69.1

[a]Italicized values are in inches. Other values are in millimeters.

$$\Delta_L = 0.36 \text{ in. } (9.1 \text{ mm})$$

Assuming nonstructural elements are installed after the composite slab has hardened, $\Delta_t + \Delta_L$ = Terms (3) + (5) + (7) + (8) + (9) = 0.05 + 0.07 + 0.37 + 0.15 + 0.36 = 1.00 in. (25.4 mm) should be a reasonable estimation. Hence,

$$0.36 \text{ in. } (9 \text{ mm}) < \ell/180 = (30)(12)/180 = 2.00 \text{ in. } (51 \text{ mm}) \quad \text{OK}$$

$$0.36 \text{ in. } (9 \text{ mm}) < \ell/360 = 1.00 \text{ in. } (25 \text{ mm}) \quad \text{OK}$$

$$1.00 \text{ in. } (25 \text{ mm}) > \ell/480 = 0.75 \text{ in. } (19 \text{ mm}) \quad \text{NG}$$

$$1.00 \text{ in. } (25 \text{ mm}) < \ell/240 = 1.50 \text{ in. } (38 \text{ mm}) \quad \text{OK}$$

Thus, the beam meets three of the four deflection criteria in Table 4.6, but does not meet the requirement for roof or floor construction supporting or attached to nonstructural elements likely to be damaged by large deflections, unless Footnotes b, d, or both are followed (or possibly unless shoring is used). This $\ell/480$ limitation is typically a rather stringent criterion.

[b]It should be noted that the total deflection of 2.72 in. (69 mm) or $\Delta_u/\ell = 2.72/360 = 1/132$ is quite large and may warrant cambering or shoring during construction. With the use of shoring, the relatively large Term 6 is significantly reduced, etc.

$$= \frac{(1)(5/48)(63{,}340 + 18{,}790)(12)(30)^2(12)^2}{(2.74 \times 10^6)(3235)} - 0.271$$

$$= 1.501 - 0.271 = 1.230 \text{ in. } (31.2 \text{ mm})$$

Term 7. *Creep deflection of the composite beam due to slab dead load:*

$$k_r C_{ul}(\Delta_i)_1 \frac{I_2}{I_c} = (0.85)(1.40)(1.230)(0.252)$$

$$= 0.37 \text{ in. } (9.4 \text{ mm})$$

Term 8. *Deflection due to differential shrinkage and creep:*

$$Q_{DS} = (Q \text{ in Eq. B.1})/2 = D_u A_1 E_1/2 \quad \text{(Eq. B.1)}$$

$$= (458)(84 \times 4.5)(3.30)/2$$

$$= 285{,}700 \text{ lb } (1{,}271{,}000 \text{ N})$$

$$y_{cs} = h_c - (y_b)_g - t/2 = 20.5 - 15.21 - 2.25$$

$$= 3.04 \text{ in. } (7.72 \text{ cm}) \quad \text{(Fig. B.1)}$$

$$\Delta_{DS} = \frac{(4/3)K_{DS} Q_{DS} y_{cs} \ell^2}{E_c (I_c)_g} \quad \text{(Eqs. B.7, B.12)}$$

$$= \frac{(4/3)(0.0528)(285{,}700)(3.04)(30)^2(12)^2}{(2.74 \times 10^6)(19{,}050)}$$

$$= 0.15 \text{ in. } (3.8 \text{ mm})$$

Term 9. *Initial or short-time live-load deflection of the composite beam:*

Using the midspan $(I_c)_{D+L} = 11{,}330$ in.^{4} (471,600 cm^4) (Table 4.12)

$$\Delta_L = \frac{K(5/48)M_L \ell^2}{E_c (I_c)_{D+L}}$$

$$= \frac{(0.925)(5/48)(81{,}820)(12)(30)^2(12)^2}{(2.74 \times 10^6)(11{,}330)}$$

$$= 0.395 \text{ in. } (10.0 \text{ mm})$$

Using the weighted average $(I_c)_{D+L} = 12{,}490$ in.^{4} (519,800 cm^4) (Table 4.12),

$$\Delta_L = (0.395)(11{,}330/12{,}490)$$

$$= 0.36 \text{ in. } (9.1 \text{ mm}).$$

The difference in the two solutions is small.

Total ultimate deflection by Eq. 4.31, which is also summarized in Table 4.13:

$$\Delta_u = \overset{(1)}{0.27} + \overset{(2)}{0.15} + \overset{(3)}{0.05} + \overset{(4)}{0.07} + \overset{(5)}{0.07} + \overset{(6)}{1.23} + \overset{(7)}{0.37} + \overset{(8)}{0.15} + \overset{(9)}{0.36}$$

$$= \underline{2.72 \text{ in. } (69.1 \text{ mm})}$$

Solution by the Simplified Eq. 4.31b

$$\Delta_u = \left(1.56 + 0.77 \frac{I_2}{I_c}\right)(\Delta_i)_2 + \left(0.19 + 0.81 \frac{I_2}{I_c}\right)(362 \times 10^{-6})K_{sh}A_{sh}\ell^2/h_p + \left(1.00 + 1.22 \frac{I_2}{I_c}\right)(\Delta_i)_1 + \Delta_{DS} + \Delta_L + (\Delta_{cp})_L \quad \text{(Eq. 4.31b)}$$

$$= (1.56 + 0.77 \times 0.252)(0.271) + (0.19 + 0.81 \times 0.252)(362 \times 10^{-6})(1/8)(1.00)(30)^2(12)^2/16 + (1.00 + 1.22 \times 0.252)(1.230) + 0.15 + 0.36 + 0 = \underline{2.74 \text{ in. } (69.6 \text{ mm})}$$

vs. 2.72 in. (69.1 mm) by the more general Eq. 4.31. The difference in the two solutions in this example was due to small differences in the creep and shrinkage factors used, as a result of differences in humidity and the fact that the creep and shrinkage size correction factors are not used in the simplified Eq. 4.31b (see values preceding Eq. 4.31a and in Table 4.11).

Solution by the rough approximate method (Eq. 4.33)

$$\Delta_D = [(\Delta_i)_1 + (\Delta_i)_2] = 1.501 \text{ in. } (38.1 \text{ mm})$$

(from Term 6, solution by Eq. 4.31)

$$(k_r T)\Delta_D(I_2/I_c) = (3.5)(1.501)(0.252)$$

$$= 1.32 \text{ in. } (33.5 \text{ mm})$$

$$\Delta_L = 0.40 \text{ in. } (10.2 \text{ mm})$$

(from Term 9, solution by Eq. 4.31 using the positive region I_e)

Total ultimate deflection by Eq. 4.33:

$$\Delta_u = \Delta_D + (k_r T)\Delta_D(I_2/I_c) + \Delta_L \quad \text{(Eq. 4.33)}$$

$$= 1.50 + 1.32 + 0.40 = \underline{3.22 \text{ in. } (82 \text{ mm})}$$

versus 2.72 in. (69.1 mm) by the more general Eq. 4.31, and 2.74 in. (69.6 mm) by the simplified Eq. 4.31b.

5

Prestressed concrete composite flexural members

DAN E. BRANSON, Ph.D., P.E.
Professor of Civil Engineering
University of Iowa
Iowa City, Iowa

5.1 INTRODUCTION

In this chapter, prestressed concrete composite flexural members are considered in detail. See Section 4.1 for the introduction to both reinforced and prestressed composite members. The references, notation, and appendixes for Chapters 4 and 5 are placed at the end of this chapter.

5.2 CAST-IN-PLACE SLAB ON PRESTRESSED CONCRETE PRECAST BEAM

5.2.1 Design of Beams and Girders

The consideration in this chapter will be limited to steam-cured pretensioned precast simple-span beams with both simple and continuous cast-in-place slabs or toppings using unshored and shored construction. As previously discussed, it will be assumed in the case of shored construction that the precast-beam supports its own dead load and the composite beam supports all other loading. In all cases, the gross-section properties will be used, as shown in Fig. B.1, although the effect of computed stresses above the cracking stress (e.g., up to the ACI Code limit of $12\sqrt{f_c'}$) on deflections will be included.

The subjects of span-depth ratios, loss of prestress, flexural stress, flexural strength, web shear, and horizontal shear are included. Also presented are sections on additional design considerations, the effect of continuity and design of negative-moment region joints, and camber and deflection control. Both design and camber-deflection examples are included for a simple-span composite beam and a four-span continuous composite girder. The effects of differential shrinkage and creep on both stresses and deflections are also summarized in Appendix B and included in the two examples for computing loss of prestress and camber.

Subscripts 1 and 2 are used to refer to the slab (or effect of the slab, such as under slab dead load) and precast beam, respectively, as shown in Fig. B.1.

5.2.1.1 Span-depth ratios

The approximate span-depth ratio limits as given by Lin[9] for side-by-side single and double tees are for simple spans (36 for floors and 40 for roofs) and for continuous spans (40 for floors and 44 for roofs). For composite beams, the total depth may be considered in computing these span-depth ratios. These ratios should be reduced by about 5% for lightweight concrete. According to Lin, for simple-span highway bridges of the I-beam type, a span-depth ratio for normal-weight concrete of 20 might be considered conservative, 22 to 24 normal, and 26 to 28 the approximate upper limit.

5.2.1.2 Loss of prestress

The method used in this section is based on the procedure of Branson[17,36,39,40,41] and includes the effect of composite action, differential shrinkage and creep, and the use of different-weight concrete. See Ref. 17 for procedures to include the effect of nonprestressed tension steel on both steel and concrete stresses. Equations 5.1 through 5.4 are given for the ultimate loss of prestress for both unshored and shored composite beams in general and simplified forms.

General equation

$$(\Delta f_s)_u = \overbrace{(nf_{ci})}^{(1)} + \overbrace{(nf_{ci})(\alpha_s C_u)\left(1 - \frac{\Delta P_s}{2P_o}\right)}^{(2)}$$

$$+ \overbrace{(nf_{ci})C_u\left[\left(1 - \frac{\Delta P_u}{2P_o}\right) - \alpha_s\left(1 - \frac{\Delta P_s}{2P_o}\right)\right]\frac{I_2}{I_c}}^{(3)}$$

$$+ \overbrace{(\epsilon_{sh})_u E_s/(1 + b_{11})_u}^{(4)} + \overbrace{(\Delta f_{sr})_u}^{(5)} - \overbrace{(mf_{cs})}^{(6)}$$

$$- \overbrace{[(mf_{cs})(\beta_s C_u)/(1 + b_{11})_u]\frac{I_2}{I_c}}^{(7)} - \overbrace{(\Delta f_{sds})_u}^{(8)} \quad (5.1)$$

Term 1 is the *prestress loss due to elastic shortening*; $f_{ci} = \frac{P_o}{A_2} + \frac{P_o e^2}{I_2} - \frac{M_2 e}{I_2}$; and n is the modular ratio at the time of prestressing (see Table 5.1 for typical values). The gross-section properties are used herein, and A_2 and I_2 refer to the precast-beam section. Trial $P_o = P_i -$ (say $2)(P_o/A_2E_{ci})A_{ps}E_s = P_i/(1 + 2n\rho_{ps})$ may be used, where $\rho_{ps} = A_{ps}/A_2$. P_i is the tensioning or jacking force, and P_o is the prestress force after transfer (after elastic loss). M_2 is the precast-beam dead-load moment, and e is the steel eccentricity.

Term 2 is the *prestress loss due to concrete creep up to the time of slab casting.* Also, nf_{ci} is the same as in Term 1. As defined in Table 5.2, α_s is the ratio of the creep coefficient (defined as *ratio of creep strain to initial strain*) up to the time of slab casting to the ultimate creep coefficient. The average ultimate creep coefficient C_u, based on the concrete age of 1 day for steam-cured precast beams, is given in Appendix A (for example, see Table A.8 plus the size-correction factor for the precast beam in Table A.6 or A.7). A typical value is C_u = 1.88 in Table A.8 when H = 70%. The expressions, $\left(1 - \frac{\Delta P_s}{2P_o}\right)$ and $\left(1 - \frac{\Delta P_u}{2P_o}\right)$, in Terms 2 and 3 are used to approximate the creep effect resulting from the variable stress history under the decreasing prestress force. See Table 5.3 for approximate values of $\Delta P_s/P_o$ and $\Delta P_u/P_o$, where ΔP_s and ΔP_u are defined as the time-dependent prestress loss (total minus initial or elastic loss) up to the time of slab casting and ultimate value, respectively.

Term 3 is the *prestress loss due to concrete creep (under prestress and precast-beam dead load) following slab casting.* The ratio, I_2/I_c, modifies the initial stress (strain) and accounts for the effect of the composite section in re-

Table 5.1. Modular ratios for steam-cured concrete.

Modular Ratios and Time Conditions		Normal Weight (w = 145 pcf, 2320 kg/m^3)		Sand-Lightweight (w = 120 pcf, 1920 kg/m^3)		All-Lightweight (w = 100 pcf, 1600 kg/m^3)	
		[a]3.5	[b]4.0	[a]3.5	[b]4.0	[a]3.5	[b]4.0
n at release of prestress		8.1	7.7	11.2	10.6	14.1	13.4
m for indicated concrete age (such as age of precast-beam concrete when slab is cast)	3 weeks	6.8	6.7	9.3	9.2	11.8	11.7
	1 month	6.7	6.7	9.3	9.2	11.8	11.6
	2 months	6.7	6.7	9.3	9.2	11.8	11.6
	3 months	6.7	6.6	9.3	9.2	11.8	11.6
	4 months	6.7	6.6	9.3	9.2	11.8	11.6

[a]Based on f'_{ci} = 3.5 ksi (24.1 N/mm^2) for steam-cured concrete prestressed at age 1 day, up to 4 month f'_c = 7100 psi (49.0 N/mm^2) (using Eq. A.3); on E_c in Eq. A.9; and on E_s = 28 × 10^6 psi (193 kN/mm^2).
[b]Based on f'_{ci} = 4.0 ksi (27.6 N/mm^2) for steam-cured concrete prestressed at age 1 day, up to 4 month f'_c = 8100 psi (55.8 k mm^2) (using Eq. A.3); on E_c in Eq. A.9; and on E_s = 28 × 10^6 psi (193 kN/mm^2).

Table 5.2. Values of α_s and β_s for steam-cured concrete.

Time Between Prestressing and Slab Casting	[a] α_s	[b] β_s
3 weeks	0.38	0.85
1 month	0.44	0.83
2 months	0.54	0.76
3 months	0.60	0.74
4 months	0.64	0.71

[a] α_s is the ratio of the precast-beam creep that takes place up to the time of slab casting to the total creep (same as in Table 4.4). From Eq. A.10: $\alpha_s = C_t/C_u = t^{0.60}/(10 + t^{0.60})$, where t = time in days between prestressing and slab casting.

[b] β_s is the creep-correction factor for the effect of loading age (same as in Table 4.4). From Eq. A.14: $\beta_s = 1.13 t_{LA}^{-0.095}$, where t_{LA} = precast-beam concrete age when the slab is cast (1-day difference between t and t_{LA} is neglected in the table).

straining additional creep curvature (strain) after the composite beam becomes effective. The other parameters are the same as described in Terms 1 and 2.

Term 4 is the *prestress loss due to shrinkage of the precast-beam concrete.* The average ultimate shrinkage strain $(\epsilon_{sh})_u$, based on the concrete age of 1 day for steam-cured precast beams, is given in Appendix A (see Table A.8 plus the size-correction factor for the precast beam in Table A.6 and A.7, for example). A typical value is $(\epsilon_{sh})_u = 546 \times 10^{-6}$ in./in. (mm/mm) in Table A.8 when $H = 70\%$. E_s is the modulus of elasticity of the prestressing steel (see Appendix A). The parameter $(b_{11})_u = n(\rho_{ps})(1 + e^2/r^2)(1 + \eta C_u)$, where $n(\rho_{ps})(1 + e^2/r^2)$ accounts for the reduction in the free shrinkage due to the stiffening effect of the steel, $(1 + \eta C_u)$ is the creep effect, and η is a relaxation coefficient. As a design simplification, $(1 + b_{11})_u = 1.25$ will be used herein.

Term 5 is the *prestress loss due to steel relaxation.* As a matter of consistency with the accuracy of creep and shrinkage loss prediction, the author recommends the use of $(\Delta f_{sr})_u = 0.075 f_{si}$ for stress relieved strand and $(\Delta f_{sr})_u = 0.035 f_{si}$ for stabilized (low-relaxation) strand in most cases, where f_{si} is the initial tensioning stress in the steel. For a discussion of this method and that of more accurate formulas, see Reference 17.

Term 6 is the *elastic prestress gain due to slab dead load (plus other loads applied at the time of slab casting or soon thereafter, such as due to diaphragms, sidewalks, partitions, ceilings, roofing, etc.).* The variable m is the modular ratio of the precast-beam concrete at the time of slab casting (see Table 5.1 for typical values); $f_{cs} = M_1 e/I$, where M_1 is the slab dead-load moment (plus other loads mentioned previously); and e, I, refer to the precast-beam section for unshored construction and the composite beam section for shored construction.

Term 7 is the *prestressed gain due to creep under slab (plus other loads mentioned previously) dead load.* (mf_{cs}) is the same as in Term 6. As defined in Table 5.2, β_s is the creep-correction factor for the age of the precast-

Table 5.3. Typical loss of prestress ratios for different-weight concrete.[a]

Prestress Loss Ratio	Normal Weight	Sand-Lightweight	All-Lightweight
$\Delta P_s/P_o$–for 3 weeks to 1 month between prestressing and slab casting	0.10	0.12	0.14
$\Delta P_s/P_o$–for 2 to 4 months between prestressing and slab casting	0.14	0.16	0.18
$\Delta P_u/P_o$	0.18	0.21	0.23

[a] The values are different for different-weight concrete because of the initial strains (due to E_c) for normal stress levels. These ratios refer only to the time-dependent (total minus elastic loss) loss of prestress.

beam concrete when the slab is cast. C_u is the same as in Term 2, $(1 + b_{11})_u$ is the same as in Term 4, and I_2/I_c is the same as in Term 3. For shored construction, delete I_2/I_c, since the composite I (or I_c) is already used in Term 6.

Term 8 is the *prestress gain due to differential shrinkage and creep.* As a design approximation. $(\Delta f_{sds})_u = 0.015 f_{si}$ will be used herein. This is discussed in detail by Branson.[17]

The effect of creep under any sustained loads (such as partial live load) applied to the composite beam and not included in Term 7 would be to contribute to the gain in prestress, and is not included in Eq. 5.1.

Other equations used herein include $(\Delta P) = (\Delta f_s) A_{ps}$ and $PL = (\Delta f_s)(100/f_{si}$, where (ΔP) is the loss of prestress force and PL is the loss of prestress in percent.

It can be seen by inspection that Eq. 5.1 for composite beams reduces to the basic equation for noncomposite prestressed beams of

$$(\Delta f_s)_u = (nf_{ci}) + (nf_{ci}) C_u \left(1 - \frac{\Delta P_u}{2P_o}\right) + (\epsilon_{sh})_u E_s/(1 + b_{11})_u + (f_{sr})_u$$

when $I_2/I_c = 1$ and Terms 6 through 8—which refer to the composite beam—are deleted. Note that Term 2 and the second part of Term 3 cancel.

The loss of prestress at the time of erection (e.g., taken either just before or just after slab casting) can be easily estimated in Eq. 5.1 and in equations similar to the simplified Eqs. 5.2 through 5.4. In this case, the steel relaxation might be taken as 55% to 65% of the ultimate value for erection times of 1 and 2 months, respectively, according to the usual log-time relation.

As discussed in Reference 17, the effect of a continuous composite slab normally has a negligible effect on the loss of prestress.

Equation 5.1 reduces to the simplified Eqs. 5.2 through 5.4 for stress relieved strand when the following representative parameters are used: $C_u = 1.88$ and $(\epsilon_{sh})_u = 546 \times 10^{-6}$ in./in. (mm/mm) from Table A.8 for $H = 70\%$ and neglecting the size-correction factors in Table A.6 or A.7. $(1 + b_{11})_u = 1.25$. $E_s = 28 \times 10^6$ psi (193 kN/mm^2), $f_{si} = 189{,}000$ psi (1300 N/mm^2). Assuming a 2-month period between prestressing and slab casting: from Table 5.2, $\alpha_s = 0.54$ and $\beta_s = 0.76$. From Tables 5.1 through 5.3: for normal-weight concrete, $n = 7.9$, $m = 6.7$, $\Delta P_s/P_o = 0.14$, and $\Delta P_u/P_o = 0.18$; for sand-lightweight concrete, $n = 10.9$, $m = 9.3$, $\Delta P_s/P_o = 0.16$, and $\Delta P_u/P_o = 0.21$; and for all-lightweight concrete, $n = 13.8$, $m = 11.7$, $\Delta P_s/P_o = 0.18$, and $\Delta P_u/P_o = 0.23$.

Simplified equation for normal-weight concrete ($w = 145$ pcf $= 2320$ kg/m^3)

$$\begin{aligned}(\Delta f_s)_u &= \overbrace{7.9 f_{ci}}^{(1)} + \overbrace{(7.9 f_{ci})(0.54 \times 1.88)(1 - 0.070)}^{(2)} \\ &+ \overbrace{(7.9 f_{ci})(1.88)[(1 - 0.090) - (0.54)(1 - 0.070)]\, I_2/I_c}^{(3)} + \overbrace{(546)(28)/1.25}^{(4)} \\ &+ \overbrace{(0.075)(189{,}000)}^{(5)} - \overbrace{6.7 f_{cs}}^{(6)} - \overbrace{[(6.7 f_{cs})(0.76 \times 1.88)/1.25]\, I_2/I_c}^{(7)} \\ &- \overbrace{(0.015)(189{,}000)}^{(8)} = \overbrace{\left(15.36 + 6.06 \frac{I_2}{I_c}\right) f_{ci}}^{(1+2+3)} \\ &+ \overbrace{23{,}570 \text{ psi } (163 \text{ N/mm}^2)}^{(4+5+8)} - \overbrace{\left(6.7 + 7.66 \frac{I_2}{I_c}\right) f_{cs}}^{(6+7)}\end{aligned} \tag{5.2}$$

Simplified equation for sand-lightweight concrete ($w = 120$ pcf $= 1920$ kg/m^3)

$$(\Delta f_s)_u = \left(19.53 + 7.56 \frac{I_2}{I_c}\right) f_{ci} + 23{,}570 \text{ psi } (163 \text{ N/mm}^2) - \left(9.3 + 10.63 \frac{I_2}{I_c}\right) f_{cs} \tag{5.3}$$

Simplified equation for all-lightweight concrete (w = 100 pcf = 1600 kg/m^3)

$$(\Delta f_s)_u = \left(26.55 + 10.22\frac{I_2}{I_c}\right)f_{ci} + 23{,}570 \text{ psi } (163 \text{ N/mm}^2) - \left(11.7 + 13.37\frac{I_2}{I_c}\right)f_{cs} \quad (5.4)$$

From the PCI Handbook[4] using 2 in. (5.1 cm) topping, I_2/I_c = 0.34 to 0.64 for normal weight 4 in. (10 cm) to 12 in. (30 cm) hollow-core slabs, 0.68 to 0.77 for normal-weight and lightweight 12 in. (30 cm) to 32 in. (81 cm) double tees, and 0.76 to 0.85 for normal weight and lightweight 24 in. (61 cm) to 36 in. (91 cm) single tees. Typical values of I_2/I_c for bridge girders with composite slabs are 0.30 to 0.55. In all of these cases, the smaller ratios apply for the shallower members and the larger ratios for the deeper members.

Various other simplified methods for computing loss of prestress may be found in the *PCI Design Handbook*,[4] AASHTO,[6] and PCI.[31] These are summarized by Branson in Ref. 17.

Examples of simplified results

For f_{ci} = 1400 psi (9.65 N/mm^2), I_2/I_c = 0.75, f_{cs} = 500 psi (3.45 N/mm^2):

Normal weight (Eq. 4.36)

$(\Delta f_s)_u$ = 45,220 psi (312 N/mm^2)/1890 = 23.9%

Sand-lightweight (Eq. 4.37)

$(\Delta f_s)_u$ = 50,220 psi (346 N/mm^2)/1890 = 26.6%

All-lightweight (Eq. 4.38)

$(\Delta f_s)_u$ = 60,610 psi (418 N/mm^2)/1890 = 32.1%

5.2.1.3 Flexural stress

The gross-section properties are used herein, with the composite section properties shown in Fig. B.1. Concrete stresses are computed at transfer by Eq. 5.5 or 5.6, and after losses with live load at positive-moment locations by Eq. 5.7 or 5.8 for both unshored and shored construction. The stress-distribution diagrams for each term and the total in Eq. 5.7 are shown in Fig. 5.1. The flexural strength method will be used for the negative-moment design under live load (plus other loads applied to the composite beam) for unshored construction, and under slab dead load (plus diaphragm, sidewalk, partition, roofing, etc., dead load) plus live load for shored construction.

Concrete stresses at transfer

$$f = \frac{P_o}{A_2} \mp \frac{P_o e y}{I_2} \pm \frac{M_2 y}{I_2} \quad (5.5)$$

$$f_{t,b} = \frac{P_o}{A_2} \mp \frac{P_o e}{(Z_2)_{t,b}} \pm \frac{M_2}{(Z_2)_{t,b}} \quad (5.6)$$

The signs for the $P_o e y/I_2$ or $P_o e/(Z_2)_{t,b}$ term are reversed when the steel cgs is above the precast-beam centroidal axis.

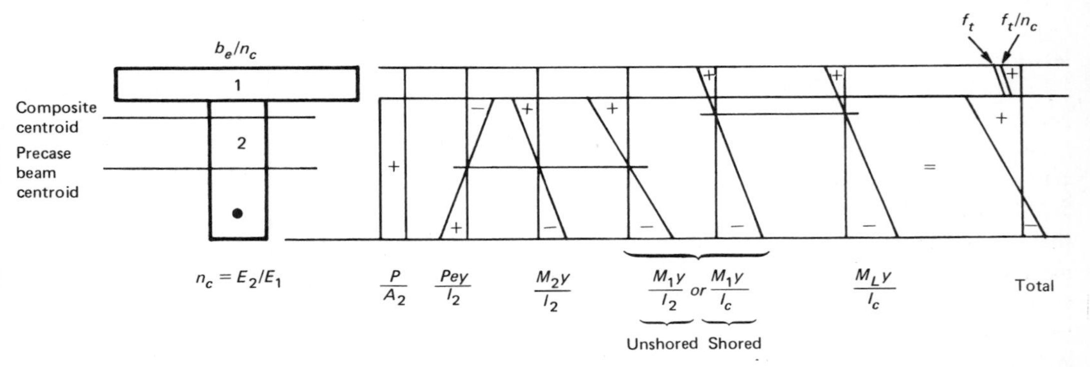

Fig. 5.1. Typical concrete stress distribution after losses with live load at positive-moment locations for unshored and shored composite prestressed concrete beams, by Eq. 4.41.

Concrete stresses after losses at positive-moment locations

$$f = \frac{P}{A_2} \mp \frac{Pey}{I_2} \pm \frac{M_2 y}{I_2} \pm [M_1(y/I)_2 \text{ for unshored case, or } M_1(y/I)_c \text{ for shored case}] \pm M_L(y/I)_c \quad (5.7)$$

$$f_{t,b} = \frac{P}{A_2} \mp \frac{Pe}{(Z_2)_{t,b}} \pm \frac{M_2}{(Z_2)_{t,b}} \pm [M_1/(Z_2)_{t,b} \text{ for unshored case, or } M_1/(Z_c)_{t,b} \text{ for shored case}] \pm M_L/(Z_c)_{t,b} \quad (5.8)$$

where the subscripts t and b refer to the top and bottom fibers, respectively; M_1 = slab (plus diaphragm, sidewalk, partition, ceiling, roofing, etc., when placed at the time the slab is cast or soon thereafter) dead-load moment; M_2 = precast-beam dead load moment; and M_L = live load (plus other loads applied to the composite beam and not included in M_1) moment.

It should be noted in Eqs. 5.7 and 5.8 for the top fiber stress that the terms with A_2, I_2, and Z_2 refer to the top of the precast section, and the terms with I_c and Z_c refer to the top of the composite section, with the latter stress divided by $n_c = E_2/E_1$ to convert back to the slab concrete stress (as shown in Fig. 5.1).

The critical sections for checking stresses are the ends and midspan at transfer, the midspan for straight and 2-point depressed strands after losses with live load, near the 0.4-point of the span or at midspan for 1-point depressed strands after losses with live load. The allowable stresses by the ACI Code and AASHTO are used in the examples herein.

5.2.1.4 *Flexural strength*

The method summarized in this section is used in the ACI Code and Commentary and the AASHTO Specifications (see Section 4.2.2). The load factors for gravity loading are given in Eq. 4.9 by the ACI Code and Eq. 5.9 by AASHTO for gravity loads not less than H 20.

$$\text{Design } M_u = 1.30(M_D + \tfrac{5}{3} M_{L+I}) \quad (5.9)$$

The following approach for underreinforced beams is the same as in Fig. 4.11 for positive-moment regions, with f_{ps} (calculated stress in prestressing steel at ultimate load) being used instead of f_y. In lieu of a strain compatibility analysis, Eq. 5.10 may be used to approximate f_{ps} for pretensioned members, provided that f_{se} (effective stress in the prestressing steel, after losses) is not less than $0.5 f_{pu}$.

$$f_{ps} = f_{pu}\left(1 - \frac{\rho_p f_{pu}}{2f_c'}\right) \quad (5.10)$$

where

f_{pu} = ultimate strength of prestressing steel = 270 ksi (1860 N/mm^2) for 270 K steel, etc.

$\rho_p = A_{ps}/bd$

A_{ps} = area of prestressing steel.

The procedure in Fig. 4.12 for negative-moment regions is also used for continuous composite prestressed beams. The contribution of compression steel to the flexural strength is neglected herein.

Underreinforced beams—ω_p, $(\omega + \omega_p)$, or $(\omega_w + \omega_{pw}) \leqslant 0.30$.

For rectangular sections, or flanged sections in which the neutral axis lies within the flange (when the flange thickness is greater than or equal to $1.4 d\rho_p f_{ps}/f_c'$, which is normally the case for composite beams):

$$M_u/\phi = A_{ps} f_{ps}(d - a/2) = A_{ps} f_{ps} d\left(1 - \frac{\rho_p f_{ps}}{1.7f_c'}\right) \quad (5.11)$$

where $\phi = 0.90$ by ACI and 1.0 by AASHTO, and the ratio 1/1.7 is rounded to 0.59 by ACI and 0.6 by AASHTO.

For flanged sections in which the neutral axis falls outside the flange:

$$M_u/\phi = A_{pw} f_{ps} d\left(1 - \frac{A_{pw} f_{ps}}{1.7bdf_c'}\right) + 0.85 f_c'(b - b')t(d - t/2) \quad (5.12)$$

where

$A_{pw} = A_{ps} - A_{pf}$

$A_{pf} = 0.85 f_c'(b - b')t/f_{ps}$.

A_{pf} and A_{pw} are those portions of the prestressing steel required to develop the compressive strengths of the overhanging flanges and the web, respectively.

Nonprestressed ordinary reinforced (mild) steel when used in combination with prestressed steel may be considered to contribute to the tensile force in a member at ultimate moment an amount equal to its area times its yield strength. For other types of nonprestressed reinforcement, a strain compatibility analysis is required (see PCA,[19] pp. 23-10, for an example using both prestressed and nonprestressed tension steel).

Overreinforced beams—ω_p, $(\omega + \omega_p)$, or $(\omega_w + \omega_{pw}) > 0.30$.

For rectangular sections, or flanged sections in which the neutral axis lies within the flange (using $\rho_p f_{ps}/f_c' = \rho_p f_{pu}/f_c' = 0.30$):

$$M_u/\phi = A_{ps} f_{ps} d\left(1 - \frac{\rho_p f_{ps}}{1.7 f_c'}\right)$$

$$= \rho_p b d f_{ps} d\left(1 - \frac{\rho_p f_{ps}}{1.7 f_c'}\right)$$

$$= f_c' b d^2 \left(\rho_p \frac{f_{ps}}{f_c'}\right)\left(1 - \frac{\rho_p f_{ps}}{1.7 f_c'}\right)$$

$$= f_c' b d^2 (0.30)(1 - 0.30/1.7)$$

$$= 0.25 f_c' b d^2 \quad (5.13)$$

For flanged sections in which the neutral axis falls outside the flange:

$$M_u/\phi = 0.25 f_c' b' d^2 + 0.85 f_c' (b - b') t (d - t/2) \quad (5.14)$$

Cracking moment

As a precaution against abrupt flexural failure resulting from rupture of the prestressing steel when failure occurs immediately after cracking, it is required by ACI and AASHTO that a minimum steel content (nonprestressed and prestressed) be provided so that the ultimate moment is at least 1.2 times the cracking moment. For the purpose of checking this requirement, use $M_u/M_{cr} \geqslant 1.2$, where M_{cr} is computed by Eq. 5.15 for both unshored and shored composite beams. See Section 5.2.2.2 for M_{cr} in deflection calculations. The tensile stress at the extreme fiber is

$$-f_r = \frac{P}{A_2} + \frac{Pe_2}{(Z_2)_b} - \frac{M_{cr}}{(Z_c)_b}$$

Solving,

$$M_{cr} = \frac{P(Z_c)_b}{A_2} + \frac{Pe_2(Z_c)_b}{(Z_2)_b} + f_r(Z_c)_b \quad (5.15)$$

5.2.1.5 Web shear

The same ACI Code strength method given by Eqs. 4.12 through 4.17 in Section 4.2.2.2 for reinforced members applies also to prestressed members, except as follows: For prestressed members, v_c (shear allowed to be carried by the concrete) is based on one of two types of failure: one in which flexural cracks develop into inclined cracks or v_{ci}, and the other in which failure results from cracking that initiates in the web or v_{cw}. These parameters are explained further elsewhere,[14, 16, 19] and only the ACI Code simplified method given by Eq. 5.16 is used in this chapter.

For members with f_{se} at least equal to $0.40 f_{pu}$:

$$v_c = 0.6\sqrt{f_c'} + 700 V_u d/M_u \quad (5.16)$$

but v_c need not be taken less than $2\sqrt{f_c'}$, nor shall v_c be greater than $5\sqrt{f_c'}$. As shown by MacGregor and Hanson,[32] and Wang and Salmon,[14] Eq. 5.16 provides a linear variation between the limits of 2 and $5\sqrt{f_c'}$. In Eq. 5.16, V_u is the maximum design shear (including load factors) at a section, and M_u is the simultaneously occurring design moment; the ratio $V_u d/M_u$ is not to exceed 1.0, and d is the distance from the extreme compressive fiber to the centroid of the prestressing steel at the section being checked.

Alternatively, for prestressed members with f_{se} at least equal to $0.40 f_{pu}$, Eq. 5.17 may be used instead of Eq. 4.14.

$$\frac{A_v}{s} = \frac{A_{ps} f_{pu}}{80 f_y d}\sqrt{\frac{d}{b_w}} \quad (5.17)$$

Also, use Eq. 5.18 for prestressed members instead of Eqs. 4.15 and 4.16.

$$s_{max} = 3h/4 \quad (5.18)$$

For many deck sections in buildings (hollow-core, double tees, etc.), only minimum or no shear reinforcement is required.

The AASHTO method for prestressed members given by Eq. 5.19 is similar to that for reinforced members by Eq. 4.13, except that the required spacing of web reinforcement for prestressed members is allowed to be twice that for reinforced members. This is based on experimental results and on experience with prestressed members having an improved shear capacity due to the precompressed concrete, and has been found to be safe for most prestressed members, except that this spacing should be decreased as the member approaches the condition of the reinforced member.

$$s = 2A_v f_y jd/(V_u - V_c) \qquad (5.19)$$

where f_y shall not exceed 60,000 psi (414 N/mm^2), and $V_c = 0.06 f'_c b_w jd$ but not more than $180 b_w jd$ (which will apply to prestressed members). Normally, use $jd = d - \rho_p f_{ps} d/1.7 f'_c$ in positive-moment regions, and $jd = d - a/2$ where a = (Slab A_s)f_y/0.85(Stem f'_c)$_b$, in negative-moment regions.

$$s_{max} = A_v f_y / 100 b_w \qquad (5.20)$$

$$s_{max} = 3d/4 \qquad (5.21)$$

The ACI Code method is also acceptable by AASHTO, provided that Eq. 5.20 is also satisfied.

The critical sections for shear in simply supported beams will usually not be near the ends of the span where the shear is maximum, but will be at some point away from the ends in a region of high moment.

For the design of web reinforcement in simply supported members carrying moving loads, it is recommended that shear be investigated only in the middle half of the span length. The web steel required at the quarter points should be used throughout the outer quarters of the span.

For continuous bridges whose individual spans consist of precast prestressed girders, web reinforcement shall be designed for the full length of interior spans and for the interior three-fourths of the exterior spans.

The load factors for shear are the same as for flexure in Eq. 5.9, and $\phi = 0.90$ for shear, according to AASHTO. Use $V_u = \frac{1.30}{\phi}\left(V_D + \frac{5}{3} V_{L+I}\right)$ with the equations for shear herein.

5.2.1.6 *Horizontal shear*

The ACI Code method for the horizontal shear design of reinforced composite beams in Section 4.2.2.3 also applies to prestressed composite beams. Essentially the same method is also recommended by AASHTO, with the following modifications:

When computing horizontal shear, use Eq. 5.22 and these allowable stresses:

$$v = V_u Q / I_c b_v \qquad (5.22)$$

The critical section locations will be assumed to be the same as for web shear in Section 5.2.1.5.

When the minimum steel tie requirements are met—75 psi (0.52 N/mm^2)

When the minimum steel tie requirements are met and the contact surfaces are clean and artificially roughened—300 psi (2.07 N/mm^2)

In addition to the preceding values, for each percent of stirrup of vertical tie reinforcement crossing the joint in excess of the minimum requirements—150 psi (1.03 N/mm^2)

Minimum tie requirements

All web reinforcement shall extend into cast-in-place decks. The minimum total area of vertical ties per linear foot of span shall be not less than the area of two No. 3 (10-mm) bars spaced at 12 in. (30 cm), or 0.22 $in.^2$/ft (4.66 cm^2/m). Web reinforcement may be used to satisfy the vertical tie requirement. The spacing of vertical ties shall not be greater than four times the average thickness of the composite flange and in no case greater than 24 in. (61 cm).

5.2.1.7 *Additional design considerations*

A brief summary of the principal design considerations required for prestressed concrete composite beams, in addition to those already described, is given in this section.

Nonprestressed tension reinforcement in top of precast beams at transfer.

When tension in the top of precast beams at transfer exceeds the allowable value at the extreme fiber, nonprestressed tension steel may be used to resist the total tension force in the concrete computed on the assumption of an uncracked section. When applying this method, the ACI Code does not state an upper limit for the computed concrete stress, and AASHTO has a maximum value for this stress of $7.5\sqrt{f'_{ci}}$.

Development length of prestressing strands

The following provision is given by both the ACI Code and AASHTO: Three- or seven-wire pretensioning strand shall be bonded beyond the critical section for a development length (in inches) not less than

$$(f_{ps} - \tfrac{2}{3} f_{se})d_b \qquad (5.23)$$

where d_b is the nominal diameter in inches, f_{ps} and f_{se} are expressed in kips/in.^{2}, and the expression in the parentheses is used as a constant without units. Equation 5.23 is an empirical relationship established by tests of Hanson and Kaar.[33]

Investigation may be limited to those cross sections nearest each end of the member that are required to develop their full strength under the specified design load. Where strand is debonded at the end of a member, the development length required above shall be doubled. Development length is usually not a problem for all but very short spans.

End blocks

Reinforcement must be provided when required in the end regions to resist bursting, horizontal splitting, and spalling forces induced by the tendons. Such reinforcement is usually not required for precast building units. However, the following provision is given by AASHTO: In pretensioned beams, vertical stirrups acting at a unit stress of 20,000 psi (138 N/mm^2) to resist at least 4% of the total prestressing force shall be placed within the distance of $d/4$ of the end of the beam, the end stirrups to be as close to the end of the beam as practicable. For at least the distance d from the end of the beam, nominal reinforcement shall be placed to enclose the prestressing steel in the bottom flange.

Spacing and cover of reinforcement

ACI Code. The clear distance between pretensioning steel at each end of the member shall be not less than three times the diameter of strands. Closer vertical spacing may be permitted in the middle portion of the span. The following minimum concrete cover shall be provided for prestressing strands:

Exposed to earth or weather:
- Joists — 1 in. (2.5 cm)
- Other members — $\frac{3}{2}$ in. (3.8 cm)

Not exposed to weather or in contact with the ground:
- Joists — $\frac{3}{4}$ in. (1.9 cm)
- Beams, girders (principal steel) — $\frac{3}{2}$ in. (3.8 cm)
- Beams, girders (ties, stirrups) — 1 in. (2.5 cm)

The cover for nonprestressed steel in prestressed members under plant control may be that given for precast members in Section 4.2.3.6.

AASHTO. The minimum clear spacing of pretensioning steel at the ends of beams shall be three times the diameter of the steel or $\frac{4}{3}$ times the maximum size aggregate, whichever is greater. The following minimum concrete cover shall be provided for prestressing and conventional steel:

Prestressing steel and
- main reinforcement — $\frac{3}{2}$ in. (3.8 cm)

Slab reinforcement
- top of slab — $\frac{3}{2}$ in. (3.8 cm)
- top of slab when deicers are used — 2 in. (5.1 cm)
- bottom of slab — 1 in. (2.5 cm)

Stirrups and ties — 1 in. (2.5 cm)

When deicer chemicals are used, drainage details shall dispose of deicer solutions without constant contact with the prestressed girders. Where such contact cannot be avoided, or in locations where members are exposed to salt water, salt spray, or chemical vapor, additional cover should be provided.

5.2.1.8 Effect of continuity

The following provisions of AASHTO are given for bridges composed of simple-span precast prestressed girders made continuous by the deck slab. The principals may also be applied to other types of continuous composite prestressed structures.

1. *General.* When structural continuity is assumed in calculating live loads plus impact and composite dead-load moments, the effects of creep and shrinkage shall be considered in the design of bridges incorporating simple-span precast, prestressed girders, and deck slabs continuous over two or more spans.

2. *Positive-moment connection at piers.* Provision shall be made in the design for the positive moments that may develop in the negative-moment region due to the combined effects of creep and shrinkage in the girders and deck slab, and due to the effects of live load plus impact in remote spans. Shrinkage and elastic shortening of the piers shall be considered when significant. Nonprestressed positive-moment connection reinforcement at piers may be designed at a working stress of $0.6 f_y$, but not to exceed 38 ksi (248 N/mm^2).

3. *Negative moments.* Negative-moment reinforcement shall be proportioned by ultimate-strength design. The effect of initial precompression due to prestress in the girders may be neglected in the negative-moment calculation of ultimate strength if the maximum precompression stress is less than $0.4 f'_c$, and the continuity steel, ρ, in the deck slab is less than 0.015; where $\rho = A_s/bd$. The ultimate negative resisting moment shall be calculated using the compressive strength of the girder concrete regardless of the strength of the diaphragm concrete.

4. *Compressive stress in girders at piers at service loads.* The compressive stress in ends of girders at piers resulting from addition of the effects of prestressing and negative live-load bending shall not exceed $0.60 f'_c$.

5. *Shear.* In continuous bridges of this type, shear reinforcement shall be designed according to Section 5.2.1.5 herein for bridges. The horizontal shear connection between the cast-in-place slab and the precast girder shall be designed in accordance with Section 5.2.1.6 herein.

Author's discussion of positive connection at piers

Positive restraint moments (tension in bottom) at the supports may occur as a result of time-dependent camber following slab casting, as described in (2) above. An approximate method for determining these moments is given in Fig. 5.2 in terms of the time-dependent camber due to prestress minus the deflection under precast beam and slab dead load—all following slab casting. See the appropriate terms in Eq. 5.24 and 5.26. These moments may then be combined with the restraint moments that accompany differential shrinkage and creep of the continuous girder, as given in Fig. B.3 and Eq. B.11, and the positive moments due to live load plus impact on remote spans. A more exact method is given by Freyermuth,[34] which also includes a detailed discussion of the positive connection steel design. For shored construction, such positive restraint moments due to live load plus impact on remote spans are decreased or eliminated by the negative restraint moments due to the initial and time-dependent effect of the slab dead load.

Experience in the 1950s showed that vertical cracks tended to develop in the joint or diaphragm concrete when no reinforcement was provided between the prestressed girders and the joint, and that these cracks seemed to be larger at the bottom with decreasing widths toward the top. This occurred even though the bottom flanges of opposing girders in some but not all cases were connected solely by bond to the joint concrete. Such cracks were believed to be a result primarily of creep and other volumetric changes occurring in the prestressed girders, and from the positive live-load moments under remote positions of the loads. Because there was no observed change in the crack sizes under traffic, it appeared that the volumetric changes in the girders were the primary cause of the cracking.

This phenomenon was apparently due in large part to the continued upward camber with time following slab and joint casting, which may have been more pronounced in girders that were conservatively designed with regard to tension in the bottom near midspan. However, in many cases, it appears to the author that the net time-dependent deformation following slab casting—primarily Terms 4, 6, and 8 in Eq. 5.24 in the case of unshored construction—will be downward and not upward (producing negative not positive restraint moments). Branson and Kripanarayanan[40] found this to be the case for

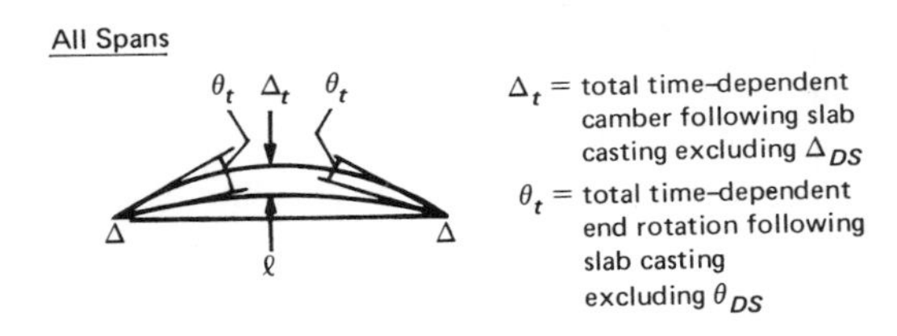

Assuming uniform load distribution under dead load and prestress (correct for parabolic tendons, otherwise approximate),

$$\Delta_t = \frac{5w\ell^4}{384EI}, \quad \theta_t = \frac{w\ell^3}{24EI} = \frac{16}{5}\frac{\Delta_t}{\ell}$$

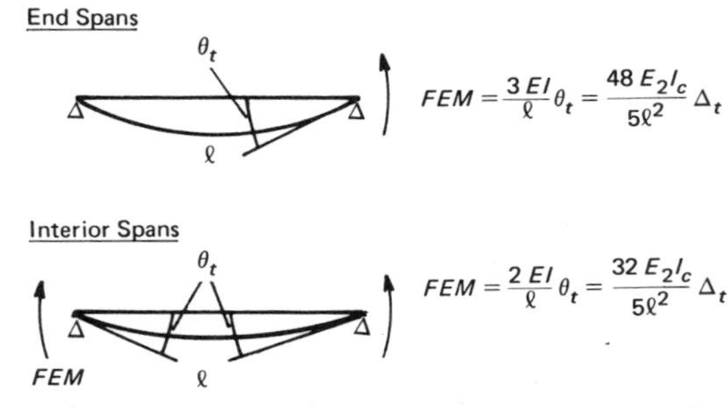

These end moments can be distributed for any multispan continuous beam, or they may be conservatively used directly in the design of positive-moment connection steel in negative-moment regions (shown in Fig. 4.6). I_c = positive section composite I.

Fig. 5.2. Approximate method for determining positive restraint moments that accompany time-dependent camber following the casting of a continuous slab on precast prestressed simple beams—in terms of the total time-dependent midspan camber following slab casting but excluding the deflection due to differential shrinkage and creep. The negative moment due to this effect may then be added from Fig. B.3 and Eq. B.11.

the measured camber-deflection versus time curves of five girders (sand-lightweight prestressed girders) of a composite highway bridge in Iowa, and for the calculated camber-deflection of the five girders as well. The same is true of Ex. 5.4 herein, as summarized in Table 5.9. In all of these cases, the initial or short-time deformation due to prestress and precast beam and slab dead load is also downward not upward (the latter result, unlike the time-dependent results, depends on well-defined material parameters), which tends to confirm the above observations because of the resulting strain distribution in the prestressed girders. Thus, it appears in most cases that the positive (bottom) connection steel between the precast prestressed girders and the joint concrete might employ only a nominal design, as illustrated in Ex. 5.2.

5.2.1.9 Example 5.1

Design of simple span, unshored, composite prestressed concrete double-T floor beam. See Fig. 5.3 for the section and other details, and Table 5.4 for the material and section properties, loads, and moments.

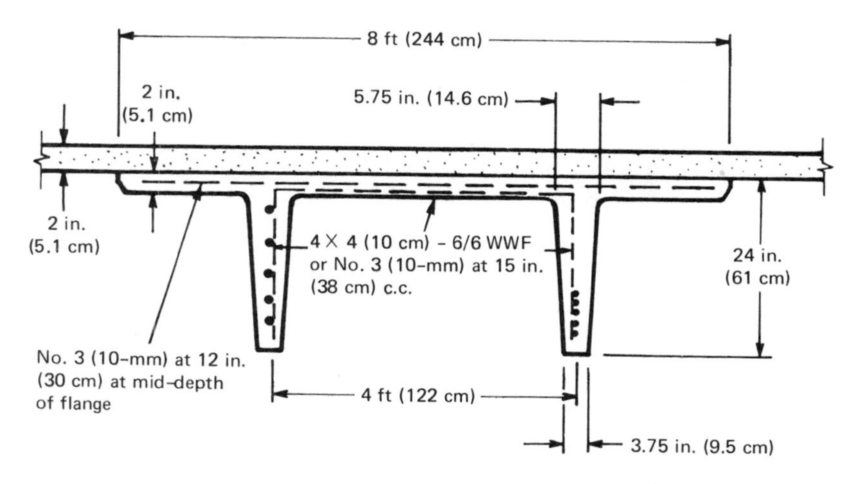

8 *LDT* 24 + 2, y_b

= 17.15 in. (43.6 cm)–*DT* $10\frac{1}{2}$-in.

(13-mm) Strands, One-Point Depression

$e_e = 6.75$ in. (17.1 cm)

$e_c = y_b - 3.00 = 14.15$ in. (35.9 cm)

Simple Span = 56 ft (17.1 m)

Fig. 5.3. Double-tee section used in Ex. 4.5 and 4.7.

Trial section and span-depth ratio

From *PCI Design Handbook*, try 8 LDT 24 + 2

Approx. ℓ/h = (0.95 for lightweight)(36) = 34 (Section 5.2.1.1)

Req. h = (56)(12)/34 = 20 in. (51 cm) OK

Loss of prestress by simplified method for stress relieved strand

$$P_i = A_{ps}f_{si} = (1.53)(189)$$

$$= 289 \text{ kips (1290 kN)}$$

$$\rho_{ps} = A_{ps}/A_g = 1.53/401 = 0.00382,$$

$$n = 11.2 \quad \text{(Table 5.1)}$$

$$\text{Trial } P_o = P_i/(1 + 2n\rho_{ps}) \quad \text{(Eq. 5.1, Term 1)}$$

$$= 298/(1 + 2 \times 11.2 \times 0.00382)$$

$$= 266 \text{ kips (1180 kN)}$$

$$\text{End } f_{ci} = \frac{P_o}{A_g} + \frac{P_o e_e^2}{I_g}$$

$$= \frac{266}{401} + \frac{(266)(6.75)^2}{20{,}985}$$

$$= 1.241 \text{ ksi } (8.56 \text{ N/mm}^2)$$

$$\text{Revised End } P_o = P_i - nf_{ci}A_{ps}$$

$$= 289 - (11.2)(1.241)(1.53)$$

$$= 268 \text{ kips (1190 kN)}$$

$$\text{Revised End } f_{ci} = (268/266)(1.241)$$

$$= 1.250 \text{ ksi } (8.62 \text{ N/mm}^2)$$

At transfer,

$$\text{End } \Delta f_s = nf_{ci} = (11.2)(1250)$$

$$= 14{,}000 \text{ psi } (96.5 \text{ N/mm}^2)$$

$$/\ 1890 = 7.4\%$$

$$0.4\text{-}Pt\ f_{ci} = \frac{P_o}{A_g} + \frac{P_o e^2}{I_g} - \frac{M_2 e}{I_g} = \frac{266}{401}$$

$$+ \frac{(266)(12.67)^2}{20{,}985} - \frac{(130.96)(12)(12.67)}{20{,}985}$$

$$= 1.749 \text{ ksi } (12.06 \text{ N/mm}^2)$$

Table 5.4. Material and section properties, loads and moments for Ex. 5.1.

8 *LDT* 24 + 2, *Unshored Composite Beam* (See Fig. 5.3)
10 - $\frac{1}{2}$ in. (13 mm) Strands, One-Point Depression

Area/Strand = 0.153 in.^{2} (0.99 cm^2)

Simple Span = 56 ft (17.1 m)

$A_{ps} = (10)(0.153) = 1.53$ in.^{2} (9.87 cm^2)

Material Properties

Steam-cured sand-lightweight double-T (DT) (120 pcf = 18.9 kN/m^3), and 2 in. (5.1 cm) normal-weight topping (145 pcf = 22.8 kN/m^3)

$f_{pu} = 270$ ksi (1860 N/mm^2)

$f_{si} = (0.70)(270) = 189$ ksi (1300 N/mm^2)

$f'_{ci} = 3500$ psi (24 N/mm^2) or specified otherwise

$f'_c = 5000$ psi (34 N/mm^2)

Topping $f'_c = 3000$ psi (21 N/mm^2)

Double-T Section Properties (*PCI Design Handbook*)

$A_g = 401$ in.^{2} (2587 cm^2)

$I_g = 20{,}985$ in.^{4} (873,400 cm^4)

$y_b = 17.15$ in. (43.6 cm)

$y_t = 6.85$ in. (17.4 cm)

$Z_b = 1224$ in.^{3} (20,060 cm^3)

$Z_t = 3064$ in.^{3} (50,220 cm^3)

$e_e = 6.75$ in. (17.1 cm)

$e_c = y_b - 3.00 = 14.15$ in. (35.9 cm)

$e_{0.4\text{-Pt}} = 6.75 + (4/5)(14.15 - 6.75) = 12.67$ in. (32.2 cm)

Composite Section Properties (*PCI Design Handbook*)

$I_g = 29{,}853$ in.^{4} (1,242,500 cm^4)

$y_b = 19.94$ in. (50.6 cm)

$y_t = 6.06$ in. (15.4 cm)

Midspan $d_c = 26 - 3 = 23.00$ in. (58.4 cm)

0.4-Pt $d_c = 2 + (y_t + e)_{DT} = 2 + 6.85 + 12.67 = 21.52$ in. (54.7 cm)

$Z_b = 1497$ in.^{3} (24,540 cm^3)

$Z_t = 4926$ in.^{3} (80,740 cm^3)

$I_2/I_c = 20{,}985/29{,}853 = 0.703$

Loads

DT dead load = (125 including steel)(401)/144 = 348 lb/ft (5080 N/m)

Topping dead load = (145)(8)(2/12) = 193 lb/ft (2820 N/m)

Ceiling load = 10 psf (479 N/m^2)

Live load = 50 psf (2390 N/m^2)

Table 5.4. (cont.)

For the unshored composite beam, all loads except the live load will be assumed to be supported by the precast beam.

Midspan Moments and 0.4-Pt *Moments*

$$M_{\text{Topping}} = M_1 = (193)(56)^2/8 = 75{,}660 \text{ ft-lb } (102.6 \text{ kN-m})$$

$$0.4\text{-Pt } M_1 = (0.96)(75{,}660) = 72{,}630 \text{ ft-lb } (98.5 \text{ kN-m})$$

$$M_{\text{DT}} = M_2 = (348)(56)^2/8 = 136{,}420 \text{ ft-lb } (185.0 \text{ kN-m})$$

$$0.4\text{-Pt } M_2 = (0.96)(136{,}420) = 130{,}960 \text{ ft-lb } (177.6 \text{ kN-m})$$

$$M_{\text{Ceiling}} = (10)(8)(56)^2/8 = 31{,}360 \text{ ft-lb } (42.5 \text{ kN-m})$$

$$0.4\text{-Pt Ceiling } M_1 = (0.96)(31{,}360) = 30{,}110 \text{ ft-lb } (40.8 \text{ kN-m})$$

$$M_L = (50)(8)(56)^2/8 = 156{,}800 \text{ ft-lb } (213 \text{ kN-m})$$

$$0.4\text{-Pt } M_L = (0.96)(156{,}800) = 150{,}530 \text{ ft-lb } (204 \text{ kN-m})$$

Revised 0.4-*Pt* P_o = 289 − (11.2)(1.749)(1.53)
= 259 kips (1150 kN)

Revised 0.4-*Pt* f_{ci} = 1.678 ksi (11.57 N/mm²)

Revised 0.4-*Pt* P_o = 260 kips (1160 kN)

Revised 0.4-*Pt* f_{ci} = 1.688 ksi (11.64 N/mm²)

Similarly,

Midspan P_o = 255 kips (1130 kN)

Midspan f_{ci} = 1.965 ksi (13.55 N/mm²)

From Eq. (5.1), Term 6 for unshored construction,

$$0.4\text{-}Pt\ f_{cs} = \frac{M_1 e}{I_2}$$

$$= \frac{(72.63 + 30.11)(12)(12.67)}{20{,}985}$$

= 0.744 ksi (5.13 N/mm²)

After losses,

$$0.4\text{-}Pt\ (\Delta f_s)_u = \left(19.53 + 7.56 \frac{I_2}{I_c}\right) f_{ci} + 23{,}570 - \left(9.3 + 10.63 \frac{I_2}{I_c}\right) f_{cs}$$

$$= (19.53 + 7.56 \times 0.703)(1688) + 23{,}570 - (9.3 + 10.63 \times 0.703)(744)$$

= 53,030 psi (366 N/mm²)/1890 = 28.1% (Eq. 5.3)

Similarly,

Midspan $(\Delta f_s)_u$ = 57,860 psi (399 N/mm²)/1890 = 30.6%

Flexural stress

At Transfer,

$$f_{t,b} = \frac{P_o}{A_2} \mp \frac{P_o e}{(Z_2)_{t,b}} \pm \frac{M_2}{(Z_2)_{t,b}} \qquad \text{(Eq. 5.6)}$$

$$\text{End } f_t = \frac{268}{401} - \frac{(268)(6.75)}{3064}$$

= 0.078 ksi (0.54 N/mm²) Comp.

$$\text{End } f_b = \frac{268}{401} + \frac{(268)(6.75)}{1224}$$

= 2.146 ksi (14.80 N/mm²) Comp.

$$\text{Midspan } f_t = \frac{255}{401} - \frac{(255)(14.15)}{3064} + \frac{(136.42)(12)}{3064} = -0.007 \text{ ksi}$$

(0.05 N/mm²) Tens.

$$\text{Midspan } f_b = \frac{255}{401} + \frac{(255)(14.15)}{1224} - \frac{(136.42)(12)}{1224} = 2.246 \text{ ksi}$$

(15.49 N/mm²) Comp.

Similarly,

0.4-Pt f_t = 0.086 ksi (0.59 N/mm^2) Comp.

0.4-Pt f_b = 2.055 ksi (14.17 N/mm^2) Comp.

Req. f'_{ci} = 2246/0.60 = 3743 psi. Use f'_{ci} = 3750 psi (26 N/mm^2).

Max. $f_{\text{Comp.}}$

= 2246 psi (15.49 N/mm^2) at Midspan

< ACI 0.60f'_{ci}

= (0.60)(3750) = 2250 psi (15.51 N/mm^2) OK

Max. f_{Tensile}

= 7 psi (0.05 N/mm^2) at Midspan < ACI 3$\sqrt{f'_{ci}}$

= (3)$\sqrt{3750}$ = 184 psi (1.27 N/mm^2) OK

After Losses,

$$f_{t,b} = \frac{P}{A_2} \mp \frac{Pe}{(Z_2)_{t,b}} \pm \frac{M_2}{(Z_2)_{t,b}} \pm \frac{M_1}{(Z_2)_{t,b}} \pm \frac{M_L}{(Z_c)_{t,b}} \quad \text{(Eq. 4.42)}$$

See the typical stress distribution for unshored construction in Fig. 5.1.

From the last term of Eq. 5.8,

Midspan Slab f_t

$$= \frac{M_L}{(Z_c)_t} = \frac{(156.80)(12)}{4926}$$

= 0.382 (2.63 N/mm^2) (low stress) Comp.

> f_t/n_c < ACI 0.45f'_c

= (0.45)(3.00) = 1.35 ksi (9.31 N/mm^2) OK

0.4-Pt $P = P_i - (\Delta f_s)_u A_{ps}$ = 289 - (53.03)(1.53)

= 208 kips (925 kN)

$$\text{0.4-Pt Precast } f_t = \frac{208}{401} - \frac{(208)(12.67)}{3064} + \frac{(72.63 + 130.96 + 30.11)(12)}{3064} + \frac{(150.53)(12)}{4926} \cdot \frac{6.06 - 2}{6.06} = 0.820 \text{ ksi } (5.65 \text{ N/mm}^2) \text{ Comp.}$$

$$\text{0.4-Pt } f_b = \frac{208}{401} + \frac{(208)(12.67)}{1224} - \frac{(72.63 + 130.96 + 30.11)(12)}{1224} - \frac{(150.53)(12)}{1497}$$

= - 0.826 ksi (5.70 N/mm^2) Tens.

Similarly,

Midspan P = 200 kips (890 kN)

Midspan Precast f_t = 0.784 ksi (5.41 N/mm^2) Comp.

Midspan f_b = - 0.833 ksi (5.74 N/mm^2) Tens.

Max. $f_{\text{Comp.}}$ = 820 psi (5.65 N/mm^2) at 0.4-Pt

< ACI 0.45f'_c = (0.45)(5000)

= 2250 psi (15.51 N/mm^2) OK

Max. $f_{\text{Tens.}}$ = 833 psi (5.74 N/mm^2) at Midspan (slightly greater than 826 psi = 5.70 N/mm^2 at 0.4-Pt in this case)

> ACI 6$\sqrt{f'_c}$ = 424 psi (2.92 N/mm^2)

and

f_r (from Table A.1)

= 7.12$\sqrt{f'_c}$ = 503 psi (3.47 N/mm^2),

but

< ACI 12$\sqrt{f'_c}$ = (12)$\sqrt{5000}$

= 848 psi (5.85 N/mm^2) (Table A.2).

Hence, the stresses are satisfactory, provided deflections are checked.

Flexural strength

0.4-Pt Design M_u

= 1.4M_D + 1.7M_L

= (1.4)(72.63 + 130.96 + 30.11) + (1.7)(150.53)

= 583 ft-k (751 kN-m)

$\rho_p = A_{ps}/bd$ = 1.53/(0.96)(21.52) = 0.000741

$$f_{ps} = f_{pu}\left(1 - \frac{\rho_p f_{pu}}{2f'_c}\right) \quad \text{(Eq. 5.10)}$$

$$= 270\left[1 - \frac{(0.000741)(270)}{(2)(3)}\right]$$

= 261 ksi (1800 N/mm^2)

1.4$d\rho_p f_{ps}/f'_c$ = (1.4)(21.52)(0.000741)(261)/3

$= 1.94$ in. (4.9 cm)

< 2 in. (5.1 cm).

Hence, use Eq. 5.11 for rectangular compression block and $f_c' = 3000$ psi (20.7 N/mm²)

$$\omega_p = \rho_p f_{ps}/f_c' = (0.000741)(261)/3$$
$$= 0.064 < 0.30.$$

Hence, use Eq. 5.11 for underreinforced case.

0.4-Pt M_u

$$= \phi A_{ps} f_{ps} d \left(1 - \frac{\rho_p f_{ps}}{1.7 f_c'}\right) \quad \text{(Eq. 5.11)}$$

$$= (0.90)(1.53)(261)(21.52)$$
$$\left[1 - \frac{(0.000741)(261)}{(1.7)(3)}\right]$$

$= 7441$ in-k $= 620$ ft-k (841 kN-m)

> 583 ft-k (751 kN-m) OK

Similarly,

Midspan Design $M_u = 607$ ft-k (823 kN-m)

Midspan Usable Capacity M_u

$= 667$ ft-k (904 kN-m) OK

0.4-Pt Design $M_u/M_u = 583/620 = 0.94$, Midspan Design $M_u/M_u = 607/667 = 0.91$. Hence, the flexural strength at midspan is less critical than at the 0.4-Pt.

Cracking moment

$$f_r = 7.12\sqrt{f_c'} = 7.12\sqrt{5000}$$
$$= 503 \text{ psi } (3.47 \text{ N/mm}^2)$$

(Table A.1)

$$\text{0.4-Pt } M_{cr} = \frac{P(Z_c)_b}{A_2} + \frac{Pe_2(Z_c)_b}{(Z_2)_b} + f_r(Z_c)_b$$

(Eq. 5.15)

$$= \frac{(208)(1497)}{401}$$

$$+ \frac{(208)(12.67)(1497)}{1224} + (0.503)(1497)$$

$= 4753$ in.-k

$= 396$ ft-k (537 kN-m)

0.4-Pt $M_u/M_{cr} = 620/396 = 1.57 > 1.2$ OK

Similarly,

Midspan $M_u/M_{cr} = 667/427 = 1.56 > 1.2$ OK

Web Shear

From *PCI Design Handbook*[4] (Table 5.2.15), Max. $(v_u - v_c)$ occurs at about 0.1-Pt.

$$e = 6.75 + (1/5)(14.15 - 6.75)$$
$$= 8.23 \text{ in. } (20.9 \text{ cm})$$

$$d_c = 2 + 6.85 + e = 8.85 + 8.23$$
$$= 17.08 \text{ in.} = 1.42 \text{ ft } (43.4 \text{ cm})$$

$$b_w = 2(3.75 + 5.75)/2$$
$$= 9.50 \text{ in. } (24.1 \text{ cm}) \quad \text{(See Fig. 5.3)}$$

$$w_u = 1.4(193 + 348 + 80)1.7(50 \times 8)$$
$$= 1550 \text{ lb/ft } (22{,}600 \text{ N/m})$$

$$V_u = w_u \ell/2 - w_u x = 1.550(28 - 5.6)$$
$$= 34.7 \text{ kips } (154 \text{ kN})$$

$$v_u = \frac{V_u}{\phi b_w d} = \frac{34{,}700}{(0.85)(9.50)(17.08)}$$
$$= 252 \text{ psi } (1.74 \text{ N/mm}^2) \quad \text{(Eq. 4.12)}$$

$$\text{Min. } v_c = (0.85 \text{ for sand-lightweight})(2)\sqrt{5000}$$
$$= 120 \text{ psi } (0.83 \text{ N/mm}^2)$$

$$\text{Max. } v_c = (0.85)(5)\sqrt{5000}$$
$$= 301 \text{ psi } (2.08 \text{ N/mm}^2)$$

$$f_{se} = P/A_{ps}$$
$$= (\text{Use 208 at 0.4-Pt, OK here})/1.53$$
$$= 136 \text{ ksi } (938 \text{ N/mm}^2) > 0.40 f_{pu}$$
$$= (0.40)(270) = 108 \text{ ksi } (745 \text{ N/mm}^2).$$

Hence, OK to use Eqs. 5.16 and 5.17.

$$M_u = w_u \ell x/2 - w_u x^2/2$$
$$= 0.775(56 \times 5.6 - 5.6^2)$$
$$= 219 \text{ ft-k } (297 \text{ kN-m})$$

$$V_u d/M_u = (34.7)(1.42)/219 = 0.225 < 1.0$$

OK

$$v_c = (0.85)(0.6)\sqrt{f_c'} + 700 V_u d/M_u$$

(Eq. 5.16)

$$= 0.51\sqrt{5000} + (700)(0.225)$$
$$= 194 \text{ psi } (1.34 \text{ N/mm}^2)$$

Use

$$\frac{A_v}{s} = \frac{(v_u - v_c)b_w}{f_y} = \frac{(252 - 194)(9.50)}{40{,}000}$$

$$= 0.0138 \text{ in. } (0.0351 \text{ cm}) \qquad \text{(Eq. 4.13)}$$

By Eq. 4.14,

$$\text{Min. } \frac{A_v}{s} = \frac{50 b_w}{f_y} = \frac{(50)(9.50)}{40{,}000}$$

$$= 0.0119 \text{ in. } (0.0302 \text{ cm})$$

By Eq. 5.17,

$$\text{Min. } \frac{A_v}{s}$$

$$= \frac{A_{ps} f_{pu}}{80 f_y d}\sqrt{\frac{d}{b_w}} = \frac{(1.53)(270)}{(80)(40)(17.08)}\sqrt{\frac{17.08}{9.50}}$$

$$= 0.0101 \text{ in. } (0.0257 \text{ cm})$$

$$s_{max} = 3h/4 = (3)(26)/4 = 19.5 \text{ in. } (49.5 \text{ cm}) \qquad \text{(Eq. 5.18)}$$

For No. 3 (10 mm) stirrups,

$$s = (2)(0.11)/0.0138 = 15.9 \text{ in. } (40.4 \text{ cm})$$

For 4 X 4 (10 cm) - 6/6 WWF,

$$A_v = (2)(0.087 \text{ in.}^2/\text{ft})$$

$$= 0.174 \text{ in.}^2/\text{ft } (3.7 \text{ cm}^2/\text{m})$$

$$> (0.0138 \text{ in.}^2/\text{in.})(12 \text{ in./ft})$$

$$= 0.166 \text{ in.}^2/\text{ft } (3.5 \text{ cm}^2/\text{m}) \quad \text{OK}$$

Use f_y = 40,000 psi (275 N/mm²) for web reinforcement. Use No. 3 (10-mm) stirrups at 15 in. (38 cm) c.c., or Use 4 X 4 (10 cm) - 6/6 WWF.

Horizontal shear

$$V_u = w_u \ell/2 = (1550)(28)$$

$$= 43{,}400 \text{ lb } (193 \text{ kN})$$

$$V_{dh} = \frac{V_u}{\phi b_v d_c} = \frac{43{,}400}{(0.85)(96)(\text{Say } 0.8h = 20.8)}$$

$$= 26 \text{ psi } (0.18 \text{ N/mm}^2) \qquad \text{(Eq. 4.18)}$$

$$< 80 \text{ psi } (0.55 \text{ N/mm}^2)$$

in Section 4.2.2.3. Hence, ties are not required.

Development length of prestressing strands

$$(f_{ps} - \tfrac{2}{3} f_{se})d = (261 - \tfrac{2}{3} \times 136)\tfrac{1}{2} \qquad \text{(Eq. 5.23)}$$

$$= 85 \text{ in. } (216 \text{ cm}) < (28)(12)$$

$$= 336 \text{ in. } (853 \text{ cm}) \quad \text{OK}$$

Transverse flange reinforcement

Based on 2-ft (0.6-m) overhang and d = 1 in. (2.5 cm).

$$w_u = 1.4(125 + 145)(2/12) + 1.7(50)$$

$$= 148 \text{ lb/ft } (2160 \text{ N/m})$$

$$M_u = w_u \ell^2/2 = (148)(2)^2/2$$

$$= 296 \text{ ft-lb } (401 \text{ N-m})$$

For No. 3 (10-mm) stirrups at 12 in. (30 cm) c.c.,

$$A_s = 0.11 \text{ in.}^2/\text{ft } (2.3 \text{ cm}^2/\text{m})$$

$$a = \frac{A_s f_y}{0.85 f_c' b} = \frac{(0.11)(40)}{(0.85)(5)(12)}$$

$$= 0.086 \text{ in. } (0.22 \text{ cm})$$

$$\text{Req. } A_s = \frac{M_u}{\phi f_y (d - a/2)}$$

$$= \frac{(296)(12)}{(0.90)(40{,}000)(1 - 0.043)}$$

$$= 0.103 \text{ in.}^2/\text{ft } (2.2 \text{ cm}^2/\text{m}) \quad \text{OK}$$

5.2.1.10 Example 5.2

Design of four-span, continuous, unshored composite prestressed concrete bridge girder. See Fig. 5.4 for the girder details, and Table 5.5 for the material and section properties.

Slab design

$$S = 8 - 1.33 = 6.67 \text{ ft } (2.0 \text{ m})$$

$$w_D = (7/12)(150) = 88 \text{ lb/ft } (1280 \text{ N/m})$$

$$\pm M_D = 0.10 w_D S^2 = (0.10)(88)(6.67)^2$$

$$= 392 \text{ ft-lb } (532 \text{ N-m})$$

$$\pm M_L = (0.80)\frac{S + 2}{32} P_{20}$$

$$= (0.8)\frac{6.67 + 2}{32}(16{,}000)$$

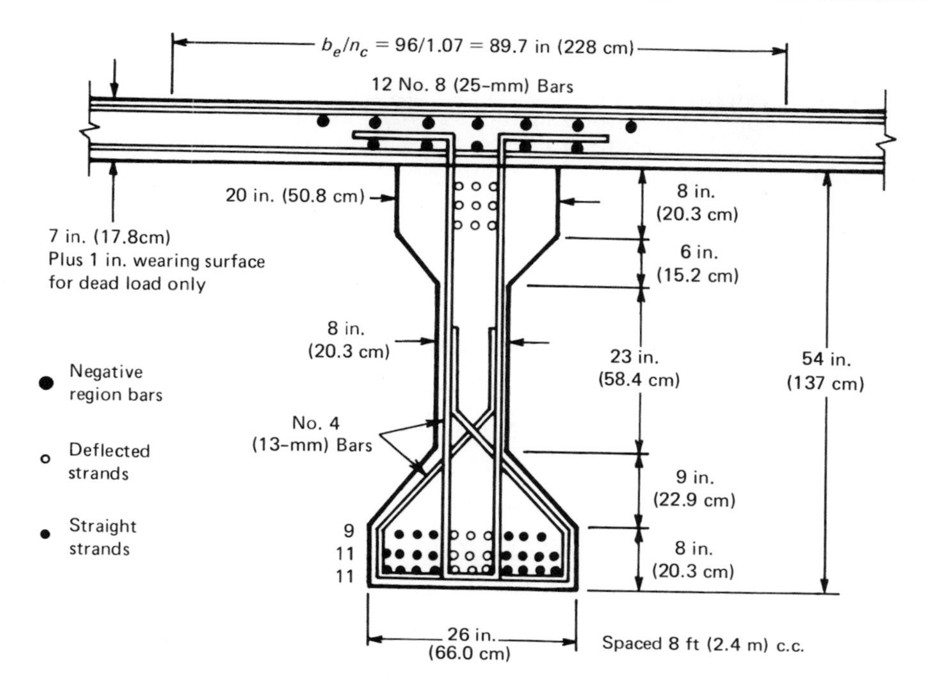

Interior Girder, AASHTO Type IV,

$(y_b)_{\text{Precast}}$

= 24.73 in. (62.8 cm), 31 $\frac{1}{2}$-in. (13-mm) Strands,

Two-Point Depression at 3rd Points

e_e = 7.50 in. (19.1 cm)

e_c = 20.86 in. (53.0 cm)

4-Span Continuous Bridge (50-90-90-50 ft, 15.2-27.4-27.4-15.2 m)

From Section 4.1.3,

b_e = Beam Span/4 = (90)(12)/4

= 270 in. (686 cm)

= Beam Spacing = (8)(12)

= 96 in. (244 cm) USE

= $12t + b_w$ = (12)(7) + 20

= 104 in. (264 cm)

Fig. 5.4. Bridge girder used in Ex. 4.6 and 4.8.

= 3470 ft-lb (4710 N-m) (AASHTO)

$$\pm M_u = 1.30(M_D + \tfrac{5}{3} M_{L+I}) \quad \text{(Eq. 5.9)}$$

= 1.30[392 + (5/3)(3470)(1.31)]

= 10,280 ft-lb (13,940 N-m)

Remainder of slab design similar to Ex. 4.2.

Also see AASHTO for distribution reinforcement, etc.

Loads and moments for interior spans (controls over end spans, as shown)

w_{Slab} = (150)(8/12)(8)

= 800 lb/ft (11,700 N/m)

Table 5.5. Material and section properties for Ex. 5.2.

AASHTO Type IV, Unshored Composite Interior Girder (See Fig. 5.4)
Four-Span Continuous Bridge (50-90-90-50 ft, 15.2-27.4-27.4-15.2 m)
Width 28 ft (8.5 m) curb to curb, Beam Spacing 8 ft (2.4 m) c.c. 31 $\frac{1}{2}$ in. (13 mm) Strands, Two-Point Depression at Third Points

$$A_{ps} = (31)(0.153) = 4.74 \text{ in. } (30.6 \text{ cm}^2)$$

Material Properties
Steam-Cured Precast Girders and Moist-Cured Slab
All Concrete—Normal Weight (See Table A.2 for E_c)

$$\text{Slab } f'_c = 4000 \text{ psi } (28 \text{ N/mm}^2)$$

$$E_c = 3.64 \times 10^6 \text{ psi } (25.1 \text{ kN/mm}^2)$$

$$\text{Precast Girder } f'_c = 5000 \text{ psi } (34 \text{ N/mm}^2)$$

$$E_c = 3.89 \times 10^6 \text{ psi } (26.8 \text{ kN/mm}^2)$$

$$f'_c = 4000 \text{ psi } (28 \text{ N/mm}^2) \text{ or specified otherwise}$$

$$n_c = 3.89/3.64 = 1.07$$

$$f_{pu} = 270 \text{ ksi } (1860 \text{ N/mm}^2)$$

$$f_{si} = (0.70)(270) = 189 \text{ ksi } (1300 \text{ N/mm}^2)$$

AASHTO Type IV Precast Girder Section Properties (PCI Design Handbook)

$$A_g = 789 \text{ in.}^2 \ (5090 \text{ cm}^2)$$

$$I_g = 260{,}741 \text{ in.}^4 \ (10{,}758{,}000 \text{ cm}^4)$$

$$y_b = 24.73 \text{ in. } (62.8 \text{ cm})$$

$$y_t = 29.27 \text{ in. } (74.4 \text{ cm})$$

$$Z_b = 10{,}544 \text{ in.}^3 \ (172{,}800 \text{ cm}^3)$$

$$Z_t = 8908 \text{ in.}^3 \ (146{,}000 \text{ cm}^3)$$

Composite Section Properties

$$y_b = \frac{(789)(24.73) + (7)(89.7)(54 + 3.5)}{789 + (7)(89.7)} = 39.25 \text{ in. } (99.7 \text{ cm})$$

$$I_g = 260{,}741 + (789)(39.25 - 24.73)^2 + (89.7)(7)^3/12 + (89.7)(7)(54 + 3.5 - 39.25)^2$$
$$= 638{,}780 \text{ in.}^4 \ (26{,}356{,}000 \text{ cm}^4)$$

$$I_2/I_c = 260{,}741/638{,}780 = 0.408$$

$$y_t = 54 + 7 - 39.25 = 21.75 \text{ in. } (55.3 \text{ cm})$$

$$Z_b = 638{,}780/39.25 = 16{,}270 \text{ in.}^3 \ (266{,}700 \text{ cm}^3)$$

$$Z_t = 638{,}780/21.75 = 29{,}370 \text{ in.}^3 \ (481{,}400 \text{ cm}^3)$$

$$w_{\text{Precast Girder}} = (150)(789)/144$$
$$= 822 \text{ lb/ft } (12{,}000 \text{ N/m})$$

$$M_1 = (0.800)(90)^2/8 + 60 \text{ (2 diaphragms)}$$
$$= 870 \text{ ft-k } (1180 \text{ kN-m})$$

$$M_2 = (0.822)(90)^2/8$$
$$= 832 \text{ ft-k } (1130 \text{ kN-m})$$

For four-span continuous interior girder (from Iowa Dept. of Transportation computer printout):

For future wearing surface of 20 psf (960 N/m^2):

Max. $+M_{FWS}$ = 60 ft-k (81 kN-m) near mid-span (17 ft-k in end span)

Max. $- M_{FWS} = 120$ ft-k (163 kN-m) at center pier (83 ft-k at first interior pier)

For HS 20 loading, including load distribution factor ($S/5.5 = 8/5.5$) and impact factor ($50/\ell + 125$, where ℓ = span for positive moment and average of adjacent spans for negative moment):

Max. $+ M_{L+I} = 733$ ft-k (994 kN-m) near midspan (truck loading controls, with lane loading $M_{L+I} = 559$ ft-k) (492 ft-k, truck, controls in end span)

Max. $- M_{L+I} = 708$ ft-k (960 kN-m) at center pier (lane loading controls, with truck loading $M_{L+I} = 563$ ft-k), (590 ft-k, lane, controls at first interior pier)

The moments due to military loading consisting of two 24-kip (107-kN) axle loads spaced 4 ft (122 cm) apart were less than the moments due to the HS 20 loading.

Required number of $\frac{1}{2}$-in. stress relieved strands for interior spans

Using AASHTO $f_b = 6\sqrt{f_c'} = 6\sqrt{5000} = 424$ psi (2.92 N/mm²), assuming 24% loss, and estimating $e_c = y_b - 4 = 20.7$ in. (52.6 cm),

$$f_b = \frac{0.76P_i}{A_2} + \frac{0.76P_i e_c}{(Z_2)_b} - \frac{M_1 + M_2}{(Z_2)_b} - \frac{M_{L+I+FWS}}{(Z_c)_b} \qquad \text{(Eq. 4.42)}$$

$$\text{Req. } P_i = \frac{1}{\dfrac{0.76}{789} + \dfrac{(0.76)(20.7)}{10{,}544}} \left[-0.424 + \frac{(870 + 832)(12)}{10{,}544} + \frac{(733 + 60)(12)}{16{,}270}\right]$$

$$= 854.4 \text{ kips (3800 kN)}$$

$$\text{Req. } A_{ps} = 854.4/189 = 4.52 \text{ in.}^2$$

$$\text{Req. No.} = 4.52/0.153 = 29.5$$

Use 31 Strands for Symm.,

$$A_{ps} = (31)(0.153) = 4.74 \text{ in.}^2 \text{ (30.6 cm}^2\text{)}$$

$$P_i = A_{ps} f_{si} = (4.74)(189)$$

$$= 896 \text{ kips (3990 kN)}$$

$$e_c = 24.73 - \frac{(11 \times 2) + (11 \times 4) + (9 \times 6)}{31}$$

$$= 20.86 \text{ in. (53.0 cm)}$$

Depressing $3 \times 3 = 9$ strands,

$$e_e = 24.73 - [(8 \times 2) + (8 \times 4) + (6 \times 6) + (3 \times 52) + (3 \times 50) + (3 \times 48)]/31$$

$$= 7.50 \text{ in. (19.1 cm)}$$

Loss of prestress by simplified method for stress relieved strand in interior spans

$$\rho_{ps} = A_{ps}/A_g = 4.74/789 = 0.00601$$

$$n = 7.7 \quad \text{(Table 5.1)}$$

$$\text{Trial } P_o = P_i/(1 + 2n\rho_{ps})$$

$$= 896/(1 + 2 \times 7.7 \times 0.00601)$$

$$= 820 \text{ kips (3650 kN)}$$

$$\text{End } f_{ci} = \frac{P_o}{A_2} + \frac{P_o e_e^2}{I_2} = \frac{820}{789} + \frac{(820)(7.50)^2}{260{,}471}$$

$$= 1.216 \text{ ksi (8.72 N/mm}^2\text{)}$$

$$\text{Revised End } P_o = P_i - n f_{ci} A_{ps}$$

$$= 896 - (7.7)(1.216)(4.74)$$

$$= 852 \text{ kips}$$

Revised End $f_{ci} = 1.261$ ksi (8.69 N/mm²)

Revised End $P_o = 850$ kips (3781 kN),

At Transfer,

$$\text{End } \Delta f_s = n f_{ci} = (7.7)(1261)$$

$$= 9710 \text{ psi (67.0 N/mm}^2\text{)}/1890 = 5.1\%$$

Midspan f_{ci}

$$= \frac{P_o}{A_2} + \frac{P_o e_c^2}{I_2} - \frac{M_2 e_c}{I_2}$$

$$= \frac{820}{789} + \frac{(820)(20.86)^2}{260{,}741} - \frac{(832)(12)(20.86)}{260{,}741}$$

$$= 1.609 \text{ ksi (11.09 N/mm}^2\text{)}$$

$$\text{Revised Midspan } P_o = P_i - n f_{ci} A_{ps}$$

$$= 896 - (7.7)(1.609)(4.74)$$

$$= 837 \text{ kips}$$

Revised Midspan $f_{ci} = 1.656$ ksi (11.42 N/mm²)

Revised Midspan $P_o = 836$ kips (3720 kN),

At Transfer,

Midspan $\Delta f_s = nf_{ci} = (7.7)(1656)$

$= 12{,}750$ psi (87.9 N/mm^2)/1890

$= 6.7\%$

From Eq. 5.1, Term 6 for unshored construction,

$$\text{Midspan } f_{cs} = \frac{M_1 e_c}{I_2} = \frac{(870)(12)(20.86)}{260{,}741}$$

$= 0.835$ ksi (5.76 N/mm^2)

After Losses,

End $(\Delta f_s)_u$

$$= \left(15.36 + 6.06\,\frac{I_2}{I_c}\right) f_{ci} + 23{,}570$$

$= (15.36 + 6.06 \times 0.408)(1261) + 23{,}570$

$= 46{,}060$ psi (318 N/mm^2)/1890

$= 24.4\%$ (Eq. 5.2)

$$\text{Midspan } (\Delta f_s)_u = \left(15.36 + 6.06\,\frac{I_2}{I_c}\right) f_{ci} + 23{,}570 - \left(6.7 + 7.66\,\frac{I_2}{I_c}\right) f_{cs}$$

$= (15.36 + 6.06 \times 0.408)(1656) + 23{,}570 - (6.7 + 7.66 \times 0.408)(835)$

$= 44{,}900$ psi (310 N/mm^2)/1890

$= 23.8\%$ (Eq. 5.2)

Flexural stress

At Transfer,

$$f_{t,b} = \frac{P_o}{A_2} \mp \frac{P_o e}{(Z_2)_{t,b}} \pm \frac{M_2}{(Z_2)_{t,b}} \quad \text{(Eq. 5.6)}$$

$$\text{End } f_t = \frac{850}{789} - \frac{(850)(7.50)}{8908}$$

$= 0.362$ ksi (2.50 N/mm^2) Comp.

$$\text{End } f_b = \frac{850}{789} + \frac{(850)(7.50)}{10{,}544}$$

$= 1.682$ ksi (11.60 N/mm^2) Comp.

$$\text{Midspan } f_t = \frac{836}{789} - \frac{(836)(20.86)}{8908} + \frac{(832)(12)}{8908}$$

$= 0.223$ ksi (1.54 N/mm^2) Comp.

$$\text{Midspan } f_b = \frac{836}{789} + \frac{(836)(20.86)}{10{,}544} - \frac{(832)(12)}{10{,}544}$$

$= 1.767$ ksi (12.18 N/mm^2) Comp.

Max. $f_{\text{Comp.}}$

$= 1767$ psi (12.18 N/mm^2) at Midspan

$<$ AASHTO $0.60 f'_{ci}$

$= (0.60)(4000)$

$= 2400$ psi (16.55 N/mm^2) OK

No $f_{\text{Tens.}} <$ AASHTO $3\sqrt{f'_{ci}} = 3\sqrt{4000}$

$= 190$ psi (1.31 N/mm^2) OK

Positive-Moment Region After Losses,

$$f_{t,b} = \frac{P}{A_2} \mp \frac{Pe}{(Z_2)_{t,b}} \pm \frac{M_1 + M_2}{(Z_2)_{t,b}} \pm \frac{M_{L+I+FWS}}{(Z_c)_{t,b}} \quad \text{(Eq. 5.8)}$$

See the typical stress distribution for unshored construction in Fig. 5.1. From the last term of Eq. 5.8,

$$\text{Midspan Slab } f_t = \frac{M_{L+I+FWS}}{n_c (Z_c)_t} = \frac{(733 + 60)(12)}{(1.07)(29{,}370)}$$

$= 0.303$ ksi (2.09 N/mm^2) (low stress) Comp.

$<$ AASHTO $0.45 f'_c = (0.45)(4.00)$

$= 1.80$ ksi (12.41 N/mm^2) OK

Also,

$$\text{End Slab } f_t = \frac{(708 + 120)(12)}{(1.07)(29{,}370)}$$

$= 0.316$ ksi (2.18 N/mm^2) Tens.

$< f_r = 0.495$ ksi (3.42 N/mm^2) (Table A.1)

Hence uncracked.

Midspan $P = P_i - (\Delta f_s)_u A_{ps}$

$= 896 - (44.90)(4.74)$

$= 683$ kips (3040 kN)

$$\text{Midspan Precast } f_t = \frac{683}{789} - \frac{(683)(20.86)}{8908} + \frac{(870 + 832)(12)}{8908} + \frac{(733 + 60)(12)}{29{,}370}$$

$$\cdot\frac{21.75-7}{21.75} = 1.779 \text{ ksi } (12.27 \text{ N/mm}^2) \text{ Comp.}$$

$$\text{Midspan } f_b = \frac{683}{789} + \frac{(683)(20.86)}{10{,}544} - \frac{(870+832)(12)}{10{,}544} - \frac{(733+60)(12)}{16{,}270}$$

$$= -0.305 \text{ ksi } (2.10 \text{ N/mm}^2) \text{ Tens.}$$

Max. Girder $f_{\text{Comp.}}$

$$= 1779 \text{ psi } (12.27 \text{ N/mm}^2) \text{ at Midspan}$$

$$< \text{AASHTO } 0.45f'_c = (0.45)(5000)$$

$$= 2250 \text{ psi } (15.51 \text{ N/mm}^2) \text{ OK}$$

Max. Girder $f_{\text{Tens.}}$

$$= 305 \text{ psi } (2.10 \text{ N/mm}^2) \text{ at Midspan}$$

$$< \text{AASHTO } 6\sqrt{f'_c} = 6\sqrt{5000}$$

$$= 424 \text{ psi } (2.92 \text{ N/mm}^2) \text{ OK}$$

Negative-Moment Region After Losses (for continuity effect),

$$\text{End } P = P_i - (\Delta f_s)_u A_{ps}$$

$$= 896 - (46.06)(4.74)$$

$$= 678 \text{ kips } (3020 \text{ kN})$$

$$\text{End Precast } f_t = \frac{678}{789} - \frac{(678)(7.50)}{8908} - \frac{(708+120)(12)}{29{,}370}\ \frac{21.75-7}{21.75}$$

$$= -0.059 \text{ ksi } (0.41 \text{ N/mm}^2) \text{ Tens.}$$

$$< \text{AASHTO } 0.190 \text{ ksi } (1.31 \text{ N/mm}^2) \text{ OK}$$

$$\text{End } f_b = \frac{678}{789} + \frac{(678)(7.50)}{10{,}544} + \frac{(708+120)(12)}{16{,}270}$$

$$= 1.952 \text{ ksi } (13.46 \text{ N/mm}^2)$$

$$< \text{AASHTO } 0.60f'_c = (0.60)(5.00)$$

$$= 3.00 \text{ ksi } (20.69 \text{ N/mm}^2) \text{ OK}$$

The flexural stresses after losses may be reduced somewhat by using the uncracked transformed section properties in Eq. 5.8 and excluding the elastic loss in the calculations. The Iowa Department of Transportation uses this procedure for stresses, for example, both at transfer and after losses. In such cases, the slab and steel are transformed into equivalent precast-beam concrete, in which case different precast girder (E_c)'s—and hence different transformed precast sections—at transfer and, e.g., for age 28 days, would logically be used. Also, differential shrinkage and creep stresses (as illustrated in Ex. 4.1 and Appendix B) might be included in any refined method, but are normally neglected in stress calculations. For these and other reasons, the author prefers to use the gross-section properties as herein, and especially for initial and time-dependent camber and deflection calculations.

Flexural strength in positive-moment region

Design Midspan M_u

$$= 1.30(M_D + \tfrac{5}{3}M_{L+I}) \qquad \text{(Eq. 5.9)}$$

$$= 1.30[(870 + 832 + 60) + (5/3)(733)]$$

$$= 3879 \text{ ft-k } (5260 \text{ kN-m})$$

$$d_c = (e_c + y_t)_{\text{Precast Girder}} + 7$$

$$= 20.86 + 29.27 + 7 = 57.13 \text{ in. } (145 \text{ cm})$$

$$\rho_p = A_{ps}/b_e d = 4.74/(96)(57.13) = 0.000864$$

$$f_{ps} = f_{pu}\left(1 - \frac{\rho_p f_{pu}}{2f'_c}\right) \qquad \text{(Eq. 5.10)}$$

$$= 270\left[1 - \frac{(0.000864)(270)}{(2)(4)}\right]$$

$$= 262 \text{ ksi } (1806 \text{ N/mm}^2)$$

$$1.4 d_c \rho_p f_{ps}/f'_c$$

$$= (1.4)(57.13)(0.000864)(262)/4$$

$$= 4.53 \text{ in. } (11.5 \text{ cm}) < 7 \text{ in. } (17.8 \text{ cm}).$$

Hence, use Eq. 5.11 for rectangular compression block and $f'_c = 4000$ psi (27.6 N/mm²)

$$\omega_p = \rho_p f_{ps}/f'_c = (0.000864)(262)/4$$

$$= 0.057 < 0.30.$$

Hence, use Eq. 5.11 for underreinforced case.

Midspan M_u

$$= \phi A_{ps} f_{ps} d_c\left(1 - \frac{\rho_p f_{ps}}{1.7f'_c}\right) \qquad \text{(Eq. 5.11)}$$

$$= (1.0)(4.74)(262)(57.13)\left[1 - \frac{(0.000864)(262)}{(1.7)(4)}\right] = 68{,}590 \text{ in.-k}$$

= 5716 ft-k (7750 kN-m)

> 3879 ft-k (5260 kN-m) OK

Cracking moment in positive-moment region

$$f_r = 7.84\sqrt{f_c'} = 7.84\sqrt{5000}$$

$$= 554 \text{ psi } (3.82 \text{ N/mm}^2)$$

(Table A.1)

Midspan M_{cr}

$$= \frac{P(Z_c)_b}{A_2} + \frac{Pe_2(Z_c)_b}{(Z_2)_b} + f_r(Z_c)_b$$

(Eq. 5.15)

$$= \frac{(683)(16{,}270)}{789} + \frac{(683)(16{,}270)}{18{,}544} + (0.554)(16{,}270)$$

= 24,150 in.-k (2013 ft-k (2730 kN-m)

Midspan M_u/M_{cr} = 5716/2013 = 2.84 > 1.2 OK

Flexural strength in negative-moment region for $L + I + FWS$

$$\text{Design } M_u = 1.30(M_D + \tfrac{5}{3}M_{L+I})$$

(Eq. 5.9)

$$= 1.30[120 + (5/3)(708)]$$

$$= 1690 \text{ ft-k } (2290 \text{ kN-m})$$

Assume steel centroid approximately at center of slab (d = 57.5 in. = 145 cm). Assume a = 3.3 in. (8.4 cm) and f_y = 40 ksi (276 N/mm²).

$$\text{Req. } A_s = \frac{M_u}{\phi f_y(d - a/2)} = \frac{(1690)(12)}{(1.0)(40)(57.5 - 1.65)}$$

$$= 9.08 \text{ in.}^2 \ (58.6 \text{ cm}^2)$$

Checking,

$$a = \frac{A_s f_y}{0.85(\text{Stem } f_c')_b} = \frac{(9.08)(40)}{(0.85)(5)(26)}$$

= 3.29 in. (8.36 cm) OK

From previous calculation,

Max. End $f_{\text{Comp.}}$ = 1952 psi (13.46 N/mm²)

< AASHTO-0.4f_c' = (0.4)(5000)

= 2000 psi (13.79 N/mm²).

Hence, the effect of precompression due to prestress in the girders may be neglected in the negative-moment calculation of ultimate strength (Section 5.2.1.8).

Use 12 NO. 8 (25-mm) bars,

$$A_s = (12)(0.79) = 9.48 \text{ in.}^2 \ (61.2 \text{ cm}^2)$$

(Fig. 5.4)

$$\rho = A_s/bd = 9.48/(26)(57.5)$$

$$= 0.0063 < 0.015 \quad \text{OK}$$

(Section 5.2.1.8)

It should be noted that only the moment due to $L + I + FWS$ was used above for the negative-moment design by AASHTO, whereas the negative moment due to the total load was used in Ex. 4.2, as discussed at the beginning of Section 4.2.2 by ACI. Examples 4.2 (sections shown in Fig. 4.14) and 4.6 (section shown in Fig. 5.4) were both considered to be unshored construction. Also, see the typical negative-moment joint in Fig. 4.6.

Reinforcement for positive (bottom) connection at piers

For the time-dependent deformation following slab casting (excluding Δ_{L+I} and Δ_{FWS}) from Ex. 5.4, Table 5.9, Δ_t = - 0.60 (Term 4 due to prestress) + 0.41 (Term 6 due to precast-beam dead load + 0.59 (Term 8 due to slab and diaphragm dead load) - 0.05 (Term 9 due to differential shrinkage and creep) = 0.35 in. (0.89 cm) downward, for slab cast 1 month after prestressing Δ_t = - 0.52 + 0.33 + 0.54 - 0.07 = 0.28 in. (0.71 cm) downward, for slab cast 2 months after prestressing.

Using Fig. 5.2 and assuming the same Δ_t/ℓ^2 in the end span (since it was not computed in the example herein) as in the interior spans,

$$\text{Int. Span FEM} = \frac{32E_2I_c}{5\ell^2}\Delta_t$$

$$= \frac{(32)(3890)(638{,}780)}{(5)(90)^2(12)^2}(0.35)$$

$$= 4772 \text{ in.-k}$$

$$= 398 \text{ ft-k } (540 \text{ kN-m})$$

$$\text{End Span FEM} = (48/32)(398)$$

$$= 597 \text{ ft-k } (810 \text{ kN-m})$$

Distributing these fixed end moments, Negative (since deflection is downward) M_t = 484 ft-k (656 kN-m) at the first interior support and 355 ft-k (481 kN-m) at the center support.

For live load plus impact on remote spans, Pos. M_{L+I} = 113 ft-k (153 kN-m) at the first interior support and 59 ft-k (80 kN-m) at the center support. Hence, these moments are more than offset by the Neg. M_t above.

Also,

$$\text{Bot. } (f_{\text{ten}})_{L+I} = \frac{\text{Pos. } M_{L+I}}{Z_b} = \frac{(113{,}000)(12)}{16{,}270}$$

$$= 83 \text{ psi } (0.57 \text{ N/mm}^2)$$

which is a small stress that is also offset by the end moments resulting from the downward time-dependent deflection following slab casting.

Thus, the author recommends the following nominal design for the positive (bottom) connection steel between the precast prestressed girders and the joint concrete:

$$A_{ps}/4 = 4.74/4 = 1.19 \text{ in.}^2 \ (7.7 \text{ cm}^2).$$

This is similar to the AASHTO requirement for continuous reinforced girders that one-fourth of the positive reinforcement be extended beyond the face of the supports.

Use the typical Iowa Department of Transportation nominal design of four prestressed strands in the bottom and four in the top ($A_s = 8 \times 0.153 = 1.22$ in.^{2} = 7.9 cm^2) extended into the joint and bent similar to Fig. 4.6, plus a $\frac{3}{4}$-in. (20-mm) coil tie connecting the bottom flange of the girder to the diaphragm and placed transverse to the direction of traffic (transverse to the girder).

Web shear

From the Iowa Dept. of Transportation computer printout, the maximum shear occurs at the first interior pier in the interior spans.

$$V_{\text{Slab}} = (0.800)(45) + 2(2 \text{ diaphragms})$$

$$= 38.0 \text{ kips } (169 \text{ kN})$$

$$V_{\text{Precast Girder}} = (0.822)(45)$$

$$= 37.0 \text{ kips } (165 \text{ kN})$$

$$V_{FWS} = 6.8 \text{ kips } (30 \text{ kN})$$

$$V_{L+I} = 60.1 \text{ kips } (267 \text{ kN}),$$

truck loading controlling, with lane loading V_{L+I} = 51.6 kips (56.6 kips, truck, controls in end span)

$$V_u = \frac{1.30}{\phi}(V_D + \tfrac{5}{3} V_{L+I})$$

$$= \frac{1.30}{0.90}[38.0 + 37.0 + 6.8 + (5/3)(60.1)]$$

$$= 263 \text{ kips } (1170 \text{ kN})$$

$$jd = d - a/2 = 57.5 - 3.29/2$$

$$= 55.85 \text{ in. } (142 \text{ cm})$$

$$V_c = 0.180 b_w jd = (0.180)(8)(55.85)$$

$$= 80 \text{ kips } (356 \text{ kN})$$

Using No. 4 (13-mm) stirrups, A_s = 0.40 in.^{2} (2.58 cm^2), and 40 ksi (276 N/mm^2) steel,

$$\text{Req. } s = 2A_v f_y jd/(V_u - V_c) \qquad \text{(Eq. 5.19)}$$

$$= (2)(0.40)(40)(55.85)/(263 - 80)$$

$$= 9.77 \text{ in. } (24.8 \text{ cm})$$

$$s_{\max} = A_v f_y / 100 b_w = (0.40)(40{,}000)/(100)(8)$$

$$= 20 \text{ in. } (51 \text{ cm}) \qquad \text{(Eq. 5.20)}$$

$$\text{End } s_{\max} = 3d/4 = (3)(57.5)/4$$

$$= 43 \text{ in. } (108 \text{ cm}) \qquad \text{(Eq. 5.21)}$$

Use No. 4 (13-mm) stirrups at 10 in. (25 cm) c.c. near the first interior piers (the first stirrup, say, is 5 in. = 13 cm from support) of the interior spans, and increase this spacing elsewhere according to the shear envelope. As noted in Section 5.2.1.5 herein, AASHTO requires that web reinforcement for continuous bridges be designed for the full length of interior spans and for the interior three-fourths of the exterior spans.

Horizontal shear

$$Q = (89.7)(7)(21.75 - 3.5)$$

$$= 11{,}460 \text{ in.}^3 \ (187{,}800 \text{ cm}^3)$$

$$v = V_u Q / I_c b_v$$

$$= (263{,}000)(11{,}460)/(638{,}780)(20) \qquad \text{(Eq. 5.22)}$$

$$= 236 \text{ psi } (1.63 \text{ N/mm}^2)$$

$$< 300 \text{ psi } (2.07 \text{ N/mm}^2)$$

Minimum tie requirements are No. 4 (13-mm) stirrups at (0.40/0.22)(12) = 22 in. (56 cm).

Hence, use No. 4 ties at 10 in. (25 cm) c.c., etc., the same as for stirrups above, with the contact surfaces clean and artificially roughened.

End spans and exterior girders

The end spans and exterior girders are usually constructed of the same section, as the interior girders in the interior spans, with the prestressing steel, negative-moment steel, and shear steel designed as previously.

End blocks

See Section 5.2.1.7 herein, and other references for details.

Development length of prestressing strands

From Eq. 5.23, the development length is easily satisfactory.

4.3.2 Camber and Deflection Control

The equations for computing camber (negative or upward movement) and deflection (positive or downward movement) of composite beams are based primarily on the work of Branson and colleagues[17,23,25,27,36-41] and includes the use of his I_e method for computing cracked deflections—which is used in the *PCI Design Handbook.*[4] The effects of composite action, sustained loads applied after the composite beam is formed, nonprestressed tension steel, differential shrinkage and creep, and different-weight concrete are also included.

The ultimate camber and deflection of prestressed concrete composite beams using unshored and shored construction are computed by Eqs. 5.24 through 5.32. For more information on the calculation of results at any time (as opposed to ultimate values), see References 17 and 41 in particular, as well as the other references cited. With the continuous material parameter equations in Appendix A, the procedures herein readily lend themselves to computer programming in which the material and time factors are read-in for a particular problem. Results by this approach for composite prestressed beams have been favorably compared with experimental data for both laboratory specimens and actual structures by Branson and Kripanarayanan.[40]

These equations could be greatly shortened by combining terms (as the simplified Eqs. 5.27 through 5.32), but are presented in the form of separate terms in order to show the individual effects or contributions to the structural deformation, such as due to the prestress force, dead load, creep, and shrinkage, that take place both before and after slab casting.

Although the loss of prestress varies along the span, as shown in Exs. 5.1 and 5.2, it is normally desirable to use the prestress force near midspan in camber calculations. In many cases, the approximate prestress loss ratios in Table 5.3 are satisfactory for computing camber and deflection.

5.2.2.1 *Camber and Deflection of Uncracked Members*

General method for unshored composite beams

$$\Delta_u = -\overbrace{(\Delta_i)_{P_o}}^{(1)} + \overbrace{(\Delta_i)_2}^{(2)} - \overbrace{\left[\frac{\Delta P_s}{P_o} + (\alpha_s k_r C_u)\left(1 - \frac{\Delta P_s}{2P_o}\right)\right](\Delta_i)_{P_o}}^{(3)} - \overbrace{\left\{-\frac{\Delta P_u - \Delta P_s}{P_o} + (k_r C_u)\left[\left(1 - \frac{\Delta P_u}{2P_o}\right) - \alpha_s\left(1 - \frac{\Delta P_s}{2P_o}\right)\right]\right\}(\Delta_i)_{P_o}\frac{I_2}{I_c}}^{(4)}$$

$$+ \overbrace{(\alpha_s k_r C_u)(\Delta_i)_2}^{(5)} + \overbrace{(1 - \alpha_s)(k_r C_u)(\Delta_i)_2 \frac{I_2}{I_c}}^{(6)}$$

$$+ \overbrace{(\Delta_i)_1}^{(7)} + \overbrace{(\beta_s k_r C_u)(\Delta_i)_1 \frac{I_2}{I_c}}^{(8)} + \overbrace{\Delta_{DS}}^{(9)} + \overbrace{\Delta_L}^{(10)}$$

$$+ \overbrace{(\Delta_{cp})_L}^{(11)} \quad (5.24)$$

Term 1 is the *initial camber of the precast*

beam due to the initial prestress moment at transfer (after elastic loss), $P_o e$. (See Appendix D for common cases of prestress moment diagrams, with formulas for computing camber.) $E_{ci} = E_c$ at transfer. For practical reasons, I_g is recommended for camber and deflection calculations.

Term 2 is the *initial dead-load deflection of the precast beam.* For prismatic simply supported precast beams, $(\Delta_i)_2 = 5M_2 \ell^2/48E_{ci}I_2$, where M_2 = midspan moment due to precast-beam dead load, $E_{ci} = E_c$ at transfer, and $I_2 = I_g$.

Term 3 *is the creep (time-dependent) camber of the precast beam, due to the prestress moment, up to the time of slab casting.* In Terms 3 and 4, ΔP_s and ΔP_u are defined as the time-dependent prestress loss (total minus initial or elastic loss) up to the time of slab casting and ultimate value, respectively. See Table 5.3 for approximate values of $\Delta P_s/P_o$ and $\Delta P_u/P_o$. The expressions,

$$\left(1 - \frac{\Delta P_s}{2P_o}\right) \quad \text{and} \quad \left(1 - \frac{\Delta P_u}{2P_o}\right),$$

in Terms 3 and 4 are used to approximate the creep effect resulting from the variable stress history under the decreasing prestress force. As defined in Table 5.2 where representative values are tabulated, α_s is the ratio of the creep coefficient (defined as *ratio of creep strain to initial strain*) up to the time of slab casting to the ultimate creep coefficient. The reduction factor k_r takes into account the effect of any nonprestressed tension steel in reducing time-dependent camber, in much the same way that compression steel reduces time-dependent deflection of the reinforced concrete beam. Based on the work of Shaikh and Branson[38] Eq. 5.25 is recommended. The average ultimate creep coefficient C_u, based on the concrete age of 1 day for steam-cured precast beams, is given in Appendix A (see Table A.8 plus the size-correction factor for the precast beam in, for example, Table A.6 or A.7). A typical value is $C_u = 1.88$ in Table A.8 when $H = 70\%$.

$$k_r = 1/[1 + (A_{ns}/A_{ps})], A_{ns}/A_{ps} \leqslant 2 \tag{5.25}$$

Term 4 is the *creep camber of the composite beam, due to the prestress moment, following slab casting.* $I_c = (I_c)_g$ of the composite section with transformed slab (normally using the 28-day E_c's for both slab and precast beam). The ratio, I_2/I_c, modifies the initial stress (strain) and accounts for the effect of the composite section in restraining additional creep curvature (strain) after the composite beam becomes effective. The other parameters are the same as described in Terms 1 and 3.

Term 5 is the *creep deflection of the precast beam up to the time of slab casting due to the precast-beam dead load.* Since creep under prestress and dead load takes place under their combined stress, the effect of any nonprestressed tension steel (k_r) is also included in Terms 5, 6, and 8 for deflection components as it was in Terms 3 and 4 for camber components.

Term 6 is the *creep deflection of the composite beam following slab casting due to the precast-beam dead load.* All parameters in this term are defined above.

Term 7 is the *initial deflection of the precast beam under slab dead load. For prismatic simply supported precast beams*

$$(\Delta_i)_1 = 5M_1\ell^2/48E_c I_2$$

where M_1 = midspan moment due to slab dead load, E_c at age 28 days will be used, and I_2 is the same as above. When diaphragms, sidewalks, partitions, ceilings, roofing, etc., are applied at the time of slab casting or soon thereafter, their dead load is also included in M_1.

Term 8 is the *creep deflection of the composite beam due to slab (plus other loads in Term 7, if any) dead load.* As defined in Table 5.2 where representative values are tabulated, β_s is the creep correction factor for the age of the precast-beam concrete at the time of slab casting. The other parameters in this term are the same as described previously.

Term 9 is the *deflection due to differential shrinkage and creep*, as presented in Appendix B. This term for prestressed composite beams is the same as Term 8 of Eq. 4.31 for reinforced composite beams, except for the selection of D_u in Table B.1.

Term 10 is the *initial or short-time live load (plus other loads applied to the composite beam and not included in Term 7) deflection of the composite beam* using Eq. C.4 for the deflection of simple and continuous members: $\Delta_L = K(5/48)M_L \ell^2/E_c I_a$, and Eq. C.5 or C.6 for K. M_L = midspan moment due to live load (plus any other loads in this term). E_c (of the precast-beam concrete, since the slab is transformed into equivalent precast-beam concrete) at age 28 days will be used. $I_a = I_c = (I_c)_g$ of the composite section with transformed slab (using the 28-day E_c's for both slab and precast beam) at midspan, unless Eqs. C.2 and C.3 are used to obtain I_a in the case of continuous members (which is normally not recommended).

Term 11 is the *partial live-load (also including other sustained loads in Term 10) creep deflection for any permanent or sustained live load applied to the composite beam:*

$$(\Delta_{cp})_L = (M_P/M_L)k_{rs}C_{up}\Delta_L$$

where M_P is the moment due to the permanent load supported by the composite beam, M_L and Δ_L are the same as in Term 10, and k_{rs} is given by Eq. C.8 to take into account any slab compression steel at midspan—unless Eqs. C.10 and C.11 are used to obtain $(k_{rs})_a$. When the longitudinal slab steel in the positive region consists of temperature or distribution steel only, $k_{rs} = 1$ is considered reasonable. C_{up} is the ultimate creep coefficient based on the slab-concrete age when the permanent load is applied to the composite beam (see Table A.8 plus the size-correction factor for the slab, for example, in Table A.6 or A.7.

It can be seen by inspection that Eq. 5.24 for composite beams reduces to the basic equation for noncomposite prestressed beams of

$$\Delta_u = -(\Delta_i)_{P_o} + (\Delta_i)_D - \left[-\frac{\Delta P_u}{P_o} + (k_r C_u)\left(1 - \frac{\Delta P_u}{2P_o}\right)\right](\Delta_i)_{P_o} + (k_r C_u)(\Delta_i)_D + \Delta_L + (\Delta_{cp})_L$$

when $I_2/I_c = 1$, subscript $2 = D$, and Terms 7 through 9—which refer to the composite beam—are deleted.

General method for shored composite beams, assuming the precast-beam supports its own dead load and the composite beam supports all other loads

$$\Delta_u = \text{Eq. 5.24,} \qquad (5.26)$$

with Terms 7 through 9 modified as follows:

Term 7 is the *initial deflection of the composite beam under slab (plus other dead load applied before the shores are removed, or soon thereafter) dead load*, using Eq. C.4 for the deflection of simple and continuous members: $(\Delta_i)_1 = K(5/48)M_1\ell^2/E_c I_a$, and Eq. C.5 or C.6 for K. M_1 = midspan moment due to the slab (etc.) dead load. E_c (of the precast-beam concrete, since the slab is transformed into equivalent precast-beam concrete) at age 28 days will be used. $I_a = I_c = (I_c)_g$ of the composite section with transformed slab (using the 28-day E_c's for both slab and precast beam) at midspan, unless Eqs. C.2 and C.3 are used to obtain I_a in the case of continuous members (which is normally not recommended).

Term 8 is the *creep deflection of the composite beam under slab (plus other loads used in Term 7, if any) dead load.* This term is the same as Term 8 of Eq. 5.24, except that the ratio, I_2/I_c, is deleted, and β_s is defined as the creep correction factor (given in Table 5.2) for the age of the precast-beam concrete when the shores are removed.

Term 9 is the *deflection due to differential shrinkage and creep.* This term is the same as Term 4 of Eq. 4.32, except for the value of the differential shrinkage and creep coefficient, D_u, in Table B.1.

Equation 5.24 for unshored construction reduces to the simplified Eqs. 5.27a through 5.29b, and Eq. 5.26 for shored construction reduces to the simplified Eqs. 5.30a through 5.32b, when the same representative parameters substituted in the simplified Eqs. 5.2 through 5.4 for loss of prestress are used.

Unshored composite beams, as by Eq. 5.24

Simplified equation for normal-weight concrete (w = 145 pcf = 2320 kg/m^3)

$$\Delta_u = \underbrace{-(\Delta_i)_{P_o}}_{(1)} + \underbrace{(\Delta_i)_2}_{(2)} - \underbrace{[-0.14 + (0.54 \times 1.88k_r)(0.93)](\Delta_i)_{P_o}}_{(3)} - \underbrace{\{-0.04 + (1.88k_r)[0.91 - (0.54)(0.93)]\}(\Delta_i)_{P_o}\frac{I_2}{I_c}}_{(4)}$$

$$+ \overbrace{(0.54 \times 1.88k_r)(\Delta_i)_2}^{(5)}$$

$$+ \overbrace{(0.46)(1.88k_r)(\Delta_i)_2 \frac{I_2}{I_c}}^{(6)} + \overbrace{(\Delta_i)_1}^{(7)} + \overbrace{(0.76 \times 1.88k_r)(\Delta_i)_1 \frac{I_2}{I_c}}^{(8)} + \overbrace{\Delta_{DS}}^{(9)} + \overbrace{\Delta_L}^{(10)} + \overbrace{(\Delta_{cp})_L}^{(11)}$$

$$= - \overbrace{\left(0.86 + 0.94k_r - 0.04 \frac{I_2}{I_c} + 0.77k_r \frac{I_2}{I_c}\right)(\Delta_i)_{P_o}}^{(1+3+4)} + \overbrace{\left(1.00 + 1.02k_r + 0.86k_r \frac{I_2}{I_c}\right)(\Delta_i)_2}^{(2+5+6)} + \overbrace{\left(1.00 + 1.43k_r \frac{I_2}{I_c}\right)(\Delta_i)_1}^{(7+8)}$$

$$+ \overbrace{\Delta_{DS}}^{(9)} + \overbrace{\Delta_L}^{(10)} + \overbrace{(\Delta_{cp})_L}^{(11)} \quad (5.27a)$$

When $k_r = 1$ (no nonprestressed tension steel), Eq. 5.27a reduces to Eq. 5.27b.

$$\Delta_u = - \overbrace{\left(1.80 + 0.73 \frac{I_2}{I_c}\right)(\Delta_i)_{P_o}}^{(1+3+4)} + \overbrace{\left(2.02 + 0.86 \frac{I_2}{I_c}\right)(\Delta_i)_2}^{(2+5+6)} + \overbrace{\left(1.00 + 1.43 \frac{I_2}{I_c}\right)(\Delta_i)_1}^{(7+8)} + \overbrace{\Delta_{DS}}^{(9)} + \overbrace{\Delta_L}^{(10)} + \overbrace{(\Delta_{cp})_L}^{(11)} \quad (5.27b)$$

Simplified equations for sand-lightweight concrete (w = 120 pcf = 1920 kg/m^3)

Δ_u = Same as Eq. 5.27a, except

$$- \left(0.84 + 0.93k_r - 0.05 \frac{I_2}{I_c} + 0.75k_r \frac{I_2}{I_c}\right)(\Delta_i)_{P_o} \quad (5.28a)$$

When $k_r = 1$, Δ_u = Same as Eq. 5.27b, except

$$\left(1.77 + 0.70 \frac{I_2}{I_c}\right)(\Delta_i)_{P_o} \quad (5.28b)$$

Simplified equations for all-lightweight concrete (w = 100 pcf = 1600 kg/m^3)

Δ_u = Same as Eq. 5.27a, except

$$- \left(0.82 + 0.92k_r - 0.05 \frac{I_2}{I_c} + 0.74k_r \frac{I_2}{I_c}\right)(\Delta_i)_{P_o} \quad (5.29a)$$

When $k_r = 1$, Δ_u = Same as Eq. 5.27b, except

$$\left(1.74 + 0.69 \frac{I_2}{I_c}\right)(\Delta_i)_{P_o} \quad (5.29b)$$

Shored composite beams, as by Eq. 5.26

Simplified equations for normal-weight concrete (w = 145 pcf = 2320 kg/m^3)

Δ_u = Same as Eq. 5.27a, except

$$(1.00 + 1.43k_r)(\Delta_i)_1 \quad (5.30a)$$

When $k_r = 1$, Δ_u = Same as Eq. 5.27b, except

$$(2.43)(\Delta_i)_1 \quad (5.30b)$$

Simplified equations for sand-lightweight concrete (w = 120 pcf = 1920 kg/m^3)

Δ_u = Same as Eq. 5.28a, except

$$(1.00 + 1.43k_r)(\Delta_i)_1 \quad (5.31a)$$

When $k_r = 1$, Δ_u = Same as Eq. 5.28b, except

$$(2.43)(\Delta_i)_1 \quad (5.31b)$$

Simplified equations for all-lightweight concrete (w = 100 pcf = 1600 kg/m^3)

Δ_u = Same as Eq. 5.31a, except

$$(1.00 + 1.43k_r)(\Delta_i)_1 \quad (5.32a)$$

When $k_r = 1$, Δ_u = Same as Eq. 5.31b, except

$$(2.43)(\Delta_i)_1 \quad (5.32b)$$

The differences in the simplified camber-and-deflection Eqs. 5.27a through 5.32b for different-weight concrete are due to the differences in the loss of prestress ratios, $\Delta P_s/P_o$ and $\Delta P_u/P_o$.

5.2.2.2 Live-load deflection of partially cracked members

The I-effective (I_e) method, initially developed by the author[22] for reinforced beams, was shown by Shaikh and Branson[38] to apply equally well to bonded prestressed beams, either with or without nonprestressed tension steel, loaded into the cracking range. The effect of nonprestressed steel is included in I_e, and the effect of the prestress force on the neutral axis location is not included in the moment of inertia calculations. This I_e procedure is used in the ACI Code for reinforced beams, slabs, and composite members, and in the *PCI Design Handbook*[4] (1971, 1978) for satisfying the ACI Code provisions on deflections of partially cracked prestressed members.

For computing the live-load deflection increment due to a live-load into the cracking range, as shown in Fig. 5.5, use Eqs. C.1 through C.6, with either of two methods described by Eqs. 5.34 through 5.36 for determining $(M_L)_{cr}/(M_L)_{max}$ in Eq. C.1. Also see Term 10 of Eq. 5.24 for comments regarding continuous uncracked members. $(M_L)_{cr}$ is the additional moment above the prestress and dead-load moment necessary to crack the beam, and $(M_L)_{max}$ is the maximum live-load moment. Computed results by the two methods are the same, as shown in Ex. 5.3. Equations 5.34 through 5.36 apply for beams containing only prestressing steel. See References 17 (Branson) and 40 (Branson and Kripanarayanan) for the corresponding formulas for beams with both prestressed steel and nonprestressed tension steel.

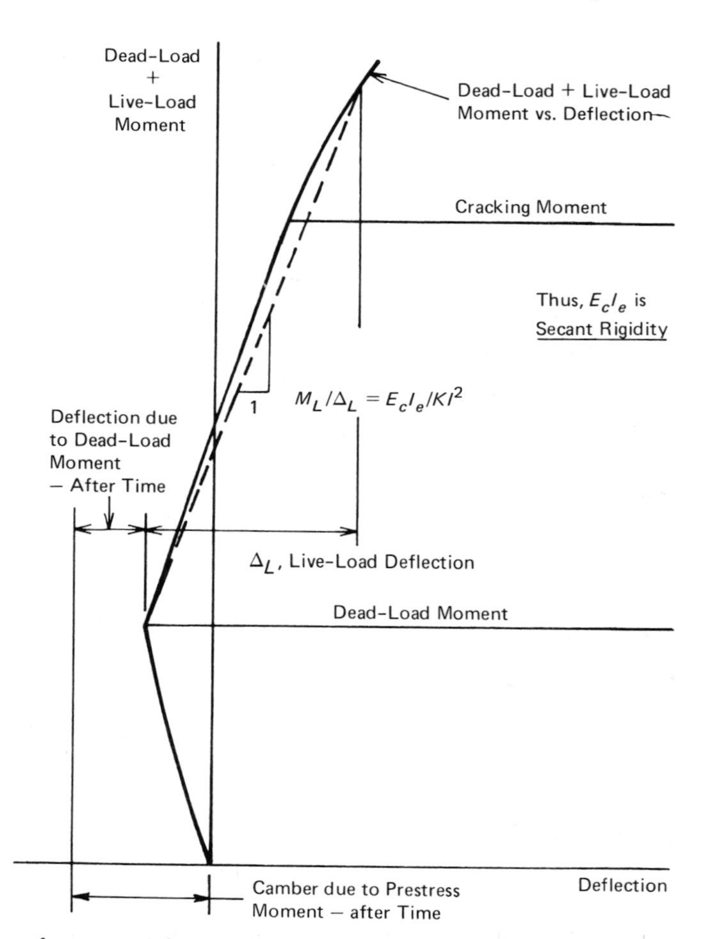

Fig. 5.5. Illustration of moment-deflection increments and flexural rigidity for partially cracked prestressed beams.

Method 1—Direct Determination of $(M_L)_{cr}$

Unshored composite beams

$$f_r = -\frac{P}{(A_g)_2} - \frac{Pe_2(y_{ten})_2}{(I_g)_2} + \frac{(M_1 + M_2)(y_{ten})_2}{(I_g)_2} + \frac{(M_L)_{cr}(y_{ten})_c}{(I_g)_c} \quad (5.33)$$

where subscripts 2 and c refer to the precast and composite sections, respectively, and M_1 and M_2 refer to the slab and precast-beam dead-load moments, respectively. Tension is assumed positive in Eq. 5.33. Solving for $(M_L)_{cr}$,

$$(M_L)_{cr} = \frac{P(I_g)_c}{(A_g)_2(y_{ten})_c} + Pe_2\frac{(y_{ten})_2(I_g)_c}{(y_{ten})_c(I_g)_2} - (M_1 + M_2)\frac{(y_{ten})_2(I_g)_c}{(y_{ten})_c(I_g)_2} + \frac{f_r(I_g)_c}{(y_{ten})_c} \quad (5.34)$$

Shored composite beams

$(M_L)_{cr}$ = [Same as Eq. 5.34, except delete M_1 in the third term] $- M_1$ (5.35)

Method 2—Stress ratio determination of $(M_L)_{cr}/(M_L)_{max}$

$$(M_L)_{cr}/(M_L)_{max} = (f_{pe} - f_D + f_r)/(f_{pe} - f_D + f_{to}) \quad (5.36)$$

where

f_r = modulus of rupture of concrete

f_{to} = total computed tensile stress (above the cracking stress, f_r, but not to exceed the maximum allowable stress, which is $12\sqrt{f'_c}$ by the ACI Code; f_{to} is a fictitious elastic stress

$f_{pe} = P/A_g + Pey_t/I_g$, in which A_g, e, y_t, and I_g refer to the precast-beam section

$f_D = (M_1 + M_2)(y_t)_2/(I_g)_2$ for unshored composite beams

$f_D = M_2(y_t)_2/(I_g)_2 + M_1(y_t)_c/(I_g)_c$ for shored composite beams

Since Eq. 5.36 is in the form of a ratio, the section modulus in both numerator and denominator cancel. Equation 5.36 was used in the development of the *PCI Design Handbook*[4] chart to be used in the calculation of I_e.

5.2.2.3 Allowable deflections

The limiting deflections in the ACI Code are shown in Table 4.6. For more detailed information on allowable deflections, see the ACI Deflection Committee (435) Report.[30]

5.2.2.4 Example 5.3. Camber and deflection of simple span, partially cracked, unshored composite double-T floor beam in Ex. 5.1

See Ex. 5.1, Fig. 5.3, and Table 5.4 for the material properties and design conditions. See Table 5.6 for a summary of the parameters and conditions used in the camber and deflection calculations. The results are summarized and compared with allowable values in Table 5.7.

Solution by the general Eq. 5.24

Initial camber due to prestress—Term 1:

From Case 3, Appendix D, for 1-point harping,

$$(\Delta_i)_{P_o} = \frac{P_o(e_c - e_e)\ell^2}{12E_{ci}I_2} + \frac{P_o e_e \ell^2}{8E_{ci}I_2}$$

$$= \frac{(260)(14.15 - 6.75)(56)^2(12)^2}{(12)(2690)(20{,}985)} + \frac{(260)(6.75)(56)^2(12)^2}{(8)(2690)(20{,}985)}$$

$= -3.038$ in. (77.2 mm)

Initial dead-load deflection of precast beam—Term 2:

$$(\Delta_i)_2 = \frac{5M_2\ell^2}{48E_{ci}I_2} = \frac{(5)(136.42)(12)(56)^2(12)^2}{(48)(2690)(20{,}985)}$$

$= 1.364$ in. (34.7 mm)

Time-dependent camber due to prestress up to the time of slab casting—Term 3:

$$-\left[-\frac{\Delta P_s}{P_o} + (\alpha_s k_r C_u)\left(1 - \frac{\Delta P_s}{2P_o}\right)\right](\Delta_i)_{P_o}$$

$= -\ [-0.16 + (0.54 \times 2.20)(1 - 0.08)]\,(3.038)$

$= -2.83$ in. (71.9 mm)

Time-dependent camber due to prestress following slab casting—Term 4:

$$-\left\{-\frac{\Delta P_u - \Delta P_s}{P_o} + (k_r C_u)\left[\left(1 - \frac{\Delta P_u}{2P_o}\right)\right.\right.$$

Table 5.6. Summary of parameters and conditions for Ex. 5.3.

From Example 5.1, Fig. 5.3, and Table 5.4

Simple Span = 56 ft (17.1 m),

$$e_e = 6.75 \text{ in. } (17.1 \text{ cm})$$

$$e_c = 14.15 \text{ in. } (35.9 \text{ cm})$$

$$DT\, A_g = A_2 = 401 \text{ in.}^2 \ (2590 \text{ cm}^2)$$

$$DT\, I_g = I_2 = 20{,}985 \text{ in.}^4 \ (873{,}400 \text{ cm}^4)$$

$$DT\, Z_b = 1224 \text{ in.}^3 \ (20{,}060 \text{ cm}^3)$$

$$\text{Composite } I_g = I_c = 29{,}853 \text{ in.}^4 \ (1{,}242{,}000 \text{ cm}^4)$$

$$I_2/I_c = 0.703$$

$$\text{Midspan } d_c = 23.00 \text{ in. } (58.4 \text{ cm})$$

$$(y_{ten})_2 = 17.15 \text{ in. } (43.6 \text{ cm})$$

$$(y_{ten})_c = 19.94 \text{ in. } (50.6 \text{ cm})$$

Midspan moments,

$$M_1 = (75.66 + 31.36) = 107.02 \text{ ft-lb } (145.1 \text{ kN-m})$$

$$M_2 = 136.42 \text{ ft-lb } (185.0 \text{ kN-m})$$

$$M_L = 156.80 \text{ ft-lb } (213 \text{ kN-m})$$

$$\text{End } P_o = 268 \text{ kips}$$

$$0.4\text{-Pt } P_o = 260 \text{ kips}$$

$$\text{Midspan } P_o = 255 \text{ kips}$$

Use P_o = 260 kips (1160 kN) in camber calculations
Also Use P = 200 kips (890 kN) in cracking-moment calculations

Other Parameters and Conditions

For steam-cured sand-lightweight DT (120 pcf = 18.9 kN/m^3) with f'_{ci} = 3750 psi (26 N/mm^2), f'_c = 5000 psi (34 N/mm^2); from Tables A.1 and A.2, and Eq. A.9:

$$f_r = 503 \text{ psi } (3.47 \text{ N/mm}^2)$$

$$E_{ci} = 2.69 \times 10^6 \text{ psi } (18.6 \text{ kN/mm}^2)$$

$$E_c = E_2 = 2.93 \times 10^6 \text{ psi } (20.2 \text{ kN/mm}^2)$$

For normal-weight topping with f'_c = 3000 psi (21 N/mm^2); from Table A.2:

$$E_c = E_1 = 3.30 \times 10^6 \text{ psi } (22.8 \text{ kN/mm}^2)$$

k_r = 1 (no nonprestressed tension steel)
Assume Average Relative Humidity = 60% and Avg. DT stem width = 4.5 in. (11.4 cm)
Size Creep $(C.F.)_T$ = 1.08 (Table A.6)
Basic C_u = 2.04 (Table A.8), Use C_u = (2.04)(1.08) = 2.20
Assuming the slab is cast 2 months after prestressing:
From Table B.1 for 70% humidity, $D_u = 375 \times 10^{-6}$ in./in. (mm/mm)
For 2-in. (5.1-cm) topping, Size Shrinkage $(C.F.)_T$ = 1.35 (Table A.6)
Also, $(C.F.)_H$ = 0.70 at 70% and 0.80 at 60% humidity (Table A.5)
Hence, Use $D_u = (1.35)(0.80/0.70)(375 \times 10^{-6}) = 579 \times 10^{-6}$ in./in. (mm/mm)
From Table 5.3, $\Delta P_s/P_o$ = 0.16 and $\Delta P_u/P_o$ = 0.21
From Ex. 5.1,

Table 5.6. (cont.)

0.4-Pt $\Delta P_u/P_o = (53.03 - 11.2 \times 1.688)(1.53)/260 = 0.20$

Midspan $\Delta P_u/P_o = (57.86 - 11.2 \times 1.965)(1.53)/255 = 0.22$

Use $\Delta P_u/P_o = \underline{0.21}$
$\alpha_s = 0.54$, $\beta_s = 0.76$ (Table 5.2)

$$- \alpha_s\left(1 - \frac{\Delta P_s}{2P_o}\right)\Bigg]\Bigg\}(\Delta_i)_{P_o}\,\frac{I_2}{I_c}$$

$$= - \{- 0.05 + (1 \times 2.20)[(1 - 0.105) - (0.54)(1 - 0.08)]\}(3.038)(0.703)$$

$$= -1.76 \text{ in. } (44.7 \text{ mm})$$

Dead-load creep deflection of the precast beam up to the time of slab casting—Term 5:

$$(\alpha_s k_r C_u)(\Delta_i)_2 = (0.54 \times 1 \times 2.20)(1.364)$$

$$= 1.62 \text{ in. } (41.1 \text{ mm})$$

Creep deflection of the composite beam following slab casting due to the precast-beam dead load—Term 6:

$$(1 - \alpha_s)(k_r C_u)(\Delta_i)_2\,\frac{I_2}{I_c}$$

$$= (1 - 0.54)(1 \times 2.20)(1.364)(0.703)$$

$$= 0.97 \text{ in. } (24.6 \text{ mm})$$

Initial deflection of the precast beam under slab plus ceiling dead load—Term 7:

$$(\Delta_i)_1 = \frac{5M_1\ell^2}{48E_cI_2} = \frac{(5)(107.02)(12)(56)^2(12)^2}{(48)(2930)(20{,}985)}$$

$$= 0.983 \text{ in. } (25.0 \text{ mm})$$

Creep deflection of the composite beam under slab plus ceiling dead load—Term 8:

$$(\beta_s k_r C_u)(\Delta_i)_1\,\frac{I_2}{I_c}$$

$$= (0.76 \times 1 \times 2.20)(0.983)(0.703)$$

$$= 1.16 \text{ in. } (29.5 \text{ mm})$$

Differential shrinkage and creep deflection—Term 9:

$$Q_{DS} = (Q \text{ in Eq. B.1})/2 = D_u A_1 E_1/2 \qquad \text{(Eqs. B.1, B.11)}$$

$$= (579)(96 \times 2)(3.30)/2$$

$$= 183{,}400 \text{ lb } (815{,}800 \text{ N})$$

$$y_{cs} = h_c - (y_b)_c - t/2 = 26 - 19.94 - 1$$

$$= 5.06 \text{ in. } (12.9 \text{ cm}) \qquad \text{(Fig. B.1)}$$

$$\Delta_{DS} = \frac{(4/3)K_{DS}Q_{DS}y_{cs}\ell^2}{E_cI_c} \qquad \text{(Eqs. B.7, B.12)}$$

$$= \frac{(4/3)(1/8)(183.40)(5.06)(56)^2(12)^2}{(2930)(29{,}853)}$$

$$= 0.80 \text{ in. } (20.3 \text{ mm})$$

Live-load deflection—Term 10:

The beam was determined to be partially cracked in Ex. 5.1. Hence, use Eqs. C.1, C.4, and 5.34 or 5.36. At midspan,

$$(M_L)_{cr} = \frac{P(I_g)_c}{(A_g)_2(y_{ten})_c} + Pe_2\,\frac{(y_{ten})_2(I_g)_c}{(y_{ten})_c(I_g)_2} - (M_1 + M_2)\frac{(y_{ten})_2(I_g)_c}{(y_{ten})_c(I_g)_2} + \frac{f_r(I_g)_c}{(y_{ten})_c}$$

$$= \frac{(200)(29{,}853)}{(401)(19.94)(12)} + (200)(14.15)\frac{(17.15)(29{,}853)}{(19.94)(20{,}985)(12)} - (107.02 + 136.42)\frac{(17.15)(29{,}853)}{(19.94)(20{,}985)} + \frac{(0.503)(29{,}853)}{(19.94)(12)}$$

$$= 115.7 \text{ ft-k } (157 \text{ kN-m}) \qquad \text{(Eq. 5.34)}$$

$$(M_L)_{cr}/(M_L)_{max} = 115.7/156.8 = 0.738.$$

Alternatively, at midspan,

$$f_{pe} = \frac{200}{401} + \frac{(200)(14.15)}{1224}$$

$$= 2.811 \text{ ksi } (19.4 \text{ N/mm}^2)$$

$$f_D = \frac{(107.02 + 136.42)(12)}{1224}$$

$$= 2.387 \text{ ksi } (16.5 \text{ N/mm}^2)$$

$$f_r = 0.503 \text{ ksi } (3.47 \text{ N/mm}^2)$$

$$f_{to} = 0.833 \text{ ksi } (5.74 \text{ N/mm}^2)$$

(Ex. 5.1)

$(M_L)_{cr}/(M_L)_{max}$

$$= (f_{pe} - f_D + f_r)/(f_{pe} - f_D + f_{to}) \quad \text{(Eq. 5.36)}$$

$$= \frac{2.811 - 2.387 + 0.503}{2.811 - 2.387 + 0.833} = 0.737$$

which is essentially the same as by the other method. Use $0.737^3 = 0.40$.

From Fig. 4.9 at midspan:

$$n_c = 2.93/3.30 = 0.89$$

$$n_{28d} = 28/2.93 = 9.6$$

$$(b_e/n_c)(y_t)^2/2 = nA_s(d_c - y_t)$$

$$(96/0.89)(y_t)^2/2 = (9.6)(1.53)(23.00 - y_t)$$

$$y_t = 2.37 \text{ in. } (6.02 \text{ cm})$$

Neglecting the effect of the 0.37 in. (0.94 cm) of precast-beam concrete having a slightly different E_c,

$$I_{cr} = (b_e/n_c)t^3/3 + nA_s(d_c - y_t)^2$$

$$= (96/0.89)(2.37)^3/3 + (9.6)(1.53)(23.00 - 2.37)^2$$

$$= 6730 \text{ in.}^4 \ (280{,}100 \text{ cm}^4)$$

$$I_e = (M_{cr}/M_a)^3 I_g + [1 - (M_{cr}/M_a)^3] I_{cr} \quad \text{(Eq. C.1)}$$

$$= (0.40)(29{,}853) + (1 - 0.40)(6730)$$

$$= 15{,}980 \text{ in.}^4 \ (665{,}100 \text{ cm}^4)$$

$$\Delta_L = \frac{(5/48)M_L \ell^2}{E_c I_e}$$

$$= \frac{(5/48)(156.80)(12)(56)^2(12)^2}{(2930)(15{,}980)}$$

$$= 1.89 \text{ in. } (48.0 \text{ mm})$$

The camber and deflection results are shown in Table 5.7.

Solution by the simplified Eq. 5.28b

$$\Delta_u = -\left(1.77 + 0.70\frac{I_2}{I_c}\right)(\Delta_i)_{P_o} + \left(2.02 + 0.86\frac{I_2}{I_c}\right)(\Delta_i)_2 + \left(1.00 + 1.43\frac{I_2}{I_c}\right)(\Delta_i)_1 + \Delta_{DS} + \Delta_L \quad \text{(Eq. 5.28b)}$$

$$= -(1.77 + 0.70 \times 0.703)(3.038) + (2.02 + 0.86 \times 0.703)(1.364) + (1.00 + 1.43 \times 0.703)(0.983) + 0.80 + 1.89$$

$$= 1.37 \text{ in. } (34.8 \text{ mm})$$

versus 1.15 in. (29.2 mm) by the more general Eq. 5.24.

The more general solution resulted in a smaller downward deflection than did the simplified solution because the net time-dependent camber due to creep was larger in the general solution, since $C_u = 2.20$ in the general solution (because of the humidity and size correction factors in Table 5.6) compared to the value of $C_u = 1.88$ used in the simplified solution.

5.2.2.5 Example 5.4. Camber and deflection of the interior spans of four-span continuous unshored composite bridge girder in Ex. 5.2

See Ex. 5.2, Fig. 5.4, and Table 5.5 for the material properties and design conditions. See Table 5.8 for a summary of the parameters and conditions used in the camber and deflection calculations. The results are summarized in Table 5.9.

Solution by the general Eq. 5.24

Initial camber due to prestress—Term 1:

From Case 9, Appendix D, for 2-point harping at 3rd points,

$$(\Delta_i)_{P_o} = \frac{P_o(e_c - e_e)}{E_{ci}I_2}\left(\frac{\ell^2}{8} - \frac{a^2}{6}\right) + \frac{P_o e_e \ell^2}{8E_{ci}I_2}$$

$$= \frac{(840)(20.86 - 7.50)}{(3640)(260{,}741)}\left[\frac{90^2}{8} - \frac{30^2}{6}\right](12)^2 + \frac{(840)(7.50)(90)^2(12)^2}{(8)(3640)(260{,}741)}$$

$$= -2.436 \text{ in. } (61.9 \text{ mm})$$

Table 5.7. Term-by-term camber, deflection in inches (mm) for Ex. 5.3.

Initial Camber Due to Prestress	Initial Dead-Load Deflection of Precast Beam	Time-Dependent Prestress Camber up to Time Slab Cast	Time-Dependent Prestress Camber after Slab Cast	Creep Dead-Load Deflection of Precast Beam up to Time Slab Cast	Creep Deflection of Composite Beam after Slab Cast Due to Precast Beam Dead-Load	Initial Deflection of Precast Beam Due to Slab + Ceiling Dead-Load	Creep Deflection of Composite Beam Due to Slab + Ceiling Dead-Load	Differential Shear and Creep Deflection	Ultimate Camber Equals Sum of Terms 1 to 9, and Not Including Δ_L
(1)	(2)	(3)	(4)	(5)	(6)	(7)	(8)	(9)	
−3.04 (77.2)	1.36 (34.7)	−2.83 (71.9)	−1.76 (44.7)	1.62 (41.1)	0.97 (24.6)	0.98 (25.0)	1.16 (29.5)	0.80 (20.3)	−0.74 (18.8)

Live-load deflection increment (see Fig. 5.5) = 1.89 in. (48.0 mm)

Initial camber = $\overset{(1)}{-3.04} + \overset{(2)}{1.36} = -1.68$ in. (42.7 mm)

Camber just prior to slab casting = $\overset{(1)}{-3.04} + \overset{(2)}{1.36} - \overset{(3)}{2.83} + \overset{(5)}{1.62} = -2.89$ in. (73.4 mm)

Camber immediately after slab casting = $-2.89 + \overset{(7)}{0.98} = -1.91$ in. (48.5 mm)

Ultimate camber + live-load deflection = $\overset{(10)}{-0.74} + 1.89 = 1.15$ in. (29.2 mm)

ACI Code (318-71) checks for floor beams (Table 4.6):

Considering the time-dependent deflection that might damage nonstructural elements to occur after slab casting,

$$\Delta_t = \overset{(4)}{-1.76} + \overset{(6)}{0.97} + \overset{(8)}{1.16} + \overset{(9)}{0.80} = 1.17 \text{ in. } (29.7 \text{ mm})$$

Hence, $\Delta_t + \Delta_L = 1.17 + 1.89 = 3.06$ in. (77.7 mm)

1.89 in. (48 mm) $\simeq \ell/360 = (56)(12)/360 = 1.87$ in. (48 mm) OK
3.06 in. (78 mm) $> \ell/480 = 1.40$ in. (36 mm) NG
3.06 in. (78 mm) $> \ell/240 = 2.80$ in. (71 mm) NG

However, the last check might be deemed close enough, especially since the final deflection (ultimate camber + Δ_L) = 1.15 in. (29.2 mm).

Initial dead-load deflection of precast beam—Term 2:

$$(\Delta_i)_2 = \frac{5M_2\ell^2}{48E_{ci}I_2} = \frac{(5)(832)(12)(90)^2(12)^2}{(48)(3640)(260{,}741)}$$

$= 1.278$ in. (32.5 mm)

Time-dependent camber due to prestress up to the time of slab casting—Term 3:

$$-\left[-\frac{\Delta P_s}{P_o} + (\alpha_s k_r C_u)\left(1 - \frac{\Delta P_s}{2P_o}\right)\right](\Delta_i)_{P_o}$$

$= -[-0.10 + (0.44 \times 1.39)(1 - 0.05)](2.436)$

$= 1.17$ in. (29.7 mm)

for slab cast 1 month after prestressing

$= -[-0.14 + (0.54 \times 1.39)(1 - 0.07)](2.436)$

$= -1.36$ in. (34.5 mm)

for slab cast 2 months after prestressing

Time-dependent camber due to prestress following slab casting—Term 4:

$$-\left\{-\frac{\Delta P_u - \Delta P_s}{P_o} + (k_r C_u)\left[\left(1 - \frac{\Delta P_u}{2P_o}\right) - \alpha_s\left(1 - \frac{\Delta P_s}{2P_o}\right)\right]\right\}(\Delta_i)_{P_o}\frac{I_2}{I_c}$$

$= -\{-0.08 + (1.39)[(1 - 0.09) - (0.44)(1 - 0.05)]\}(2.436)(0.408)$

$= -0.60$ in. (15.2 mm)

Table 5.8. Summary of parameters and conditions for Ex. 5.4.

From Ex. 5.2, Fig. 5.4, and Table 5.5

Four-Span Continuous Bridge (50-90-90-50 ft, 15.2-27.4-27.4-15.2 m)

$$e_e = 7.50 \text{ in. } (19.1 \text{ cm})$$

$$e_c = 20.86 \text{ in. } (53.0 \text{ cm})$$

2-Point Depression at 2-Point Dep. at 3rd Points

$$A_2 = 789 \text{ in.}^2 \ (5090 \text{ cm}^2)$$

$$I_2 = 260{,}741 \text{ in.}^4 \ (10{,}758{,}000 \text{ cm}^4)$$

and

$$I_c = 638{,}780 \text{ in.}^4 \ (26{,}356{,}000 \text{ cm}^4)$$

for both positive and negative region values (composite girder is uncracked in both regions, from Ex. 5.2), and also for I_a.

$I_2/I_c = 0.408$

$$M_1 = 870 \text{ ft-k } (1180 \text{ kN-m})$$

$$M_2 = 832 \text{ ft-k } (1130 \text{ kN-m})$$

$$\text{Pos. } M_{L+I} = 733 \text{ ft-k } (994 \text{ kN-m})$$

$$\text{Pos. } M_{FWS} = 60 \text{ ft-k } (81 \text{ kN-m})$$

$$\text{Neg. } M_{L+I} = 708 \text{ ft-k } (960 \text{ kN-m})$$

$$\text{Neg. } M_{FWS} = 120 \text{ ft-k } (163 \text{ kN-m})$$

$$\text{End } P_o = 850 \text{ kips } (3780 \text{ kN})$$

$$\text{Midspan } P_o = 836 \text{ kips } (3720 \text{ kN}).$$

Use P_o = 840 kips (3740 kN) in camber calculations.
Midspan $\Delta P_u/P_o = (44.90 - 12.75)(4.74)/836 = 0.18$ (same as Table 5.3).

Other Parameters and Conditions

For steam-cured normal-weight precast girder and normal-weight slab with

$$\text{girder } f'_{ci} = \text{slab } f'_c = 4000 \text{ psi } (28 \text{ N/mm}^2)$$

$$\text{girder } f'_c = 5000 \text{ psi } (34 \text{ N/mm}^2)$$

from Table A.2:

$$\text{girder } E_{ci} = \text{slab } E_c = 3.64 \times 10^6 \text{ psi } (25.1 \text{ kN/mm}^2)$$

$$\text{girder } E_c = 3.89 \times 10^6 \text{ psi } (26.8 \text{ kN/mm}^2)$$

$k_r = 1$ (no nonprestressed tension steel), Assume Average Relative Humidity = 70%
For $v/s = 4.8$ in. (12.2 cm) (*PCI Design Handbook*), Size Creep $(C.F.)_T = 0.74$ (Table A.7), Basic girder $C_u = 1.88$ (Table A.8), Use girder $C_u = (0.74)(1.88) = 1.39$
For slab $t = 8$ in. (20.3 cm), Shrinkage size $(\text{C.F.})_T = 0.94$ and Creep size $(\text{C.F.})_T = 0.96$ (Table A.6)
Assume future wearing surface (FWS) is applied at or after the slab age of 90 days, and basic slab C_{up} = 1.39 (Table A.8).
Hence, use slab $C_{up} = (0.96)(1.39) = 1.33$
For slab cast 1 month after prestressing:

$$D_u = 270 \times 10^{-6} \text{ in./in. (mm/mm)} \qquad \text{(Table B.1)}$$

Use $D_u = (0.94)(270 \times 10^{-6}) = 254 \times 10^{-6}$ in./in. (mm/mm)
$\alpha_s = 0.44$, $\beta_s = 0.83$ (Table 5.2), $\Delta P_s/P_o = 0.10$ (Table 5.3)

For slab cast 2 months after prestressing:

$$D_u = 375 \times 10^{-6} \text{ in./in. (mm/mm)} \qquad \text{(Table B.1)}$$

Table 5.8. (cont.)

Use $D_u = (0.94)(375 \times 10^{-6}) = 353 \times 10^{-6}$ in./in. (mm/mm)
$\alpha_s = 0.54$, $\beta_s = 0.76$ (Table 5.2), $\Delta P_s/P_o = 0.14$ (Table 5.3)

---

for slab cast 1 month after prestressing

$$= -\{-0.04 + (1.39)[(1 - 0.09) - (0.54)(1 - 0.07)]\}(2.436)(0.408)$$

$$= -0.52 \text{ in. } (13.2 \text{ mm})$$

for slab cast 2 months after prestressing

Dead-load creep deflection of the precast beam up to the time of slab casting—Term 5:

$$(\alpha_s k_r C_u)(\Delta_i)_2 = (0.44 \times 1.39)(1.278)$$

$$= 0.78 \text{ in. } (19.8 \text{ mm})$$

for slab cast 1 month after prestressing

$$= (0.54 \times 1.39)(1.278)$$

$$= 0.96 \text{ in. } (24.4 \text{ mm})$$

for slab cast 2 months after prestressing

Creep deflection of the composite beam following slab casting due to the precast-beam dead load—Term 6:

$$(1 - \alpha_s)(k_r C_u)(\Delta_i)_2 \frac{I_2}{I_c}$$

$$= (1 - 0.44)(1.39)(1.278)(0.408)$$

$$= 0.41 \text{ in. } (10.4 \text{ mm})$$

for slab cast 1 month after prestressing

$$= (1 - 0.54)(1.39)(1.278)(0.408)$$

$$= 0.33 \text{ in. } (8.4 \text{ mm})$$

for slab cast 2 months after prestressing

Initial deflection of the precast beam under slab plus diaphragm dead load—Term 7:

$$(\Delta_i)_1 = \frac{5M_1\ell^2}{48E_cI_2} = \frac{(5)(870)(12)(90)^2(12)^2}{(48)(3890)(260{,}741)}$$

$$= 1.251 \text{ in. } (31.8 \text{ mm})$$

Creep deflection of the composite beam under slab plus diaphragm dead load—Term 8:

$$(\beta_s k_r C_u)(\Delta_i)_1 \frac{I_2}{I_c} = (0.83 \times 1.39)(1.251)(0.408)$$

$$= 0.59 \text{ in. } (15.0 \text{ mm})$$

for slab cast 1 month after prestressing

$$= (0.76)(1.39)(1.251)(0.408)$$

$$= 0.54 \text{ in. } (13.7 \text{ mm})$$

for slab cast 2 months after prestressing.

The effect of the composite section is included in Terms 4, 6, and 8 (which are time-dependent terms following slab casting) in the same way as for simple spans (without any effect of continuity), because the initial strains and curvatures in these terms are induced in the precast girder as simple spans. However, the effect of continuity must be included in the following Terms 9 through 11.

Differential shrinkage and creep camber (interior spans)—Term 9:

$$y_{cs} = h_c - (y_b)_c - t/2 = 62 - 39.25 - 4$$

$$= 18.75 \text{ in. } (47.6 \text{ cm}) \quad \text{(Fig. B.1)}$$

$$Q_{DS} = (Q \text{ in Eq. B.1})/2 = D_u A_1 E_1/2 \quad \text{(Eqs. B.1, B.11)}$$

$$= (254)(96 \times 8)(3.64)/2$$

$$= 355{,}000 \text{ lb } (1579 \text{ kN})$$

for slab cast 1 month after prestressing

$$= (353)(96 \times 8)(3.64)/2$$

$$= 493{,}400 \text{ lb } (2195 \text{ kN})$$

for slab cast 2 months after prestressing

Even though the spans are unequal, $K_{DS} = -0.011$ at the 3rd point (from Fig. B.3) for the interior spans may be used in this case as an approximation at midspan.

$$\Delta_{DS} = \frac{(4/3)K_{DS}Q_{DS}y_{cs}\ell^2}{E_cI_c} \quad \text{(Eqs. B.7, B.12)}$$

$$= \frac{(4/3)(-0.011)(355.0)(18.75)(90)^2(12)^2}{(3640)(638{,}780)}$$

Table 5.9. Term-by-term camber, deflection in inches (mm) for Ex. 4.8.

	Initial Camber Due to Prestress	Initial Dead-Load Deflection of Precast Beam	Time-Dependent Prestress Camber up to Time Slab Cast	Time-Dependent Prestress Camber after Slab Cast	Creep Dead-Load Deflection of Precast Beam up to Time Slab Cast	Creep Deflection of Composite Beam after Slab Cast Due to Precast-Beam Dead-Load	Initial Deflection of Precast Beam Due to Slab + Diaphragm Dead-Load	Creep Deflection of Composite Beam Due to Slab + Diaphragm Dead-Load	Differential Shear and Creep Camber	Initial Deflection of Composite Beam Due to FWS	Creep Deflection of Composite Beam Due to FWS	Ultimate Deflection Equals Sum of Terms 1 to 11, Not Including Δ_{L+I}
	(1)	(2)	(3)	(4)	(5)	(6)	(7)	(8)	(9)	(10b)	(11)	
Slab Cast 1 Month After Prestressing	−2.44 (61.9)	1.28 (32.5)	−1.17 (29.7)	−0.60 (15.2)	0.78 (19.8)	0.41 (10.4)	1.25 (31.8)	0.59 (15.0)	−0.05 (1.2)	0.02 (0.5)	0.03 (0.8)	0.10 (2.5)
Slab Cast 2 Months After Prestressing	Same	Same	−1.36 (34.5)	−0.52 (13.2)	0.96 (24.4)	0.33 (8.4)	Same	0.54 (13.7)	−0.07 (1.8)	Same	Same	0.02 (0.5)
Live load plus impact deflection (Term 10a) = 0.37 in. (9.4 mm)												

Initial camber $= \underset{(1)}{-2.44} + \underset{(2)}{1.28} = -1.16$ in. (29.5 mm)

Camber just prior to slab casting $= \underset{(1)}{-2.44} + \underset{(2)}{1.28} - \underset{(3)}{1.17} + \underset{(5)}{0.78} = -1.55$ in. (39.4 mm) for slab cast 1 month after prestressing
$= -2.44 + 1.28 - 1.36 + 0.96 = -1.56$ in. (39.6 mm) for slab cast 2 months after prestressing

Camber immediately after slab casting $= -1.55 + \underset{(7)}{1.25} = -0.30$ in. (7.6 mm) for slab cast 1 month after prestressing
$= -1.56 + 1.25 = -0.31$ in. (7.9 mm) for slab cast 2 months after prestressing

Ultimate deflection $+ \Delta_{L+I} = 0.10 + \underset{(10a)}{0.37} = 0.47$ in. (11.9 mm) for slab cast 1 month after prestressing
$= 0.02 + 0.37 = 0.39$ in. (9.9 mm) for slab cast 2 months after prestressing

$= -0.049$ in. (1.2 mm)

for slab cast 1 month after prestressing

$= (493.4/355.0)(-0.049)$

$= -0.07$ in. (1.8 mm)

for slab cast 2 months after prestressing

Live load plus impact deflection—Term 10a:

From Ex. 5.2, the bridge girder (both precast and composite sections) remains uncracked in both positive- and negative-moment regions. Hence, the gross section $I = I_c$ of the composite section is used as the average I.

The deflection is computed at midspan (using Eq. C.4) based on the loading position for maximum positive moment and hence deflection (and not including the negative-moment envelope). Use

M_m = Pos. M_{L+I} = 733 ft-k

and

M_o = (lane load moment for 90-ft simple span/2

$= 1344 \text{ ft-k}/2)(8/5.5)\left(1 + \frac{50}{90 + 125}\right)$

= 1205 ft-k (1630 kN-m)

$$K = 1.20 - 0.20 M_o/M_m \qquad \text{(Eq. C.5)}$$

for uniform loading and used here as an approximation

$= 1.20 - (0.20)(1205/733) = 0.87$

compared to $K = 1$ for simple spans and $K = 0.60$ for fixed-fixed spans under uniform loading.

$$\Delta_{L+I} = \frac{K(5/48)M_{L+I}\ell^2}{E_c I_c} \qquad \text{(Eq. C.4)}$$

$$= \frac{(0.87)(5/48)(733)(12)(90)^2(12)^2}{(3890)(638{,}780)}$$

= 0.37 in. (9.4 mm)

Future wearing surface initial deflection—Term 10b:

M_m = Pos M_{FWS} = 60 ft-k

$M_o = w\ell^2/8 = (0.020 \times 8)(90)^2/8$

= 162 ft-k

$K = 1.20 - 0.20 M_o/M_m$

$= 1.20 - (0.20)(162/60) = 0.66$

(Eq. C.5)

$$(\Delta_i)_{FWS} = \frac{K(5/48)M_{FWS}\ell^2}{E_c I_c} \qquad \text{(Eq. C.4)}$$

$$= \frac{(0.66)(5/48)(60)(12)(90)^2(12)^2}{(3890)(638{,}780)}$$

= 0.02 in. (0.5 mm)

Creep deflection due to future wearing surface—Term 11:

$(\Delta_{cp})_{FWS}$ = (Slab C_{up})$(\Delta_i)_{FWS}$

$= (1.33)(0.02) = 0.03$ in. (0.8 mm)

The camber and deflection results are shown in Table 5.9.

Solution by the simplified Eq. 4.61b

$$\Delta_u = -\left(1.80 + 0.73\frac{I_2}{I_c}\right)(\Delta_i)_{P_o} + \left(2.02 + 0.86\frac{I_2}{I_c}\right)(\Delta_i)_2 + \left(1.00 + 1.43\frac{I_2}{I_c}\right)(\Delta_i)_1 + \Delta_{DS} + \Delta_{L+I+FWS} + (\Delta_{cp})_{FWS}$$

(Eq. 5.27b)

$= -(1.80 + 0.73 \times 0.408)(2.436) + (2.02 + 0.86 \times 0.408)(1.278) + (1.00 + 1.43 \times 0.408)(1.251) - (0.05 \text{ or } 0.07, \text{ use } 0.06) + (0.37 + 0.02) + 0.03$

= 0.26 in. (6.6 mm)

versus 0.47 in. (11.9 mm) and 0.39 in. (9.9 mm) by the more general Eq. 5.24 for slabs cast at 1 and 2 months, respectively, after prestressing.

The more general solution resulted in a larger downward deflection than did the simplified solution because the net time-dependent camber due to creep was smaller in the general solution, since $C_u = 1.39$ in the general solution (because of the size-correction factor in Table 5.8) compared to the value of $C_u = 1.88$ used in the simplified solution.

REFERENCES

1. ACI-ASCE Committee 333, Chairman, I. M. Viest, Tentative Recommendations for Design of Com-

posite Beams and Girders for Buildings, *ACI Journal, Proceedings*, Vol. 57, No. 6, Part 1, Dec. 1960, pp. 609–628; also *Journal of the Structural Division, Proceedings*, ASCE, Vol. 86, No. ST 12, Dec. 1960.

2. American Concrete Institute, *Building Code Requirements for Reinforced Concrete*, pp. 1–144, and *Commentary on Building Code Requirements for Reinforced Concrete*, pp. 1–91, (ACI 318-63), Detroit, Michigan, 1963.
3. American Concrete Institute, *Building Code Requirements for Reinforced Concrete*, pp. 1–78, and *Commentary on Building Code Requirements for Reinforced Concrete*, pp. 1–96, (ACI 318-71), Detroit, Michigan, 1971.
4. Prestressed Concrete Institute, *PCI Design Handbook*, Chicago, Illinois, 1971, pp. 1-1 to 12-8.
5. British Standards Institution, *Code of Practice for The Structural Use of Concrete, Part 1. Design, Materials, Workmanship*, CP 110, 1972, pp. 1–155; Cement and Concrete Association, *Handbook on the Unified Code for Structural Concrete*, London, 1972, pp. 1–153.
6. American Association of State Highway and Transportation Officials, *Standard Specifications for Highway Bridges*, Washington, D.C., 1973, pp. 1–469.
7. Wang, C. K. and Salmon, C. G., *Reinforced Concrete Design*, First Edition, International Textbook Company, Scranton, Pennsylvania, 1965, pp. 1–754.
8. Evans, R. H. and Bennett, E. W., *Pre-Stressed Concrete, Theory and Design*, John Wiley & Sons, New York, 1958, pp. 1–294.
9. Lin, T. Y., *Design of Prestressed Concrete Structures*, Second Edition, John Wiley & Sons, New York, 1963, pp. 1–614.
10. Preston, H. K. and Sollenberger, N. J., *Modern Prestressed Concrete*, McGraw-Hill Book Company, New York, 1967, pp. 1–337.
11. Khachaturian, N. and Gurfinkel, G., *Prestressed Concrete*, McGraw-Hill Book Company, New York, 1969, pp. 1–460.
12. Winter, G. and Nilson, A. H., *Design of Concrete Structures*, Eighth Edition, McGraw-Hill Book Company, New York, 1972, pp. 1–615.
13. Ferguson, P. M., *Reinforced Concrete Fundamentals*, Third Edition, John Wiley & Sons, New York, 1973, pp. 1–750.
14. Wang, C. K. and Salmon, C. G., *Reinforced Concrete Design*, Second Edition, Intext Press, New York, 1973, pp. 1–934.
15. Libby, J. R., *Modern Prestressed Concrete*, Van Nostrand Reinhold, New York, 1971, pp. 1–516.
16. Libby, J. R., Prestressed Concrete, *Handbook of Concrete Engineering*, Chapter 9, edited by M. Fintel, Van Nostrand Reinhold, New York, 1974, pp. 249–286.
17. Branson, D. E., *Deformation of Concrete Structures*, McGraw-Hill International Book Company, Center for Advanced Publishing, Dusseldorf and New York, 1977, pp. 1–546.
18. Neville, A. M. and Dilger, W., *Creep of Concrete: Plain, Reinforced, and Prestressed*, North-Holland Publishing Company, Amsterdam, 1970, pp. 1–622.
19. Portland Cement Association, Notes on ACI 318-71, *Building Code Requirements with Design Applications*, Portland Cement Association, Skokie, Illinois, 1972, pp. 1–1 to 26–27.
20. Gergely, P. and Lutz, L. A., Maximum Crack Width in Reinforced Concrete Flexural Members, Causes, Mechanism, and Control of Cracking in Concrete, *SP-20, American Concrete Institute*, Detroit, 1968, pp. 1–17.
21. Building Code Subcommittee, ACI Committee 435; D. E. Branson, Subcommittee Chairman; G. M. Sabnis, Committee 435 Chairman, *Revised ACI Code and Commentary Provision on Deflections, Section 9.5*, Dec. 1975, pp. 1–26.
22. Branson, D. E., *Instantaneous and Time-Dependent Deflections of Simple and Continuous Reinforced Concrete Beams*, HPR Publication No. 7, Part 1, Alabama Highway Department, Bureau of Public Roads, Aug. 1963 (1965), pp. 1–78.
23. Branson, D. E., Time-Dependent Effects in Composite Concrete Beams, *ACI Journal, Proceedings*, Vol. 61, No. 2, Feb. 1964, pp. 213–230.
24. ACI Committee 435, D. E. Branson, Chairman, Deflections of Reinforced Concrete Flexural Members, *ACI Journal, Proceedings*, Vol. 63, No. 6, Part 1, June 1966, pp. 637–674. Republished in *ACI Manual of Concrete Practice*, Part 2, 1967, 1972, 1974, 1976.
25. Branson, D. E., Design Procedures for Computing Deflections, *ACI Journal, Proceedings*, Vol. 65, No. 9, Sept. 1968, pp. 730–742.
26. Branson, D. E., Compression Steel Effect on Long-Time Deflections, *ACI Journal, Proceedings*, Vol. 68, No. 8, Aug. 1971, pp. 555–559.
27. Subcommittee 2, ACI Committee 209 (D. E. Branson, Subcommittee 2 Chairman; J. R. Keeton, Committee 209 Chairman), *Prediction of Creep, Shrinkage, and Temperature Effects in Concrete Structures*, Designing For Effects of Creep, Shrinkage and Temperature in Concrete Structures, *ACI Publication SP 27-3*, 1971, pp. 51–93.
28. Kripanarayanan, K. M. and Branson, D. E., Short-Time Deflections of Beams Under Single and Repeated Load Cycles, *ACI Journal, Proceedings*, Vol. 69, No. 2, Feb. 1972, pp. 110–117.
29. Kripanarayanan, K. M. and Branson, D. E., *Some Experimental Studies of Time-Dependent Deflections of Noncomposite and Composite Reinforced Concrete Beams*, ACI Publication SP 43-16, Deflections of Concrete Structures, 1974, pp. 409–419.
30. Subcommittee 1, ACI Committee 435; R. S. Fling, Subcommittee 1 Chairman; D. E. Branson, Committee 435 Chairman; Allowable Deflections, *ACI*

Journal, Proceedings, Vol. 65, No. 6, June 1968, pp. 433–444. Republished in *ACI Manual of Concrete Practice*, Part 2, 1972, 1974, 1976.

31. PCI Committee on Prestress Losses (H. K. Preston, Chairman), Recommendations for Estimating Prestress Losses, *PCI Journal*, Vol. 20, No. 4, July-Aug. 1975, pp. 43–75.
32. MacGregor, J. G. and Hanson, J. M., Proposed Changes in Shear Provisions for Reinforced and Prestressed Concrete Beams, *ACI Journal, Proceedings*, Vol. 66, Apr. 1969, pp. 276–288.
33. Hanson, N. W. and Kaar, P. H., Flexural Bond Tests of Pretensioned Prestressed Beams, *ACI Journal, Proceedings*, Vol. 55, Jan. 1955, pp. 783–802.
34. Freyermuth, C. L., Design of Continuous Highway Bridges with Precast Prestressed Concrete Girders, *PCI Journal*, Vol. 14, No. 2, Apr. 1969, pp. 14–36. Also, *Engineering Bulletin*, EB014.01E, Portland Cement Association, Aug. 1969.
35. Mattock, A. H., Precast-Prestressed Concrete Bridges: 5. Creep and Shrinkage Studies, *PCA Journal*, Vol. 3, No. 2, May 1961, pp. 32–66.
36. Branson, D. E. and Ozell, A. M., Camber in Prestressed Concrete Beams, *ACI Journal, Proceedings*, Vol. 32, No. 12, June 1961, pp. 1549–1574.
37. Subcommittee 5, ACI Committee 435; A. C. Scordelis, Subcommittee Chairman; D. E. Branson, Committee 435 Chairman; Deflections of Prestressed Concrete Members, *ACI Journal, Proceedings*, Vol. 60, No. 12, Dec. 1963, pp. 1697–1728. Republished in *ACI Manual of Concrete Practice*, Part 2, 1967, 1972, 1974, 1976.
38. Shaikh, A. F. and Branson, D. E., Non-Tensioned Steel in Prestressed Concrete Beams, *PCI Journal*, Vol. 15, No. 1, Feb. 1970, pp. 14–36.
39. Branson, D. E., Meyers, B. L., and Kripanarayanan, K. M., *Time-Dependent Deformation of Noncomposite and Composite Prestressed Concrete Structures*, Highway Research Record, No. 324, Symposium on Concrete Deformation, 1970, pp. 15–43.
40. Branson, D. E. and Kripanarayanan, K. M., Loss of Prestress, Camber, and Deflection of Noncomposite and Composite Prestressed Concrete Structures (four papers presented by D. E. Branson on various aspects at the 6th Congress, Federation Internationale de la Precontrainte, Prague, June 1970; and at Design Seminars sponsored by the Prestressed Concrete Institute, Chicago, Jan. 1971 and June 1971; and by the California Division of Highways and Ceramic Lightweight Aggregate Association, Sacramento, Mar. 1971), *PCI Journal*, Vol. 16, No. 5, Sept.-Oct. 1971, pp. 22–52.
41. Branson, D. E., The Deformation of Noncomposite and Composite Prestressed Concrete Members, *ACI Publication SP 43-4*, Deflections of Concrete Structures, 1974, pp. 83–127.
42. Branson, D. E. and Christiason, M. L., Time-Dependent Concrete Properties Related to Design—Strength and Elastic Properties, Creep and Shrinkage, Designing for Effects of Creep, Shrinkage, and Temperature in Concrete Structures, *ACI Publication SP 27-13*, 1971, pp. 257–277.
43. Pauw, A., Static Modulus of Elasticity of Concrete as Affected by Density, *ACI Journal, Proceedings*, Vol. 57, No. 6, Dec. 1960, pp. 679–687.
44. Meyers, B. L., Branson, D. E., and Schumann, C. G., Prediction of Creep and Shrinkage Behavior for Design from Short-Term Tests, *PCI Journal*, Vol. 17, No. 3, May–June 1972, pp. 29–45.
45. Meyers, B. L. and Branson, D. E., Design Aid for Predicting Creep and Shrinkage of Concrete, *ACI Journal, Proceedings*, Vol. 69, No. 9, Sept. 1972, pp. 551–555.
46. Comite Europeen Du Beton—Federation Internationale de la Pre-Contrainte, International Recommendations for the Design and Construction of Concrete Structures, *Cement and Concrete Association*, London, June 1970, pp. 1–80.
47. Subcommittee 7, ACI Committee 435 (R. S. Fling and A. F. Shaikh, Subcommittee Chairmen; B. L. Meyers and G. M. Sabnis, Committee 435 Chairmen), Deflections of Continuous Concrete Beams, *ACI Journal, Proceedings*, Vol. 70, No. 12, Dec. 1973, pp. 781–787. Republished in *ACI Manual of Concrete Practice*, Part 2, 1976.
48. Branson, D. E., Chapter 5—"Design for Deflections," *Metric Design Handbook for Reinforced Concrete Elements*, Canadian Portland Cement Association, Ottawa, Canada, Editor, M. Fintel, 1978, pp. 5-1 to 5-61.
49. Branson, D. E., *Deflexiones de Estructuras de Concreto Reforzado y Presforzado*, Serie Concreto Estructural, CE-1, Instituto Mexicano del Cemento y del Concreto, Mexico City, 1978, pp. 1–130.
50. Building Code Subcommittee, ACI Committee 435; D. E. Branson, Subcommittee Chairman; J. R. Libby, G. M. Sabnis, Committee 435 Chairmen; "Proposed Revisions by Committee 435 to ACI Buiding Code and Commentary Provisions on Deflections," *ACI Journal*, *Proceedings* Vol. 75, No. 6, June 1978, pp. 229–238.

NOTATION

A = cross-sectional area

A = effective tension area of concrete surrounding the main tension reinforcing bars and having the same centroid as that reinforcement, divided by the number of bars, in.^{2} When the main reinforcement consists of several bar sizes, the number of bars shall be computed as the total steel area divided by the area of the largest bar used

A_g = area of gross section, neglecting the steel

A_{ns} = area of nonprestressed tension steel in prestressed member

A_{pf} = portion of prestressing steel required to develop the compressive strength of the overhanging flanges

A_{ps} = area of prestressing steel

A_{pw} = portion of prestressing steel required to develop the compressive strength of the web

A_s = area of tension steel in reinforced member

A_{sf} = area of steel to develop compressive strength of overhanging flanges

A_{sh} = shrinkage coefficient

A'_s = area of compression steel in reinforced member

A_v = area of shear steel

A_{vf} = area of shear-friction steel

A_1 = area of slab

A_2 = area of precast beam

a = depth of equivalent rectangular stress block

b = width of compression face of member

b_e = effective flange width

b_v = width of the cross section being investigated for horizontal shear

b_w = web width

b' = minimum width of web of a flanged member

b_{11} = parameter used in prestress loss equation

C = creep coefficient defined as the ratio of creep strain to initial strain

C = compressive force

$C.F.$ = correction factor for creep and shrinkage

$(C.F.)_H$ = correction factor for humidity

$(C.F.)_{LA}$ = correction factor for loading age

$(C.F.)_T$ = correction factor for member size or average thickness

C_t = creep coefficient at any time t

C_u = ultimate (in time) creep coefficient

C_{u1} = ultimate creep coefficient based on the age of the precast-beam concrete when the slab is cast

C_{u2} = ultimate creep coefficient based on the concrete age when the precast-beam dead-load moment is applied

C_{up} = ultimate creep coefficient based on the slab-concrete age when the permanent load is applied to the composite beam

C_{us} = ultimate creep coefficient based on the slab-concrete age when the shores are removed

D = differential shrinkage and creep strain or coefficient

D_u = ultimate differential shrinkage and creep coefficient

d = effective depth of section = distance from extreme compression fiber to centroid of tension steel

d_b = nominal diameter of bar, wire, or strand

d_c = effective depth of composite section

d_c = thickness of concrete cover measured from the extreme tension fiber to the center of the bar located closest thereto

d_p = effective depth of precast section

E = modulus of elasticity

E_c = modulus of elasticity of concrete

E_{ci} = modulus of elasticity of concrete at the time of initial loading

E_s = modulus of elasticity of steel

E_1 = modulus of elasticity of slab concrete

E_2 = modulus of elasticity of precast-beam concrete

e = eccentricity of steel

e_c = eccentricity of steel at center of span

e_e = eccentricity of steel at end of span
f = concrete or steel stress
f_c = concrete stress
f_{ci} = initial concrete stress (concrete stress due to initial loading, such as at the steel centroid)
f_{cs} = concrete stress at the steel centroid for any loading time
f_{ct} = splitting tensile strength of concrete
f_{pe} = compressive stress in concrete due to prestress only after all losses, at the extreme fiber of a section at which tensile stresses are caused by applied loads
f_{ps} = stress in prestressing steel at design (ultimate) load
f_{pu} = ultimate strength of prestressing steel
f_r = modulus of rupture of concrete
f_s = steel stress
f_{se} = effective stress in prestressing steel after losses
f_{si} = initial or tensioning stress in prestressing steel
f_{to} = total computed tensile stress in concrete (as after cracking in a prestressed member, but based on the uncracked–gross –section)
f_y = yield point or yield strength of steel
f'_c = compressive strength of concrete
f'_{ci} = compressive strength of concrete at the time of initial loading
$\sqrt{f'_c}$, $\sqrt{f'_{ci}}$ = square root of concrete compressive strength—same units as f'_c, f'_{ci} (psi) in empirical expressions
$(f'_c)_t$ = compressive strength of concrete at any time t
$(f'_c)_{7d}$, $(f'_c)_{28d}$ = compressive strength of concrete at age 7, 28, days, etc.
$(f'_c)_u$ = ultimate (in time) compressive strength of concrete
f'_t = direct tensile strength of concrete
Δf_s = loss of stress in prestressing steel (loss of prestress)
$(\Delta f_{sds})_u$ = ultimate loss of prestress due to differential shrinkage and creep
$(\Delta f_{sr})_u$ = ultimate loss of prestress due to steel relaxation
$(\Delta f_s)_u$ = ultimate loss of prestress
H = ambient relative humidity in percent
h = overall thickness of member
I = moment of inertia (second moment of the area) of a section
I_a = average moment of inertia
I_c = moment of inertia of composite section with transformed slab. The slab width is divided by E_2/E_1
I_{cr} = moment of inertia of cracked transformed section
I_e = effective moment of inertia
I_g = moment of inertia of gross concrete section, neglecting the steel
I_1 = moment of inertia of slab
I_2 = moment of inertia of precast beam
I_{2s} = moment of inertia of cracked precast beam, as defined in Term 6 of Eq. 4.31 and in Eq. 4.33
jd = internal lever arm by the elastic theory
K = deflection coefficient
K_{DS} = differential shrinkage and creep-deflection coefficient
K_{sh} = shrinkage warping deflection coefficient
kd = depth of neutral axis by the elastic theory
k_r = reduction factor
$(k_r)_a$ = average value of k_r
k_{rs} = reduction factor for slab compression steel (if any)
ℓ = span length
ℓ_n = clear span
M = bending moment
M_a = maximum service-load moment (unfactored moment) at

the stage for which deflections are being considered
M_{CL} = temporary construction live-load moment
M_{cr} = cracking moment
M_D = dead-load moment
M_L = live-load moment
M_{L+I} = live-load plus impact moment
M_m = net moment at midspan
M_o = beam statical moment
M_P = moment due to permanent load supported by composite beam
M_u = ultimate moment
M_1 = moment under slab dead load, etc.
M_2 = moment under precast-beam dead load, etc.
m = modular ratio at the time of additional load application, such as the time of slab casting for a composite beam
$m = f_y/0.85f'_c$
n = modular ratio at the time of loading, such as the time of prestressing or at 28 days—$n = E_{steel}/E_{stem} = E_s/E_{ci}$ or E_s/E_c
$n_c = E_{stem}/E_{slab}$
P = prestress force after losses
P_i = initial tensioning force
P_o = prestress force at transfer (after elastic loss)
ΔP = loss of prestress force
ΔP_s = total loss of prestress at the time of slab casting minus the initial elastic loss
ΔP_u = total ultimate loss of prestress minus the initial elastic loss
p = tension steel percentage = $100A_s/bd$
p' = compression steel percentage = $100A'_s/bd$
Q = differential shrinkage and creep force
Q = statical moment of the transformed area outside the slab-beam interface about the centroidal axis of the composite section
$R = f_cjk/2$
r = radius of gyration = $\sqrt{I/A}$
S = section modulus
S_b = section modulus for bottom fiber stress
S_t = section modulus for top fiber stress
S = clear span for slab design, and center-to-center beam spacing for lateral load distribution
s = spacing
T = tensile force
T = multiplier for additional long-time deflections due to time-dependent effects, such as creep and shrinkage
t = time
t = flange thickness of T-beam
V = shear force
V_D = dead-load shear force
V_L = live-load shear force
V_u = ultimate shear force
v = shear stress
v_c = nominal permissible shear stress carried by concrete
v_h = horizontal shear stress
v_{dh} = design horizontal shear stress in strength method
v/s = volume-surface ratio
v_u = ultimate shear stress
w = uniformly distributed load
w = unit weight of concrete
y = vertical distance of cross section measured from the centroidal axis to any fiber
y_b = distance from the centroidal axis to the bottom fiber
y_{cs} = distance from the centroidal axis of the composite section to the centroid of the slab
y_t = distance from the centroidal axis to the top fiber
y_{ten} = distance from the centroidal axis (of gross section when computing cracking moment) to the extreme fiber in tension
Z = section modulus
Z_b = section modulus for bottom fiber stress
Z_t = section modulus for top fiber stress

z = a parameter limiting distribution of flexural steel (a crack-control parameter)
α (alpha) = ratio of creep coefficient at any time to ultimate creep coefficient = C_t/C_u
α_s = ratio of creep coefficient up to the time of slab casting to the ultimate creep coefficient
$(\alpha_s)_{cp}$ = ratio of precast-beam creep that takes place up to the time of slab casting to the total creep
$(\alpha_s)_{sh}$ = ratio of precast-beam shrinkage that takes place between the time of dead-load application and slab casting to the ultimate shrinkage
β_s (beta) = creep-correction factor for precast-beam concrete age when a sustained load (such as the slab dead load for a prestressed composite beam) is applied
(gamma) = ratio of differential shrinkage and creep that takes place after the shores are removed to the ultimate value
Δ (delta) = deflection or camber
Δ_{cp} = creep deflection
$(\Delta_{cp})_L$ = live-load creep deflection
Δ_{DS} = differential shrinkage and creep deflection
Δ_i = initial deflection or camber
$(\Delta_i)_1$ = initial deflection under slab dead load
$(\Delta_i)_2$ = initial deflection under precast-beam dead load
$(\Delta_i)_D$ = initial dead-load deflection
$(\Delta_i)_{P_o}$ = initial camber due to the initial prestress moment $P_o e$
Δ_L = live-load deflection
Δ_{sh} = shrinkage warping deflection
Δ_t = time-dependent deflection or camber
Δ_u = ultimate deflection or camber
(epsilon) = unit strain
$(\epsilon_{sh})_t$ = shrinkage strain at any time t
$(\epsilon_{sh})_u$ = ultimate (in time) shrinkage *stri*
η (eta) = relaxation coefficient
θ (theta) = slope of deflection curve
μ (mu) = coefficient of friction
ρ (rho) = tension steel ratio = A_s/bd
ρ' = compression steel ratio = A_s'/bd
ρ_b = steel ratio producing a balanced condition
$\rho_f = A_{sf}/b_w d$
ρ_p = prestressing steel ratio = A_{ps}/bd. Also ρ of precast beam
$\rho_{ps} = A_{ps}/A_g = A_{ps}/A_p = A_{ps}/A_2$
$\rho_w = A_s/b_w d$
σ (sigma) = unit stress
ϕ (phi) = curvature
ϕ = capacity reduction factor
ω (omega) = $\rho f_y/f_c'$
$\omega_p = \rho_p f_{ps}/f_c'$
ω_w, ω_{pw} = reinforcement indices for flanged sections computed as for ω, and ω_p, except that b shall be the web width, and the steel area shall be that required to develop the compressive strength of the web only

Subscripts

1 = cast-in-place slab of a composite beam or the effect of the slab, such as due to slab dead load
2 = precast-beam section of a composite beam
a = average value
b = bottom fiber. Also balanced value
c = composite section. Also concrete. Also center of beam (as at midspan)
ci = initial concrete value
cp = concrete creep
cr = cracking
D = dead load
DS = differential shrinkage and creep
e = end of beam
FWS = future wearing surface
g = gross section
H = relative humidity
I = impact
i = initial value. Also slab-beam interface fiber

L = live load
LA = loading age
m = midspan
ns = nonprestressed steel in prestressed member
P = permanent load applied to composite section
p = precast. Also prestressing steel
ps = prestressing steel
s = slab. Also steel. Also distance to composite slab centroid, as y_{cs}
sh = shrinkage
T = thickness as average thickness for creep and shrinkage correction factors
t = time-dependent. Also top fiber. Also time in days in equations
ten = tension
u = ultimate value

Appendix A

MATERIAL PROPERTIES

Concrete Compressive Strength

Based on some 253 data points for different-weight concrete, Eqs. A.1 through A.4 have been recommended for computing compressive strength at any time, t (Branson,[17] Branson and Christiason[42]).

Moist-cured concrete, Type 1 or normal cement

$$(f_c')_t = \frac{t}{4.00 + 0.85t}(f_c')_{28d} \qquad \text{(A.1)}$$

For example, $(f_c')_{7d} = 0.70(f_c')_{28d}$, and $(f_c')_u = 1.18(f_c')_{28d}$.

Moist-cured concrete, Type III or high-early strength cement

$$(f_c')_t = \frac{t}{2.30 + 0.92t}(f_c')_{28d} \qquad \text{(A.2)}$$

For example, $(f_c')_{7d} = 0.80(f_c')_{28d}$, and $(f_c')_u = 1.09(f_c')_{28d}$.

Steam-cured concrete, Type I or normal cement

$$(f_c')_t = \frac{t}{1.00 + 0.96t}(f_c')_{28d} \qquad \text{(A.3)}$$

For example, $(f_c')_{1d} = 0.51\ (f_c')_{28d}$, and $(f_c')_u = 1.04(f_c')_{28d}$.

Steam-cured concrete, Type III or high-early strength cement

$$(f_c')_t = \frac{t}{0.70 + 0.98t}(f_c')_{28d} \qquad \text{(A.4)}$$

For example, $(f_c')_{1d} = 0.60(f_c')_{28d}$, and $(f_c')_u = 1.02(f_c')_{28d}$.

In Eqs. A.1 through A.4, t is the age of concrete in days and $(f_c')_u$ refers to an ultimate (in time) value. These equations have also been recommended by ACI Committee 209.[27]

Concrete Tensile Strength

Equation A.5 has been suggested by ACI Committee 209[27] for computing the direct tensile strength of different-weight concrete.

$$f_t' = (1/3)\sqrt{wf_c'}\text{, psi; } w \text{ in pcf and } f_c' \text{ in psi} \qquad \text{(A.5)}$$

Representative values of Eq. A.5 are given in Table A.1.

Concrete Modulus of Rupture

The ACI Code recommends Eqs. A.6 for computing the modulus of rupture of different-weight concrete.

Normal-Weight Concrete

$$f_r = 7.5\sqrt{f'_c},$$

Sand-Lightweight Concrete

$$f_r = 6.4\sqrt{f'_c}, \text{ psi}; f'_c \text{ in psi}$$

All-Lightweight Concrete

$$f_r = 5.6\sqrt{f'_c}, \tag{A.6}$$

Representative values of Eq. A.6 are given in Table A.1. The above equations for sand-lightweight and all-lightweight concrete are to be used when the tensile splitting strength, f_{ct}, is not specified. Otherwise, for lightweight aggregate concretes, the preceding equation for normal-weight concrete shall be modified by substituting $f_{ct}/6.7$ for $\sqrt{f'_c}$, but the value of $f_{ct}/6.7$ used shall not exceed $\sqrt{f'_c}$.

The corresponding factors in Eq. A.6 by AASHTO are 7.5, 6.3, and 5.5 instead of 7.5, 6.4, 5.6, respectively.

Based on some 332 data points for different-weight concrete, Branson[17] recommends the slightly improved and less conservative Eq. A.7 for computing the modulus of rupture.

$$f_r = 0.65\sqrt{wf'_c}, \text{ psi}; w \text{ in pcf and } f'_c \text{ in psi} \tag{A.7}$$

Representative values of Eq. A.7 are given in Table A.1.

Table A.1. Representative values of concrete tensile strength, f_t, and concrete modulus of rupture, f_r.[a]

Parameter	Compressive Strength, f'_c, psi, N/mm^2						
	2500 *17.2*	3000 *20.7*	4000 *27.6*	5000 *34.5*	6000 *41.4*	7000 *48.3*	8000 *55.2*
Concrete Tensile Strength							
$f'_t = (1/3)\sqrt{wf'_c}$–Eq. A.5							
Normal Weight ($w = 145$), $f'_t = 4.01\sqrt{f'_c}$	201 *1.39*	220 *1.52*	254 *1.75*	284 *1.96*	311 *2.14*	336 *2.32*	359 *2.48*
Sand-Lightweight ($w = 120$), $f'_t = 3.65\sqrt{f'_c}$	183 *1.26*	200 *1.38*	231 *1.59*	258 *1.78*	283 *1.95*	305 *2.10*	326 *2.25*
All-Lightweight ($w = 100$), $f'_t = 3.33\sqrt{f'_c}$	167 *1.15*	182 *1.25*	211 *1.45*	235 *1.62*	258 *1.78*	279 *1.92*	298 *2.05*
Concrete Modulus of Rupture							
ACI Code–Eq. A.6							
Normal Weight $f_r = 7.5\sqrt{f'_c}$	375 *2.59*	411 *2.83*	474 *3.27*	530 *3.65*	581 *4.01*	628 *4.33*	671 *4.63*
Sand-Lightweight $f_r = 6.4\sqrt{f'_c}$	320 *2.21*	351 *2.42*	405 *2.79*	453 *3.12*	496 *3.42*	535 *3.69*	572 *3.94*
All-Lightweight $f_r = 5.6\sqrt{f'_c}$	280 *1.93*	307 *2.12*	354 *2.44*	396 *2.73*	434 *2.99*	469 *3.23*	501 *3.45*
Concrete Modulus of Rupture							
Author's Recommendation–							
$f_r = 0.65\sqrt{wf'_c}$–Eq. A.7							
Normal Weight ($w = 145$), $f_r = 7.83\sqrt{f'_c}$	392 *2.70*	429 *2.96*	495 *3.42*	554 *3.82*	607 *4.19*	655 *4.52*	700 *4.83*
Sand-Lightweight ($w = 120$), $f_r = 7.12\sqrt{f'_c}$	356 *2.45*	390 *2.69*	450 *3.10*	503 *3.47*	552 *3.81*	596 *4.11*	637 *4.39*
All-Lightweight ($w = 100$), $f_r = 6.50\sqrt{f'_c}$	325 *2.24*	356 *2.45*	411 *2.83*	460 *3.17*	503 *3.47*	544 *3.75*	581 *4.01*

[a]Values such as 2500 and 201 are in psi. Italicized values such as *17.2* and *1.39* are in N/mm^2.

Concrete Modulus of Elasticity

The ACI Code recommends Pauw's[43] Eq. A.8 for computing the modulus of elasticity of different-weight concrete.

$$E_c = 33\sqrt{w^3 f'_c}, \text{ psi}; w \text{ in pcf and } f'_c \text{ in psi} \tag{A.8}$$

Representative values of Eq. A.8 are given in Table A.2.

Based on some 274 data points for different-weight concrete, the author[17] recommends the somewhat improved (especially for higher strengths above, for example, 5000 psi or 34.5 N/mm^2) and the more conservative Eq. A.9 for computing the modulus of elasticity.

$$E_c = a_1\sqrt{w^3 f'_c}, \text{ psi}; w \text{ in pcf and } f'_c \text{ in psi} \tag{A.9}$$

where $a_1 = 39.0 - 0.0015f'_c$. Representative values of Eq. A.9 are given in Table A.2.

Steel Modulus of Elasticity

The ACI Code recommends: $E_s = 29 \times 10^6$ psi (200 kN/mm^2) for nonprestressed steel, and the modulus of elasticity of prestressing steel to be determined by tests or supplied by the manufacturer. Values commonly used[9] for prestressing steel are: $E_s = 29 \times 10^6$ psi (200 kN/mm^2) for wires and bars; $E_s = 28 \times 10^6$ psi (193 kN/mm^2) for 270-K seven-wire strand;

Table A.2. Representative values of concrete modulus of elasticity, E_c.[a]

Parameter	Compressive Strength, f'_c, psi, N/mm^2						
	2500 *17.2*	3000 *20.7*	4000 *27.6*	5000 *34.5*	6000 *41.4*	7000 *48.3*	8000 *55.2*
Concrete Modulus of Elasticity ACI Code—Eq. A.8 $E_c = 33\sqrt{w3f'_c}$							
Normal Weight ($w = 145$), $E_c = 57{,}620\sqrt{f'_c}$	2.88 *19.9*	3.16 *21.8*	3.64 *25.1*	4.07 *28.1*	4.46 *30.8*	4.82 *33.2*	5.15 *35.5*
Sand-Lightweight ($w = 120$), $E_c = 43{,}380\sqrt{f'_c}$	2.17 *15.0*	2.38 *16.4*	2.74 *18.9*	3.07 *21.2*	3.36 *23.2*	3.63 *25.0*	3.88 *26.8*
All-Lightweight ($w = 100$), $E_c = 33{,}000\sqrt{f'_c}$	1.65 *11.4*	1.81 *12.5*	2.09 *14.4*	2.34 *16.1*	2.55 *17.6*	2.76 *19.1*	2.95 *20.3*
Concrete Modulus of Elasticity Author's Recommendation— $E_c = a_1\sqrt{w^3f'_c}$—Eq. A.9							
$a_1 = 39.0 - 0.0015f'_c$	35.25	34.5	33.0	31.5	30.0	28.5	2.70
Normal Weight ($w = 145$) E_c =	3.08 *21.2*	3.30 *22.8*	3.64 *25.1*	3.89 *26.8*	4.06 *28.0*	4.16 *28.7*	4.22 *29.1*
Sand-Lightweight ($w = 120$), E_c =	2.32 *16.0*	2.48 *17.1*	2.74 *18.9*	2.93 *20.2*	3.05 *21.0*	3.13 *21.6*	3.17 *21.9*
All-Lightweight ($w = 100$), E_c =	1.76 *12.1*	1.89 *13.0*	2.09 *14.4*	2.23 *15.4*	2.32 *16.0*	2.38 *16.4*	2.41 *16.6*
Maximum Computed Tensile Stress Allowed by the ACI Code for Partially Prestressed Beams— $12\sqrt{f'_c}$, psi; f'_c in psi N/mm^2	600 *4.14*	657 *4.53*	759 *5.23*	849 *5.85*	930 *6.41*	1004 *6.92*	1073 *7.40*

[a]Compressive strength, f'_c, such as 2500 in psi; with italicized values such as *17.2* in N/mm^2. Modulus of elasticity such as $E_c = 2.88 \times 10^6$ psi; with italicized values such as $E_c =$ *19.9* kN/mm^2.

$E_s = 27 \times 10^6$ psi (186 kN/mm^2) for 250-K seven-wire strand; and $E_s = 25 \times 10^6$ psi (172 kN/mm^2) for large strand.

Concrete Creep and Shrinkage

The following procedure of Branson and colleagues[17,27,39–42,44,45] (1970 to 1977) is recommended for computing creep and shrinkage in the form of standard values for standard conditions, with correction factors used for other conditions. This information is based on 470 creep data points (in form of creep coefficient defined as ratio of creep strain to initial strain) and 356 shrinkage data points for different-weight concrete in which no consistent distinction was found between normal-weight and lightweight concrete. However, it is noted that higher shrinkage has been found in some cases for lightweight concrete as compared to normal-weight concrete of comparable strength and quality.

Standard equations and standard conditions—ambient relative humidity ≤40%, average thickness of member (part under consideration) 6 in. (15 cm)

Creep coefficient for loading age 7 days for moist-cured concrete and 1 day for steam-cured concrete

$$C_t = \frac{t^{0.60}}{10 + t^{0.60}} C_u \qquad \text{(A.10)}$$

Upper Limit $C_u = 4.15$, Average $C_u = 2.35$, Lower Limit $C_u = 1.30$. From Table A.5 for $H =$ 70%, for example, Average $C_u = 0.80(2.35) =$ 1.88.

Shrinkage after age 7 days for moist-cured concrete

$$(\epsilon_{sh})_t = \frac{t}{35 + t} (\epsilon_{sh})_u \qquad \text{(A.11)}$$

Shrinkage after age 1 day for steam-cured concrete

$$(\epsilon_{sh})_t = \frac{t}{55 + t} (\epsilon_{sh})_u \qquad \text{(A.12)}$$

For both Eqs. A.11 and A.12:

Upper Limit $(\epsilon_{sh})_u$

$= 1070 \times 10^{-6}$ in./in. (mm/mm)

Average $(\epsilon_{sh})_u$

$= 780 \times 10^{-6}$ in./in. (mm/mm)

Lower Limit $(\epsilon_{sh})_u$

$= 415 \times 10^{-6}$ in./in. (mm/mm)

From Table A.5 for $H = 70\%$, for example, Average $(\epsilon_{sh})_u = 0.70\ (780 \times 10^{-6}) = 546 \times 10^{-6}$ in./in. (mm/mm).

In Eqs. A.10 to A.12, t is time in days after loading for creep, and, for shrinkage, time after initial shrinkage is considered (age 1 day for steam-cured and 7 days for moist-cured concrete. These equations are tabulated in normalized form (time-ratio term versus time) in Table A.3. This procedure has also been recommended by ACI Committee 209.[27]

Correction factors

For loading ages later than 7 days for moist-cured concrete and later than 1 day for steam-cured concrete

Use Equations A.13 and A.14 for the creep-correction factors.

Creep $(C.F.)_{LA}$

$= 1.25 t_{LA}^{-0.118}$, for moist-cured concrete (A.13)

Creep $(C.F.)_{LA}$

$= 1.13 t_{LA}^{-0.095}$, for steam-cured concrete (A.14)

where t_{LA} is the loading age in days. Representative values of Eqs. A.13 and A.14 are given in Table A.4.

For shrinkage considered from other than 7 days for moist-cured concrete and other than 1 day for steam-cured concrete

Determine the differential in Eq. A.11 or A.12 for any period starting after this time. That is, for example, shrinkage between 28 days and 1 year would be equal to 7-day to 1-year shrinkage minus 7-day to 28-day shrinkage. For shrinkage of moist-cured concrete from day 1 (can be used to determine differential shrinkage in composite beams, for example), use Shrinkage $C.F. = 1.20$. In this procedure for moist-cured concrete, the concrete is

Table A.3. Time-ratio values for the creep coefficient and shrinkage strain from Eqs. A.10 through A.12.

Creep, Shrinkage Ratios	Time								
	28-day	3-month	6-month	1-year	2-year	5-year	10-year	20-year	30-year
C_t/C_u (A.10)	0.42	0.60	0.69	0.78	0.84	0.90	0.93	0.95	0.96
$(\epsilon_{sh})_t/(\epsilon_{sh})_u$ (A.11)	0.44	0.72	0.84	0.91	0.95	0.98	0.99	1.00	1.00
$(\epsilon_{sh})_t/(\epsilon_{sh})_u$ (A.12)	0.34	0.62	0.77	0.87	0.93	0.97	0.99	0.99	1.00

assumed to have been cured for approximately 5 to 7 days.

For greater than 40% ambient relative humidity
Use Eqs. A.15 through A.17 for the creep and shrinkage correction factors.

Creep $(C.F.)_H$

$$= 1.27 - 0.0067H, \text{ when } H \geqslant 40\% \quad (A.15)$$

Shrinkage $(C.F.)_H$

$$= 1.40 - 0.010H, \text{ when } 40\% \leqslant H \leqslant 80\% \quad (A.16)$$

Shrinkage $(C.F.)_H$

$$= 3.00 - 0.030H, \text{ when } 80\% \leqslant H \leqslant 100\% \quad (A.17)$$

where H is ambient relative humidity in percent. Representative values of Eqs. A.15 through A.17 are given in Table A.5.

For average thickness of member (part under consideration) greater than 6 in. (15 cm)
Use Eqs. A.18 through A.21 by the average-thickness method or Eqs. A.22 and A.23 by the volume-surface ratio method for the creep and shrinkage correction factors.

Average-thickness method (recommended for average thicknessses up to about 12 *to* 15 in. *or* 30 *to* 38 cm)

Creep $(C.F.)_T$

$$= 1.14 - 0.023T, \text{ for} \leqslant 1 \text{ year loading} \quad (A.18)$$

Creep $(C.F.)_T$

$$= 1.10 - 0.017T, \text{ for ultimate values} \quad (A.19)$$

Shrinkage $(C.F.)_T$

$$= 1.23 - 0.038T, \text{ for} \leqslant 1 \text{ year drying} \quad (A.20)$$

Shrinkage $(C.F.)_T$

$$= 1.17 - 0.029T, \text{ for ultimate values} \quad (A.21)$$

where T is the average thickness of the part of the member under consideration in inches. Representative values of Eqs. A.18 to A.21 are given in Table A.6. It may be noted in Table A.6 that the correction factors are closer to unity (smaller size effect) for ultimate values than for 1-year values because of the longer

Table A.4. Creep correction factors for various nonstandard loading ages, computed by Eqs. A.13 and A.14.

Loading Age, t_{LA}, days	Creep $(C.F.)_{LA}$, moist-cured (A.13)	Creep $(C.F.)_{LA}$, steam-cured (A.14)
7	1.00	0.97
10	0.95	0.90
20	0.87	0.85
28	0.84	0.83
60	0.77	0.76
90	0.74	0.74

Table A.5. Creep and shrinkage correction factors for nonstandard relative humidity, computed by Eqs. A.15 through A.17.

Relative Humidity H, %	Creep $(C.F.)_H$ (A.15)	Shrinkage $(C.F.)_H$ (A.16, A.17)
40% or less	1.00	1.00
50%	0.94	0.90
60%	0.87	0.80
70%	0.80	0.70
80%	0.73	0.60
90%	0.67	0.30
100%	0.60	0.00

time for internal adjustment in the case of the ultimate values.

Volume-surface ratio method (recommended for average thicknesses greater than about 12 *to* 15 in. *or* 30 *to* 38 cm)

Creep $(C.F.)_T$

$$= 1.12 - 0.08(v/s),\ v/s \geqslant 1.5 \text{ in. } (3.8 \text{ cm}) \tag{A.22}$$

Shrinkage $(C.F.)_T$

$$= 1.14 - 0.09(v/s),\ v/s \geqslant 1.5 \text{ in. } (3.8 \text{ cm}) \tag{A.23}$$

where v/s is the volume-surface ratio of the member in inches. Representative values of Eqs. A.22 and A.23 are given in Table A.7.

For average thickness of member (part under consideration) less than 6 in. (15 cm)

Creep and shrinkage may be noticeably higher for very thin members, for example, 2 in. (5 cm) and 3 in. (8 cm) thick, as compared to 6 in. (15 cm). For members less than 6 in. (15 cm) thick, the correction factors in Table A.6 should be used.

Table A.6. Creep and shrinkage correction factors for average thickness of member ⩾ 2 in. (5 cm), computed by Eqs. A.18 through A.21 above 6 in. (15 cm) thick.[a]

Average Thickness of Member, T (Part Being Considered)		Creep $(C.F.)_T$		Shrinkage $(C.F.)_T$	
		⩽ 1 yr	ultimate value	⩽ 1 yr	ultimate value
in.	cm	A.18	A.19	A.20	A.21
2[b]	5.1[b]	1.30	1.30	1.35	1.35
3[b]	7.6[b]	1.17	1.17	1.25	1.25
4[b]	10.2[b]	1.11	1.11	1.17	1.17
5[b]	12.7[b]	1.04	1.04	1.08	1.08
6	15.2	1.00	1.00	1.00	1.00
8	20.3	0.96	0.96	0.93	0.94
10	25.4	0.91	0.93	0.85	0.88
12	30.5	0.86	0.90	0.77	0.82
15	38.1	0.80	0.85	0.66	0.74
20	50.8	0.68	0.76	0.47	0.59

[a]This method is recommended for average thicknesses up to about 12 to 15 in. (30 to 38 cm).
[b]The creep and shrinkage correction factors for members less than 6 in. (15 cm) thick were obtained from CEB.[46]

Table A.7. Creep and shrinkage correction factors for volume-surface ratios ⩾ 1.5 in. (3.8 cm), computed by Eqs. A.22 and A.23.[a]

Volume-Surface Ratio, v/s[b]		Creep $(C.F.)_T$	Shrinkage $(C.F.)_T$
in.	cm	(A.22)	(A.23)
1.5	3.8	1.00	1.00
2	5.1	0.96	0.96
3	7.6	0.88	0.87
4	10.2	0.80	0.78
5	12.7	0.72	0.69
6	15.2	0.64	0.60
7	17.8	0.56	0.51
8	20.3	0.48	0.42

[a]This method is recommended for average thicknesses greater than about 12 to 15 in. (30 to 38 cm).
[b]Examples: For rectangular sections 6 by 12 in. (15.2 by 30.5 cm) and 18 by 36 in. (45.7 by 91.4 cm), v/s = 2.0 in. (5.1 cm) and 6.0 in. (15.2 cm), respectively.

Ultimate values of creep and shrinkage are tabulated in Table A.8, in which the humidity and age correction factors (but not the size-correction factors) are already included.

The correction factors considered previously were due to age effects, humidity, and member size. Other effects described in References 17 and 42 can also be of significance in certain cases, but for most practical purposes the effects included above are sufficient.

Simple equations have been provided for all the material parameters in Appendix A for ready programming, with results also tabulated in Tables A.1 through A.8 for hand calculations.

In addition to the material parameters in Appendix A, multipliers for the ultimate long-time deflection of reinforced concrete flexural members in Eq. C.13 are given in Table A.9 by Branson[25] and Table A.10 by AASHTO[6] for different conditions, as $T_u = 2.0$ in the ACI Code, and by Eqs. A.24 and A.25 for nominal calculations according to ACI Committee 435[21] and Branson.[17]

For Beams and One-Way Slabs

$$T_u = 2.5 \tag{A.24}$$

For Two-Way Construction

$$T_u = 3.0 \tag{A.25}$$

Table A.8. **Summary of ultimate (in time) creep coefficients and shrinkage strains, including the effects of relative humidity, moist and steam curing, and concrete age effects.**[a–c]

age[c]	Average Relative Humidity, Moist or Steam Curing, Ultimate Creep Coefficient or Shrinkage Strain																							
	≧ 90%				80%				70%				60%				50%				≦ 40%			
	Moist		Steam		Moist		Steam		Moist		Steam		Moist		Steam		Moist		Steam		Moist		Steam	
(days)	C_u	d	C_u	d	C_u	d	C_u	d	C_u	d	C_u	d	C_u	d	C_u	d	C_u	d	C_u	d	C_u	d	C_u	d
1	-	281	1.57	234	-	562	1.72	468	-	655	1.88	546	-	749	2.04	624	-	842	2.21	702	-	936	2.35	780
7	1.57	234	1.53	209	1.72	468	1.66	418	1.88	546	1.82	487	2.04	624	1.98	557	2.21	702	2.14	626	2.35	780	2.28	696
10	1.50	182	1.42	198	1.63	364	1.54	396	1.79	425	1.69	462	1.94	485	1.84	528	2.10	546	1.99	594	2.23	607	2.12	660
20	1.37	149	1.34	172	1.49	298	1.46	343	1.64	347	1.60	400	1.78	397	1.74	458	1.92	447	1.88	515	2.05	496	2.00	572
28	1.32	130	1.31	155	1.44	260	1.42	310	1.58	303	1.56	362	1.72	347	1.70	414	1.86	390	1.83	465	1.97	433	1.95	517
60	1.21	86	1.20	112	1.32	172	1.30	224	1.45	201	1.43	261	1.57	230	1.55	298	1.70	259	1.68	336	1.81	287	1.79	373
90	1.17	66	1.17	89	1.27	131	1.27	178	1.39	153	1.39	207	1.51	175	1.51	237	1.63	197	1.63	266	1.74	218	1.74	296

[a] See Tables A.6 and A.7 for correction factors due to the effect of member size. The other principal effects (due to age and humidity) are included in this table. These are average values that apply nominally to different weight and strength concretes, and Types I and III cement.
[b] See Eqs. A.10 through A.12 for multipliers to obtain values at times other than ultimate (such as at 28 days, 3 months, 6 months, 1 year, 5 years, etc., as given in Table A.3.
[c] Concrete age referred to in the table–loading *age* for creep, and period from the *age* indicated to ultimate for shrinkage.
[d] Ultimate shrinkage micro-strain, $(\epsilon_{sh})_u$, as 281×10^{-6} in./in. (mm/mm).

Table A.9. Multiplier, T_u, for computing the additional long-time deflection of reinforced concrete flexural members due to creep and shrinkage for both normal-weight and lightweight concrete members of common types and sizes.[25 a, b]

Concrete Strength, f'_c, at Age 28 Days	Average Relative Humidity, Age When Loaded								
	100%			70%			50%[b]		
	⩽7*d*	14*d*	⩾28*d*	⩽7*d*	14*d*	⩾28*d*	⩽7*d*	14*d*	⩾28*d*
2500 to 4000 psi (17 to $28\ N/mm^2$)	2.0	1.5	1.0	3.0	2.0	1.5	4.0	3.0	2.0
> 4000 psi ($> 28\ N/mm^2$)	1.5	1.0	0.7	2.5	1.8	1.2	3.5	2.5	1.5

[a]For the period: 1 *month or less* 3 *months* 1 *year* 5 *years or more*
Use: 25% 50% 75% 100% of table values

[b]The 50% humidity values may normally be used for lower relative humidities, such as in heated buildings.

Table A.10. Multiplier, $k_r T_u$, recommended by AASHTO.[6]

Humidity Condition	$A'_s = 0$	$A'_s = 0.5A_s$	$A'_s = A_s$
Climate of high humidity	2.5	1.8	1.5
Climate of average humidity	3.0	2.2	1.8
Climate of low humidity	3.5	2.5	2.0

Appendix B

DIFFERENTIAL SHRINKAGE AND CREEP STRESSES AND DEFLECTIONS

Stress and Deflection Equations for a Simple Span–Elastic Theory

Referring to Figs. B.1 and B.2a for an uncracked section,[17]

$$Q = DA_1E_1 \tag{B.1}$$

where Q is the slab shrinkage force (see Eq. B.11 for practical value of Q as Q_{DS}) and D is the differential shrinkage and creep strain or coefficient. Applying this force to the slab and then to the composite section, as in Fig. B.2b, yields Eqs. B.2 to B.5 for the fiber stresses of an uncracked (including steel) section, as shown in Fig. B.2c.

$$\sigma_{1t} = \frac{Q}{A_1} + \left(-\frac{Q}{A_c} - \frac{Qy_{cs}y_{ct}}{I_c}\right)\frac{E_1}{E_2} \tag{B.2}$$

$$\sigma_{1b} = \frac{Q}{A_1} + \left(-\frac{Q}{A_c} \pm \frac{Qy_{cs}y_{ci}}{I_c}\right)\frac{E_1}{E_2} \tag{B.3}$$

$$\sigma_{2t} = -\frac{Q}{A_c} \pm \frac{Qy_{cs}y_{ci}}{I_c} \tag{B.4}$$

$$\sigma_{2b} = -\frac{Q}{A_c} + \frac{Qy_{cs}y_{cb}}{I_c} \tag{B.5}$$

The plus-minus term in Eqs. B.3 and B.4 is minus when the composite centroid is below the slab and plus when the composite centroid is within the slab.

For the constant moment, Qy_{cs}, along a simple span, the midspan deflection is given by Eq. B.8.

$$\Delta = \frac{Qy_{cs}\ell^2}{8E_2I_c} \tag{B.6}$$

With reference to Fig. 4.8 or 4.9, the following parameters are used for a cracked section in Eqs. B.1 to B.7:

$$y_{cs} = y_t - t/2$$

for both centroid cases (composite centroid either below or within the slab) in Figs. 4.8 and 4.9

$$y_{ct} = y_t$$

$$y_{ci} = y_t - t$$

when the centroid is below the slab and $t - y_t$ when the centroid is within the slab

$$y_{cb} = y_b = d_{c3} - y_t$$

$$A_1 = b_e t$$

(for example) rather than (beam spacing) (t) in computing the shrinkage force, etc., for both centroid cases

Subscript Notation

1, cast-in-place slab
2, precast beam or steel beam
b, bottom fiber of section
c, composite section with transformed slab
i, slab-beam interface
s, slab centroid
t, top fiber of section

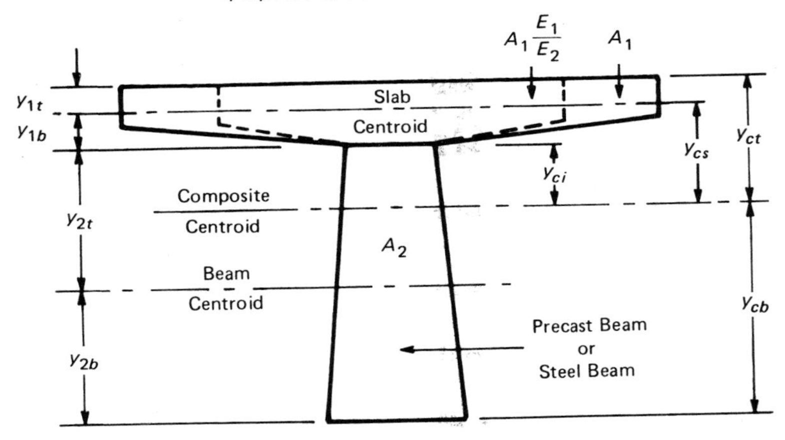

$$A_c = A_1 \frac{E_1}{E_2} + A_2$$

$$I_c = I_1 \frac{E_1}{E_2} + I_2 + A_1 \frac{E_1}{E_2} (y_{cs})^2 + A_2 (y_{cb} - y_{2b})^2$$

Fig. B.1. Geometry, nomenclature, and section properties for an uncracked concrete-concrete composite beam section.

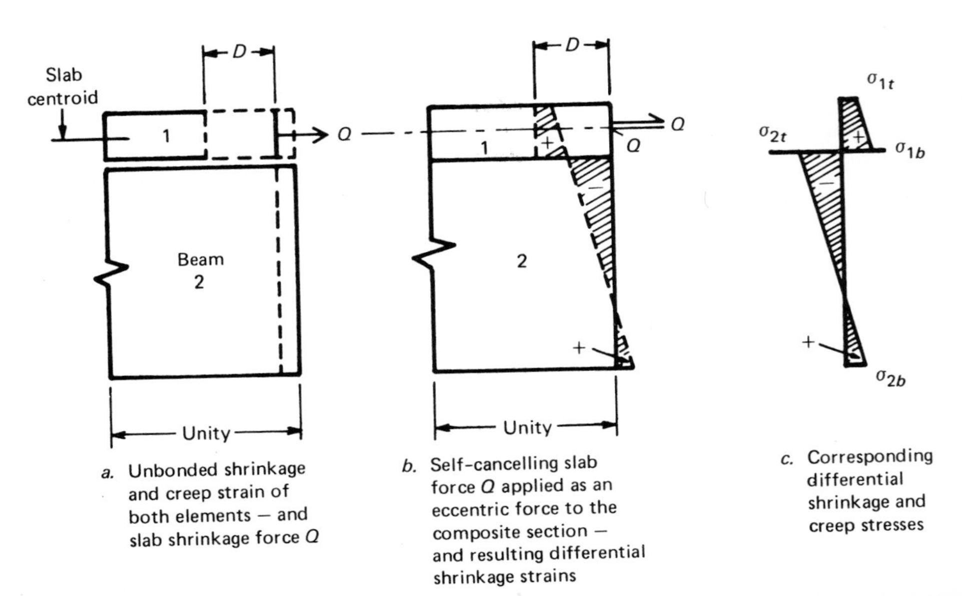

Fig. B.2. Shrinkage force *Q* applied to slab and to composite section, with resulting differential shrinkage and creep stresses.

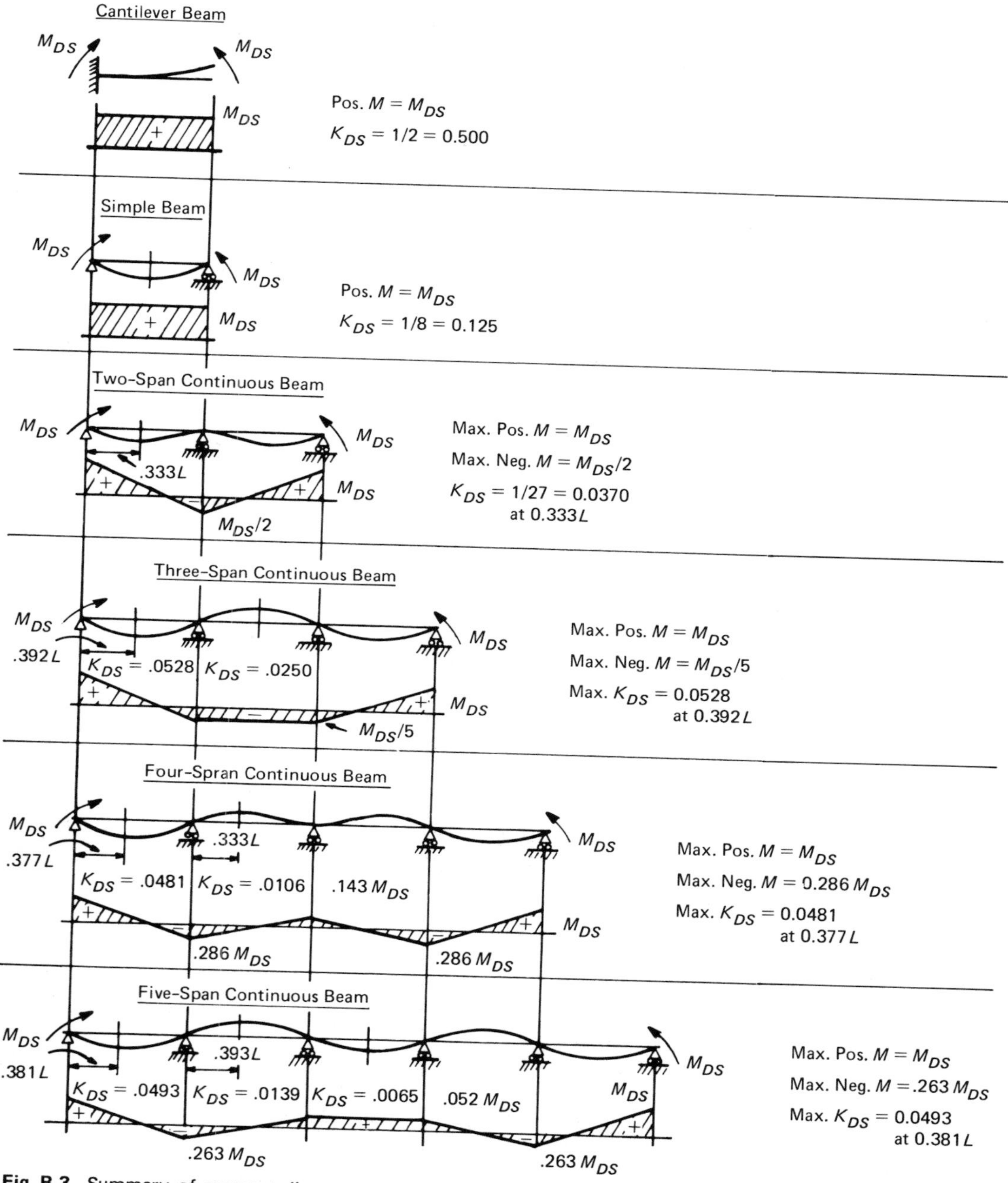

Fig. B.3. Summary of moment diagrams and deflection coefficients due to differential shrinkage and creep for cantilever, simple, and two- to five-span (equal spans) continuous beams.

$$A_c = (b_e/n_c)(t) + nA_s$$

when the centroid is below the slab

$$A_c = (b_e/n_c)(y_t) + nA_s$$

when the centroid is within the slab

$$I_c = I_{cr}$$

in Fig. 4.8 or 4.9.

Moments and Deflections for Cantilever, Simple, and Continuous Spans–Elastic Theory

In the previous simple-span elastic equations, $M_{DS} = Qy_{cs}$. For cantilever, simple, and two- to five-span (equal spans) continuous beams, the differential shrinkage and creep moment diagrams (in terms of M_{DS}) and deflection coefficients (K_{DS}) by the elastic theory have been derived by Branson[17] and are given in Fig. B.3. Using these moments, stresses are calculated by the previous equations in the positive-moment regions and by the usual procedure for elastic calculations in the negative-moment regions. Equation B.7 is the general elastic deflection equation.

$$\Delta = \frac{K_{DS} Q y_{cs} \ell^2}{E_2 I_c} \qquad \text{(B.7)}$$

Equations B.8 *through* B.12 *Are Recommended for Any Time After Slab Concrete Age of* 28 *Days, Including Ultimate Values*[17]

For Steel Beam E_s,

$$E_2 = E_s \qquad \text{(B.8)}$$

For Slab E_c,

$$E_1 = (E_c)_{28d} \qquad \text{(B.9)}$$

For Precast Concrete Beam E_c,

$$E_2 = (E_c)_{28d} \qquad \text{(B.10)}$$

For Q in Eqs. B.1 through B.7,

$$\text{Use } Q_{DS} = (Q \text{ in Eq. B.1})/2 \qquad \text{(B.11)}$$

For Deflection of Concrete-Concrete Composite Beams,

$$\Delta_{DS} = (4/3)$$

$$(\Delta \text{ in Eq. B.6 or B.7, with Eqs. B.9 through B.11}) \qquad \text{(B.12)}$$

The 1/2-factor in Eq. B.11 is a modification factor for creep relaxation effects due to differential shrinkage. The factor of 4/3 in Eq. B.12 takes into account the additional reduced

Table B.1. Recommended ultimate differential shrinkage and creep strains or coefficients, D_u, due to shrinkage plus creep under gravity loads, etc. (from Branson[17]).[a,b]

Beam Type	Unshored or Shored Construction	Age of Precast-Beam Concrete When Slab Cast (months)	Recommended (Average) Values of D_u, in./in. (mm/mm)
Steel-Concrete Composite Beam	Either Unshored or Shored	-	*$D_u = 655 \times 10^{-6}$*
Prestressed Concrete Composite Beam	Shored	1	$D_u = 310 \times 10^{-6}$
		2[b]	*$D_u = 415 \times 10^{-6}$*
		3	$D_u = 470 \times 10^{-6}$
	Unshored	1	$D_u = 270 \times 10^{-6}$
		2[b]	*$D_u = 375 \times 10^{-6}$*
		3	$D_u = 435 \times 10^{-6}$
Reinforced Concrete Composite Beam	Shored	1	$D_u = 290 \times 10^{-6}$
		2[b]	*$D_u = 395 \times 10^{-6}$*
		3	$D_u = 455 \times 10^{-6}$
	Unshored	1	$D_u = 260 \times 10^{-6}$
		2[b]	*$D_u = 355 \times 10^{-6}$*
		3	$D_u = 410 \times 10^{-6}$

[a]The numerical values in the table are based on the results in Appendix A for the conditions: 70% humidity and 6-in. (15-cm) slab thickness. Correction factors are thus required for other humidity from Table A.5 and slab thickness from Table A.6.

[b]The 2-month period between precast-beam and slab casting is recommended for routine calculations. These values are italicized.

E_c-effect in deflection calculations due to creep of the precast-beam concrete.

Differential Shrinkage and Creep Strains or Coefficients

Based on the creep and shrinkage relations and values in Appendix A, the author recommends the ultimate differential shrinkage and creep coefficients, D_u, given in Table B.1, for unshored and shored construction of steel-concrete, prestressed concrete, and reinforced concrete composite beams. Other than ultimate values may be estimated or computed using the procedures in Reference 17.

Appendix C

APPROACH AND FORMULAS FOR COMPUTING DEFLECTIONS

The following approach by Branson[17,22-26] for computing deflections of reinforced concrete beams was recommended by ACI Committee 435.[21] These procedures are approximate in nature. Because of the variability of concrete structural deformation, it appears to be not only feasible but essential that relatively simple procedures be used so that engineers will guard against placing undue reliance on the computed results.

Effective Moment of Inertia

The effective moment of inertia for a simple beam, or between inflection points for a continuous beam, is given by Eq. C.1, but not greater than I_g.

$$I_e = (M_{cr}/M_a)^3 I_g + [1 - (M_{cr}/M_a)^3]\, I_{cr} \tag{C.1}$$

where M_a is the maximum service-load moment (unfactored moment) at the stage for which deflections are being considered, $M_{cr} = f_r I_g / y_{ten}$, y_{ten} refers to the extreme tension fiber, and f_r is given in Appendix A.

For different load levels, the deflection should be computed in each case using Eq. C.1 for the total load being considered, such as dead load or dead plus live load. The incremental deflection, such as live load, is then computed as the difference in these values, as shown in Fig. C.1.

For continuous beams of constant depth under uniform loading, the appropriate average I_e is given by Eqs. C.2 and C.3.

One End Continuous,

$$I_a = 0.85 I_m + 0.15(I_{\text{cont. end}}) \tag{C.2}$$

Both Ends Continuous,

$$I_a = 0.70 I_m + 0.15(I_{e1} + I_{e2}) \tag{C.3}$$

where I_m refers to the midspan section, and I_{e1}, I_{e2} to the beam ends. However, ACI Committee 435[47] and Branson[17] have found that the use of I_a = midspan I_e will be satisfactory in most cases for continuous prismatic beams. Moment-envelope values should be used in computing both positive and negative values of I_e.

For a single heavy concentrated load at midspan, use I_a = midspan I_e. For a cantilever beam, use $I_a = I_e$ at the face of the support.

Table C.1. Values of A_{sh} for calculating shrinkage curvature.[a]

p \ p'	0.00	0.25	0.50	0.75	1.00	1.25	1.50	1.75	2.00	2.25	2.50	2.75	3.00
0.25	0.44	0.00	–	–	–	–	–	–	–	–	–	–	–
0.50	0.56	0.31	0.00	–	–	–	–	–	–	–	–	–	–
0.75	0.64	0.45	0.26	0.00	–	–	–	–	–	–	–	–	–
1.00	0.70	0.55	0.39	0.22	0.00	–	–	–	–	–	–	–	–
1.25	0.75	0.63	0.49	0.35	0.20	0.00	–	–	–	–	–	–	–
1.50	0.80	0.69	0.57	0.45	0.32	0.18	0.00	–	–	–	–	–	–
1.75	0.84	0.74	0.64	0.53	0.42	0.30	0.17	0.00	–	–	–	–	–
2.00	0.88	0.79	0.69	0.60	0.50	0.39	0.28	0.16	0.00	–	–	–	–
2.25	0.92	0.83	0.74	0.65	0.56	0.47	0.37	0.26	0.15	0.00	–	–	–
2.50	0.95	0.87	0.79	0.71	0.62	0.53	0.44	0.35	0.25	0.14	0.00	–	–
2.75	0.98	0.91	0.83	0.75	0.67	0.59	0.51	0.42	0.33	0.24	0.13	0.00	–
3.00	1.00	0.94	0.87	0.79	0.72	0.64	0.57	0.49	0.40	0.32	0.23	0.13	0.00
3.25	1.00	0.97	0.90	0.83	0.76	0.69	0.62	0.54	0.47	0.39	0.31	0.22	0.12
3.50	1.00	1.00	0.94	0.87	0.80	0.74	0.67	0.60	0.52	0.45	0.37	0.29	0.21
3.75	1.00	1.00	1.00	0.90	0.84	0.78	0.71	0.64	0.58	0.51	0.44	0.36	0.28
4.00	1.00	1.00	1.00	1.00	0.88	0.81	0.75	0.69	0.62	0.56	0.49	0.42	0.35

[a] $p = 100A_s/bd$, $p' = 100A'_s/bd$. When $p' > p$, interchange p' and p to obtain corresponding solution in opposite direction.

Initial Deflection

The midspan deflection of simple and continuous beams is given by Eq. C.4. In the case of continuous beams, the midspan deflection may normally be used as an approximation of the maximum deflection.

$$\Delta_i = \Delta_m = K(5/48)M_m\ell^2/E_cI_a \quad (C.4)$$

where M_m is the midspan moment.

For uniform loading in Eq. C.4, $K = 1$ for simple spans, $K = 0.800$ for fixed-hinged beams ($K = 0.738$ when using the maximum moment and computing the maximum deflection), $K = 0.600$ for fixed-fixed beams, and for continuous spans,

$$K = 1.20 - 0.20M_o/M_m \quad (C.5)$$

where M_o = statical moment at midspan = $w\ell^2/8$.

For a midspan concentrated load in Eq. C.4, $K = 0.800$ for simple spans, $K = 0.560$ for fixed-hinged beams ($K = 0.5725$ when using the maximum moment and computing the maximum deflection), $K = 0.400$ for fixed-fixed beams, and for continuous spans,

$$K = 1.20 - 0.40M_o/M_m \quad (C.6)$$

where M_o = statical moment at midspan = $P\ell/4$.

The end deflection of cantilever beams under uniform loading or an end concentrated load is given by Eq. C.7.

$$\Delta_i = \Delta_e = K(5/48)M_e\ell^2/E_cI_a \quad (C.7)$$

where

M_e is the moment at the face of the support
$K = 12/5$ for uniform loading
$K = 16/5$ for an end concentrated load

When beams are cantilevered from other than a fully fixed support, the deflection due to rotation at the support must also be included.

Since deflections are logically computed for a given continuous span based on the loading pattern for maximum positive moment and deflection, Eq. C.4 together with the values given for K, is believed to be the most convenient form of a deflection equation. These are theoretically correct elastic equations (assuming a constant EI in the statically indeterminate cases) based on the loading and the positive-moment envelope value at midspan. Thus, it is seen that negative moments are not required when computing maximum deflections of continuous beams, unless Eqs. C.2 and C.3 are used to compute I_a.

Effect of Compression Steel in Reducing Time-Dependent Deflections

For creep deflections,

$$k_r = 0.85/(1 + 50\rho') \quad \text{(C.8)}$$

For creep plus shrinkage (time-dependent) deflections,

$$k_r = 1.00/(1 + 50\rho') \quad \text{(C.9)}$$

where $\rho' = A_s'/bd$.

For continuous beams, the same weighted average values as in Eqs. C.2 and C.3 for I_e may be used, since both relate to curvatures.

One End Continuous,

$$(k_r)_a = 0.85(k_r)_m + 0.15(k_r)_{\text{cont. end}} \quad \text{(C.10)}$$

Both Ends Continuous,

$$(k_r)_a = 0.70(k_r)_m + 0.15[(k_r)_{e1} + (k_r)_{e2}] \quad \text{(C.11)}$$

However, the use of k_r based on the midspan section ρ' only will be satisfactory in most cases for continuous beams.

Creep and Creep Plus Shrinkage (Time-Dependent) Deflection

$$\Delta_{cp} = k_r C_t \Delta_i \quad \text{(C.12)}$$

$$\Delta_t = \Delta_{cp} + \Delta_{sh} = K_r T \Delta_i \quad \text{(C.13)}$$

where C_t is given by Eq. A.10 and the procedure of Appendix A (see Table A.8), and T may be obtained in Table A.9 or A.10. For average conditions, $C_u = 1.6$ and $T_u = 2.5$ for beams and one-way slabs may be used.

Shrinkage Deflection

$$\Delta_{sh} = K_{sh}\phi_{sh}\ell^2 \quad \text{(C.14)}$$

$$\phi_{sh} = A_{sh}\epsilon_{sh}/h \quad \text{(C.15)}$$

where A_{sh} (based on the author's shrinkage curvature equations) is obtained from Table C.1 or Fig. C.2, and ϵ_{sh} is given by Eqs. A.11 and A.12 and the procedure of Appendix A (see Table A.8). This procedure for computing shrinkage warping is used in the 1972 British Standard Code of Practice. For average conditions, Ult. $\epsilon_{sh} = 400 \times 10^{-6}$ in./in. (mm/mm)

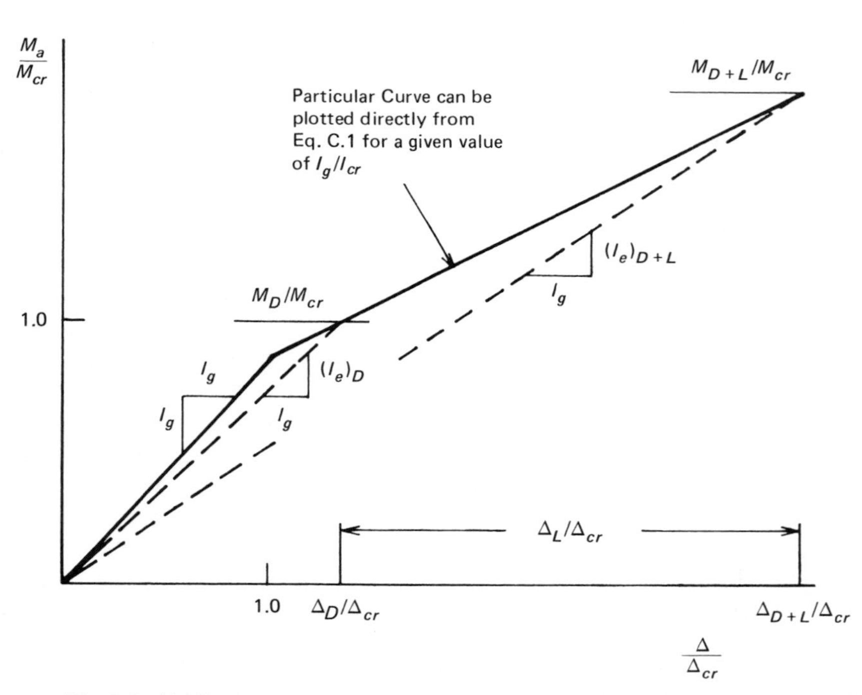

Fig. C.1. Idealized short-time moment versus deflection diagram using Eq. C.1.

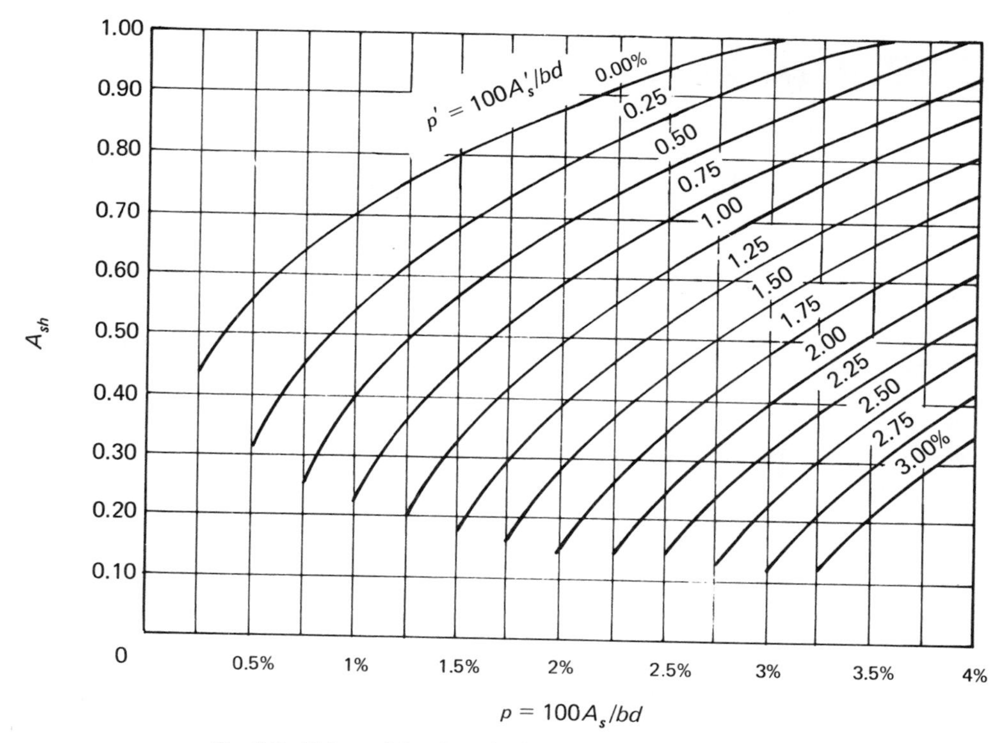

Fig. C.2. Values of A_{sh} for calculating shrinkage curvature in Eq. C.15.

may be used. It is recommended that the steel precentages refer to the support section of cantilevers and the midspan section of simple and continuous beams.

Based on the assumption of equal positive and negative shrinkage curvatures with an inflection point at the quarter-point of continuous spans, the following shrinkage deflection coefficients were derived by Branson[17]:

Cantilever Beams—

$$K_{sh} = \frac{1}{2} = 0.500 \qquad \text{(C.16)}$$

Simple Beams—

$$K_{sh} = \frac{1}{8} = 0.125 \qquad \text{(C.17)}$$

Two-Span Continuous Beams—

$$K_{sh} = 0.084 \qquad \text{(C.18)}$$

Continuous Beam With Three or More Spans—
End Span (When Discontinuous at One End),

$$K_{sh} = 0.090 \qquad \text{(C.19)}$$

Interior Spans or Spans Continuous at Both Ends,

$$K_{sh} = 0.065 \qquad \text{(C.20)}$$

Beam or Frame With One or More Spans Continuous at Both Ends,

$$K_{sh} = 0.065 \qquad \text{(C.21)}$$

Appendix D

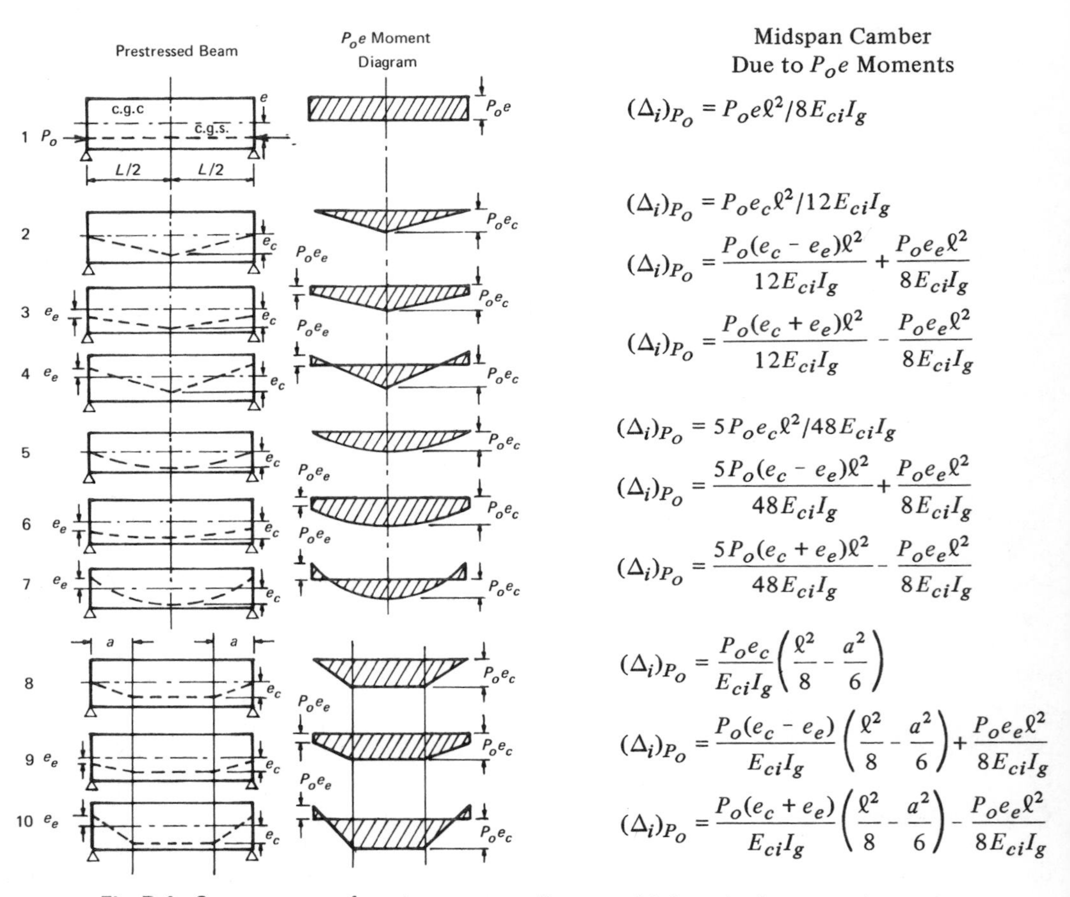

Fig. D.1. Common cases of prestress moment diagrams with formulas for computing camber.

For practical reasons, I_g is recommended.

6

Steel-concrete composite columns

RICHARD W. FURLONG, Ph.D., P.E.
Professor of Civil Engineering
University of Texas at Austin
Austin, Texas

6.1 WHAT IS A COMPOSITE COLUMN?

A composite column is a compression member that is constructed with load-bearing concrete plus steel in any form different from reinforcing rods. Several examples of cross sections of composite columns are shown in Fig. 6.1. Concrete-filled pipe (Fig. 6.1a) and tubing (Fig. 6.1b) represent a familiar form of a composite column. Quite often, the drilling "mud" or mortar of caissons can serve to help carry load as suggested by Fig. 6.1c. Load-bearing concrete has been used as fireproofing in the past (see cross sections, Fig. 6.1d and e), but sprayed-on fireproofing has become more common since the 1930s. As the unit cost of steel increases, the use of load-bearing fireproofing may become necessary once again. The use of structural shapes to protect exposed corners of concrete columns in dock and traffic areas creates another form of composite column, as in Fig. 6.1f. The section shown in Fig. 6.1g represents a seismic resistant optimization of shear strength from structural shapes and ductility from a spirally reinforced core to stabilize post-yield modes of local buckling of the shapes.

The first "reinforcements" that were used for concrete columns were structural shapes. Shapes tied together with batten plates and rivets formed excellent reinforcing members. Obviously, the use of wired rods required less fabrication expense, and reinforcement with rods was recognized as more economical after steel producers developed rods and bars that had yield strengths considerably higher than those of rolled shapes (SSRC Guide 1977).

Nevertheless, there remain numerous applications for which some forms of composite column provide excellent solutions to structural problems. When contrasted with plain steel columns, the composite column has superior load retention at increasing temperatures, considerably more resistance to local buckling, and greater stiffness and abrasion resistance. When compared with reinforced concrete, the composite column possesses much better strength and ductility in shear and generally in flexure also. Concrete-filled pipe or tube requires no column formwork. In earthquake-prone regions, the composite column can provide excellent ductility and load retention even after extensive concrete damage. The damage can be repaired if the overall structure can survive.

Reinforcing a concrete compression member with structural shapes or tubes does permit the designer the use of a greater percentage of steel in cross sections than the maximum (8%) permitted in conventionally reinforced concrete

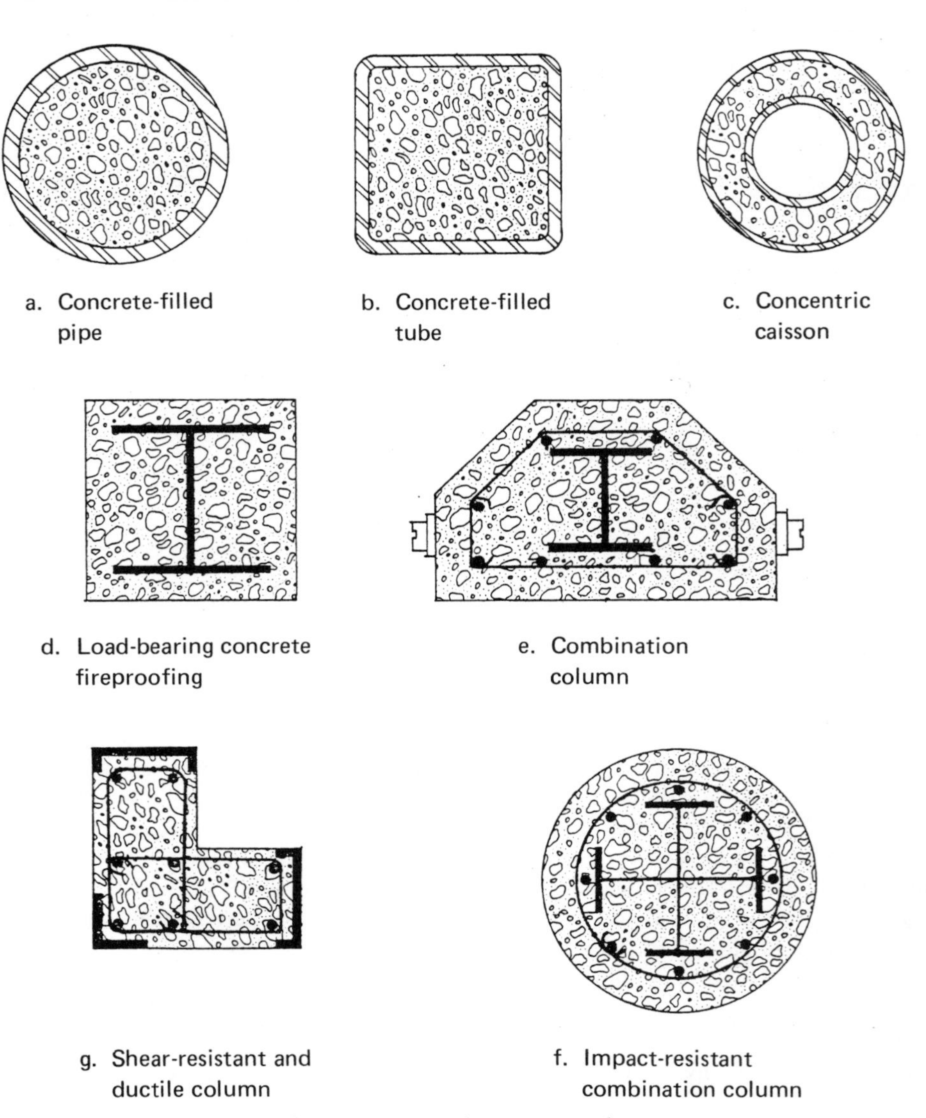

Fig. 6.1. Composite-column cross sections.

columns (ACI 318-71). The composite column can support more thrust per unit area than can any other reinforced concrete column of the same dimensions. The structural shape core of a concrete-encased column may support 80% of the thrust on a composite column, but the stiffening effect of the concrete encasement does permit the useful stress level on the core to be set at a level higher than that permitted on the bare steel shape. The concrete encasement thereby not only carries its portion of thrust, but it stiffens the structural core steel shape enough to make the core itself more effective against local and overall buckling.

Composite columns are a rather natural and logical structural form for compression members in fire-resistant construction. Encasing a structural shape in load-bearing concrete accomplishes a double benefit of providing the steel core with insulation from heat while supporting enough load to permit a reduction in the required size of the core. Lightweight concrete serves best for load retention of composite columns at high temperatures because the kiln-produced aggregate remains stable at elevated temperatures (Malhotra and Stevens 1964). Concrete-filled steel tubes can continue to support service loads at high temperatures only if the concrete filling is strong enough to support the imposed loading *without help from the softened tube*. The concrete filling can conduct some heat away from the steel tubing exposed to fire, but the delay time prior to steel softening is rather insignificant.

6.2 CHARACTERISTIC STRUCTURAL BEHAVIOR OF COMPOSITE COLUMNS

The behavior of steel-concrete compression members can be deduced on the basis of the following characteristics for steel and for concrete materials:

1. Steel is eight to 10 times as stiff as concrete until the steel is strained beyond its yield point, generally near 0.12% to 0.18% for commercial grades of structural steel.
2. Steel tends to buckle locally after it yields in compression.
3. Concrete cannot resist much tension strain without fracturing at strains less than 0.1%.
4. Concrete will sustain loads that create compression strains higher than 0.16%, but in the absence of some lateral confining pressure, concrete will display no stiffness when compressed to strains above 0.2%. In the absence of a strain gradient (adjacent fibers are strained less than surface fibers) or lateral confining pressure, concrete strained beyond 0.2% will spall and "flow" laterally, typically with an explosive failure. In the presence of strain gradients, spalling and explosive failure can be delayed until surface strains from 0.2% to 0.5% occur, the maximum increasing with the gradient.
5. Concrete strained less than 0.1% exhibits a Poisson ratio only one-third to one-half that of steel at the same strain. Concrete strained more than 0.16% exhibits a Poisson ratio greater than that of unyielded steel.

6.2.1 Concentrically Loaded Columns

Truly concentric loads can exist only instantaneously for composite columns. The heterogeneous nature of concrete invariably permits one set of particles to respond to load more stiffly than others, and concentricity is destroyed. Under virtually concentric loads, commercial grades of structural steel reach their yield strength before concrete has exhausted its compressive strength. Concrete inside steel tubing prohibits post-yield wall buckling of steel by pushing outward against the tubing. Concrete encasement around structural shapes will prohibit local buckling of yielded steel. Consequently, strains on composite columns can be increased without loss of load after steel yields, but longitudinal stiffness to resist more concentric load is virtually exhausted after steel yields. In the absence of neighboring particles to share overloads on virtually concentrically loaded columns, concrete will spall at strains near 0.2% unless concrete is confined laterally. Concrete confined laterally (as inside steel tubing) at strains above 0.3% eventually forces a post-yield outward buckling mode to occur in the tube wall.

6.2.2 Eccentrically Loaded Composite Columns

In the presence of eccentric compression loads, there are always some fibers subjected to less strain than that imposed on edge fibers. Much of the concrete and steel in eccentrically loaded composite columns will separate from one another due to their differences in Poisson's ratio. Nevertheless, it is probable that both the steel and the concrete experience a planar variation of strain, even though the steel plane may be different from the concrete plane of strain. Concrete does fracture in tension, and only the compressed concrete in flexure contributes significantly to resist longitudinal strain. Concrete-filled steel tubes fail under eccentric compression much the same as they do in concentric compression except that only one face or side of the eccentric compression specimen experiences outward buckling of the tube wall when concrete spalls at the same location. In all cases, the surface strains will reach at least 0.3% before concrete spalls against the steel encasement.

Concrete-encased structural shapes can be said to "fail" when surface concrete spalls. In the presence of only a slightly eccentric thrust, it is possible that the surface concrete will fail at strains lower than the 0.3% generally expected because of the Poisson-ratio separation between steel and concrete at lower strains. Lateral binding reinforcement less effective than a column spiral cannot restrain such failures of surface concrete. As the eccentricity of thrust increases, the probability of attaining surface strains above 0.3% is high enough that strength estimates based on such surface strain are accurate. Concrete-encased structural shapes may require special treatment of the thin segment of concrete adjacent to the flat face of a flange or box steel core as shown in Fig. 6.2a. If the width-to-thickness ratio of the edge concrete (b_f/t_c) is greater than 4, the edge concrete could become unstable and spall shortly after surface strains reach 0.2%. Until laboratory studies reveal better techniques, a mechanical attachment such as one or more longitudinal reinforcing steel bars welded to the flat face as suggested in Fig. 6.2b is encouraged.

6.3 STRUCTURAL CAPACITY OF COMPOSITE COLUMNS

Concentric Thrust Capacity

Under the action of a concentric compression force, a composite column should shorten in length, and it seems logical to assume that all particles of each cross section experience the same amount of shortening per unit length. If all elements of a cross section experience identical strains, the analysis of response to increasing amounts of load could be derived from the simple superposition of forces that are caused by stresses associated with increases in overall strain. Some characteristic stress-strain curves for steel and for concrete are shown in Fig. 6.3a. Just below Fig. 6.3a, the forces associated with the stresses at each strain for the assumed concrete and steel composite cross section are illustrated in Fig. 6.3b. When the forces on concrete are added to forces on steel, the graph of total force can be derived. The composite-

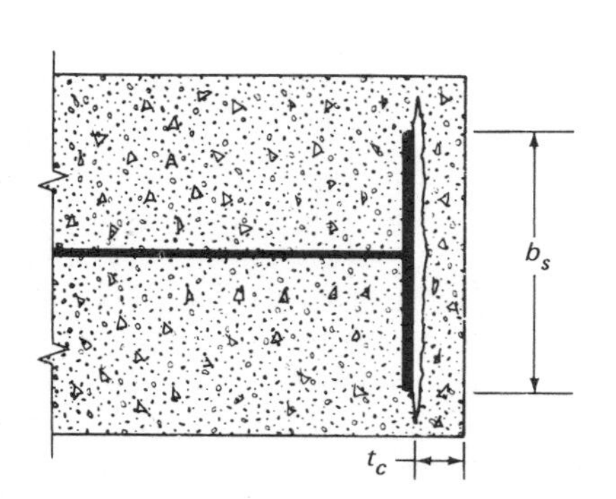

a. Weak adhesion of concrete cover

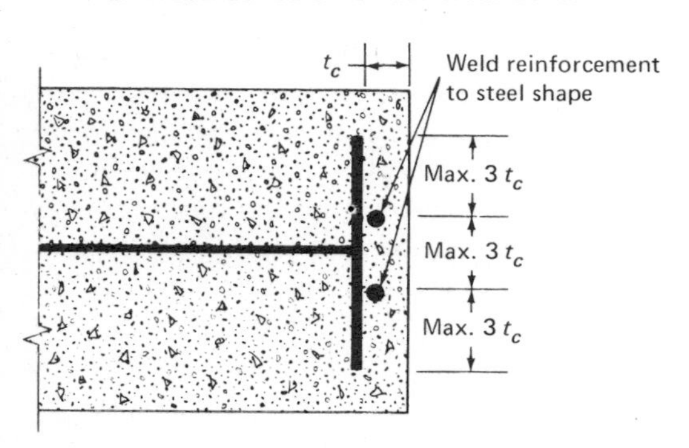

b. Mechanical aid to adhesion

Fig. 6.2. Concrete adhesion to steel surfaces.

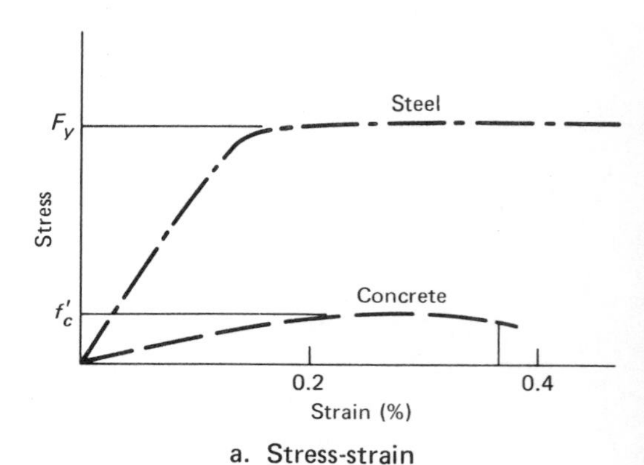

a. Stress-strain

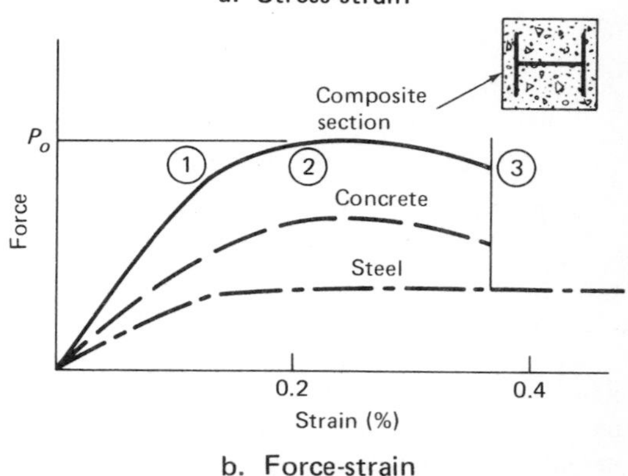

b. Force-strain

Fig. 6.3. Axially concentric column load.

section graph of Fig. 6.3b contains a relatively linear portion from the origin to the strain value at which the proportional limit of steel is reached. After the steel begins to yield, the curve by its reduced slope reflects the loss of steel stiffness in the region from (1) to (2). After concrete reaches its maximum strength and microcracking around aggregates progresses, there is a loss of total force and a measurable decrease in stiffness until the concrete fractures or spalls off entirely. Thereafter, only the yielded strength of the steel core is available (at zero stiffness) to resist further compression strains.

Early investigators were able to observe the initial linear behavior as suggested to the left of (1) on Fig. 6.3b. There was considerably less reporting of load-deformation behavior beyond the yield region (to the right of [1]) of composite columns. The nature of residual stress in steel shapes and the tendency of concrete to creep when strains exceed 0.1% impaired the reproducibility of the few load-deformation functions that had been reported. Even the early investigators realized that the peak value of force, shown as P_o in Fig. 6.3b, could be evaluated reasonably well simply by adding together the separate capacities of concrete and steel. The maximum thrust or the squash load, P_o, on composite columns can be estimated by adding the products of steel area A_s and steel yield strength f_y to the product of concrete area A_c and concrete strength $\beta f'_c$. A coefficient β has been used to modify laboratory or material control values of concrete compression strength f'_c. If 10-cm cubes are used for concrete control, the values of β vary somewhat with f'_{cu}, the measured strength of each cube, but the quantity $\beta = 0.75$ is used frequently. If 6-in. (15-cm) diameter cylinders, 12 in. (30 cm) long, are used for measuring f'_c, the coefficient $\beta = 0.85$ provides a reliable lower bound value for concrete capacity. The maximum compression strength or squash load is then expressed:

$$P_o = A_s f_y + 0.85 f'_c A_c \tag{6.1}$$

Some tests of short lengths of concrete-filled pipe sections have revealed the possible development of loads higher than P_o if the steel pipe after yielding longitudinally can continue to provide effective lateral confining pressures to the contained concrete. However, for columns more than a few diameters in height, the loss of longitudinal stiffness in the steel wall leads to subsequent buckling instability, and thrusts greater than P_o cannot be realized.

The buckling stability of composite columns can be analyzed at any strain level if the basic stress-strain behavior of the materials is known. A composite tangent-modulus form of strength-slenderness behavior represents a lower limit to column strength P_c. A buckling load P_c can be expressed (Knowles and Park 1969):

$$P_c = \pi^2 EI_{\tan}/(kl)^2 \tag{6.2}$$

The value of the quantity kl represents the effective length of the compression member (distance between points of inflection in the buckled shape of the column), and the quantity $EI_{\tan}$ represents an effective tangent-modulus stiffness for the cross section. The effective tangent-modulus stiffness can be established by tests, but more commonly it must be estimated from characteristic stress-strain properties of steel and concrete.

For purposes of design, stress-strain functions for steel and for concrete, once expressed in analytic form, can be differentiated to obtain the slope of each as a function of strain. These slopes represent tangent moduli for each material. Then for any strain level, products of tangent-modulus stiffness, E_{st} and E_{ct}, and flexural shape factors, I_s and $\frac{1}{2}I_c$, can be combined to give an effective stiffness quantity $EI_{\tan}$

$$EI_{\tan} = E_{st}I_s + \tfrac{1}{2}E_{ct}I_c \tag{6.3}$$

Only half the moment of inertia of concrete is suggested here because it should be assumed that something less than the full concrete cross section can remain uncracked in pure flexure. At the same strain level for which $EI_{\tan}$ is evaluated, the corresponding thrust can be evaluated simply by adding together the products of material stress, f_s and f_c, and material areas, A_s and A_c:

$$P_{cr} = f_s A_s + f_c A_c \tag{6.4}$$

After values of $EI_{\tan}$ and P_{cr} are evaluated, the effective length kl can be determined:

$$kl = \pi \sqrt{\frac{EI_{\tan}}{P_{cr}}} \tag{6.5}$$

The designer must evaluate Eq. 6.5 for enough different strain levels to define an adequate strength-slenderness design curve. The calculations and graph for such a curve are given in Fig. 6.4 in order to illustrate the type and sequence of calculation necessary. The exponential equation that was used to represent A36 steel tubing for Fig. 6.4 will give misleadingly high values for f_s at very high strains, and an upper limit value of P_o as expressed by Eq. 6.1 should be observed.

Note that the thrust-slenderness curve of Fig. 6.4 has an S-type graph that is typical for any compression member. The graph starts with a thrust of P_o at zero slenderness, and the thrust capacity decreases as the slenderness increases. The rate of decrease begins to lessen for ultimate thrusts less than 0.5 P_o.

Two familiar analytic functions, a parabola and a hyperbola, have been used to approximate column strength curves. The fundamental parameters for such approximations involve only the effective long-column stiffness EI and the squash load P_o. If P_{cr} is less than 0.5 P_o, long-column behavior is assumed, and $EI_{\tan}$ is taken simply as the low-stress initial value of flexural stiffness:

$$EI = E_s I_s + \tfrac{1}{2} E_c I_c \tag{6.6}$$

The ACI Building Code recommends that concrete stiffness be computed from the cylinder strength f'_c in psi and concrete density w_c in pcf; thus:

$$E_c = w_c^{1.5}\, 33 \sqrt{f'_c} \quad \text{in lb units}$$
$$E_c = w_c^{1.5}\, 900 \sqrt{f'_c} \quad \text{in kg/m}^3 \text{ and MPa units} \tag{6.7}$$

Then the effective length kl_c at which long-column action commences can be computed from Eq. 6.5 as:

$$kl_c = \pi \sqrt{\frac{EI}{0.5 P_o}} \tag{6.8}$$

The thrust capacity for P_{cr} values greater than 0.5 P_o is made to fit the intermediate column portion by the relationship:

$$P_{cr} = P_o \left[1 - \frac{1}{2}\left(\frac{kl}{kl_c}\right)^2\right] \qquad kl < kl_c \tag{6.9a}$$

$$P_{cr} = \pi^2 \frac{EI}{(kl)^2} \qquad kl > kl_c \tag{6.9b}$$

Generalized strength curves useful for design can be constructed from Eqs. 6.2 and 6.9 if Eq. 6.6 is accepted for effective flexural stiffness, and if safety factors are incorporated into the design curves. The strength-slenderness curves for $f'_c = 4$ ksi (28 MPa), concrete-filled, A36 (248 MPa) steel, round tubing are derived and displayed in Fig. 6.5. The dotted line of the graph in Fig. 6.5 was obtained from the theoretically more precise curve of Fig. 6.4. Correspondence between the "precise" and the general curves indicates that the simplifications of Eqs. 6.6 through 6.9 introduce negligible error into the procedure for developing generalized strength curves. For Fig. 6.4, the curves are parabolas with a horizontal tangent at P_o. These parabolas can be constructed to pass through the coordinate point $\frac{kl_c}{D}$ and $\frac{P_o}{2D}$. To the right of $\frac{kl_c}{D}$ values, the curves are hyperbolas, and convenient points on each hyperbola were determined in order to construct the graph. At a slenderness ratio of 1.5 $\frac{kl_c}{D}$, ordinates will be equal to $\frac{P_o}{4.5 A_g}$, and, at 2 $\frac{kl_c}{D}$, the ordinates are $\frac{P_o}{8 A_g}$.

Similar sets of generalized thrust-slenderness graphs can be constructed for any combination of steel yield strength, concrete strength, and cross-section shape. The graphs are applicable only for estimating thrust capacities for concentrically loaded composite columns. A column can be considered to be concentrically loaded only if it is a part of a structure that is laterally braced against horizontal forces and only if connections that introduce thrusts to columns are not capable of introducing significant bending to the columns also.

Development of Load-Slenderness Curve

12-in. (305-mm) standard pipe, A36 steel (f_y = 248 MPa), filled with concrete

$$f'_c = 4 \text{ ksi (28 MPa)}, \qquad E_c = 3600 \text{ ksi (24,800 MPa)}$$

For steel, use

$$\epsilon_s = \frac{f_s}{29{,}000}[1 + 0.1(f_s/30)^9] \qquad E_s = 29{,}000 \text{ ksi (200,000 MPa)}$$

$$E_s = 29{,}000/[1 + (f_s/30)^9]$$

For concrete, use

$$f_c = 0.85 f'_c \frac{\epsilon_c}{0.0019}(2 - \epsilon_c/0.0019) \qquad \epsilon_c < 0.0019$$

$$f_c = 0.85 f'_c \quad \text{for} \quad \epsilon_c > 0.0019$$

$$E_c = 3600(1 - \epsilon_c/0.0019) > 0.$$

Cross section:

A_s = 14.58 in.^{2} (9400 mm^2) $\quad$ A_c = 113.1 in.^{2} (73,000 mm^2) $\quad$ $P_o = 0.85 f'_c A_c + A_s f_y$

I_s = 279.3 in.^{4} (116 × 10^6 mm^4) $\quad$ I_c = 1018 in.^{4} (424 × 10^6 mm^4) $\quad$ = 0.85(4)113.1

+ 14.58(36)

= 909 kips

(4040 kN)

f_s (ksi)	$\epsilon_s = \epsilon_c$ (%)	E_s (ksi)	f_c (ksi)	E_c (ksi)	Eq. 4 P_{cr} (k)	Eq. 3 EI_{tan} (k in.^{4})	Eq. 5 kL in.
36	0.1882	4708	3.40	34	909	1332000	120.
33	0.1406	8336	3.17	936	839	2805000	182.
30	0.1138	14500	2.85	1444	760	4785000	249.
25	0.0879	24290	2.42	1934	638	7769000	347.
20	0.0691	28260	2.02	2291	520	9059000	415.
15	0.0517	28940	1.60	2620	400	9417000	482.
10	0.0345	29000	1.12	2946	272	9599000	590.

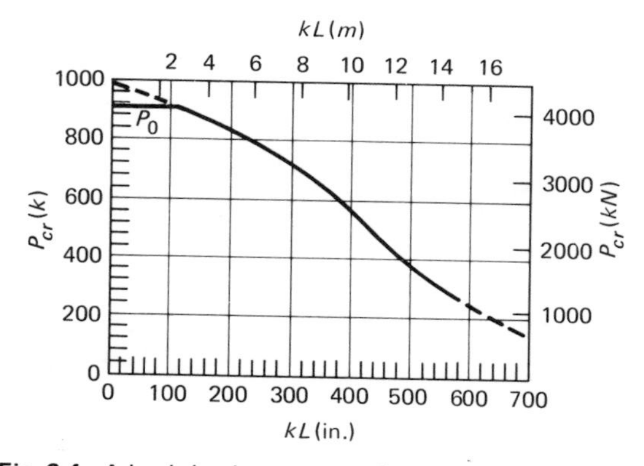

Fig. 6.4. A load-slenderness curve for concrete-filled pipe.

Development of Strength-Slenderness Curves

D = outside diameter of tube

t = thickness of tube wall

$$A_g = \frac{\pi D^2}{4} \qquad I_g = \frac{\pi D^4}{64}$$

$$A_s = \frac{\pi D^2}{4}\left[1 - \left(1 - \frac{2t}{D}\right)^2\right] \qquad I_s = \frac{\pi D^4}{64}\left[1 - \left(1 - \frac{2t}{D}\right)^4\right]$$

$$A_c = A_g - A_s \qquad I_c = I_g - I_s$$

$$\frac{P_o}{A_g} = 0.85 f'_c \left[1 - 4\left(\frac{t}{D} - \frac{t^2}{D^2}\right)\right] + f_y\left[4\left(\frac{t}{D} - \frac{t^2}{D^2}\right)\right]$$

$$\frac{EI}{I_g} = E_c\left[1 - 8\left(\frac{t}{D} - \frac{3t^2}{D^2} + \frac{4t^3}{D^3} - \frac{2t^4}{D^4}\right)\right] + 8E_s\left(\frac{t}{D} - \frac{3t^2}{D^2} + \frac{4t^3}{D^3} - \frac{2t^4}{D^4}\right)$$

$$\frac{kl_c}{D} = \frac{\pi}{4}\sqrt{\frac{2EI}{I_g} \times \frac{A_g}{P_o}} \qquad \text{say} \qquad Z = 8\left(\frac{t}{D} - \frac{3t^2}{D^2} + \frac{4t^3}{D^3} - \frac{2t^4}{D^4}\right)$$

$\frac{t}{D}$	$4\left(\frac{t}{D} - \frac{t^2}{D^2}\right)$	Z	$\frac{P_o}{A_g}$ ksi	$\frac{EI}{I_g}$ ksi	$\frac{kl_c}{D}$	$\frac{P_o}{2A_g}$ ksi	$\frac{P_o}{4.5A_g}$ ksi	$\frac{1.5kl_c}{D}$	$\frac{P_o}{8A_g}$ ksi	$\frac{2kl_c}{D}$
0.015	0.0591	0.1147	5.327	4930.	33.8	2.66	1.18	50.6	0.666	67.5
0.030	0.1164	0.2192	7.195	7762.	36.5	3.58	1.60	54.7	0.90	73.0
0.050	0.1900	0.3439	9.594	11150.	37.9	4.80	2.13	56.8	1.20	75.7
0.075	0.2775	0.4780	12.45	14800.	38.3	6.22	2.77	57.4	1.56	76.6
0.100	0.3600	0.5296	15.14	16200.	36.3	7.57	3.36	54.5	1.89	72.7
0.125	0.4375	0.6836	17.66	20390.	37.7	8.84	3.93	56.6	2.21	75.5
0.150	0.5100	0.7599	20.03	22470.	37.2	10.01	4.45	55.8	2.50	74.4

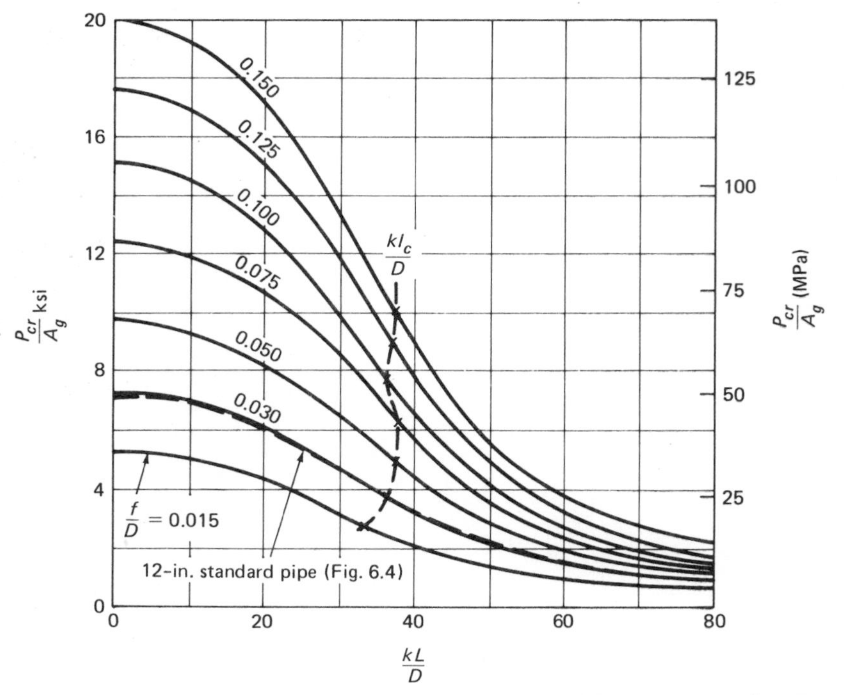

Fig. 6.5. Strength-slenderness curves for A36 round tubes filled with concrete f'_c = 4 ksi.

6.4 ECCENTRIC THRUST CAPACITY: AXIAL LOAD PLUS FLEXURE

The strength of composite cross sections under the action of thrust P_u and bending moment M_u has been described in the form of interaction diagrams. Combinations of thrust capacity and corresponding moment capacity are determined analytically on the basis of an assumed plane of strain, a limiting strain for concrete, and thereafter an integration of stresses that correspond with stress-strain characteristics for each material. In the absence of moment, the maximum thrust is, of course, the squash load P_o. In the absence of thrust, the "pure" moment capacity M_o is something larger than the plastic moment capacity for the steel alone in the cross section. Except for the case of bending about the weak axis of a concrete-encased *W*, *H*, or *I* steel shape, an acceptable lower bound to pure bending capacity can be taken as the plastic flexural strength of steel alone. For weak-axis bending of encased steel shapes, an acceptable lower bound value can be taken as the plastic bending strength of the steel flanges plus the flexural strength of a cross section reinforced only by the web of the steel shape (Furlong 1976). Values of thrust P_u in the presence of bending M_u can be estimated from linear, parabolic, or elliptical functions that include the estimated points P_o and M_o. A convenient and reliable analytic function that will be illustrated with design charts takes the form:

$$\left(\frac{P_u}{P_o}\right)^2 + \frac{M_u}{M_o} = 1 \qquad (6.10)$$

A more precise though not demonstrably more accurate estimate of P_u and M_u values can follow the logic of reinforced concrete theory (Ferguson 1972). That logic involves

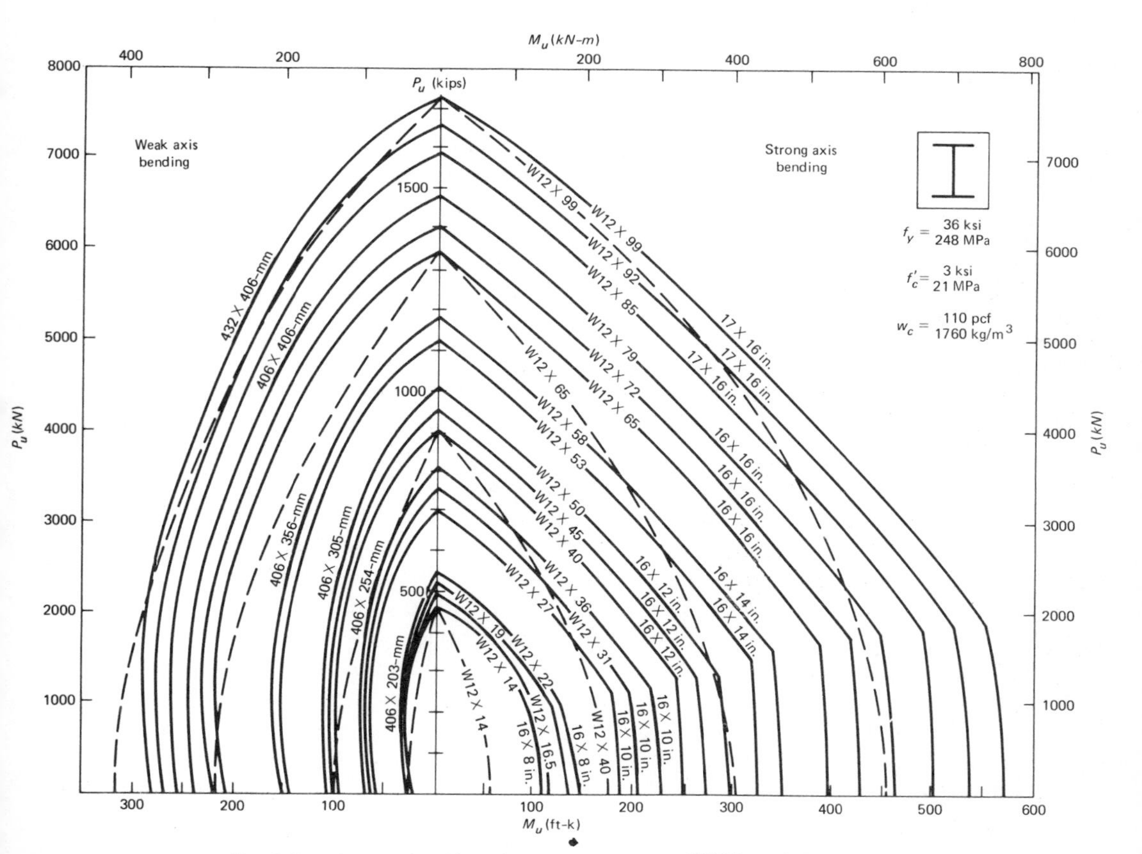

Fig. 6.6a. Interaction chart for concrete-encased W12 steel shapes.

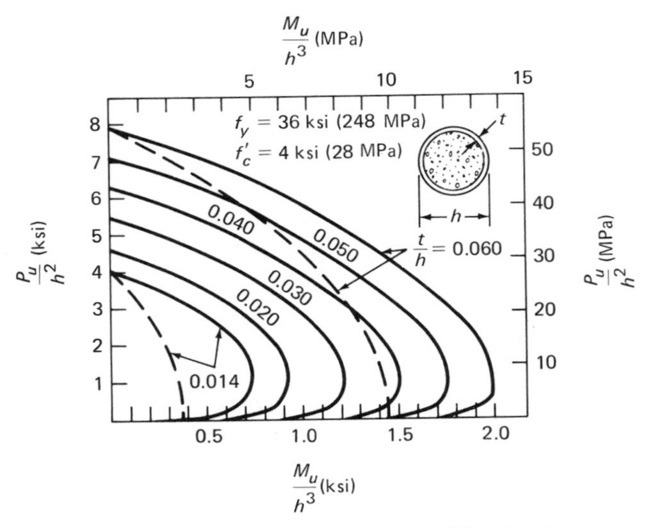

Fig. 6.6b. Interaction chart for concrete-filled steel tubes.

the use of a limiting concrete strain, usually 0.3%, as the limit of concrete strength. A selection of a neutral axis plus the designation of an extreme fiber at 0.3% strain defines a failure plane of strain for the selected neutral axis. Stresses on all steel elements and on concrete elements can be computed on the basis of characteristic stress-strain properties, such as a trapezoidal function for steel and a parabola or equivalent rectangular stress block for compressed concrete. Generally, the rectangular stress block is used for concrete. Some "precise" interaction diagrams for AISC 12-in. (305-mm) rolled shapes (AISC 1970) encased in lightweight load-bearing concrete are shown in Fig. 6.6a (Furlong 1974). Dashed lines of Fig. 6.6a show for some shapes a set of parabolic functions that pass through P_o values and estimated lower-bound values of M_o. The estimates of M_o for dashed-line graphs give moment capacities that are considerably lower than the values obtained from the more accurately derived functions. When dashed-line graphs are used for design, the resulting cross sections will be safer than cross sections that are designed on the basis of the solid-line graphs. Figure 6.6b contains a generalized interaction graph for round tubes filled with concrete. Again, some dashed-line graphs illustrate the conservatively safe values given by Eq. 6.10.

6.5 COMBINED AXIAL LOAD PLUS BENDING OF SLENDER COLUMNS

Analytic techniques for relating the thrust capacity for various proportions of slenderness of composite columns have been described. In the absence of thrust, composite columns do not experience a corresponding reduction in flexural capacity as slenderness increases. Structural shapes in flexure can develop lateral-torsional buckling modes that reduce their flexural capacity as slenderness increases, but the encasement of steel shapes in concrete effectively curtails the potential for developing lateral-torsional buckling of the composite member. Concrete-filled steel tubes possess enough inherent torsional stiffness to curtail the probability for lateral torsional buckling. Until physical tests reveal better definitions of limiting proportions of rectangular shapes that will not fail in a lateral-torsional mode, it should be conservative and safe to ignore such a possibility if the width-to-breadth ratio of cross sections is greater than $\frac{2}{3}$ and less than 1.5.

When thrust is combined with flexural forces on relatively slender members, the thrust tends to increase the lateral displacement of the flexed or bent member as illustrated in Fig. 6.7a. The column in Fig. 6.7a supports a thrust P that is applied with an eccentricity e at one end and αe at the other end. Consequently, the moment at one end will be the product Pxe at one end, and at the other end the moment is the product $Px\alpha e$ as illustrated in Fig. 6.7b. If the column is slender, moment diagrams indicated in Fig. 6.7c can develop. The long dashed line of Fig. 6.7c represents the moment diagram of an intermediate slenderness for which the maximum moment Pe remains at one end. The solid line of Fig. 6.7c is the moment diagram for a slender column, and the maximum moment δPe exists at some point between the ends of the column.

The amount of extra moment created by the lateral displacement of a slender compression member depends on the ratio α between end moments, and depends also on the member slenderness as reflected by the column buckling capacity P_c. A moment magnification factor δ can be estimated as

$$\delta = \frac{0.6 + 0.4\alpha}{1 - \dfrac{P}{P_c}} \geqslant 1 \qquad (6.11)$$

where

$$P_c = \pi^2 \frac{EI}{(kl)^2}$$

The sense of α can be negative if moments at each end create compression on opposite faces of the member. A minimum value $\alpha = -0.5$ is permitted by most building codes (ACI-318-71, AISC 1970).

The strength effect of slenderness can be estimated simply by checking against the short-column strength the combination of thrust P_u and slenderness-magnified moment δM_u. If values of δ as computed from Eq. 6.11 are less than 1, the calculations indicate simply that the end moment remains greater than any other moment between ends of the member as illustrated by the long dashed line of Fig. 6.7c. The column must support the thrust plus end moment, obviously, so factors *less* than 1 cannot be used.

Thrust-moment interaction graphs are perhaps the most precise description of strength against which the required thrust P_u and magnified moments δM_u can be compared. Most composite-column interaction graphs can fit adequately some linear analytic functions, and useful equations that relate thrust capacity and flexural capacity even for slender composite columns can be expressed with magnified moment values that are used for interaction ratios in the form of Eq. 6.10 to give Eq. 6.12:

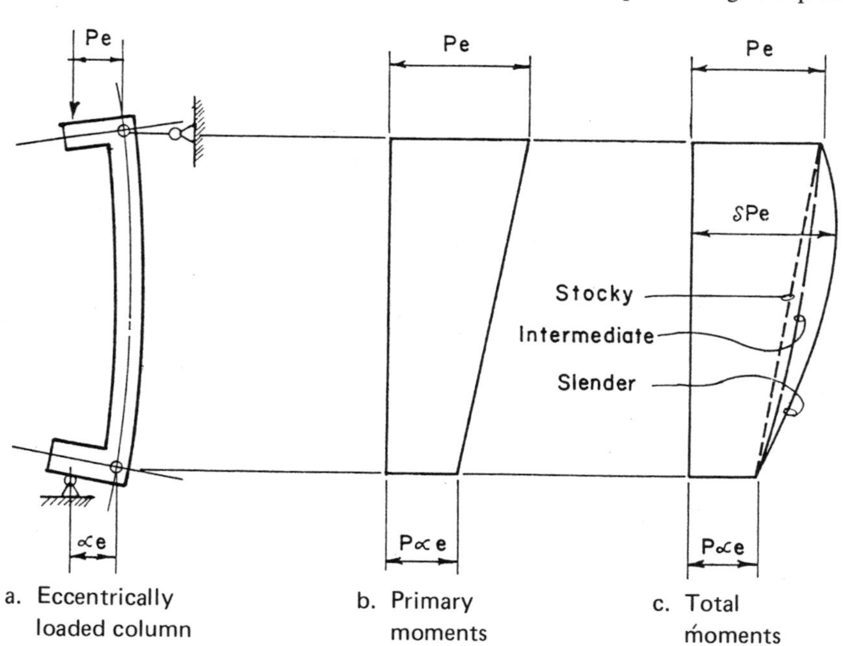

Fig. 6.7. Moments caused by eccentric thrust.

$$\left(\frac{P_u}{P_{cr}}\right)^2 + \frac{\delta M_u}{M_o} = 1 \qquad (6.12)$$

In Eq. 6.12, the magnification factor δ would be computed with the ultimate thrust P_u as the value of P in Eq. 6.11. Values of P_{cr} should be computed from Eq. 6.9a or 6.9b, and values of M_o should be taken as the plastic bending capacity for steel alone except for the special case of weak-axis bending of encased shapes.

The analysis of capacity for composite columns is a vital but not a complete part of the design process. Safe design also requires that some allowances be made for probable discrepancies between real and anticipated material properties, member dimensions, and indeterminate frame forces. Finally, even the lower bound estimates of capacity must be adequate to accommodate loads that exceed the largest probable service loads.

6.6 STRUCTURAL SAFETY: LOAD FACTORS AND MATERIAL RELIABILITY

In order to employ a simple and virtually failsafe criterion for structural proportioning, most engineers believe that at least 100% overload should be sustained before a structural system collapses. There are wide variations among real ratios between collapse load and design load even for separate parts or elements of the same well-designed structure. Surely one is more willing to accept lower margins of safety for local "failure" that is signaled by the sagging of a beam than one might be to accept overall collapse from the sudden, buckling-spalling failure of a column. Consequently, the load factors and material reliability factors recommended for composite columns will tend to permit more than 100% overload prior to collapse of a composite column.

A generally comfortable margin of overload capacity should exist if the *maximum* anticipated level of design load is simply augmented by a factor of 50% to 67% before it is checked against a *minimum* capacity estimate for the member being designed. In North American practice, the steel industry advocates a load factor of 1.7 for gravity-type loading and 1.3 for wind- or earthquake-induced loads (AISC 1970). The American Concrete Institute requires design for 1.4 permanent and 1.7 intermittent gravity-type loads, and it permits the use of 75% of these values when the effects of wind or earthquake are superimposed (ACI 318-71). General load factors of 5/3 gravity-caused loads and 5/4 wind-, dynamic-, or volume-change-induced loads are recommended as a simple, reasonable pair of values.

Real safety requires rational recognition that the actual member capacities can be lower than those that would exist if material strength and member dimensions were at least as good as the structural specification demands. Material reliability factors that are less than 100% can be assigned to concrete and to steel in composite columns on the basis of a statistical measure when adequate coupon or test data are available. In the absence of such data, the judgment of the designer must be used. The material reliability factors μ given in Table 6.1 are offered only as suggestions that will provide for at least as much safety in composite columns as that presently obtained through applications of AISC designs for steel alone. The μ factors should be multiplied by property numbers F_y, f'_c, and E_s in Eqs. 6.1 through 6.3. The net factor of safety will tend to increase with slenderness because the effect of a softened set of E_s and E_c values will become larger as slenderness increases. Laterally bound concrete, as used in Table 6.1, would be concrete within steel pipe or tubing, or a closely spaced spiral.

In the design process, structural safety is taken into account by insuring that service loads augmented by load factors are never greater than the lowest reliable strength estimates. Design aids can be constructed with the material reliability factors μ incorporated into each in order to describe lowest reliable capacities. Then the largest anticipated design loads must be augmented by load factors in order to obtain ultimate loads. The ultimate loads must be less than the lowest reliable capacities. For example, a short length of a composite cross section that contains 10 in.^{2} (645 mm^2) of

Table 6.1. Material reliability factors μ.

Material	Property	Type or Use	Symbol	Suggested Value
Steel	Yield strength, F_y	Rolled shapes	μ_{ss}	0.95
		Pipe or tube	μ_{st}	0.90
	Stiffness, E_s	All steel	μ_e	0.85
Field-cast concrete	Cylinder strength, f'_c	Laterally unbound	μ_{cu}	0.65
		Laterally bound	μ_{cc}	0.85
Factory-cast concrete	Cylinder strength, f'_c	Laterally unbound	μ_{cuf}	0.80
		Laterally bound	μ_{ccf}	0.95

A36 steel (f_y = 248 mPa) encased in 100 in.^{2} (64,500 mm^2) of f'_c = 4 ksi (28 mPa) concrete has a specified thrust capacity

$$\text{Ideal } P_o = f_y A_s + 0.85 f'_c A_c$$
$$= (36)10 + 0.85(4)100 = 700\text{k}$$

but a design aid should reflect only the reliable value of

$$P_o = (\mu_{ss} f_y) A_s + 0.85(\mu_{cu} f'_c) A_c$$
$$= (0.95 \times 36)10 + 0.85(0.65 \times 4)100$$
$$= 563\text{k}$$

using μ values from Table 6.1. A service load of 335k would be permissible if the load factor were $\frac{5}{3}$ because $\frac{5}{3} \times 335 = 558\text{k}$ is less than 563k.

6.7 BUILDING CODE OF THE AMERICAN CONCRETE INSTITUTE

The ACI Building Code defines the failure of compression members in terms of a 0.3% strain limit for any concrete fiber. That failure strain at a concrete surface can be used with a set of different neutral axis locations to define failure strains for each neutral axis. Stresses that are compatible with each failure strain are then integrated to determine a failure thrust and moment for each neutral axis location in order to describe the type of interaction curves illustrated in Fig. 6.6. Traditionally, the ACI Building Code specified a minimum value for the eccentricity of column thrusts, but now in lieu of that requirement, an upper limit on column thrust is specified. Present limits require that maximum thrust can be no greater than 85% of the squash load for concrete-filled tubes.

Material reliability factors are not used by the ACI Building Code, but somewhat similar capacity reduction factors ϕ are applied to estimates of member strength. The ϕ factor for concrete-filled tubes is 0.75, and the ϕ factor for concrete-encased shapes is 0.70. Load factors in the Building Code are 1.4 for permanent loads and 1.7 for transient or variable live load. All load factors can be reduced by 25% if wind or seismic forces comprise a part of the design loading.

Slenderness effects for columns must be analyzed in terms of moment magnification factors, as suggested by Eq. 6.11. The ACI Building Code requires the moment magnifier to be computed for the ultimate (augmented service load) thrust condition of loading, and values of EI must be evaluated for the "softest" possible circumstance of column behavior in framed structures. It is permissible to use the EI value from either of the following two equations:

$$EI_1 = \frac{0.4 E_c I_g}{1 + \beta_d} \tag{6.13a}$$

$$EI_2 = \frac{0.2 E_c I_g + E_s I_s}{1 + \beta_d} \tag{6.13b}$$

with

E_c = modulus of elasticity for concrete (Eq. 6.7)
E_s = modulus of elasticity for steel (29,000 ksi or 200,000 MPa)
I_g = moment of inertia of gross cross section
β_d = creep factor equal to the ratio between design dead-load moment and total design moment.

Cross-section capacity as defined by interaction curves must be diminished by capacity reduction factors ϕ, and the reduced strengths must exceed ultimate thrusts and magnified ultimate moments.

Lower limits to the thickness t of steel tubing for concrete-filled composite columns are set as:

$$t \geqslant h\sqrt{\frac{f_y}{8E_s}} \text{ for circular sections of diameter } h \tag{6.14a}$$

or

$$t \geqslant b\sqrt{\frac{f_y}{3E_s}} \text{ for each face of width } b \tag{6.14b}$$

6.8 EXAMPLES OF COMPOSITE COLUMN ANALYSIS AND DESIGN

EXAMPLE 6.1

How much live-load thrust can be supported by an A36 steel (f_y = 248 MPa) tube with 0.25-in. (10-mm) wall thickness and 10-in. (254-mm) diameter filled with f'_c = 4 ksi (28 MPa) concrete, if the effective length of the column is 16.4 ft (5.00 m) and the dead-load thrust is 120k (534 kN)? Moments are negligible.

Solution

1. Determine section properties and E_c.

$$A_s = \frac{\pi}{4}(10^2 - 9.5^2) = 7.66 \text{ in.}^2 \ (4940 \text{ mm}^2)$$

$$A_c = \frac{\pi}{4}(9.5)^2 = 70.9 \text{ in.}^2 \ (45{,}700 \text{ mm}^2)$$

$$I_s = \frac{\Pi}{64}(10^4 - 9.5^4)$$

$$= 90.9 \text{ in.}^4 \ (37.8 \times 10^6 \text{ mm}^4)$$

$$I_c = \frac{\Pi}{64}(9.5)^4 = 400 \text{ in.}^4 \ (166 \times 10^6 \text{ mm}^4)$$

$$E_c = 33W_c^{1.5}\sqrt{\mu_{cc}f'_c}$$

$$= 33(145)^{1.5}\sqrt{0.85 \times 4000}$$

$$= 3.36 \times 10^6 \text{ psi } (23.2 \times 10^6 \text{ MPa})$$

2. Compute EI, P_o, kl_c, and P_{cr} using Eqs. 6.1, 6.6, 6.8, and 6.9.

$$EI = (\mu_e E_s)I_s + \tfrac{1}{2}(E_c)I_c$$

$$= (0.85 \times 29{,}000)90.9 + \tfrac{1}{2}(3360)400$$

$$= 2.91 \times 10^6 \text{ k-in.}^2 \ (8.35 \text{ MN-m}^2)$$

$$P_o = (\mu_{st}f_y)A_s + 0.85(\mu_{cc}f'_c)A_c$$

$$= (0.9 \times 36)7.66 + 0.85(0.85 \times 4)70.9$$

$$= 453\text{k } (2015 \text{ kN})$$

$$kl_c = \pi\sqrt{\frac{EI}{0.5P_o}} = \pi\sqrt{\frac{2.91 \times 10^6}{0.5 \times 453}}$$

$$= 356 \text{ in. } (9.04 \text{ m})$$

$$P_{cr} = P_o\left[1 - \frac{1}{2}\left(\frac{kl}{kl_c}\right)^2\right]$$

$$= 453\left[1 - \frac{1}{2}\left(\frac{16.4 \times 12}{356}\right)^2\right]$$

$$= 384\text{k } (1700 \text{ kN})$$

3. Using a load factor of $\frac{5}{3}$, compute the safe amount of live-load thrust P_l such that the total service load augmented by the load factor remains less than P_{cr}.

$$\tfrac{5}{3}(120 + P_l) < 384$$

$$P_l < 111\text{k } (493 \text{ kN}) \qquad \text{ANSWER}$$

Comments

The empty tube would be allowed to support a 136k (605 kN) service load or 16k (71 kN) of live load if AISC rules were observed for steel alone. ACI regulations are meaningless for axially loaded columns, and no comparison will be attempted for this example.

EXAMPLE 6.2

Select a cross section with load-bearing lightweight concrete encasement around an A36 steel shape to support an axial load of 640k (2850 kN) with no moment under service-load conditions, and an effective length of 126 in. (3.2 m). Assume $f'_c = 3$ ksi (21 MPa) with $w_c =$ 110 pcf (1760 kg/m³).

1. Estimate the gross area of the required cross section as if steel occupied about 10% of the gross area. An estimating formula for concrete-encased shapes becomes:

$$A_g \geqslant \frac{2P}{0.5f'_c + 0.1f_y} \geqslant \frac{2 \times 640}{0.5 \times 3 + 0.1 \times 36}$$

$$= 251 \text{ in.}^2 \ (162{,}000 \text{ mm}^2)$$

Try a cross section that is 18 in. (460 × 350 mm) deep and 14 in. wide ($A_g = 252$ in.²) with a W14X61 core.

2. Compute P_o for the trial section

$$A_s = 17.9 \text{ in.}^2 \ (11{,}550 \text{ mm}^2)$$

$$A_c = 18 \times 14 - 17.9$$

$$= 234 \text{ in.}^2 \ (151{,}000 \text{ mm}^2)$$

$$P_o = \mu_{ss} f_y A_s + \mu_{cu} f'_c A_c$$

$$= (0.95 \times 36)17.9 + (0.65 \times 3)234$$

$$= 1068\text{k} \ (4750 \text{ MN})$$

The ultimate load of $\frac{5}{3} \times 640 = 1067$k (4746 MN), and it is probable that slenderness will make this trial section inadequate. Try the next larger core shape (W14 × 68).

Compute P_o for the new trial section:

$$A_s = 20.0 \text{ in.}^2 \ (12{,}900 \text{ mm}^2)$$

$$A_c = 18 \times 14 - 20 = 232 \text{ in.}^2 \ (150{,}000 \text{ mm}^2)$$

$$P_o = (0.95 \times 36)20.0 + (0.65 \times 3)232$$

$$= 1136\text{k} \ (5053 \text{ kN})$$

3. Weak-axis buckling is more critical than strong-axis buckling; determine P_{cr}:

$$I_s = I_y = 121 \text{ in.}^4 \ (50.4 \times 10^6 \text{ mm}^4)$$

$$I_c = \frac{18}{12}(14)^3 - 121$$

$$= 3995 \text{ in.}^4 \ (1663 \times 10^6 \text{ mm}^4)$$

$$E_c = 33w^{1.5}\sqrt{\mu_{cu} f'_c}$$

$$= 33(110)^{1.5}\sqrt{0.65 \times 3000}$$

$$= 1.68 \times 10^6 \text{ psi } (11{,}600 \text{ MPa})$$

$$EI = (\mu_e E_s)I_s + \tfrac{1}{2}E_c I_c$$

$$= (0.9 \times 29{,}000)121 + \tfrac{1}{2}(1.68 \times 10^6)3995$$

$$= 6{,}510{,}000 \text{ k-in.}^2 \ (18{,}680 \text{ kN-m}^2)$$

$$kl_c = \pi\sqrt{\frac{EI}{0.5P_o}} = \pi\sqrt{\frac{6{,}510{,}000}{0.5 \times 1136}}$$

$$= 336 \text{ in. } (8.54 \text{ m})$$

$$P_{cr} = P_o\left[1 - \frac{1}{2}\left(\frac{kl}{kl_c}\right)^2\right] = 1136\left[1 - \frac{1}{2}\left(\frac{126}{336}\right)^2\right]$$

$$= 1056\text{k} \ (4697 \text{ kN})$$

That value is close enough to the required value of $\frac{5}{3}P = 1067$k (4746 kN) to permit the second trial section to be used.

Use W14X68 core in 18 × 14-in. encasement

(460 × 350 mm)

ANSWER

EXAMPLE 6.3

Determine the moment capacity of a concrete-filled steel caisson of 24-in. (610-mm) outside diameter and 0.50-in. (13-mm) thickness if the ultimate thrust $P_u = 1390$ K (6180 kN), $f'_c = 4$ ksi (28 MPa), and $f_y = 36$ ksi (248 MPa).

Solution

Figure 6.6b can be used, as it was constructed for the appropriate material properties. Compute the ratios

$$\frac{t}{h} = \frac{0.50}{24} = 0.021$$

and

$$\frac{P_u}{h^2} = \frac{1390}{24^2} = 2.41 \text{ ksi } (166 \text{ MPa})$$

For these ratios, the chart gives $\frac{M_u}{h^3} = 0.45$ ksi (3.10 MPa).

Compute

$$M_u = 0.45h^3 = 0.45(24)^3$$

$= 6220$ in.-k (700 kN-m)

ANSWER

Alternate Solution

If Fig. 6.6b had not been available, use Eq. 6.10 to estimate the moment capacity.

1. Compute

$$P_o = 0.85 f'_c A_c + f_y A_s$$

$$= 0.85 f'_c (h - 2t)^2 \frac{\pi}{4} + f_y (h - t)\pi t$$

$$= 0.85 \times 4(23)^2 \frac{\pi}{4} + 36(23.5)\frac{\pi}{2}$$

$= 2740$k (12,200 kN)

2. Compute

$$M_o = f_y Z = f_y \left(\frac{3}{4} h^3\right)\left(\frac{t}{h} - 2\frac{t^2}{h^2} + \frac{4}{3}\frac{t^3}{h^3}\right)$$

$$= 36\left(\frac{3}{4} \times 24^3\right)\left(\frac{0.50}{24} - \frac{2 \times 0.50^2}{24^2} + \frac{4}{3} \times \frac{0.50^3}{24^3}\right)$$

$= 7460$ in.-k (10,100 kN-m)

3. Solve Eq. 6.10 for M_u:

$$\left(\frac{P_u}{P_o}\right)^2 + \frac{M_u}{M_o} = 1$$

$$M_u = M_o \left[1 - \left(\frac{P_u}{P_o}\right)^2\right]$$

$$= 7460 \left[1 - \left(\frac{1390}{2740}\right)^2\right]$$

$= 5540$ in.-k (7500 kN-m)

ANSWER

EXAMPLE 6.4

If a service-load thrust of 834k (3710 kN) is to be supported by a caisson of 24-in. (610-mm) diameter and 0.50-in. (13-mm) wall thickness with steel of $f_y = 40$ ksi (276 MPa) quality, how much service-load moment can be permitted if the tube is filled with concrete at $f'_c = 4.7$ ksi (32 MPa)?

Solution

1. The general load factor of $\frac{5}{3}$ will be used, together with material reliability factors μ of Table 6.1. Application of the material reliability factors would require that a design chart be constructed for a steel yield strength of $\mu_{st} f_y = 0.90 \times 40 = 36$ ksi (248 MPa) and a concrete compressive strength $\mu_{cc} f'_c = 0.85 \times 4.7 = 4.0$ ksi (28 MPa). Figure 6.6b can be used again.
2. Compute

$$P_u = P \times \text{Load Factor} = 834 \times \frac{5}{3}$$

$= 1390$k (6180 kN).

3. Repeat Ex. 6.3 to obtain $M_u = 6220$ in.-k (700 kN-m).
4. Compute service-load moment

$$= \frac{M_u}{\text{Load Factor}}$$

$$= \frac{6220}{5/3} = 3730 \text{ in.-k (421 kN-m)}$$ ANSWER

EXAMPLE 6.5

Determine the amount of wind force that can be permitted on a frame if wind force were limited by the strength of 12 × 12 in. (305 mm), encased, W8X31, A36 (248 MPa) steel columns, and elastic analysis indicated that gravity-type service loads generated 162k (720 kN) column thrust and 360 in.-k (40.7 kN-m) moment about the major axis, while 1k (4.45 kN) of wind produced 23 in.-k (2.60 kN) of moment about the major axis. The effective length of the columns was computed to be 212 in. (5.38 m), and the natural aggregate encasement ($w_c = 145$ lb/ft^3 = 2320 kg/m^3) has a specified value $f'_c = 4$ ksi (28 MPa). There are no moments about the minor axis, and the effective length for minor-axis buckling is only 114 in. (2.90 M).

Solution

Let it be assumed that no design aid is available, and Eq. 6.12 will be used in lieu of more precise analysis. AISC compactness criteria will not be observed for the beam-column response of the encased shape, and inspection reveals that minor-axis capacity need not be checked.

1. Compute section and member properties:

$A_s = 9.12$ in.^{2} (5880 mm^2)

$I_s = 110$ in.^{4} (45.8×10^6 mm^4)

$Z_s = 30.4$ in.^{3} (498,000 mm^3)

$A_c = 12^2 - 9.1 = 135$ in.^{2} (87,100 in.^{2})

$$I_c = \frac{12^4}{12} - 110 = 1618 \text{ in.}^4 \ (673 \times 10^6 \text{ mm}^4)$$

Using material reliability factors from Table 6.1,

$$E_c = 33w^{1.5}\sqrt{\mu_{cu}f_c'} \qquad \text{Eq. (6.7)}$$

$$= 33(145)^{1.5}\sqrt{0.65 \times 4000}$$

$$= 2.94 \times 10^6 \text{ psi} = 2940 \text{ ksi } (20{,}300 \text{ MPa})$$

$$EI = \mu_e E_s I_s + \tfrac{1}{2} E_c I_c \qquad \text{Eq. (6.6)}$$

$$= 0.85 \times 29{,}000(110) + \tfrac{1}{2} \times 2940(1618)$$

$$= 5.07 \times 10^6 \text{ k-in.}^2 \ (14.55 \times 10^6 \text{ N-m}^2)$$

$$P_o = \mu_{ss} f_y A_s + 0.85\mu_{cu} f_c' A_c \qquad \text{Eq. (6.1)}$$

$$= 0.95 \times 36(9.12) + 0.85(0.65 \times 4.0)135$$

$$= 505\text{k } (2250 \text{ kN})$$

$$M_o = \mu_{ss} f_y Z_s = 0.95 \times 36(30.4)$$

$$= 1040 \text{ k-in. } (117 \text{ kN-m})$$

2. Compute ultimate forces using the load factor of $\frac{5}{4}$ since wind is to be a part of the loading. Let x equal the number of kips of allowable service-load wind force.

$$P_u = \text{L.F.} \times P = \tfrac{5}{4} \times 162 = 202\text{k } (960 \text{ kN})$$

$$M_u = \text{L.F.} \times M$$

$$= \tfrac{5}{4}(360 + 23x)$$

$$= 450 + 28.8x \text{ in.-k } (50.8 + 3.25x \text{ kN-m})$$

3. Compute kl_c, P_{cr}, P_c, and δ

$$kl_c = \pi\sqrt{\frac{EI}{0.5P_o}} \qquad \text{Eq. (6.8)}$$

$$= \pi\sqrt{\frac{5.07 \times 10^6}{0.5 \times 505}} = 445 \text{ in. } (11.30 \text{ m})$$

$$P_{cr} = P_o\left[1 - \frac{1}{2}\left(\frac{kl}{kl_c}\right)^2\right] \qquad \text{Eq. (6.9)}$$

$$= 505\left[1 - \frac{1}{2}\left(\frac{212}{445}\right)^2\right] = 448\text{k } (1990 \text{ kN})$$

$$P_c = \frac{\pi^2 EI}{(kl)^2} \qquad \text{Eq. (6.2)}$$

$$= \frac{\pi^2(5.07 \times 10^6)}{(212)^2} = 1110\text{k } (4950 \text{ kN})$$

The value of $\alpha = 1$ for frames with columns that resist sway.

$$\delta = \frac{0.6 + 0.4\alpha}{1 - \dfrac{P_u}{P_c}} = \frac{1}{1 - \dfrac{202}{1110}} = 1.22$$

Eq. (6.11)

4. Use Eq. 6.12 and solve for the value of x, the allowable wind force.

$$\left(\frac{P_u}{P_{cr}}\right)^2 + \frac{\delta M_u}{M_o} = 1 \qquad \text{Eq. (6.12)}$$

$$\left(\frac{202}{448}\right)^2 + \frac{1.22(450 + 28.8x)}{1040} = 1$$

$$28.8x = \frac{1040}{1.22}\left[1 - \left(\frac{202}{448}\right)^2\right] - 450$$

$$x = 7.96\text{k } (35.4 \text{ kN}) \qquad \text{ANSWER}$$

EXAMPLE 6.6

Select a cross section for a concrete-encased column to support service-load thrusts of 128k (569 kN) dead load and 82k (365 kN) of live load together with 32 ftk (142 kN-m) of dead-load moment and 36 ftk (160 kN-m) of live-load moment if slenderness can be neglected and if requirements of the American Concrete Institute are to be observed. Use $f_c' = 3$ ksi (21 MPa) and $f_y = 36$ ksi (248 MPa). All moments act about the major axis of the member.

Solution

1. Compute P_u and M_u

Ultimate force = 1.4 × (dead load) + 1.7 × (live load) (ACI 9.3.1)

$$\text{Ultimate thrust} = 1.4 \times 128 + 1.7 \times 82$$

$$= 319\text{k } (1420 \text{ kN})$$

Ultimate moment = 1.4 X 32 + 1.7 X 36

= 106 ft-k (144 kN-m)

2. The graph of Fig. 6.6a is constructed for specified material properties and member capacities. Instead of reducing ordinates by the ACI capacity reduction factors ϕ, increase the design loads by multiplying by the inverse of ϕ factors ($\phi = 0.70$ for encased shapes):

Design ultimate thrust P_u

$$= \frac{319}{0.70} = 456\text{k (2028 kN)}$$

Design ultimate moment M_u

$$= \frac{106}{0.70} = 151 \text{ ft-k (205 kN-m)}$$

3. Use Fig. 6.6a with $P_u = 456$k and $M_u =$ 151 ft-k to select a core size and column shape. The W12X36 in 16 X 10 in. of concrete is almost enough, but specify the W12X40 in 16 X 12 in. (406 X 305 mm) of concrete. ANSWER

Comments

The dashed line for a W12X40 shape in 16 X 12 in. of concrete falls inside of the coordinate point $P_u = 456$k and $M_u = 151$ ft-k, indicating that the approximate value of M_o for Eq. 6.12 would be too low for the selected cross section to satisfy that equation. A W12X50 shape in a 16 X 12 in. concrete section has a value of $P_o =$ 1008k (4480 kN) and an approximate $M_o =$ 218 ft-k (295 kN). These values in Eq. 6.12 give:

$$\left(\frac{456}{1008}\right)^2 + \frac{151}{218} = 0.90$$

which satisfies that relationship which should be used if no design aid charts were available to refine the selection.

6.9 LIST OF SYMBOLS

a Depth of cross-section area on which rectangular stress block of concrete is assumed to act.

A_c Area of concrete in a cross section.

A_g Area of an entire cross section bounded by gross exterior boundaries.

A_s Area of steel in a cross section.

A_w Area of the web of a steel beam.

b Width of a cross section.

b_s Width of a steel flange.

D Outside diameter of a cross section.

e Eccentricity of thrust on a cross section.

E Modulus of elasticity.

E_c Modulus of elasticity of concrete.

E_{ct} Tangent modulus of elasticity of concrete.

E_s Modulus of elasticity of steel.

E_{st} Tangent modulus of elasticity of steel.

EI Flexural stiffness quantity.

EI_{tan} Tangent to load vs. flexural curvature diagram.

f_c Stress on concrete.

f'_c Index compressive strength of standard concrete cylinders.

f'_{cu} Index compressive strength of standard concrete cubes.

f_s Stress on steel.

f_y Yield strength of steel.

h Total thickness or depth of a cross section.

I Moment of inertia of a cross section.

I_g Moment of inertia of gross cross section.

I_s Moment of inertia of steel in a cross section.

I_x Moment of inertia about the x axis of a cross section.

I_y Moment of inertia about the y axis of a cross section.

k Coefficient of length (story height) of a column.

kl Effective length of a column, taken as the distance between points of inflection in the deflected shape of the column at buckling.

kl_c Characteristic length of a column above which elastic buckling should be expected and below which inelastic buckling occurs.

l Length of a member.

L.F. Load factor.

M Moment.

M_o Moment capacity of a cross section in the absence of thrust.

M_u Ultimate moment.

P Thrust.

P_c Elastic buckling thrust capacity of a cross section.

P_{cr} Thrust at which a column buckles.

P_l Live-load thrust.

P_o Thrust capacity of a cross section in the absence of moment (squash load).

P_u Ultimate thrust.

t Thickness of segment.

t_c Thickness of concrete cover.

w_c Density of concrete.

Z_s Plastic section modulus for steel shape or tube.

α Ratio between moments at opposite ends of a beam column.

β Coefficient relating the compressive strength of concrete in members to the compressive strength of control cylinders or cubes.

β_d A coefficient for concrete creep effects used by ACI Building Code; taken as the ratio between design dead-load moment and total design load moment.

δ Moment magnification factor.

e Strain.

e_c Strain in concrete.

e_s Strain in steel.

μ Material reliability factor.

μ_{cc} Material reliability factor for confined concrete.

μ_{ccf} Material reliability factor for factory produced confined concrete.

μ_{cu} Material reliability factor for unconfined concrete.

μ_{cuf} Material reliability factor for factory produced unconfined concrete.

μ_e Material reliability factor for stiffness of steel.

μ_{ss} Material reliability factor for steel shapes.

μ_{st} Material reliability factor for steel pipe or tubing.

ϕ Member capacity reduction factor.

6.10 REFERENCES

ACI 318-71 (1971), *Building Code Requirements for Reinforced Concrete*, American Concrete Institute, Detroit, Michigan.

AISC (1970), *Manual of Steel Construction*, 7th Ed., American Institute of Steel Construction, New York.

Ferguson, Phil M. (1972), *Reinforced Concrete Fundamentals*, 3rd Edition, John Wiley & Sons, New York.

Furlong, Richard W. (1976), AISC Design Logic Makes Sense for Composite Columns, Too, *J. American Institute of Steel Construction*, 1st Quarter.

Furlong, Richard W. (1974), Concrete Encased Steel Columns–Design Tables, *ASCE Journal of the Structural Division*, Vol. 100, No. ST9, September.

Knowles, R. B., and Park, R. (1969), Strength of Concrete Filled Steel Tubular Columns, *ASCE Journal of the Structural Division*, Vol. 95, No. ST12, December.

Malhotra, H. L., and Stevens, R. F. (1964), Fire Resistance of Encased Steel Stanchions, *Proc. Inst. of Civil Engr.*, Vol. 27, January.

SSRC Guide (1977), *Guide for the Design of Metal Compression Members*, Structural Stability Research Council (formerly Column Research Council), 3rd Edition.

7

Mixed steel-concrete high-rise systems

SRINIVASA H. IYENGAR, P.E.
Partner
Skidmore, Owings and Merrill
Chicago, Illinois

7.1 INTRODUCTION

Steel-concrete composite systems for tall buildings are composed of concrete components that interact with structural steel components within the same system. These two components, by their integral behavior, meet all the requirements of strength, stiffness, and stability of the overall high-rise system. These systems are also termed *mixed steel-concrete systems* or *steel-concrete combination systems.*

The steel components may take the form of floor framings, Vierendeel or diagonally braced truss subsystems, composite steel claddings, or structural steel embedments of various types. The concrete components may consist of exterior concrete framed-tubes, shear walls of various shapes, precast concrete wall or slab panels, precast concrete cladding panels, and others.

The type and degree of interaction are dependent on a particular combination of steel and concrete components. In essence, the composite system is formulated to utilize the most desirable characteristics and assets of each material to their best advantage so that the combination may result in a higher order of overall system efficiency.

Composite members as individual elements of a system have been in use for a considerable number of years. They consist of composite beams or trusses, encased or filled composite columns, and steel-deck-reinforced composite slabs. These members are generally used in steel structures, and their development as composite members is based on utilizing the concrete that would normally be required for floor slabs or the concrete required for fire-protective encasements for steel beams or columns. As a result, the developments have been restricted to individual members only.

In most instances, the contribution of a composite member, which is developed to support only the gravity loads, has been ignored in the overall system resistance for lateral loads of wind or earthquake. The development of an overall systems approach, where reinforced concrete and structural steel components can be used as effective contributing parts of the total system, is relatively recent, and considerable potential exists for evolving a variety of new structural system compositions. Many examples of equivalent tubular systems and core-braced systems for high-rise buildings currently exist in the mixed form, and they will be described herein. The potential of these systems to be used in different forms will also be discussed. Also presented will be background information on steel and concrete systems, out of which grew many current forms of mixed systems, together with discussions on advantages and disadvantages of various steel and concrete components. A method of evaluation of various system combinations will also be discussed.

The choice of a steel, concrete, or composite system for any particular project depends not only on system efficiency, material availability and cost, construction technology, and labor, but also on planning, architectural, and aesthetic criteria. It is thus impossible to reach definitive conclusions just on the basis of a structural system evaluation. What is therefore presented here is a structural discussion of relative merits, efficiencies, and advantages of different systems, which could then be used as part of a total building evaluation including other factors. Obviously, experience and engineering judgments are inseparable parts in this evaluation process.

7.2 STEEL AND CONCRETE HIGH-RISE SYSTEMS

The beginnings of composite steel-concrete systems for tall buildings can be directly linked to developments of steel or concrete systems. Therefore, a brief review of these developments is presented here.

The most common form of steel system is a steel skeleton or steel frame building where vertical Vierendeel frames are arranged in a rectilinear fashion in two directions. Partial or full moment connections at beam-column joints of the frames provide all the lateral load resistance. In some early versions, knee braces were provided at beam-column joints to increase joint rigidity. The frame system tends to be inefficient for longer spans and for taller buildings beyond about 20 to 25 stories. The efficiency of these frame systems is increased by the introduction of vertical diagonalized trusses interconnecting columns generally provided in the centralized core of buildings (Iyengar 1972). The truss system may provide all the required lateral load resistance, or it could be combined with Vierendeel frames resulting in a shear truss-frame interacting system. A further refinement in this interacting system, which is to link the vertical trusses with a one-story perimeter belt truss provided at various building levels, with one-story trusses outriggering from the core trusses at the same levels (see Khan 1970, Taranath 1974). The induced participation of exterior columns results in a partial equivalent-tube concept. Various types of exterior diagonally braced systems are possible, which result in equivalent-tube systems. An outstanding example of this is the John Hancock Center in Chicago, which represents a high-efficiency tube system; (see Khan et al. 1966). Many systems consisting of framed-tubes are also possible in lieu of diagonalized tubes, and they involve closely spaced, wide columns, which are rigidly connected to deep beams generally provided as fascia frames around the building. The bundled framed-tube system used for Sears Tower (height: 1454 ft or 443 m) in Chicago represents a new innovative application of framed-tubes for ultratall building systems. (See Iyengar and Khan 1973).

Similar systems as developed can also be traced for concrete buildings (Iyengar and Khan 1973). The concrete frame building uses either a beam-and-slab arrangement or waffle-slab arrangement for the floor system. Such systems tend to be massive, which limits their range of applicability to about 10 to 20 stories. The shear-wall system has been the most commonly used system for concrete high-rise buildings in the last 20 years. Initially, the shear walls were linear elements interconnecting columns that were usually located inside the buildings. Later versions, usually for office buildings, have used shear walls organized around the building core, resulting in a cantilever-tube core. In earlier versions, the shear wall provided all the lateral load resistance, which was later improvised into shear wall–frame interacting systems (see Khan and Sbarounis 1964). An efficient form of the shear wall–frame interacting system involves fascia building frames interacting with core walls. In this system, the fascia plane frame parallel to the direction of lateral load was considered interacting with the core walls. The shear wall–frame interaction brought about substantial structural benefits in terms of added stiffness and strength. Many outstanding examples of this system currently exist. Subsequent developments in concrete have centered on utilizing the exterior three-dimensional form of the building for structural resistance. The column spacing and member proportions of the fascia frame were viewed as if they were solid walls with minimum openings for windows. This transformed the exterior form into an equivalent tube with punched openings for windows whose predominant behavior was that of a vertical cantilever (Amin and Khan 1973). In some cases where a larger lateral load stiffness was required or where walls were the efficient form of interior support, an interior shear-wall tube was added, which resulted in the so-called tube-in-tube system.

Later modifications in concrete high-rise systems have centered on restructuring the geometry of the tube into a "bundled tube" form to improve structural efficiency and to create an arrangement of modular spaces similar to the bundled-tube system of Sears Tower.

7.3 ADVANTAGES OF STEEL OR CONCRETE CONSTRUCTION

In formulating a certain composite system, it is necessary to recognize the advantages and merits offered by each material and each subsystem as they relate to architecture, system efficiency, simplicity of construction, and speed of construction. Concrete possesses an inherent capability of being molded to any shape, thereby offering a certain freedom for architectural expressions; it is durable in different weather conditions, thereby offering opportunities for exposed concrete; it has inherent fire protection ability; it can yield monolithic and continuous joints readily; and it allows walls of different thicknesses to be cast in any shape with or without openings for windows and other purposes. Further, concrete could be precast or site cast. These assests make them suitable for the construction of framed-tubes, shear walls, wall panels, and cladding panels. However, concrete is at a distinct disadvantage when it comes to floor framings, which tend to be massive and therefore not easily spannable over long spans. Concrete construction has a higher labor factor because of formwork and shoring and is generally slower to construct. Interior columns in tall buildings tend to become massive and space consuming. Core walls, although relatively efficient as structural elements, impose rigidities in core planning. For extremely tall buildings, these walls may become extremely massive and space consuming if relied on exclusively for lateral load resistance.

Steel construction, on the other hand, has the advantage of spanning longer spans with relatively lighter members. Steel columns have a distinct advantage in terms of size as compared to concrete columns. Simple floor framings and steel columns offer maximum flexibility for space planning, particularly in the core. Steel construction represents a more industrialized product in that elements are prefabricated in plants under mechanized and controlled conditions. All this contributes to a higher speed of construction, which is an influential factor in most tall-building construction. Among the disadvantages of steel buildings is the need for applied fire protection, the higher cost of cladding and window wall, and the inherent cost in providing additional steel and welded moment connections or diagonalized connections to resist wind forces.

7.4 STEEL AND CONCRETE COMPONENTS

As a means of examining different possible combinations of subsystems and components and to recognize and determine a valid combination, a Systems Component Table is presented in Table 7.1 which is helpful in reviewing different system components and their specific roles. The components of a high-rise building system can be classified into: (1) lateral load resisting system or subsystem, (2) floor framing, (3) floor slabs, (4) columns, and (5) cladding and window wall. In the table, the different possibilities for each subsystem or component in steel, concrete, and composite members are listed separately.

Table 7.1. Mixed system combination table.

	Lateral Load-Resisting Subsystem	Floor Framing	Slab	Columns	Cladding
Concrete	Cast-in-place frame	Flat slab	Solid one-way slab	Concrete Tied or spiral column	Solid Masonry infill
	Shear walls	Joist slab	Solid two-way slab	Precast tied or spiral column	Precast concrete cladding
	Shear wall–frame interaction	Waffle slab	Precast plank and fill		Precast composite formwork cladding
	Exterior framed tube	Precast beams	Precast concrete floor panels		Stone cladding
	Tube-in-tube				
	Modular tube				
	Wall panels				
Composite	Encased steel frames	Concrete-encased composite steel beams	Metal deck and fill-composite	Encased-steel column	Composite-steel cladding
	Steel reinforced concrete construction	Unencased composite steel beams	Metal deck and fill composite-reinforcement for negative bending	Filled-tube column	
	Wall panels with steel embedments	Encased beams with miscellaneous steel embedments		Encased columns with miscellaneous steel embedments	
	Steel bracings with precast concrete columns				
Steel	Welded moment-resistant frame	Rolled beam–noncomposite	Metal deck and fill-noncomposite	Rolled or built-up steel column–exterior	Noncomposite steel cladding
	Shear truss	Truss or joist–noncomposite		Rolled or built-up steel column–interior	Aluminum cladding
	Shear truss–frame interaction	Welded two-way grid		All columns–steel built-up or rolled	
	Shear truss–belt truss system				
	Framed tube				
	Diagonalized tube				
	Modular tube				

7.4.1 Lateral Load Resisting System or Subsystems

Lateral load resisting subsystems are the primary subsystems on which the efficiency of the total system depends. All the possibilities listed may or may not be suitable types to be used in a mixed steel-concrete form. The following review of components follows the sequential order presented in Table 7.1.

Cast-in-place concrete frames, which utilize integral floor systems, such as a waffle slab or flat slab or flat plate, are not suitable in a mixed form. Concrete frames, which use isolated beams at frame lines that can be separated from floor framing, can be useful in a mixed form, especially if such frames are used only on the exterior fascia frames. One can then combine the concrete frame with its cast-in-place columns and beams with a structural steel framing that spans between the concrete beams.

Shear walls represent a very effective subsystem for use in a mixed form. Their ability to be cast in any shape makes them extremely versatile. Their massiveness, together with their effective cantilever behavior, makes them a highly efficient lateral load resisting element. Open or closed, linear or curved shapes can be utilized inside or on the exterior of buildings. Penetrations and openings for doors and heating, ventilating, and air-conditioning distribution can be readily incorporated into the wall. Shear walls can be cast as separate reinforced concrete elements or as elements interconnecting and encasing steel columns. They form efficient enclosures for centralized building cores, stairways, and other shafts of reasonably large size. In most situations, shear walls can be built first, and the rest of the structure can be built at a later time.

Shear wall–frame interacting subsystems can be effective if the floor system can be separately built. A particular form involving an exterior concrete frame and concrete core walls appears feasible. Structural steel framing can then be supported on the exterior frame and core walls. The exterior concrete frame is advantageous if exposed concrete or stone cladding is the preferred material for architectural expression.

Exterior framed-tubes, which are conceived as equivalent punched-tube forms, are highly effective for use in a mixed form. As with shear walls, they are versatile in terms of possible shape modulations. The framed-tube is best achieved in reinforced concrete whereby the basic rigidity at the joints is obtained from the natural monolithic character of concrete. Architecturally, the concrete of the vertical and horizontal surfaces of the tube can be molded to any reasonable shape to bring the desired articulations and can be exposed if desired. Figure 7.1 shows the elevation of a 37-story composite tubular system with exposed concrete, which typifies the basic character of the exterior tube. The punched-tube wall generally results in a reduction of glass area on the facade and therefore offers savings in the perimeter air systems because of reduced heating and cooling loads. Another advantage of the closely spaced column form of the framed-tube relates to the simplification of window wall details. Since each window is framed all around by the elements of the tube, the window glazing is directly attached to the tube by simple gasket details.

Tube-in-tube and modular tubes are efficient types for ultra-high-rise buildings. The tubes can be built in concrete with structural-steel-floor framing filled into the system.

Wall panels of reinforced concrete, especially in the precast form, are efficient elements. They can be used as infilled panels in a steel frame or as a panelized wall system in shear walls or as precast cladding panels on a steel frame. In each case, they act as bracing panels by virtue of their in-plane shear rigidity for lateral load resistance.

Encased steel frames where the encasement may be required for fire protection, although effective as a structural form, have limited potential because of the high cost of formwork.

Steel reinforced concrete construction, which uses encased built-up steel column and beam elements, is extensively used in Japan. The system tends to be ineffective where the high costs of formwork, labor, and time are controlling parameters.

Wall panels with embedded steel bracings are composite panels that facilitate a more direct

Fig. 7.1. Gateway III Building, Chicago, Illinois.

transmission of panel shear forces to the frame, thus enhancing their efficiency as infilled shear panels.

Steel bracings used with precast concrete columns are composite vertical trusses in which the columns are used as chords. They appear to have been used extensively in the USSR. This type of truss has limited potential for use in high-rise buildings because of the need for complicated axial tension load connections in columns.

Welded moment-resistant steel frames can be effective in the mixed form when augmented by panel or masonry infillings or composite claddings.

Shear trusses, shear truss-frame interacting system, shear truss-belt truss system, framed-tube, diagonalized tube, and modular tube systems in steel are uniquely suited for steel construction, and their use in the mixed form has little potential. Some possibility, however, exists for minimizing the amount of structural steel used in the overall system by replacing floor framing with precast concrete systems. Even for this, the potential is extremely small.

7.4.2 Floor Framings and Slabs

Flat slabs, joist slabs, and waffle slabs are cast-in-place concrete floor systems, which are uniquely suited for concrete buildings; their use in the mixed form offers no potential.

Precast concrete beams are supported on steel framing; although possible in some situations, they offer limited potential.

Concrete-encased composite beams represent

an obsolete form of floor framing because of the high cost of formwork. They are used only where integral heavy-duty fire protection of this type is required, and its application is therefore limited to special cases.

Steel floor framing consisting of composite or noncomposite beams or trusses, whose main function is to support the floor and its gravity load, has a great advantage over other types of floor framings. The benefits are spannability and lightness, reduction of labor due to prefabrication, faster construction, and the ability to be connected to other concrete or steel elements with simple details. The steel framing may consist of solid-web standard rolled-beam sections or open-web steel trusses applicable for spans up to 80 ft (24 m). This aspect of spannability makes them suitable for use with concrete framed-tubes, and tube-in-tube and modular tube systems where a longer span between tubular walls or from the interior tube to the exterior tube is generally desired. The metal deck, which is often cellularized for floor electrical and power distribution, is commonly used as the slab element with a concrete topping. The metal deck, apart from supporting the floor loads, also serves as a construction platform on steel framing. Composite design for the steel beams as well as for the deck, utilizing the floor slab concrete, is commonly used. In these instances, the metal deck acts as a positive reinforcment in the composite slab assembly. The composite beam behavior is established by welding shear studs through the metal deck, and the composite action for the deck is established by side embossments on the metal deck or other suitable mechanical devices.

The composite slab-and-beam arrangement is beneficial in establishing an integral floor diaphragm, which is essential in tying the steel and concrete component parts of the mixed system together. This diaphragm is essential in establishing overall system stability and for transferring wind or earthquake lateral shear forces to respective resistive elements.

Cast-in-place solid concrete slabs, which are supported on steel framing, can be used as part of a mixed system, but are less effective than the metal decked slabs described previously. The solid slabs can be one-way or two-way slab supported on steel beams.

Precast planks or floor panels supported on steel framing can be used as parts of slabs. Generally, a concrete floor fill is required, and composite action with the precast planks or panels can be established. Composite design for the steel beam with the planks or panel and fill has also been used in some instances. This system can be used where the availability of the metal deck is limited. In general, the planks perform the same function as the metal deck.

Metal decks, composite or noncomposite, offer considerable potential as discussed in this section under Steel Floor Framing.

7.4.3 Columns

Concrete columns of various types are used in the mixed system as vertical elements of concrete framed-tube or other concrete subsystems. Precast concrete columns can also be used, either as part of a precast framed-tube or as a nonlateral load-carrying interior column in lieu of steel columns.

Encased-steel columns can be part of an encased steel frame or can be used as composite columns in an otherwise concrete building to reduce the size of the concrete columns.

Filled steel tube columns have no significant application in the mixed system.

Steel columns of various types are used in mixed systems. They can be part of a moment frame braced by walls, panels, or claddings. Steel columns are also used as part of a simple frame subsystem that is braced by shear walls or concrete framed-tubes.

7.4.4 Claddings

Claddings are building enclosures that are attached to the structure either integrally or nonintegrally with flexible connections. If the cladding material is used integrally, considerable potential exists for using the enclosure as a stressed skin for wind-resistance purposes. Even though many earlier buildings consisted of heavy masonry-infill claddings, their contribution was generally ignored; however, the stiffening effect of such cladding has been well recognized. The use of contemporary cladding materials, such as the precast concrete or steel-plate cladding, enhance opportunities for composite claddings. The need for expansion joints in some cladding systems often renders them ineffective for wind-bracing purposes.

Masonry-infilled cladding in a steel frame can be used for wind-load resistance. However, the potential is limited to about five stories unless the infillings are extensively reinforced and special details are used for attachments.

Precast concrete cladding panels offer the most potential for their use as a wind-resisting element in a steel-framed structure. Various types and forms are possible, and some even incorporate the function of formwork if cast-in-place concrete members or encasements are required for fascia structural members.

Stone cladding cannot generally be used as composite cladding because of the need for expansion joints. However, some types of composite stone claddings where a precast concrete backing is required can be engineered to be used in the same manner as precast concrete claddings.

Composite-steel claddings have been used in stiffening the fascia frames of steel-frame systems, and their effectiveness is based on the need for architecturally exposed steel cladding.

Aluminum cladding systems do not offer opportunities for cladding participation because of the need for extensive expansion joints.

7.5 MIXED SYSTEM COMBINATIONS

As a first step, it is essential to examine each system combination to determine its potential for complying with the basic structural requirements. The integral behavior of all components of the mixed system taken together must satisfy all the requirements of the system. For high-rise buildings, this could be generally stated as that required to satisfy strength, stiffness, stability, and ductility of the system subjected to gravity and lateral loads.

Lateral load resistance is perhaps the most controlling subsystem into and around which are fitted other subsystems required for gravity loads. The lateral load subsystem must therefore: (1) support its share of gravity forces, (2) resist lateral load shears and overturning moments, (3) provide the stiffness required for lateral sway or drift limitations, and (4) provide the stiffness required for overall system stability under gravity loads.

The gravity load subsystems include floor slabs, floor framing, and vertical supports, which do not participate in lateral load resistance and are generally dependent on the lateral load subsystem to provide the overall system stability. The floor diaphragm, which consists of the floor slab that is stiffened by the floor framing, provides the essential link between these two subsystems and therefore plays a key role. The continuity of the diaphragm and its connections to the lateral load subsystem to transfer the in-plane wind shear forces are of extreme importance.

Investigation of the mixed system to satisfy ductility requirements for seismic resistance is essential, in addition to lateral load resistance. Connection details, in particular those between steel and concrete elements, should undergo critical review for ductility and ability to absorb large deformations produced under seismic conditions.

Construction efficiency of a certain mixed form relates to (1) level of industrialization and use of prefabricated elements which will minimize shop and field labor, (2) simplicity of connection details between elements, (3) construction sequence and its effect on speed of construction, and (4) feasibility of using larger capacity lifting equipments. The overall construction economy is controlled by such factors as (1) material and labor availability, (2) availability and capability of facilities to produce prefabricated elements in steel and/or concrete, (3) experience and familiarity with a certain construction methodology, and (4) labor-group requirements associated with different trades.

It is obvious that one has to use considerable judgment in the evaluation of a certain combination. It is clear that what may be efficient and economical under one set of circumstances may be invalid or ineffective under others.

Some common mixed systems generally suitable for construction in the United States are discussed in the remainder of this chapter.

7.5.1 Composite Tubular System

The composite tubular system combines the essential properties of an exterior equivalent framed-tube system in reinforced concrete with simply connected structural steel framing on the interior. This concept is shown in a schematic form in Fig. 7.2, which indicates the concrete framed-tube envelope on the periphery and the steel floor framing and steel columns on the interior. A typical cross section of the spandrel beam and exterior column is also shown.

The steel floor framing consists of composite-steel beams or trusses, all simply connected for shear and designed only for gravity loads. These steel members are connected to a small-size steel column, which is incorporated into the vertical members of the framed-tube and serves the function of an erection column required for steel erection. A typical composite slab is used as the slab element, which is connected to the concrete spandrel beams by means of steel dowels as shown in Fig. 7.2.

In principle, the framed-tube subsystem resists all lateral loads of wind and/or earthquake,

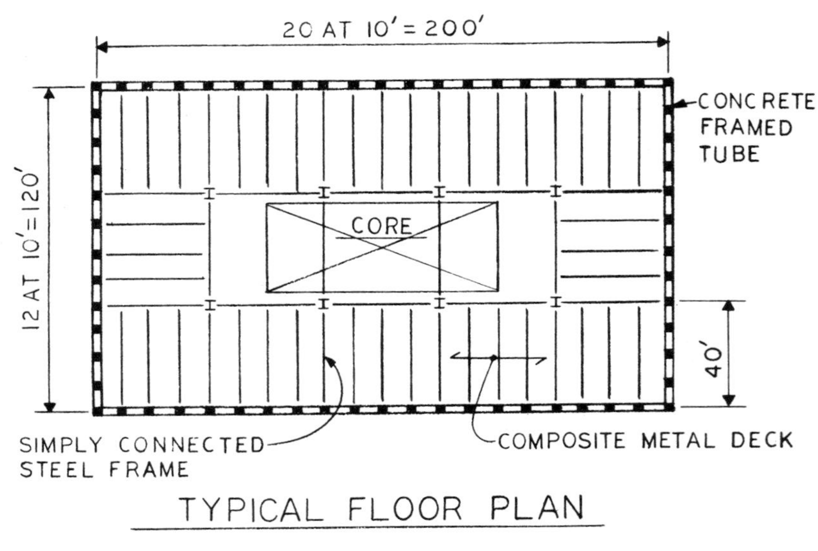

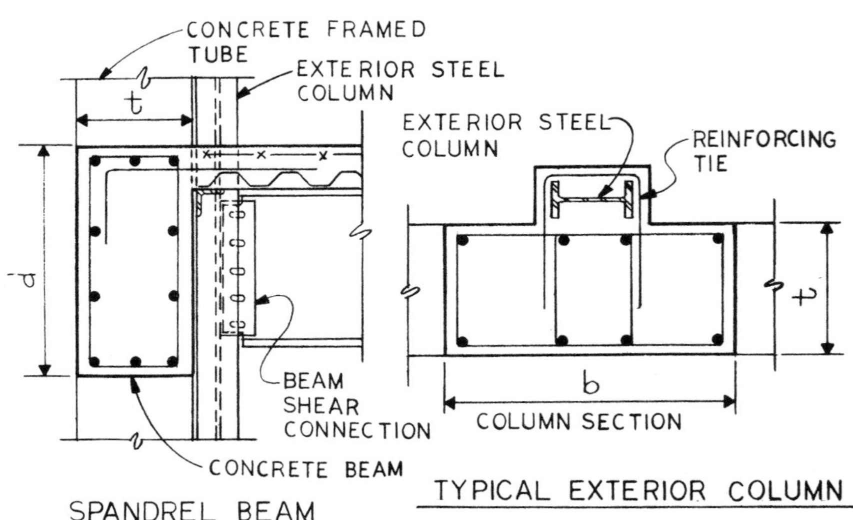

Fig. 7.2. Composite framed tube.

whereas the interior steel subsystem is required to carry only its share of gravity loads. The use of steel framing on the interior offers all the advantages of a steel system including its faster speed of construction.

7.5.1.1 Framed-Tube Member Proportions

The horizontal and vertical elements of the framed-tube can be considered as beams and columns. The spacing of columns ranges from 4 to 10 ft (1.2 to 3 m) on center. Generally, 30% to 40% of this spacing is covered by the depth of these columns. Similarly, the beam element may range from 2 to 6 ft (60 cm to 1.8 m) in depth. Determination of member proportions is related to (1) minimum member depths required to maintain a high structural efficiency as an equivalent cantilever form, and (2) member depths required to achieve a certain proportion of openings on the facades. The member thicknesses are determined to satisfy the structural requirements at various heights in the building.

7.5.1.2 Framed-Tube Analysis

For the structural analysis, all the lateral load resistance is assigned to the framed-tube neglecting any nominal resistance offered by the interior steel framing. The behavior of a framed-tube system under lateral load involves a combination of a cantilever mode and a shear-frame mode. The cantilever overturning moments under lateral load are resisted by the entire three-dimensional tube form causing compressive and tensile axial forces in the columns and also causing associated member moments as a result of relative column shortenings. The shear from the lateral load is resisted by bending of beams and columns from the shear-frame action of the two side frames parallel to the direction of the lateral load. The preliminary analysis can be approached on the basis of separate computation of cantilever forces and the shear-frame forces. The shear-frame component can be readily determined by an approximate procedure such as the portal-frame method. However, the analysis of the cantilever component requires assumptions regarding cantilever effectiveness to account for the shear-lag effect of the discretized tube as opposed to a solid-walled tube. For a preliminary estimate of the cantilever effectiveness of the tube, the configuration of the tube can be reduced to two equivalent end channels involving the web frame and only part of the flange frame. This accounts for the fact that columns in the center portions of the sides of the tube are least effective in resisting overturning forces. A more rational procedure for the determination of this cantilever effectiveness is presented by Khan and Amin (1973), which is quite useful as a practical guide.

A more complete structural analysis of the tube can be performed by utilizing three-dimensional frame-analysis computer programs involving all members and joints and including column axial deformations. A significant simulation is with respect to joint fixity, a factor related to additional restraint offered by the large finite width of members at the joints as compared to centerline analysis methods. Either the effect of "joint fixity" can be evaluated separately and incorporated into the computerized frame analysis by adjustment of member properties, or a computerized analysis based on a panelized frame utilizing finite-element methods can be performed.

Similarly, the analysis for gravity loads can be carried out either by estimation of column loads assuming a certain load distribution between columns or by use of computerized frame-analysis programs.

7.5.1.3 Framed-Tube Design

The design and detailing of the members of the framed-tube are performed using the latest ACI design rules. The strength-design method is used with appropriate load factors for various loadings. The exterior concrete column is treated as a composite column including the steel columns. In many cases, the contribution of the steel column has been negligible and therefore neglected in design. The design of this steel column is generally controlled by steel erection conditions.

7.5.1.4 Construction Coordination

The interior steel-frame construction proceeds in advance of reinforced concrete tube for a

Fig. 7.3. Construction sequence, One Shell Square Building, New Orleans, Louisiana.

certain number of stories. This frame is then enveloped by the reinforced concrete tube as shown in Fig. 7.3. This is made possible by incorporating a small steel column as part of the steel frame, which is later encased by the tube as shown in the column detail of Fig. 7.2. A separation of eight to 10 stories between steel and framed-tube working levels is generally required. Metal deck installation follows behind steel at close interval. Concrete on the deck is placed next, completing the floor-slab diaphragm. The completed slab is also used as a construction platform for lifting prefabricated reinforcement cages and gang forms for the construction of the framed tube. The steel frame above the completed framed-tube at any time during construction is required to have adequate lateral load resistance for that portion.

7.5.1.5 Example

As an example, the structural material quantities of a composite tubular system are compared with a steel-braced system, and a concrete shear-wall system in Table 7.2 for a 600 ft (180 m) tall, 43-story structure, with a plan dimension of 200 by 120 ft (61 by 36.6 m). The steel system involves a perimeter moment connected frame on a 20-ft (6.1-m) span interacting with core trusses. The floor framing and the composite slab for the steel system are the same as those for the composite system. The concrete system is a shear wall–frame interacting system. The shear walls are provided in the core and the frame on the perimeter uses 20-ft (6.1-m) column centers. A one-way and two-way concrete joist system is used for the 40-ft (12.2-m) span floor system. The design information for the composite tubular system is summarized in Table 7.3.

The amount of structural steel used in the composite tubular scheme is reduced to less than half that of the steel scheme. This reduction is to be counted against the addition of concrete, formwork, and reinforcing steel for the concrete tubular frame. Since reinforcing steel operates at higher stress levels than structural steel and that concrete is effective in compression and shear, this replacement results in the addition of a smaller amount of reinforcing steel. Costs of materials and labor vary for different regions and different times. However, potential cost savings for the composite tubular system as compared to the steel system is from 30% to 50% (Belford 1972).

Many completed examples of this system currently exist. The first application was in the Gateway III Building (Fig. 7.1) in Chicago, a 35-story office tower where the members of the framed-tube are architecturally exposed. The exterior columns are spaced at 9 ft (2.7 m) centers with 27 ft (8.2 m) interior bays. The framed-tube expression can be clearly observed. The 52-story One Shell Square Building (Fig. 7.4) in New Orleans also uses a composite system where the framed-tube is clad in travertine. The columns are spaced at 10 ft (3 m) centers with the exterior corners of the column chamfered at 45° to create an architectural articulation. Interior steel columns are placed around the building core, with floor-framing beams spanning 40 ft (12.2 m) to the exterior. A combination of K-bracing between the interior columns and cables was used during erection to withstand hurricane wind loads. Figure 7.3

Table 7.2. Comparison of steel, concrete, and composite systems.

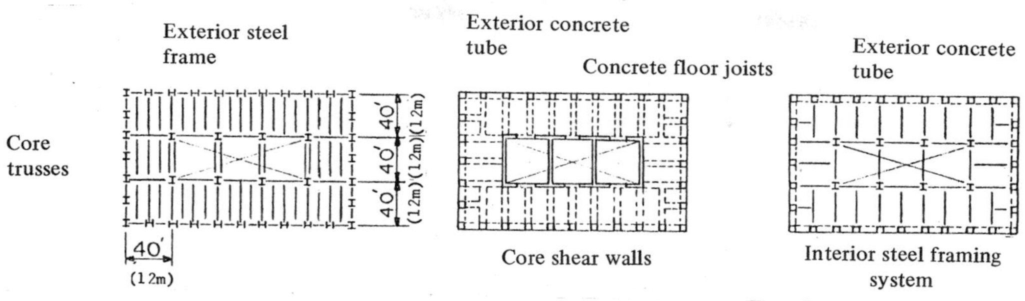

	Braced Steel Frame	Exterior Concrete Frame with Shear Wall	Composite Tubular System
Structural steel	21 lb/ft^2 (1006 Pa)	0.00	11 lb/ft^2 (527 Pa)
Concrete-floor system	0.33 ft^3/ft^2 (0.1 m^3/m^2)	0.85 ft^3/ft^2 (0.26 m^3/m^2)	0.33 ft^3/ft^2 (0.1 m^3/m^2)
Concrete-columns and walls	0.00	0.5 ft^3/ft^2 (0.15 m^3/m^2)	0.30 ft^3/ft^2 (0.09 m^3/m^2)
Total reinforcing steel	0.6 lb/ft^2 (28.7 Pa)	10.5 lb/ft^2 (503 Pa)	4.00 lb/ft^2 (192 Pa)
Formwork	1.0 ft^2/ft^2(1.0 m^2/m^2)	2.00 ft^2/ft^2 (2.0 m^2/m^2)	0.50 ft^2/ft^2 (0.50 m^2/m^2)
Metal deck	1.0 ft^2/ft^2 (1.0 m^2/m^2)	0.00	1.0 ft^2/ft^2 (1.0 m^2/m^2)
Fireproofing	0.9 ft^2/ft^2 (0.9 m^2/m^2)	0.00	0.6 ft^2/ft^2 (0.6 m^2/m^2)
Construction time	1.0 yr	2 yr	1.25 yr
Cladding	Metal/glass	Exposed concrete or stone	Exposed Concrete or stone

Table 7.3. Design example—composite tubular system.

No. stories	43	Office live load	50 lb/ft^2 (2400 Pa)
Building height	600 ft (183 m)	Partition load	20 lb/ft^2 (958 Pa)
Average wind pressure	40 lb/ft^2 (1915 Pa)	Translatory period	5 sec
Lateral load drift	9 in. (0.23 m)	Torsional period	1.5 sec
Floor-plan dimension	See Fig. 7.2		

Slab

2-in. (50 mm) composite metal deck with $3\frac{1}{4}$-in. (83-mm), 4000-psi structural lightweight concrete topping

Steel Floor Framing

W21×44, A36 typical composite-steel beams at 10-ft centers

Exterior Framed-Tube Dimensions

Floors	*b*	*d*	*t*	f_c' (psi)(kPa)
0–10	36 in. (0.9 m)	42 (1.07 m)	24 (0.6 m)	5000 (35)
10–20	36 in.	42	24	5000 (35)
20–30	36 in.	42	21 (0.53 m)	4000 (28)
30–42	36 in.	42	18 (0.45 m)	4000 (28)

Fig. 7.4. One Shell Square Building, New Orleans, Louisiana.

shows the building partially complete, and the advance erection of steel can be clearly observed.

7.5.2 Other Composite Tubular Forms

7.5.2.1 Flexibility of Shape

The exterior framed-tube form offers considerable flexibility for variations in shapes without significant loss of structural efficiency for lateral load resistance. Figures 7.5a–c show some examples of different shapes that have been studied. In fact, the form can be used in almost a free-form concept as in Fig. 7.5c as long as the asymmetry is reasonable. The inherently large torsional stiffness of the exterior framed tube form makes it acceptable for some asymmetry.

7.5.2.2 Partial Framed-Tube Systems

Partial or incomplete framed-tube components can be effectively used on the exterior facades of buildings in various forms as shown in Fig. 7.6. These systems are generally applicable for high-rise buildings in a low-to-medium height range (15 to 30 stories) where a full tubular system may not be needed. In addition to the examples in Fig. 7.6, a variety of other forms are possible. The framed-tube elements behave as individual equivalent cantilevers that are tied together by shear diagrams at each floor. The shape and disposition of these wall-frame elements must develop adequate torsional resistance in the horizontal plane in addition to normal lateral load resistance.

7.5.2.3 Cluster of Tubes

Figure 7.7 shows a concept developed on a cluster of framed tubes. Each tube can be square or rectangular in the dimensional range of 60 to 80 ft (18 to 24 m) for the sides of the tube. Interconnections between the tubes may be needed to gain access into the tubes. The floor framing generally consists of composite-

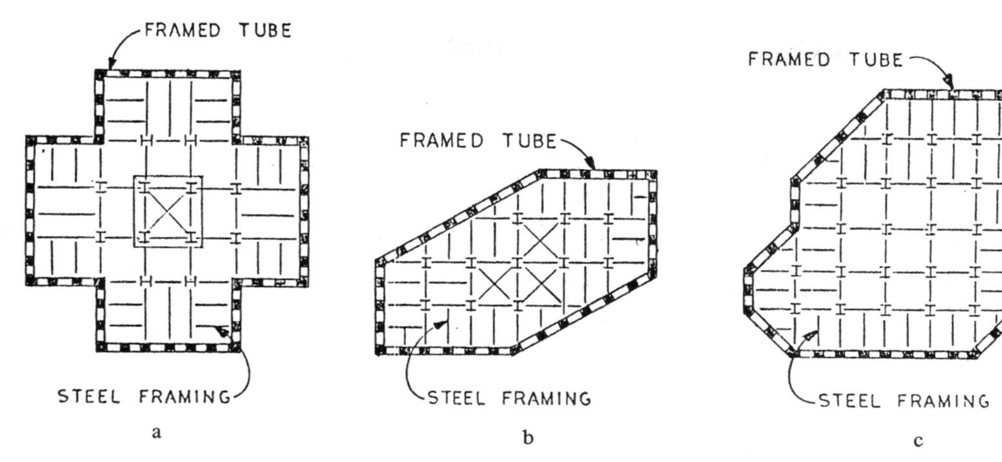

Fig. 7.5. Examples of shape variations.

steel trusses that span between tubular walls without any interior columns. The tubes are individually self-sufficient in terms of structural stiffness and strength, and the interlinking between the tubes (if provided) is required to transmit axial and shear forces in a horizontal plane as a diaphragm. This is to prevent relative lateral sway between the tubes under wind-loading conditions and also to absorb forces of dynamic coupling between the units. It is therefore necessary to investigate any particular arrangement of tubes for possible forces in the link members and design for them accordingly.

The concrete tubes can be constructed either by conventional methods or by a continuous slip-form technique. The construction sequence would be to construct the tubes for a certain number of stories, which are then framed with structural steel at a later time. Anchor plates cast in concrete are used for the shear connection of trusses as shown in Fig. 7.7. The clustering of tubes offers an opportunity for evolving different arrangements to suit different site configurations and variations in massing. Some arrangements of clustering are also shown in Fig. 7.7.

7.5.2.4 Bundled Composite Tubes

The bundled-tube system consists of an integrated cluster of framed tubes all functioning together as one overall tube. The system is highly efficient and is capable of developing large lateral load stiffnesses which make them suitable for ultra-high-rise structures. The bundled-tube system also provides an organization of modular areas, which are capable of being terminated at various levels to create floors of different shapes and sizes. The size of each modular tube is generally in the range of 60 to 80 ft (18 to 24 m). In this system, all the framed-tube lines, both interior and exterior, are of reinforced concrete and steel trusses frame between two tubular lines as shown in Fig. 7.8. This system is generally constructed with the concrete tubes built ahead of the filler steel framing. The use of this concept for the 25-story Ohio National Bank Building in Columbus, Ohio, is shown in Fig. 7.9, which clearly indicates the nature of the concrete tubes. The tubes are 60 ft (18.3 m) long, and the cluster consists of six tubes that are terminated at different levels. The floor framing consists of 60 ft (18.3 m) span composite trusses at 15 ft (4.6 m) centers, which are spanned by a composite slab with a 3 in. (75 mm) metal deck. The trusses are attached to the concrete tubes by a typical anchor-plate detail.

7.5.3 Concrete Core Braced System

Figure 7.10 shows a common form of concrete core braced mixed system. For convenience, a rectilinear floor shape with a rectilinear closed-form core is shown, although other shapes are possible. Whatever the shape, the basic concept

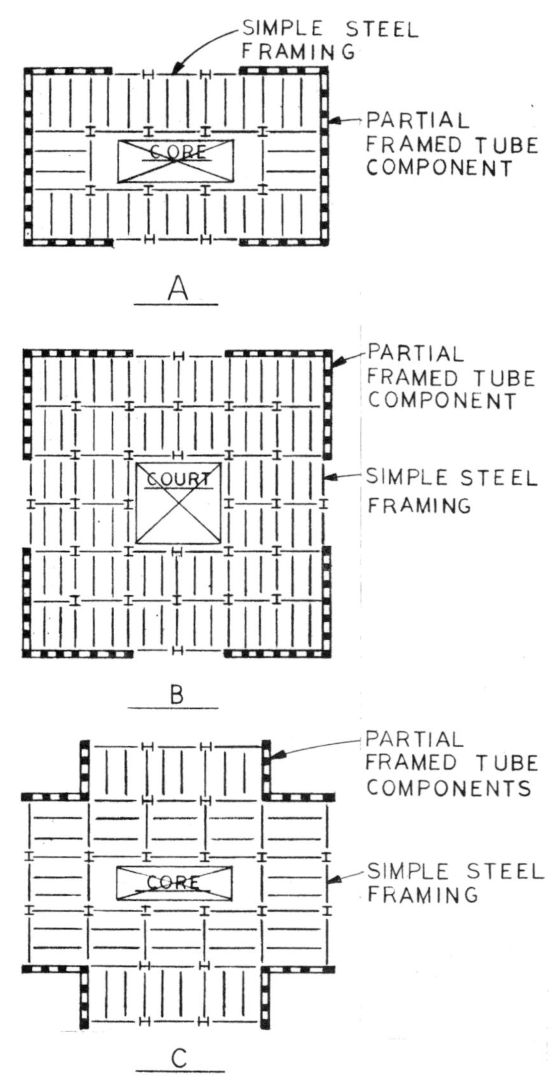

Fig. 7.6. Partial framed-tube forms.

is for the concrete core to assume all lateral forces with the exterior steel framing filled in around the core. The structural steel framing is generally designed only for gravity forces with simple shear connections. All steel framing follows the standard details of steel construction. The steel beams are connected to the core walls either by a typical corbel detail (as shown in Figure 7.10), or by bearing in a wall pocket or by an anchor shear-plate detail. The diaphragm action is established with the use of a typical arrangement of composite slabs on composite beams. Reinforcing dowels are provided from the core walls to the slab to engage the slab diaphragm. Steel columns on the exterior faces—and in some cases on the interior—complete the steel part of the framing. The basic advantage of this combination of steel and concrete is in the simplification of the steel-framing part that is not required to resist lateral loads.

Among the disadvantages are the relative inflexibility of the core walls for core planning and the loss of space due to the presence of walls themselves. Even though the system is economically effective for buildings up to about 45 stories, large inefficiencies, particularly related to core walls, result when this concept is applied to taller structures.

As far as cladding and window-wall details are concerned, the same systems that are used with a steel building are applicable.

In high seismic areas, the steel-frame part would be required to resist some part of the earthquake loads as a moment frame. The core walls may also require steel columns embedded in them to resist overturning and to provide for load-carrying capability when the shear wall is damaged.

7.5.3.1 Core Planning

The building core generally involves centralization of various building service elements, such as elevators, fire stairs, mechanical-electrical rooms and shafts, toilets, closets of all types, and laundry rooms. The primary element affecting the structural planning of the core is the elevator system, which is usually provided in several banks serving different floor ranges. As one bank terminates, the shear walls for this part may have to be discontinued, thus resulting in an eccentric core that must be considered in core-wall analysis and design. The floor framing inside the core can be either cast-in-place concrete, precast concrete, or structural steel. Precast concrete and steel framing offer an efficient alternative to cast-in-place concrete framing since formwork and shoring are not needed.

7.5.3.2 Core-Wall Analysis and Design

The concrete core must generally provide all the required torsional and flexural rigidity and

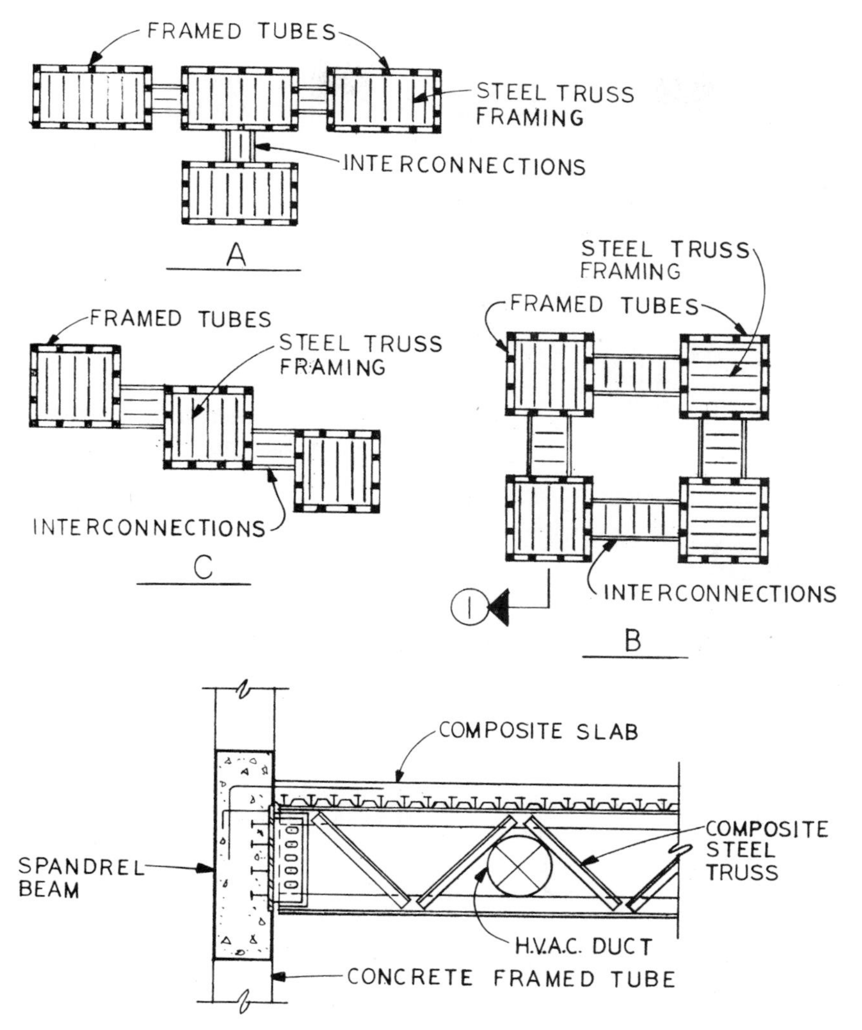

Fig. 7.7. Clusert of framed tubes.

strength with no participation from the steel system. The torsional resistance is also required for possible asymmetry of the wall system where eccentricities of lateral load with respect to the center of rigidity exist. Conceptually, the core system should be treated as a cellular wall system of a tubular form with punched openings for access to result in an effective cantilever behavior. Beam interconnections between different wall units are necessary to establish the cantilever behavior.

The plan dimensions of the core should be selected so as to produce optimum flexural stiffness of the cantilever tube for a particular volume of the building. The floor-framing span outside the core should be such as to distribute

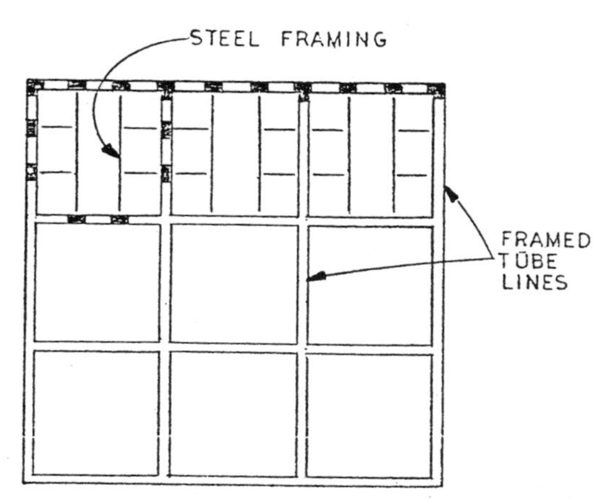

Fig. 7.8. Composite bundled tube.

Fig. 7.9. Ohio National Bank Building, Columbus, Ohio.

enough gravity loads to the core walls so that their design is controlled by compressive stresses even under wind loads. The geometric location of the core should be selected so as to minimize eccentricities for lateral load.

Simple cantilever wall analysis may be sufficient in most cases. However, if the total form is used as a wall-tube, it may be necessary to perform a more sophisticated analysis to include the effect of openings or to evaluate the effectiveness of interconnections. An analysis of this type may also be required to evaluate torsional stresses when the vertical profile of the core-wall assembly is asymmetrical. Such analysis is best performed by use of computerized finite-element programs so that the walls can be simulated as panelized plate elements.

Steel columns are proportioned for gravity loads using an effective length factor $K = 1$ on the assumption that the steel frame is braced by the walls.

The ACI design rules are generally followed for the design of walls under gravity and wind stresses. The design procedure is identical to the design of shear walls in a typical concrete structure.

7.5.3.3 Diaphragm Behavior

The diaphragm also plays a significant role in terms of providing lateral support to the steel frame, which lacks the inherent capacity to resist frame instability. The required bracing force to stabilize a steel column under axial load can generally be computed from $P\Delta/h$, where P is the column axial load, Δ is the maximum expected sway plus erection tolerance, and h is the story height. The diaphragm stresses due to wind forces are generally small and can be readily resisted by the slab. Longitudinal and corner diagonal reinforcements in addition to slab dowels are provided in the slab strip on the exterior periphery of the walls to resist hoop tensions that might possibly develop in the diaphragm.

7.5.3.4 Differential Shortening Effect

The core walls are generally proportioned for relatively smaller equivalent compressive stresses as compared to exterior steel columns, which are designed for higher stresses. High-strength steel is often utilized for these columns. The result is that the steel columns shorten more than the concrete core under gravity load, even including the long-term shortening effect due to creep. Corrections to steel columns are generally necessary to maintain floor levels within acceptable tolerances. This may be accomplished by specifying overlength columns or by providing shims at column splices.

7.5.3.5 Construction Sequence

The concrete cores are normally built up to a certain number of stories, after which the steel framing is started and filled around the core. Figure 7.11 shows a photograph of the system under construction and clearly indicates this sequence of construction. Since the lead time for delivery of structural steel is usually longer than that for reinforcing steel, significant saving in time can be affected by building the concrete core ahead during the differential lead time period.

7.5.3.6 Examples

A typical design example is illustrated in Table 7.4 which is self-explanatory. Many other examples of concrete-core braced systems

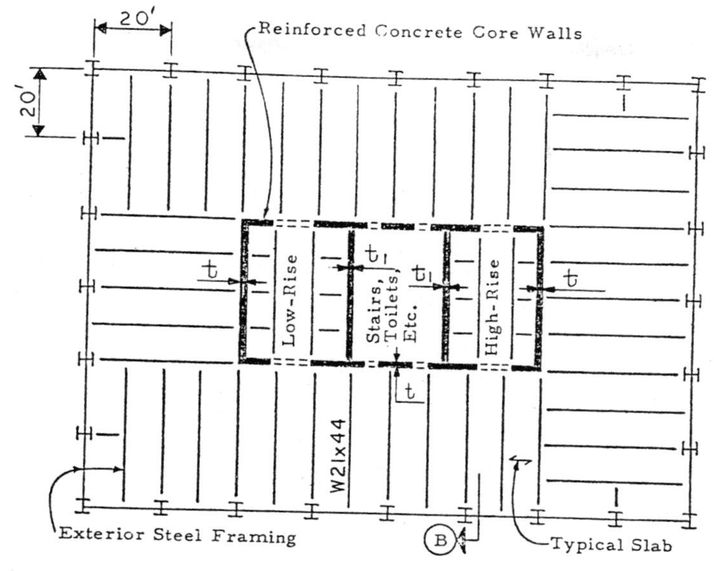

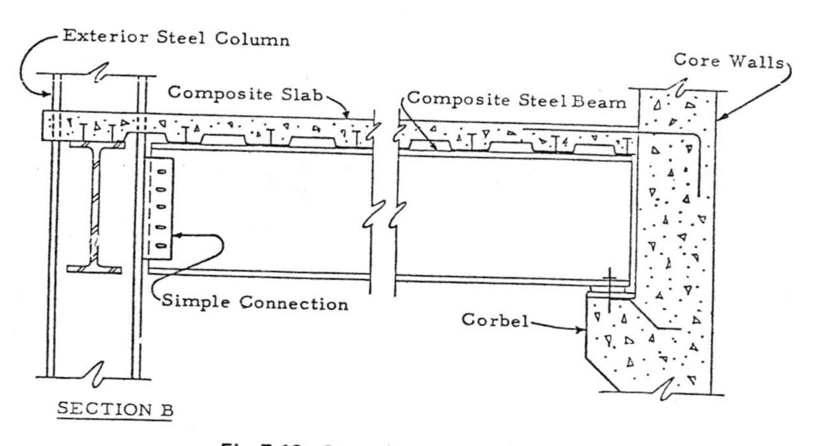

Fig. 7.10. Central core braced system.

exist in various parts of the world. In principle, they are all similar in that the centralized core system assumes all lateral load. The state-of-the-art paper by Kozak (1972) is generally useful in indicating the variations in the outside steel structure. The paper also describes, and includes photographs of, the 27-story, 100 m (328 ft) television building in Bratislava, which is typical of this type of system. Another outstanding example is the Maine-Montparnasse Building (1973) in Paris, France. The office tower consists of 58 stories for a total height of 210 m (689 ft) and perhaps represents the tallest structure for which the concrete-core-braced system has been used. The reinforced concrete core was slip-formed and was built in two stages. The core was built halfway before the steel work was started. The second-stage core was started after completion of the steel for the first stage.

An example in the United States is the 40-story Denver Square Office Building in Denver, Colorado, and many office towers in Canada in the 25- to 35-story range in the cities of Edmonton, Calgary, and Toronto, also use this system.

Fig. 7.11. Central core braced system under construction.

7.5.4 Multiple Concrete-Core Support Systems

Multiple-core systems are characterized by vertical concrete cores that are used for vertical support of steel floor framing; the cores also provide all the required lateral resistance for wind forces. A review of several existing systems indicates that each one has been evolved to respond to the particular architectural and planning needs of a project, and therefore they are not to be construed as systems for general application. The inspiration to use these systems stems from the need for large open spaces at ground level and the viability of concrete cores as an expressive architectural form. Functionally, these hollow cores house service elements, such as stairs, elevators, and mechanical shafts, which are now decentralized to occur at various locations of the floor.

An example of this type of system for the Fourth Financial Bank Building in Wichita, Kansas, is shown in Fig. 7.12. Hollow exposed concrete pylons, 15 by 15 ft (4.6 by 4.6 m) square, are arranged on a 80-by-80-ft (24.4-by-24.4-m) grid with three bays in each direction. The building is eight stories tall, and the building enclosure includes a huge interior court space and an L-shaped office floor space. The pylon is of reinforced concrete with a wall thickness of 10 in. (25 cm), which is designed for vertical loads of floor and for lateral loads

Table 7.4. Design example—Central core braced system.

No. stories	39	Office live load	50 psf (2400 Pa)
Structure height	503 ft (153 m)	Partition load	20 psf (958 Pa)
Average wind pressure	30 psf (1436 Pa)		
Drift limit index @ 30 psf (1436 Pa)	H/500		
Translatory period	5-sec short direction		
Torsional period	2 sec		
Floor-plan dimension	See Fig. 7.10		
Slab			
2-in. (5-mm) composite metal deck with $3\frac{1}{4}$-in. (83-mm), 4000-psi (28-kPa) structural lightweight concrete topping			
Steel floor framing			
W21X44–A36 typical composite-steel beams at 10-ft (3-m) centers			
Core walls			
t = 1'-6" (0.45 m), t_1 = 1'-0 (0.3 m) up to 25th floor		f_c' = 5000 psi (35 kPa)	
t = 1'-3" (0.38 m), t_1 = 1'-0 (0.30 m) above 25th floor		f_c' = 4000 psi (28 kPa)	
Material quantities			
Average structural steel		10 lb/ft^2 (479 Pa)	
Core wall reinforcing quantity		1.5 lb/ft^2 (71.8 Pa)	

with each pylon acting as an individual cantilever bending about its own axis. The floor system consists of 40-in.-deep composite trusses at 15-ft (4.6-m) centers supporting a composite slab unit consisting of a 3-in.-deep (75-mm) cellularized blended metal deck and 2.5 in. (63 mm) of semilightweight concrete topping. This composite system is similar to the one described by Iyengar and Zils (1973). Figures 7.13a and b show a truss and girder connection with the pylons.

Many other examples of multiple concrete-core braced high-rise systems exist, and so a detailed description here is not provided.

Fig. 7.12. Fourth Financial Bank Building under construction.

A

B

Fig. 7.13. Member end connections. *A*, Girder to pylon connection; *B*, truss to pylon connection.

The 23-story Knights of Columbus Building (1970) in New Haven, Connecticut, involves slip-formed cylindrical cores and vertical post tensioning.

7.5.5 Composite-Steel Frames

In a steel-frame system, the girders along column lines function as part of a continuous frame whereby reverse bending and consequently negative moments, are induced due to gravity and lateral loads. This steel frame can be transformed into a composite form by concrete encasement of beams and columns. In order to truly take advantage of the concrete encasement, the steel part should be reduced and reinforcing steel bars should be added where required. This form tends to become uneconomical because of high cost of formwork and labor, and therefore it is not commonly used in the United States.

In cases where this concrete encasement is used for fire protection of steel beams and columns, the composite properties can be considered in the stiffness computation to control frame drift, while the steel section alone is designed to resist the generated forces and moments.

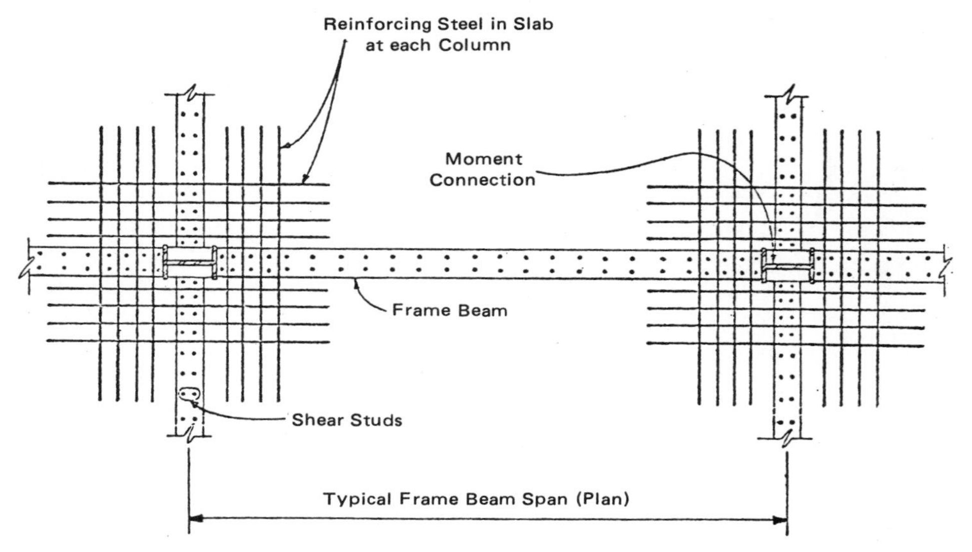

Fig. 7.14. Composite frame plan with additional reinforcing.

A more feasible form of an unbraced composite-steel frame utilizes the normal construction of a rigid frame with unencased composite beams and uses reinforcements in the negative moment areas as shown in the frame plan in Fig. 7.14. Shear studs are provided over the entire beam length to develop the positive and negative regions of the beam. Two types of composite frames of this form are possible. In one type, nominal reinforcements are provided only to control cracks in the negative moment areas, in which case the composite properties are utilized to satisfy only the lateral load stiffness of the frame with the beam section itself capable of resisting the forces from a strength point of view. A second type utilizes the reinforcements for strength development in addition to stiffness contribution. For the second type to be practical, the spans and the number of stories will have to be reasonably small.

Design rules are currently not available for design of composite frames. Design information pertaining to flexural stiffness of encased members, effective slab widths of T-sections in positive and negative regions, the extent of shear connection required, effectiveness of negative reinforcement in the slab, and other related information needs to be developed through analytical and experimental investigations.

7.5.6 Composite Steel-Plate Cladding

A commonly used steel plate cladding, shown in Fig. 7.15, consists of flat carbon steel plates attached to the steel frame by means of concrete backing behind the cladding. Shear studs are provided for attachment between the concrete and steel plates. The plate thickness varies from about 0.25 to 0.375 in. (6 to 8 mm), and the plate segments are generally welded to each other to formulate a continuous cladding. Because of fire-protection requirements, the composite properties cannot be included in strength calculations; however, their stiffening effect can be used in frame-drift calculations. The amount of stiffening as a result of this composite action can range from 10% to 50% depending on the overall geometry of the building and the composition and stiffness of individual frames (see Arndt and Scalzi 1972). It should be noted that composite action is established primarily for the two exterior frames parallel to the direction of the wind.

Figure 7.15 shows two wind-drift curves for a 25-story steel-frame building, one using only the structural steel for stiffness and the other including composite properties for fascia frames and for exterior columns of the other frames. The dimensions of the frame and of the steel-plate cladding for exterior columns and spandrel beams are also shown. A uniform wind pressure of 30 psf (1436 Pa) was assumed. The curves illustrate the stiffening effect of the composite cladding, which for this example amounts to a substantial reduction in drift at the roof level of the building.

The design of the steel frame itself can be formulated on the basis of proportioning the steel frames for gravity loads only and then performing a drift analysis including composite properties for members of the fascia frames. In most instances, the additional steel required to satisfy drift limits can either be saved or reduced. Therefore, for those building frame systems that are influenced by drift limitations, the inclusion of the fascia plate cladding, if provided for aesthetic reasons, results also in some cost savings in the structural system.

7.5.7 Interior Panel Braced Frames

The panel bracings generally take the form of concrete shear panels infilled or attached to the steel frame, which, by virtue of their in-plane shear rigidity, contribute to the overall lateral load stiffness and strength of the steel frame. These panels, which generally serve as nonstructural interior partitions in the core areas or as room-dividing partitions in apartment and hotel-type occupancies, can also be used to perform the structural stiffening function. In particular, the amount of additional steel required to satisfy wind-load requirements is either eliminated or reduced together with simplification of steel details in a typical steel-frame structure.

The panels require a bounding steel frame to which they are attached by side connections to columns and top and bottom connections

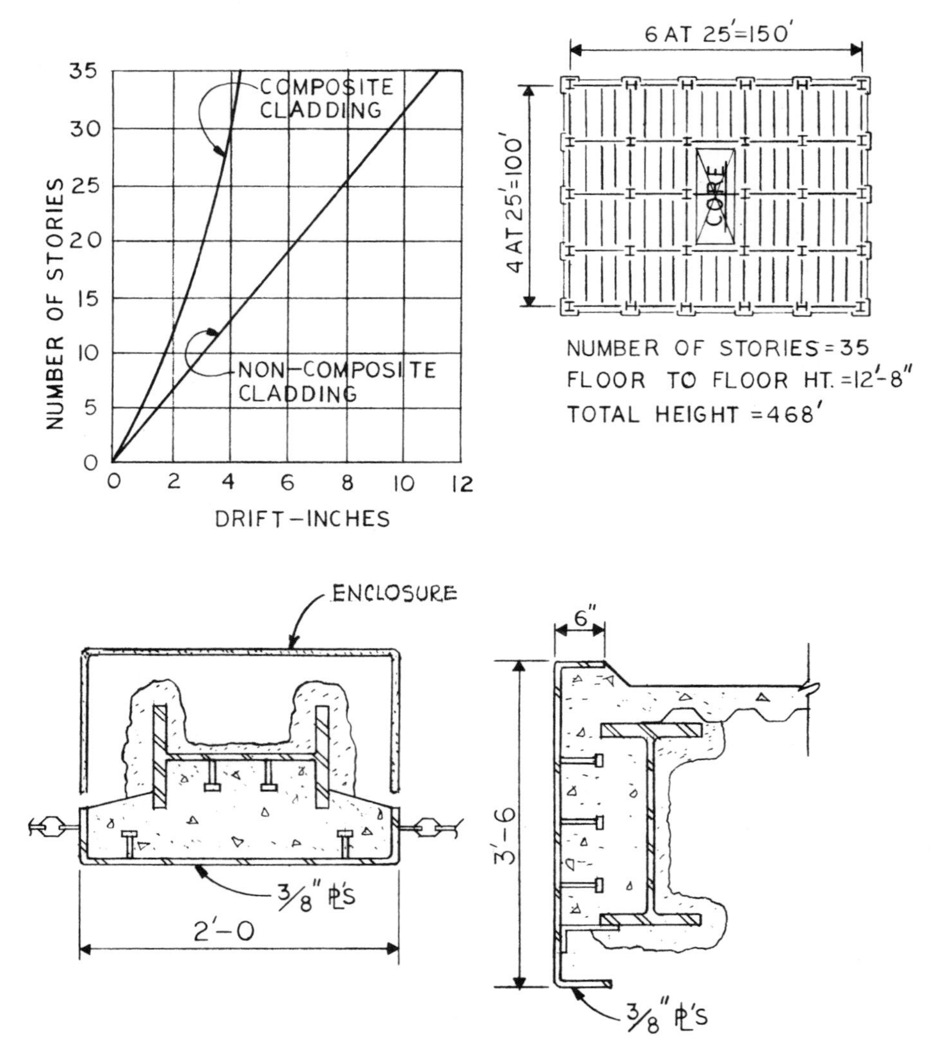

Fig. 7.15. Composite steel-plate cladding.

to steel beams as shown in Fig. 7.16a. The panels are reinforced, and in some cases direct transfer of forces is provided by steel embedments in the form of diagonalized bracings that are directly attached to the steel frame as shown in Fig. 7.16b and c.

The panelization of centralized cores of office buildings with bounding steel frames and concrete panels is generally ineffective because of the irregularities created by various openings in the core. However, apartments and hotel systems generally consist of a regular arrangement of room-dividing partitions as shown in Fig. 7.17, and the use of a concrete panel in a bounding steel frame offers greater possibilities. Even though such systems are uncommon in the United States, a considerable number of these systems are built in Japan under the so-called HPC System for apartment buildings in the 10- to 15-story range.

The basic behavior of the shear panel under lateral load is shown in Fig. 7.17. The principal mode of deformation of the panel is in shear. If the height of the panel is h, the length ℓ, and thickness t, then the shear stress τ in the panel subjected to a lateral shear of V is equal to:

$$\tau = \left(\frac{h}{\ell}\right)\frac{V}{ht} \tag{7.1}$$

and the shear strain $\gamma = \tau/G$, where G is the

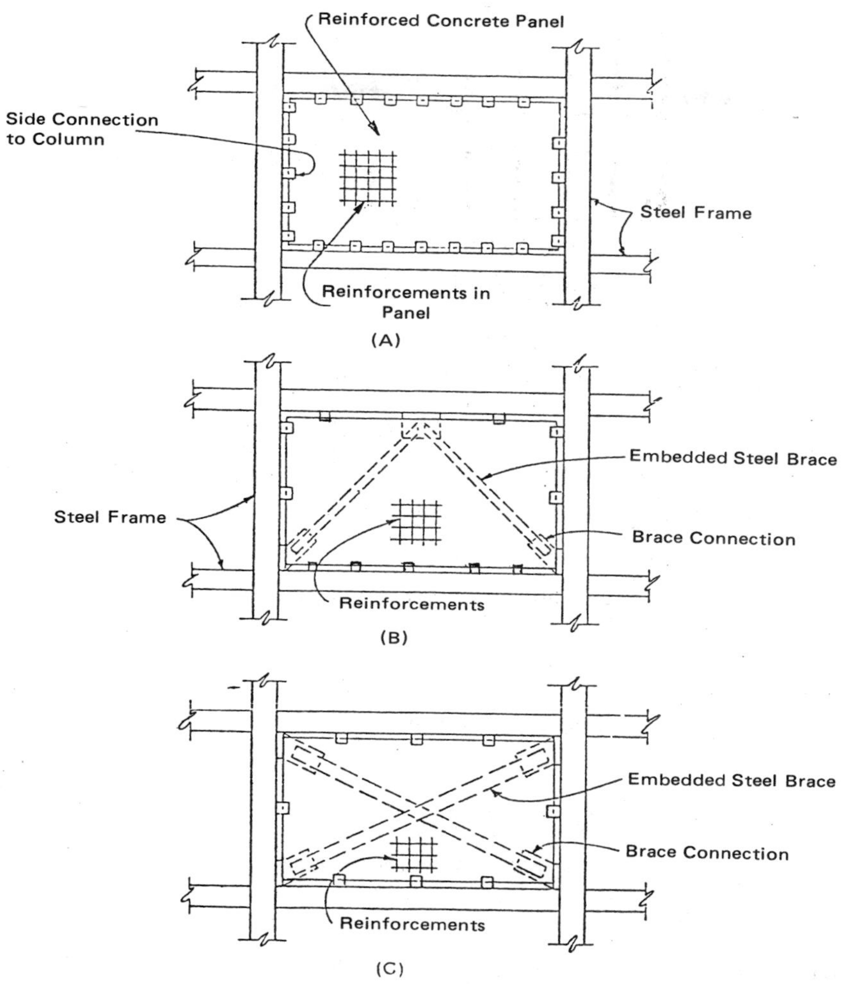

Fig. 7.16. Reinforced concrete infill panel.

shear modulus of panel concrete. The contribution of shear-panel distortion to story drift is equal to γH, where H is the floor-to-floor height.

In addition to panel-shear distortion, there is the contribution of column axial strain, usually called the *column-shortening component.* This component can be computed by assuming the panel-braced frame as a vertical cantilever resisting the overturning moment of the lateral forces. The equivalent moment of inertia of this cantilever is computed using the steel columns as chords held at their geometric location and bent about their neutral axis as indicated in Fig. 7.17.

The lateral load analysis of the total building system, consisting of several panel-braced frames within it, can be approached on the basis of assigning a portion of the total lateral load-overturning moment to each panel-braced frame in proportion to its equivalent moment of inertia at that particular floor. The resulting lateral load shear in each panel-braced frame is then deduced from the overturning moments at consecutive floors. This method of analysis is approximate, but yields results that are acceptable for design purposes. A detailed analysis based on finite-element methods can be used for more accurate results (Weidlinger 1973). The shear forces are assigned to the panels and the axial force due to overturning, to the steel columns. The total drift can be computed by

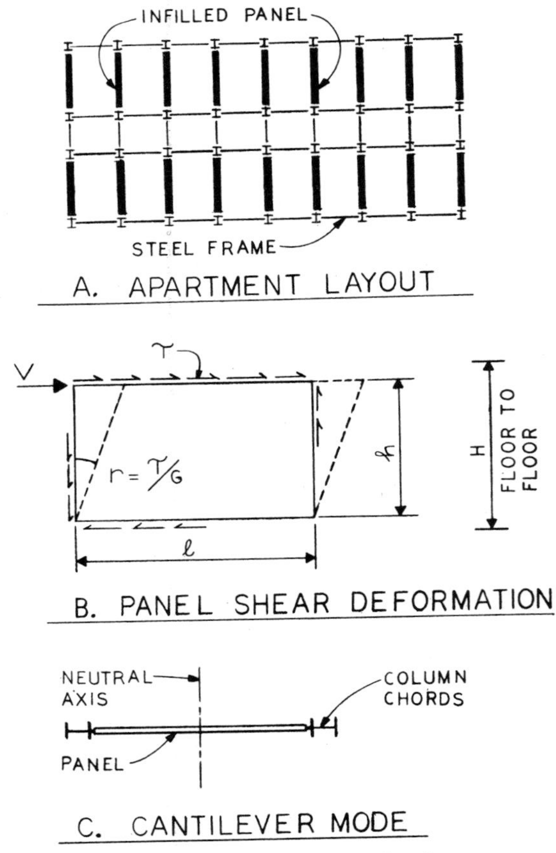

Fig. 7.17. Interior panel bracing.

superposition of shear and cantilever components of drifts. It should be noted that shear drift is computed on a story basis and accumulated from the bottom to the top of the building, whereas the cantilever component is computed for a cantilever fixed at the ground using the total height of the braced frame.

The panels must be reinforced to resist wind forces from either of two directions resulting in two-way reinforcing. Tension participation of concrete is ignored, leaving the concrete in the panel to provide compression resistance and shear rigidity. Depending on the type of panel connections, it may be necessary to evaluate the flexibility of the connections of the shear panels. Experimental values for different types of attachments will be required for this evaluation.

The participation of the panel in the gravity load should also be investigated. Since the panel is attached to the columns, the axial column strain under gravity load is also transmitted to the panel. If the panel in addition is also attached to the beam, floor loads will also be directly transferred to the panel. The panel under this condition will act like a bearing wall and the connections should be suitably designed for this gravity-load transfer. Additional gravity-load capacity is provided by embedded bracings, which are shown in Fig. 7.16a and b. If the transfer of gravity loads tends to exceed the capacity of the panels to resist these forces, connections with the beam element can be designed to be flexible in the vertical plane, but rigid for transfer of the shear in a horizontal plane. However, such connections tend to be cumbersome and impractical, and their use should be evaluated from this point of view.

7.5.8 Exterior Panel-Braced Frames

A logical evolution in the use of reinforced concrete wind-bracing panels is to visualize the exterior cladding enclosure in a panelized form. Figure 7.18 shows some possible types of exterior wind-bracing panels attached to steel frames. In its most effective form, the panel shape corresponds to an approximate square with one window opening that is attached to a fascia frame; this will result in relatively closely spaced steel columns. This is shown in Fig. 7.18a. The closely spaced columns of the exterior frame generally represent the structural character needed for an ultra-high-rise building system in the range of 40 to 80 stories; as such, panel bracings of this type offer considerable potential. This type of panelization can be construed as a stressed skin of an equivalent tubular form, which is reinforced by the fascia steel frame. In this case, the fascia steel frame primarily functions as a gravity and wind axial load carrier, while the reinforced concrete cladding panel acts as a shear panel. A continuous connection or discreet closely spaced connections between the steel frame and the panel

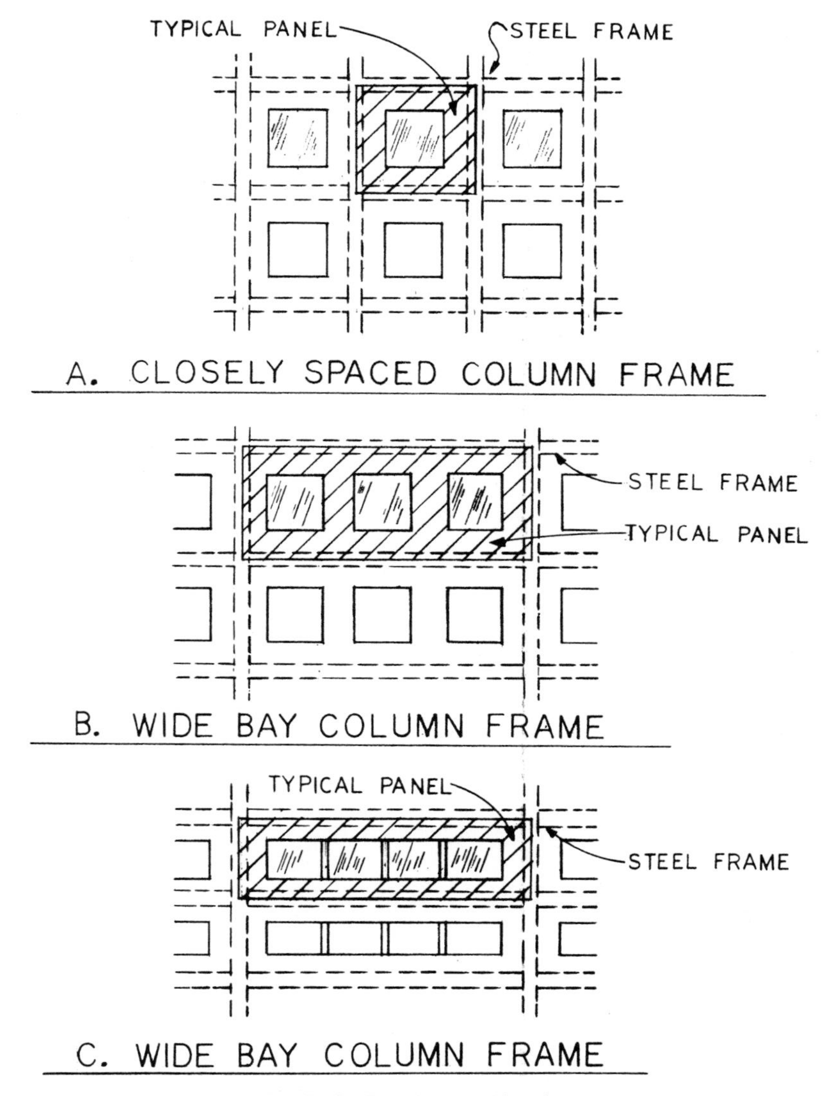

Fig. 7.18. Exterior panei bracing.

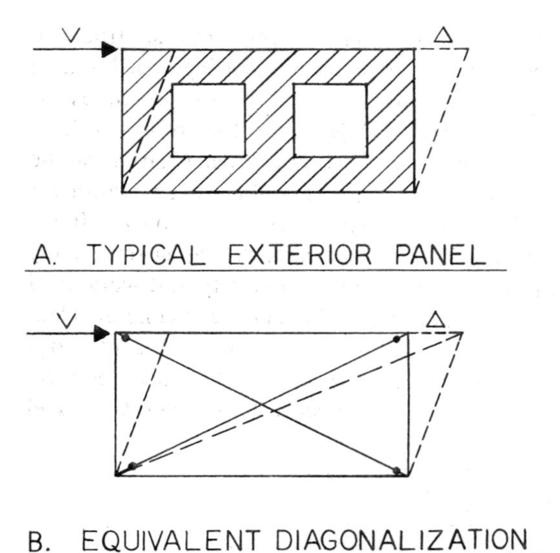

Fig. 7.19. Equivalent truss element.

are generally required. Panel-frame beam connections that are flexible in the vertical plane, but rigid for shear resistance in the horizontal plane, are required to promote gravity load flow into the steel columns.

Figure 7.18b and c show cladding panels for an exterior frame with wide bay spacing. The panel could involve multiple window openings as in Fig. 7.18b or wide single opening as in Fig. 7.18c. The transportable size of the panels as well as the lifting weights tend to govern the feasibility of these wide bay panels.

The presence of openings in the panel results in considerable reduction of panel-shear stiffness. The shear stiffness is influenced by the area and size of penetrations and also by the number and size of such penetrations. Useful information on the effect of different penetration variables on the panel stiffness is not

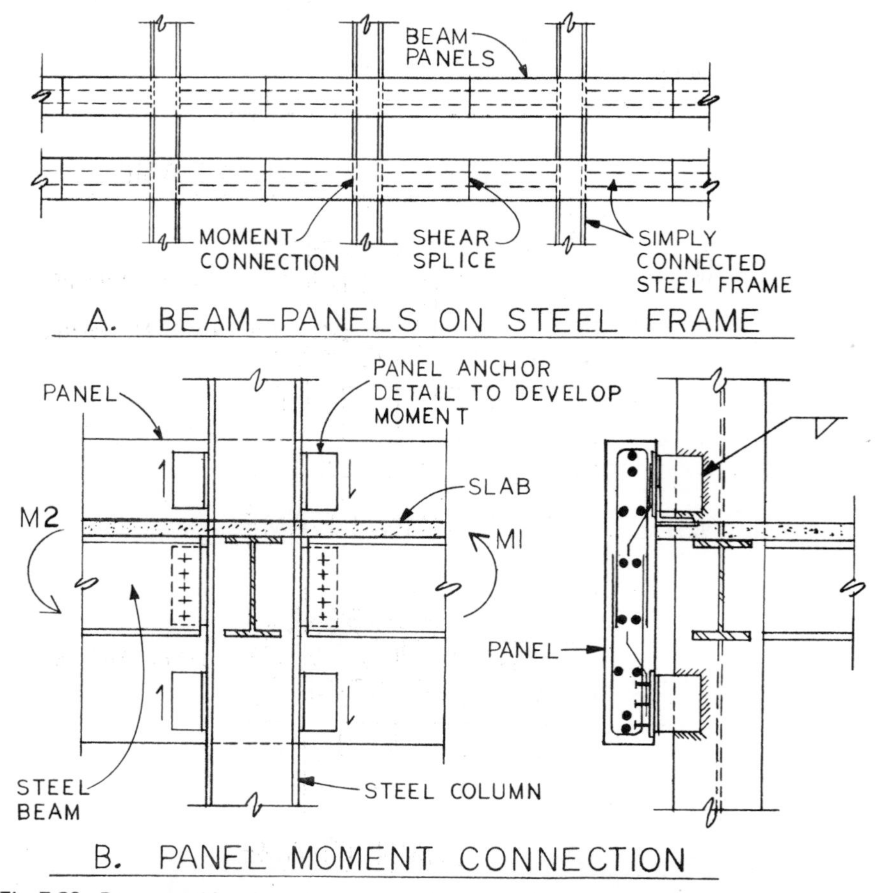

Fig. 7.20. Beam-panel bracing. A, Beam-panels on steel frame. B, Panel moment connection.

available, and therefore each case will have to be analyzed individually (Weidlinger 1973).

For purposes of structural analysis, the panel can be replaced by an equivalent pin-connected diagonal system as shown in Fig. 7.19. The areas of the diagonals of the equivalent truss system can be determined from Truss Analogy to simulate shear stiffness of the actual panel. This panel-shear stiffness can be determined from computerized finite element analysis of the panel including the penetrations. If the panel-shear stiffness is S_p, the connection stiffness S_c, and the frame stiffness S_f, then the equivalent truss stiffness S_t can be determined from:

$$\frac{1}{S_t} = \frac{1}{S_c} + \frac{1}{S_p + S_f} \tag{7.2}$$

Since the plane of cladding and that of the frame does not generally coincide, the panel-connection details will have to outrigger from the frame and therefore tend to be more flexible than the infilled panel anchorages. This brings about the necessity for determining the connection stiffness (S_c) by tests on typical anchor connections.

Panel bracings may also take the form of precast reinforced concrete panel beams that act as wind bracing, as shown in Fig. 7.20. In this case, the panel-beams augment only the spandrel beam stiffness, and all wind forces are resisted by this composite fascia frame. This can be an effective form of bracing if the panel-beam is relatively deep in the range of 30% to 60% of the height of the floor. This is consistent with some of the current energy conservation designs, which need deep spandrel beams to minimize glazed areas. Increased column stiffness will also result because of the reduction of the length of the unbraced column in the story. In order for the beam-panel to perform its stiffening function, the unbalanced moment at the beam-panel joint is required to be transferred to columns. The effectiveness with which this connection can be worked out in a simple fashion determines the viability of this form of bracing. In order to facilitate this, it would be helpful if the beam-panel is continuous at the joint and the panels themselves are spliced at midspan. Since the center of the span generally corresponds to a point of contraflexure due to reverse bending of the beam, the predominant splice force is one of transverse shear. Some possible connection details are also shown in Fig. 7.20.

Steel reinforcing bars or embedded steel plates are required in the concrete panel to develop the panel moment. Other transverse and longitudinal reinforcing will also be required as shown for shear, shrinkage, and temperature stresses.

REFERENCES

Arndt, A. P. and Scalzi, J. B. (1972), *Plate Wall Cladding*, State of the Art Report 3B, TC17, Conference Preprint Addendum, ASCE-IABSE International Conference Preprints, Volume A.

Belford, D. (1972), Composite Steel-Concrete Building Frame, *Civil Engineering*, July.

Iyengar, S. H. (1972), *Preliminary Design and Optimization of Steel Building Systems*, State of the Art Report No. 3, Technical Committee No. 14: Elastic Design, American Society of Civil Engineers–International Association for Bridge and Structural Engineering Joint Committee on Tall Buildings, August.

Iyengar, S. H. and Khan, F. R. (1972), *Optimization Approach for Concrete High-Rise Buildings*, Presented at the American Concrete Institute Convention in Dallas, Texas, March.

Iyengar, S. H. and Khan, F. R. (1973), *Structural Steel Design of World's Tallest Building–Sears Tower*. Presented at the Australian Conference on Steel Developments, Australian Institute of Steel Construction, Newcastle, Australia, May 21–25.

Iyengar, S. H. and Zils, J. (1973), Composite Floor System for Sears Tower, *American Institute of Steel Construction Engineering Journal*, Third Quarter.

Khan F. R. (1970), *New Structural Systems in Steel*, AISC Annual Convention Pittsburgh, Pennsylvania, April.

Khan, F. R. and Amin, N. R. (1973), Analysis and Design of Framed Tube Structures for Tall Concrete Buildings, *American Concrete Institute Journal* SP36.

Khan, F. R. and Sbarounis, J. A. (1964), Interaction of Shear Walls and Frames, *J. American Society of Civil Engineers*, June.

Khan, F. R., Iyengar, S. H., and Colaco, J. P. (1966), Computer Design of the 100-Story John Hancock Center, *J. American Society of Civil Engineers*, December.

Knights of Columbus Building, New Haven, Connecticut, *Progressive Architecture Magazine*, September 1970.

Kozak, J. (1972), *Structural Systems of Tall Buildings with Core Structures*, State of the Art Report No. 8, TC3, Planning and Design of Tall Buildings, ASCE-IABSE Conference, August 21-26.

Maine-Montparnasse, the Highest Tower in Europe: Its Construction in Pictures, *Acier-Stahl Magazine*, June 1973.

Taranath, B. S. (1974), Optimum Belt Truss Locations for High-Rise Structures, *American Institute of Steel Construction, Engineering Journal*, Vol. II, First Quarter.

Weidlinger, P. (1973), Shear Field Panel Bracing, *J. Structural Division, Proceedings, American Society of Civil Engineers*, Vol. 99, ST 7, July.

8

Application of composite construction in bridges

ARTHUR L. ELLIOTT, P.E.
Bridge Engineer
Sacramento, California

8.1 GENERAL

Composite action in its broadest meaning relates to the interaction of the deck slab with its supporting girder to make the two act as a unified structural system, each supporting its share of the load. Frequently, the composite concept is considered to apply only to concrete slabs on steel beams, however, in bridge construction, the composite principle is used with a number of other materials and also for some other purposes, such as: composite piles of steel, wood, or concrete; partial slabs; and various concrete-on-concrete beam systems. Concrete slabs on timber beams are often made to act compositely. Another common composite application is a poured-in-place concrete slab on precast concrete beams that may or may not have been prestressed. Although these are similar materials, positive steps must still be taken to assure composite action between the two elements. In slab construction, a portion of the slab may be precast with the full tension reinforcement or, if the slab is to be post-tensioned, with just enough prestress to support the dead load. These half-slabs act as lower forms for the deck slab. When the upper portion of the slab is poured in place, it acts compositely with the precast half-slab. In this case, various keying devices must be employed either in the concrete itself or as additional hardware to assure that the two portions of the slab will work compositely. Even the construction of an ordinary T-beam concrete bridge entails certain attention to insure composite action. It is common to place the beams and slab in separate pours. The construction joint between pours must be roughened, keyed, or mechanically connected to make the beam act compositely.

Another application of composite action found in bridges is in piles made up of two or more materials but commonly called *composite piles*. The upper portion of the pile is often of concrete, often prestressed to provide extra durability in the area aboveground and in the critical tidal range when the structure is in water. The lower portion of the pile may be a steel H-section or a timber pile. Various joining devices are used to give the pile rigidity across the joint and make it act compositely.

The advent of prestressing introduced many other possibilities for incorporating prestressed elements into cast-in-place construction to make the entire system act compositely. All of these are examples of composite action in bridges. The composite action discussed in this chapter will deal principally with the composite action of concrete slabs on steel girders.

Although positive steps must be taken to assure composite action between two dissimilar materials, or even between similar materials fastened together, there is composite action in many cases where none was intended by the designer. In research projects to study comparative stresses and deflections with and without composite action, considerable difficulty is experienced in preventing the beam and slab from acting compositely. Usually the mating surfaces must be specially smoothed and greased to be certain that they do not interact. In normal stress ranges, it is likely that ordinary roughnesses such as mill scale, welds, rivet or bolt heads, or the roughness of a set concrete surface will provide enough shear restraint to make a beam act compositely. However, as the beams approach their ultimate capacity, these casual and accidental shear resistors will fail, and slippage will occur between the two elements. Thus, since most beams are designed to be effective far beyond their working range and up to their ultimate capacity, shear connectors provided will assure the necessary composite action regardless of what natural shear restraint might be present.

Evidence of unanticipated composite action is usually apparent whenever studies are made of noncompositely designed beams to compare the calculated deflections with those actually measured in the field. The calculated deflections are often greater than the actual measured deflections, indicating that there is composite action making the beam stronger than it calculates to be.

8.2 CURVED BRIDGE GIRDERS

The use of curved girders has seen a marked increase in recent years because of the greater number of highway interchange structures and

Fig. 8.1. Example of curved composite bridge—Elysian viaduct structures. (*Courtesy California Department of Transportation.*)

Fig. 8.2. South Fork Eel River Bridge. (*Courtesy California Department of Transportation.*)

the decreasing amount of available land. This has led to a large number of curved ramps crossing at different levels, making possible a compactness not possible with the older designs.

A number of different approaches to the design of curved girders have been developed. It is obvious from their geometry of curvature and superelevation that they are subjected to both flexural and torsional stresses. The designer should be well aware of these forces as well as the rather complex interaction of the various elements in curved bridges. Box sections, which are particularly efficient in resisting torsion and shear, are frequently used to achieve both attractive appearance and economical design for these long, looping spans. The boxes may be either of concrete or steel. The concrete boxes are usually cast in place and usually post-tensioned. Steel boxes or steel curved girders are erected in the field, and then a cast-in-place concrete deck is added to provide additional lateral stability. Care must be taken with steel structures to maintain the geometry of the sections until the concrete is added to provide the additional stiffness. This means that the curved steel girders are usually erected in pairs with strong cross-bracing to maintain their curvature and vertical alignment. Steel boxes must also be rigidly supported across their top flanges to hold them in position until the concrete is placed.

Approach to the analysis of such structures ranges from solutions of the classical elasticity equations of bending, shear, and torsion, to the modern finite-element analysis techniques supplemented by tests often on scaled models of curved bridges. Following the general theories by Vlassov (1961), Timoshenko and Woinowsky-Krieger (1959), and others, Heins (1976) gives considerable information on the analysis and some tests. He presents design aids for box bridges to allow rapid evaluation of the required spacing of interior diaphragms to minimize the induced stresses as caused by distortion of the section under load. Aneja and Roll (1969) and Hsu (1968a,b) have reported tests on models as well as full-scale tests of bridges made of both steel-concrete and concrete-concrete composite curved bridges.

Two examples of curved girder bridges, located in California, are shown in Fig. 8.1 and 8.2. Figure 8.1 shows a composite steel-concrete curved bridge continuous over two spans as it crosses a flood-control channel in Los Angeles. Figure 8.2 shows two multispan continuous concrete box-girder bridges with spans near 100 ft (30 m) in an interchange in Sacramento.

8.3 ADVANTAGES OF COMPOSITE CONSTRUCTION IN BRIDGES

The composite principle has been in use in bridge construction almost since the beginning of scientific design. However, modern technology has made many special uses of the composite idea to make construction faster and more economical. The primary appeal of composite construction is probably economy since loads may be carried with fewer pounds of material when they are acting compositely. However, there are other advantages to the composite approach. The structures are somewhat stiffer so they have less movement—which is sometimes a definite advantage. The structure may be made somewhat shallower than it would have to be if it were not acting compositely. A few inches less depth of superstructure is of great importance in highway separation structures. The saving becomes very evident when one considers the additional height of structure and the large quantities of approach fill that a few inches of greater structure depth may require. In large interchanges where a number of levels may be crossing over and under, a few inches saved in each superstructure depth to meet the required clearance can effect a very considerable saving of cost in the structure itself as well as its approaches. Therefore, it becomes of great importance to design the shallowest superstructure possible, and, short of prestressing, this is usually a composite section.

Composite construction also makes other construction advantages possible. Composite construction is often thought of as merely the joining of a steel beam with a concrete deck and arranging them to act together as a T-beam. However in a broader sense, composite construction is also the joining of any two members together, even members of similar material, and then making them act together as though they had been placed monolithically. This too is composite action, and it is utilized in bridge construction in a variety of ways to save time and money.

8.3.1 Saving Formwork

One of the chief expenses in any concrete construction is the formwork. Many clever procedures have been developed to eliminate or minimize formwork. One system used with either steel or concrete beams is the half-depth slab. The half-slabs are made to fit between the supporting beams, spanning the gap and acting as a form for the deck slab. This eliminates the need for forms and their supports as well as the need for stripping them. This idea has been successfully used in buildings, wharves, and for highway bridge decks. The slabs, usually about 4 in. (10 cm) thick, are usually cast in a yard. Quality control is often better in a plant operation, and the finish on the underside of the slab will be good enough so that no field attention or work will be required. Sometimes these half-slabs are pretensioned to give them greater strength and stiffness. The upper surface of the slabs is intentionally left very rough, and various reinforcing arrangements are provided to give anchorage between the precast portion and the upper part of the slab, which will be placed *in situ*.

The anchorage devices between the upper and lower slabs may be dowel bars, hairpins of reinforcing steel, zigzag bars that alternate between the upper and lower slabs, spiral bars with half of the spiral in each section, concrete keys, and exposed aggregate, which will bond into the concrete top slab—to name a few. These may not be thought of as composite beams in the usual sense, but the action is composite in just the same manner as found in a steel beam–concrete slab system.

Use of half-slabs as forms has been a frequent practice in bridges where there are a large number of similar spans, and the saving in forms will pay for setting up a plant to make the half-slabs. Where the haul is not too long, the half-slabs may be made in a commercial plant. A large refinery dock for Union Oil Company in San Francisco was built in this manner by the Ben C. Gerwick Company, and worked very satisfactorily. The half-slabs eliminate the need for forming between the cap girders (forming is always difficult to support on the concrete piles), and also eliminate form removal in the low clearance under the deck and over the water. This system works better than many totally precast systems because the final surface is cast in place and consequently has a smooth, joint-free surface. This can also eliminate cracks that come up through the wearing surface at the joints between the precast units when used full depth and surfaced with asphalt.

8.3.2 Economy

Composite construction is common in bridges because it makes the principal elements of the bridge share in carrying the load. Using a heavy concrete slab for nothing but a wearing surface is an expensive luxury. Composite action makes it carry its share of the load.

8.3.3 Reduction in Deflection

Because the composite beams are stiffer, they deflect less under load. This gives a more solid, less bouncy structure—which is especially desired when there is considerable pedestrian traffic. In cases where precast half-slabs are used, the lower portion of the slab may be well cured before use, which results in a reduction in plastic flow (creep) of the finished deck slab, which has been detailed in Chapter 4.

Thus, the overall advantages of composite construction in bridges are that it is more economical, it often makes the bridge easier to build, and the resulting structure is stiffer, stronger, and more satisfactory.

8.4 COMPOSITE DESIGN CONSIDERATIONS

The main requirement of composite construction is, of course, that the girder and slab shall be tied together so that they will act as a T-beam. In steel-concrete combination, the concrete slab rests on the top flange of the steel beam. The concrete is usually extended down on the sides of the beam by the thickness of the top flange to form a lateral key but the top flange is not encased in concrete. Thus, the shear between the two elements must be developed by providing mechanical connectors.

Three different load conditions must be considered for composite bridge design. The system may be designed so the composite action applies to (1) live load only; (2) live load plus dead load, and (3) intermediate conditions where some of the dead load is carried by the girder alone and some by the

Fig. 8.3. Devil's Canyon Bridge. (*Courtesy California Department of Transportation.*)

composite system. The live loads are always carried by the composite section. Unless special procedures are used to insure composite participation in carrying the dead load, the dead load in place before the deck concrete sets is assumed to be carried by the girders alone.

While the effect of temperature change and shrinkage and expansion of concrete on the composite action are generally not considered in bridge design, the plastic flow or creep of the concrete can have an important effect and must be considered. The dead load causes a constant stress, and, under its sustained application, the concrete will flow plastically—which in composite action, reduces its ability to resist stress and carry its share of the load. This is compensated for by using a more conservative value of n (ratio of E_s/E_c) when computing the stresses caused in the composite action by dead loads.

8.5 COMPOSITE FOR DEAD AND LIVE LOAD, OR LIVE LOAD ONLY

Normally, where the steel beams are set in position and then the concrete deck is placed on them, the beams carry the dead load without any composite action. In order to make the system composite for both dead load and live load, it is necessary to fully support the steel beam until the concrete deck has fully set, i.e., until concrete has enough strength to support itself. In this manner, the steel beam is in a no-load condition until the concrete has set, after which the supports are removed and the entire composite section acts to support both the dead and the live loads. Although the spans should be fully supported, or at least supported at the quarter points to effectively maintain the steel beam in a no-load condition, apparently satisfactory results have been obtained by a single jacking bent placed at midspan. The beams were jacked somewhat higher than a straight line across between bearings and held there until the concrete of the deck had set. Then, when the jacks were removed, both the beams and the deck combined to support the dead load compositely. This is a risky business, however, because it is highly indeterminate and it is impossible to find what stresses are being induced.

There have been cases where the center of the span was jacked somewhat higher than a no-load position, and then, after the concrete had set, when the jacks were lowered, the concrete went into a higher state of compression than would result from only composite action, resulting in a sort of prestress. This may or may not be desirable and, if overdone, could result in crushing of the concrete. Thus, if this prestress were to be induced intentionally, extra shear connectors should be provided to hold the two elements together. The whole procedure is highly indeterminate and should not be attempted without adequate research and more complete knowledge of the stresses being induced followed by strict field control to see that the desired results are obtained. Furthermore, the savings that would result from inducing some overstress into the concrete deck are so small that they do not seem worth the effort and risk. Besides the procedure itself, there are many other variables that make the results uncertain such as the amount of support to be provided, the height of the jacking, the individual characteristics of the steel beams (which will vary in sag, camber, and straightness), and finally the character of the concrete.

8.6 AASHTO BRIDGE SPECIFICATIONS

The bridge specifications used throughout the United States and also in many other parts of the world are those published by the American Association of State Highway and Transportation Officials (AASHTO): *Specifications for Highway Bridges*. They have been developed over the past 50 years as a reflection of the actual experience and research of highway-oriented organizations throughout the country and are therefore the most reliable highway bridge specifications available. Their coverage of composite construction is quite thorough, and anyone designing composite highway bridges should be familiar with these specifications.

The following paragraphs are excerpts from the AASHTO Specifications (1973, 1974)*,

*AASHTO Specifications referred to in this chapter are those in the 1973 and 1974 Supplement (Interim Specifications).

Fig. 8.4. McBean Parkway, O.C. (*Courtesy California Department of Transportation.*)

with comments, noting some of the more important requirements governing the design of composite girders for highway bridges.

8.6.1 General (AASHTO Specifications 1.7.96)

The specifications apply only to steel girders with concrete slabs connected by shear connectors. The value of n assuming an $E_s = 29{,}000$ ksi (200,000 MPa) and $f_c' =$ 28-day cylinder strength of concrete, is then assumed to be (E_s/E_c):

$f_c' = 2000\text{–}2400$ psi (13.7–16.5 MPa) $n = 15$

$= 2500\text{–}2900$ psi (17.2–20.0 MPa) $= 12$

$= 3000\text{–}3900$ psi (20.7–26.9 MPa) $= 10$

$= 4000\text{–}4900$ psi (27.6–33.8 MPa) $= 8$

$= 5000$ or more (34.5 MPa or more) $= 6$

Because the dead-load stresses are constant and of long duration, the creep of the concrete will affect the steel-concrete relationship. Therefore, if the composite section is designed to carry dead loads, the stresses and horizontal shears produced by the dead loads should be computed with an n value multiplied by 3. Thus, if the structure has been supported or otherwise arranged for full composite action, (dead load plus live load), then the $3n$ value should be used for computing all dead-load stresses. If the composite action applies only to the live loads and the girders are not supported during the deck placement, then the $3n$ value would be used only for those dead loads that were superimposed after the deck slab had been placed and the span swung free of in-span support.

Because of the unpredictable effect of expansive cement, AASHTO warns that composite design should be used with caution if an expansive cement is to be specified.

A composite section should be proportioned so that the neutral axis lies below the top of the steel beam. Any concrete that is on the tension side of the neutral axis is neglected in computing anything but deflection.

It is of prime importance in assuring composite action that there be positive mechanical anchorage between the tension and compression sections to prevent slippage along the joining plane. This mechanical attachment is required even though it is likely the natural roughness will induce composite action between the two sections.

The AASHTO Specifications, reflecting a number of disastrous experiences, warn that the designer must be certain that the steel beams will be stable during the time when the concrete deck is being placed and until the time when the concrete has reached its strength and can provide the anticipated lateral diaphragm support. Bridges on a superelevation have had all the steel beams roll over downhill under a load of wet concrete. The beams, whether they be steel or precast concrete, should be laterally supported until the concrete has hardened and become a part of the structural system.

8.6.2 Shear Connectors (AASHTO Specifications 1.7.97)

The purpose of the shear connectors is to transmit the horizontal shear between the two beam elements. To successfully do this, they must be strong enough to resist both horizontal and vertical forces and must also be of a shape that easily permits the concrete to surround them. In the case of steel beams, the shear connector is usually welded to the top flange. Probably the most popular shear connector is the Nelson stud bolt, which is electric-welded to the beam flange by a welding gun that automatically produces the proper-size weld. The shear connectors are usually applied to the beams in the shop where labor is usually cheaper and working conditions more convenient.

There are several practical aspects governing the choice of shear-connector types. Steel erectors dislike working with beams bristling with studs and other shear connectors because they are hard to walk on and have a sinister ability to snare pant legs. More care must be taken with the shear-lugged beams, and it is more difficult to move them around without bending or breaking the shear connectors. These beams are generally more bothersome to ship, handle, and erect. These difficulties have on occasion led to a decision to place the shear connectors after the beams are erected. Another difficulty arises with the placing of the deck-reinforcing steel. Some connectors, such as hooks and spirals, require that the deck steel be threaded through the connectors. This greatly increases the difficulty of reinforcing steel placing. Larger connectors, such as tees and channels, occupy so much space that they interfere with the bars and disrupt their spacing. Ease of application and minimum interference with the deck-reinforcing steel are probably the main reasons why the studs are the most popular shear connectors.

For design purposes, the available shear area of a connector, if it is a stud, is based upon its cross-sectional area, not the area through the throat of the weld. On the other hand, if welded angles or channels are used as shear connectors, they must have at least $\frac{3}{16}$-in. (50-mm) fillet welds. For these connectors, the available shear area is based upon a cross section taken through the throat of the fillet weld.

It is important that there be adequate concrete cover around the shear lugs to avoid cracking. There should be at least 2 in. (5 cm) of concrete over the tops of the shear lugs, and

Fig. 8.5. Pico Lyons Bridge. *(Courtesy California Department of Transportation.)*

the shear lugs must penetrate at least 2 in. (5 cm) into the concrete slab above the surface of the top flange of the steel beam. Also, the clear distance between the edge of the girder flange and the edge of the shear connector must not be less than 1 in. (25 mm).

8.6.3 Effective Flange Width (AASHTO Specifications 1.7.98)

Composite construction converts a beam and slab into an effective T-beam unit. The effective width of the flange of the T-beam is limited to the smallest of:

1. one-fourth of the span length of the girder
2. the center-to-center distance of girders.
3. twelve times the least thickness of the slab.

When a girder has a flange on one side only, the effective flange width must not exceed the smallest of:

1. one-twelfth the span length of the girder
2. six times the thickness of the slab
3. one-half the center-to-center distance of the next girder.

8.6.4 Stresses (AASHTO Specifications 1.7.99)

If supports or other arrangements are not provided to make the girder act compositely for both live and dead loads, then the maximum tensile and compressive stresses are the sum of the stresses produced by the dead load acting on the girder alone and the stresses produced by the superimposed dead and live loads acting on the composite girder. If supports or other arrangements are provided and remain until the deck is fully cured and able to take its share of the composite load, then both the dead- and live-load stresses may be computed on the basis of the composite section.

In continuous spans, the positive moment portion of the beams may be designed with composite sections as in simple beams.

Whether shear connectors are to be provided in the negative moment area depends upon the design of the beam. If the reinforcing steel embedded in the concrete is considered a part of the composite section, then shear connectors should be provided in the negative moment area. However, if the reinforcing steel is not used in computing the beam properties for negative moments, the shear connectors need not be provided in the negative moment area. In that event, additional shear connectors will be required at the point of dead-load contraflexure.

8.6.5 Shear (AASHTO Specifications 1.7.100)

In positive moment areas, the spacing of shear connectors should not exceed 24 in. (60 cm). They must be designed for fatigue as suggested by Slutter and Green (1966) and checked for ultimate strength of the beam.

8.6.6 Fatigue

The range of horizontal shear on the beam is computed by the formula:

$$S_r = \frac{V_r Q}{I}$$

where

S_r = range of horizontal shear per linear inch (2.54 cm) at the junction of the slab and girder at the point in the span under consideration

V_r = range of shear due to live loads and impact (at any section, the range shall be taken as the difference in the maximum and minimum shear evelopes—excluding dead load)

Q = statical moment about the neutral axis of the composite section of the transformed compressive concrete area or the area of reinforcement embedded in the concrete for negative moment

I = moment of inertia of the transformed composite girder in the positive moment region or the moment of inertia provided by the steel beam including or excluding the area of reinforcement embedded in the concrete in the negative moment regions.

The allowable range of horizontal shear (Z_r) in pounds on an individual connector is:

For channels: $Z_r = (B)(w)$

Fig. 8.6. Bridge on Eel River Redwood Highway, US 101. *(Courtesy California Department of Transportation.)*

For welded studs: $Z_r = \alpha d^2$ (where $H/d > 4$)

where

w = length of the channel shear connector in inches measured in a transverse direction on the flange of the girder
d = diameter of the stud (in.)
α = 13,000 for 100,000 cycles
= 10,600 for 500,000 cycles
= 7850 for 2,000,000 cycles
B = 4000 for 100,000 cycles
= 3000 for 500,000 cycles
= 2400 for 2,000,000 cycles
H = height of stud in inches.

The spacing of the shear connectors is obtained by dividing the allowable range of horizontal shear of all connectors at one transverse girder cross-section (ΣZ_r) by the horizontal range of shear (S_r) per linear inch. Over the intermediate supports of continuous beams, it is permissible to modify the spacing to avoid placing connectors at points of high stress in the tension flange, provided the total required number of connectors is supplied.

8.6.7 Ultimate Strength (AASHTO Specifications 1.7.100 [A] [2])

The ultimate strength of the beam must be determined and the shear connectors checked to be certain there are enough provided to attain the failure limit. It is required that the number of shear connectors (N) between the points of maximum positive moment and the end supports or dead-load points of contraflexure shall equal or exceed the number given by:

$$N = \frac{P}{\phi S_u} \qquad (8.1)$$

where

S_u = the ultimate strength of the shear connector
ϕ = a reduction factor = 0.85

P = force in the slab defined as P_1, P_2, or P_3.

At points of maximum moment, the force in the slab is taken as the smaller of the values of P_1 or P_2:

$$P_1 = A_s F_y; \quad P_2 = 0.85 f_c'(b)(c) \quad (8.2)$$

where

A_s = total area of steel section including cover plates
F_y = specified minimum yield strength of steel being used
f_c' = compressive strength of the 28-day concrete
b = effective flange width
c = thickness of the concrete slab.

At points of maximum negative moment, the force in the slab is:

$$P_3 = A_s^r F_y^r \quad (8.3)$$

where

A_s^r = total area of reinforcing steel at the interior within the effective flange width
F_y^r = specified minimum yield of the reinforcing steel.

The ultimate strength of a shear connector (S_u) is given by:

$$\text{Channels:} \quad S_u = 550(h + t/2)(w)\sqrt{f_c'}$$

$$\text{Welded studs:} \quad S_u = 930 d^2 \sqrt{f_c'} \quad (8.4)$$

where

S_u = ultimate strength of individual shear connector (lb)
h = the average flange thickness of the channel flange (in.)
t = thickness of the web of the channel (in.)
w = length of channel shear connector (in.)
f_c' = compressive strength of 28-day concrete (psi)
d = diameter of the stud (in.)

Use 45.67 and 77.22 in place of 550 and 930, respectively, if f_c' is given in MPa, and h, t, d, and w in cms.

8.6.8 Additional Connectors to Develop Slab Stress (AASHTO 1.7.100 [A] [3])

It is also required that there be additional shear connectors provided at points of contraflexure to develop the slab stress when the reinforcement embedded in the concrete is not used in computing the strength of the section in resisting negative moments. If one thinks of the two beam elements as independently expanding and contracting members depending upon their state of tension or compression, it becomes obvious that the main function of the shear connector is to lock the two parts together so that they will act in unison. Beams deficient in shear connectors will pull apart and either hump up or crack open.

The number of additional shear connectors required assuming no help from the deck steel is:

$$N_c = \frac{A_r f_r}{Z_r} \quad (8.5)$$

where

N_c = number of additional connectors for each beam at the point of contraflexure
A_r = total area of slab reinforcement steel for each beam over interior support
f_r = range of stress due to live load plus impact in the slab reinforcement over the support (unless this range has been computed, it may safely be assumed to be 10,000 psi)
Z_r = allowable range of horizontal shear on an individual shear connector.

These additional shear connectors (N_c) are to be placed adjacent to, on either side of, or centered on the point of dead-load contraflexure within a distance equal to $\frac{1}{3}$ of the effective slab width.

8.6.9 Vertical Shear (AASHTO Specifications 1.7.100 [B])

For purposes of design, it may be assumed that the web of the supporting steel girder carries the total external shear, neglecting the effect of the steel flanges and the concrete deck slab. The shear is assumed to be distributed uniformly throughout the gross area of the web.

8.6.10 Deflection (AASHTO Specifications 1.7.101)

The deflection of any vehicular span should be less than 1:800 of the span length. Where pedestrians are to use the bridge, the better ratio is 1:1000. The deflection of cantilever arms should be limited to 1:300, raising this to 1:375 where pedestrians are concerned.

Conformance with these AASHTO deflection limits will not necessarily guarantee that the structure will not be objectionably limber. Long-span pedestrian structures may meet the requirement for static load deflection and still be subject to objectionable oscillations under a pulsating load, such as pedestrian traffic. A far safer approach is to determine both the frequency and amplitude of the structure and make sure that these values are within the ranges tolerated by humans. Wright and Green (1959, 1964) consider in detail human tolerance to vibration. Unless a designer has had experience with long and slender structures, it is possible to meet all the static requirements and still have a structure that gallops or sways under pulsating loads such as pedestrians or wind. If there is doubt, the structure should be checked to see that it has adequate stiffness to resist undulation.

When computing deflection, the gross cross-sectional area of the beam or girder is used, with the composite section considered as resisting the live loads. Until the concrete deck has achieved at least 75% of its 28-day strength, all loads are to be considered as carried by the girder alone.

8.7 SHEAR CONNECTORS

Although it is known from practical experience that composite action occurs many times where it is not expected and might be considered as fairly easy to achieve when composite action is relied upon in design, sufficient shear connectors must still be provided to positively develop the required horizontal shear force. Without the shear connectors, the bond between the slab and girder gradually loosens with movement and age, and the composite action might disappear entirely.

Shear connectors can take on a variety of forms depending upon the type of construction. For steel beams with concrete slab decks, Fig. 8.7 shows a variety of steel devices that have been used as shear connectors. All of these are welded to the steel top flange. Gravel embedded in epoxy has also been used on top flanges to create shear resistance between the beam and the concrete slab. Although this would seem to have advantages, it also has drawbacks. True, it does permit the beams to be erected without the nuisance of the shear connectors, but the application of the epoxy just before the reinforcing steel is placed is a messy job. Also, to insure the embedment of the gravel, considerable care must be exercised; thus, it seems likely that the apparent saving over welded connectors is not as much as would first appear.

For concrete beams with concrete slabs, shear devices also must be provided. If the beams have been cast smooth, then the shear devices used are usually either embedded reinforcing bars: dowels, half-buried spirals, or ends of stirrups. It used to be common practice to create keys in the top of the beams, which were usually cast before the deck slabs, by pushing two-by-fours (5 × 10 cm), about 1 foot (30 cm) long, down into the wet concrete. When removed later, these left a depressed key. Beams also used to be castellated on top by placing intermittent blocks across the full width of the forms. It was found, however, that these heavy keys were not necessary, and a good roughness such as that developed by sandblasting or wire-brushing the soft concrete to expose the coarse aggregate will provide adequate shear resistance. Care must be taken that the top surface of the concrete beam is free of laitance, dust, form oil, grease, and other debris when the concrete deck is placed, because these contaminants destroy bond.

In timber construction, where the concrete deck slab is placed on timber beams, the timber beams are often sawed down a few inches into the top surface at 12-to-18-in. (30-to-45-cm) intervals and the alternate blocks cut out to create a castellated pattern to develop the shear. On simple bridge construction, the shear has been developed by driving a number of spikes into the top of the timber beam and placing the concrete around them.

Where partial-thickness slabs are used as both form and the bottom half of the slab, there must be provision for adequate shear transfer between the two halves of the slab. Frequently, steel spirals are used, burying half of the diameter in the lower slab and leaving the upper half exposed to go into the subsequently placed upper portion of the slab. Exposed stirrups, truss bars, and other devices have been used. Also, there must be positive efforts to roughen the top surface of the bottom half of the slab to enhance the shear bond between the two sections. While it is probably true that the roughness of the concrete alone is enough to transmit shear in the ordinary stress ranges, it is important that the devices also tie the two elements together. Shrinkage and drying will almost invariably cause a crack along the construction joint and, without a positive hold-down restraint, the two sections of the slab may separate.

Probably the principal cause of failure of shear connectors is fatigue. It is therefore important that the connectors be properly designed to resist the fatigue stresses. It will frequently be found that the fatigue requirements will govern the design. Therefore, it is not possible to compute the stresses in the shear connectors from theory based upon static strength and slip criteria alone. Even though it appears that the connectors will be operating at a very low stress level, it is no assurance that fatigue is not a problem. Should the possibility of fatigue be neglected, there could very likely be a fatigue failure of the shear connectors near midspan even though the calculations might show the total number of shear connectors provided to be adequate.

Paragraph 1.7.100 of the AASHTO Specifications gives detailed directions for the calculations related to shear connectors both in fatigue conditions and to meet the ultimate strength of the beam. The fatigue design is accomplished by limiting the magnitude of the maximum shear in the connectors in the working-load range. This magnitude of maximum shear is determined as the algebraic difference between the minimum and maximum shear experienced at any given point in the beam. If an envelope curve is drawn for the maximum and minimum shears for a simple span up to about 140 ft (43 m), the variation in the algebraic difference across the span will be quite small. Therefore, in the common span lengths for highway bridges, which are most frequently around 80 ft (24 m), it is practical to use the maximum shear range calculated for the entire length of span. The required spacing may then be checked at midspan and near the supports, and, if the difference is small, it may be economical to use the same spacing as calculated for the support area across the entire span. Small variations in shear-connector spacing will only increase the fabrication costs—hence, many designers prefer to use a uniform spacing of connectors throughout the length of simple span beams. If a greater degree of economy is desired, dividing the beam into three regions—the two end quarters and the center span half—should be adequate. Spacing should never be greater than 24 in. (61 cm), except over interior supports of continuous beams where wider spacing may be used to avoid placing shear connectors at locations of high stress in the tension flange.

8.8 SHEAR-CONNECTOR SPACING IN CONTINUOUS SPANS

In continuous spans, as in simple spans, the shear-connector spacing is computed from the range of horizontal shears. In the outer quarters of the positive-moment region, the range of shears calculated for the support points or points of contraflexure are used. The connector spacing for the center portion of the positive-moment region is computed using the range of horizontal shear at the midpoint. The spacing should not exceed 24 in. (61 cm) in the positive-moment region.

In negative-moment areas, the range of horizontal shears at the points of contraflexure and over the interior supports should be calculated. When the design is governed by the fatigue reduction of stresses in the top flanges, the shear connectors should be eliminated from the areas immediately adjacent to the girder supports. In this case, the total number of connectors provided becomes more important than their spacing. The same total number of connectors must

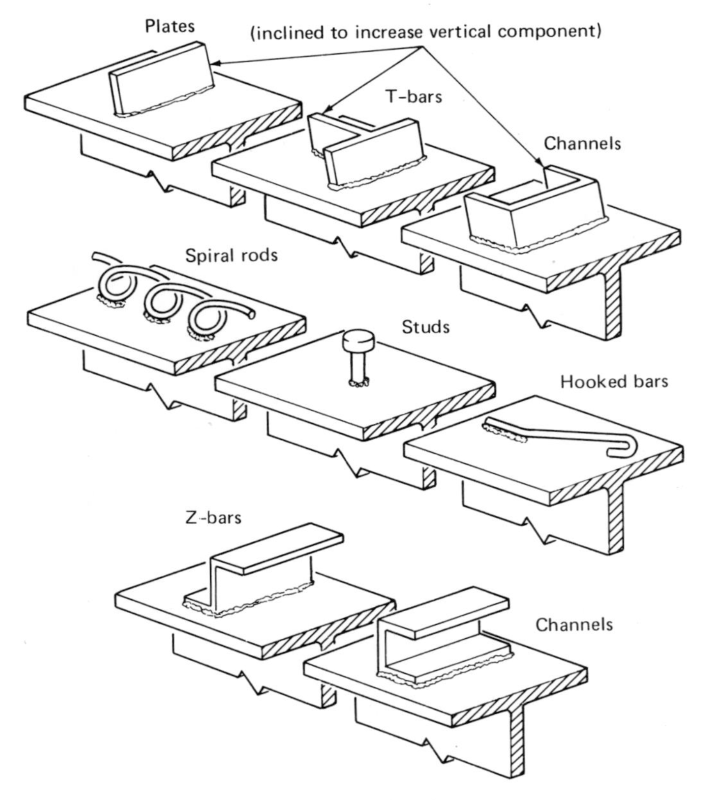

Fig. 8.7. Some devices used as shear connectors.

be provided but they are spaced back from the highly stressed tension areas in the top flange. However, this should not result in going to the other extreme and making a large grouping of connectors around the inflection points.

Under the AASHTO Specifications, the fatigue reduction for ASTM A-514 or A-517 steel is so severe that the same number of shear connectors required would be the same as required for ASTM A-36 steel. Thus, there is little advantage in designing a high-strength steel beam to act compositely in the negative-movement areas.

8.8.1 Example of a Composite Continuous Girder Design

This example explains the procedures for designing a three-span continuous steel girder bridge: Spans 80-82-80 ft (24-25-24 m), 38′-6″ (11.7 m) roadway, four girders spaced 10′-6″ (3.2 m), AASHTO H20-44 loading, 500,000 cycles for fatigue range.

Assume top flange width = 12 in. (30 cm)
Effective span length of deck slab = c/c span − $\frac{1}{2}$ top flange width
$= 10.5 - \frac{1}{2}(1.0) = 10.0$ ft (305 cm)

From standard slab design tables, the following details are determined: slab thickness = 8 in. (20 cm), top bars = #6 @ 12 in. (30 cm), bottom bars = #6 @ 12 in., truss bars = #6 @ 12 in. (30 cm).

Calculating the cantilever moment:

$$M_{\text{rail}} = (0.36)(3.42) \qquad = 1.23 \text{ kft}$$
$$M_{\text{slab}} = (0.58)(4.25)(2.125)(0.15) \qquad = 0.79$$

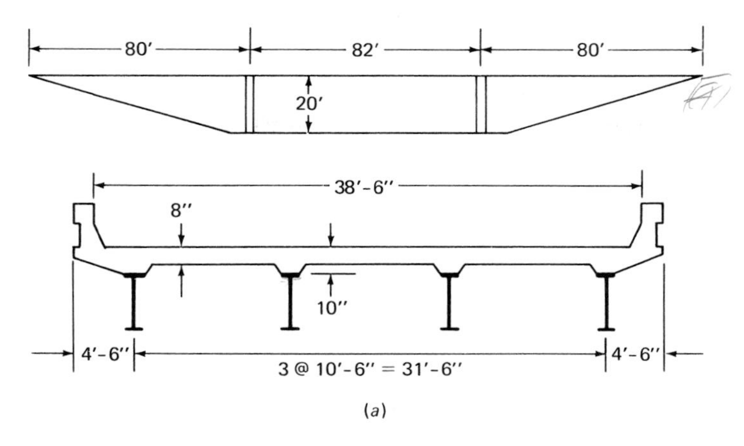

Effective span (S) = $10.5 - \frac{1}{2}$ Top flange width

$S = 10.5 - \frac{1}{2}(1.0) = 10.0''$ Deck thickness = 8″

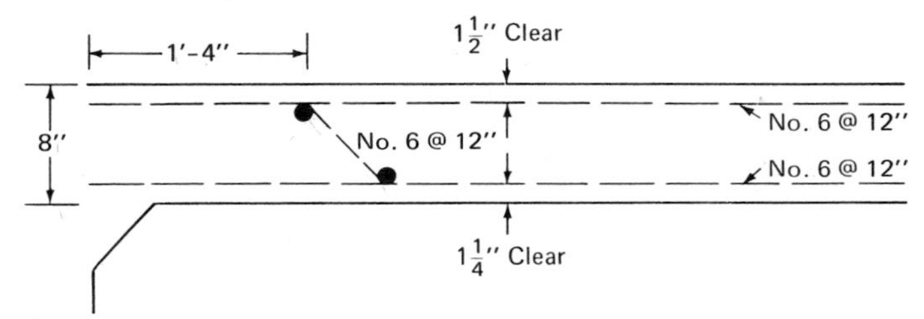

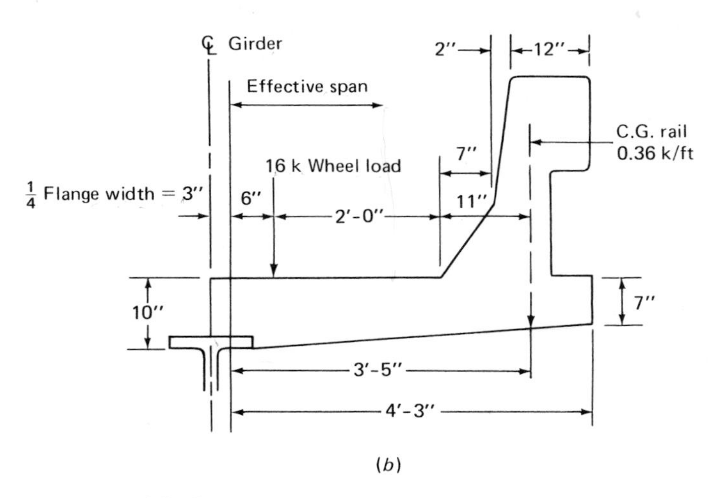

Fig. 8.8. Example of composite highway bridge design.

$$= (0.25)(4.25)(4.25/(3)(2))(0.15) = 0.11$$

$$M_{LL+I} = \frac{(16)(1.3)(0.5)}{(0.8)(0.5) + 3.75} = 2.51$$

Cantilever moment = 4.64 kft (6.29 mkN)

Required steel = $A_s = M/f_s jd$ = (4.64)(12)/(0.875)(20)(7.5) = 0.424 in.^{2}/ft (0.09 cm^2/cm)

Two #6 bars/ft. = 0.88 in.^{2}/ft (0.186 cm^2/cm) (well above required).

Girder design:

Interior girder dead load:

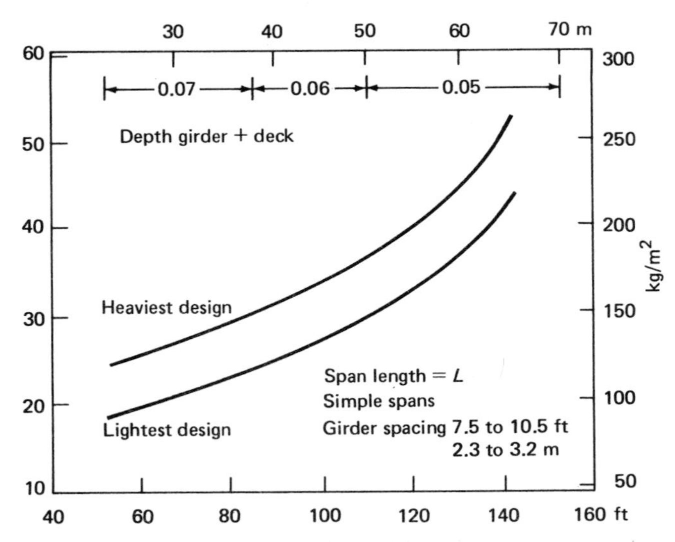

Fig. 8.9. Estimated weight per foot of composite welded plate girders including stiffeners, cross frames, diaphragms, and shear connectors.

Slab = (0.67)(10.5)(0.150) = 1.055 k/ft
Fillet = (0.17)(1.33)(0.150) = 0.034
Rail ($\frac{1}{4}$ to each girder) = 0.36/2 = 0.180
Allowance for welded girder (Fig. 8.4) = 0.273
(0.026)(10.5)
1.542 k/ft (22.5 kN/m)

Exterior girder dead load:

Slab = (0.667)(5.25)(0.150) = 0.525 k/ft
Fillet = (0.167)(1.333)(0.150) = 0.033
Overhang = (0.583 + 0.833)(4.25)(0.150)/2 = 0.451
Allowance for welded girder *DL* (Fig. 8.4) = 0.273
Rail ($\frac{1}{4}$ to each girder) = 0.36/2 = 0.180
1.462 k/ft (21.3 kN/m)

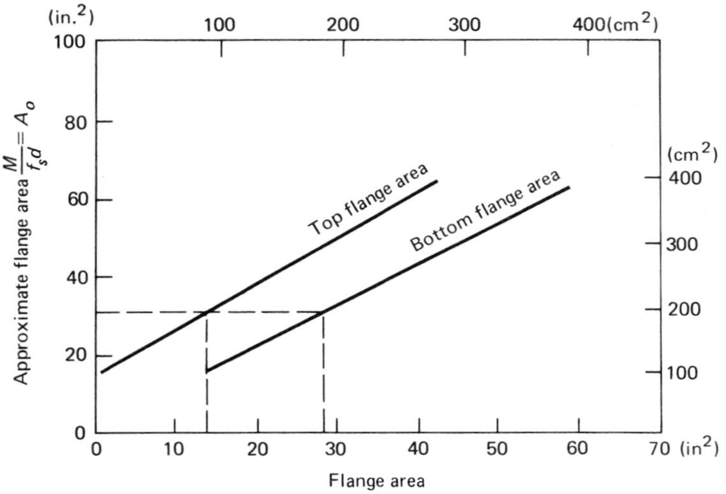

Fig. 8.10. Approximate flange areas—Composite girders.

Live load:

S/5.5 wheel lines = S/11 lanes

Impact = 50/L + 125 = 50/207 = 0.242; (midspan) = 50/41 + 125 = 0.300 (near support)

	Interior Girder L.L.	Exterior Girder L.L.
Lanes per girder =	10.5/11 = 0.955	9.50/11 = 0.864

Center span moments:

	Interior		Exterior	
$M_{\text{slab+girder}}$	= 1.363(82²)/8	= 1145	1.282(82²)/8	= 1078
M_{rail}	= 0.18(82²)/8	= 151	0.18(82²)/8	= 151
$M_{\text{LL+I}}$	= 1201(1.242)(0.955)	= 1425	1201(1.242)(0.864)	= 1289
		2721 kft		2518 kft

Use the interior girder moment = 2721 kft (3690 mkN)

Effective flange width (AASHTO 1.7.98) = $12t = 12(8) = 96$ in. (240 cm)
Equivalent flange width ($n = 30$) = Effective width/n = 96/30 = 3.2 in. (8 cm) (for dead load)
Equivalent flange area = Equivalent slab width(t) = 3.2(8) = 25.6 in.² (160 cm²)
I of equivalent slab area = $w(t^3)/12 = 3.2(8^3)/12 = 136.53$ in.⁴ (5333 cm⁴)
Equivalent flange width ($n = 10$) = Effective width/n = 96/10 = 9.6 in. (24 cm) (for live load)
Equivalent flange area = Equivalent flange width(t) = 9.6(8) = 76.8 in.² (480 cm²)
I of equivalent flange area = Equivalent flange width(t^3)/12 = $9.6(8^3)/12 = 409.6$ in.⁴ (16,000 cm⁴)
Equivalent flange width ($n = 10$) = Effective width/n = 96/10 = 9.6 in. (24 cm) for live load
Equivalent flange area = Equivalent flange width (t) = 9.6(8) = 76.8 in.² (495 cm²)
I of equivalent flange area = Equivalent flange width(t^3)/12
= $9.6(8^3)/12 = 409.6$ in.⁴ (17,000 cm⁴)

Preliminary Design

Depth of girder/span length = $D/S = 0.06$, (Fig. 8.4). Span = 82 ft (25 m)
$D = 0.06(82) = 4.92$ ft. Say 5.0 ft = 60 in. (150 cm)
Girder with 50-in. web + 10-in. deck and fillet = 60 in.

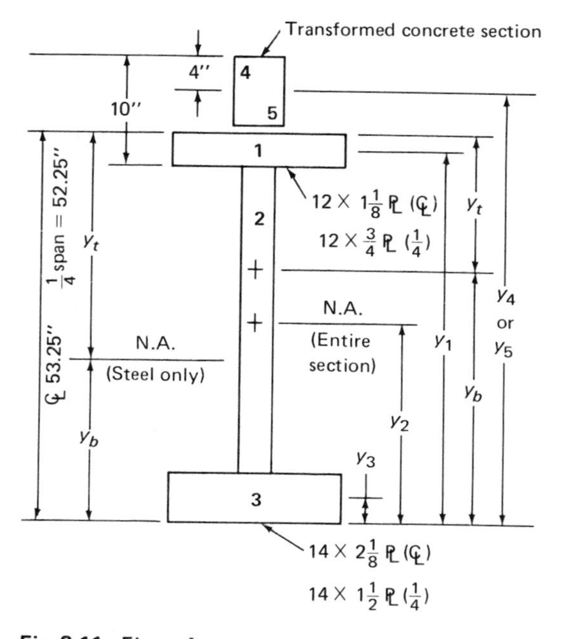

Fig. 8.11. Figure for example—Transformed section.

Area of flange (A_f) = $M/f_s d$ (Approximate design formula)

$$M = M_{DL} + M_{LL+I} = 2721 \text{ kft } (3690 \text{ mkN})$$

$A_f = (M)(12{,}000)/f_s d = (2721)(12{,}000)/20{,}000(53) = 30.80$ in.^{2}
Then from Fig. 8.8, $A_{\text{top flange}} = 13.5/12 = 1.13$ in Try $12 \times 1\frac{1}{8}$-in. plate (30 × 3 cm)
$A_{\text{bottom flange}} = 28.0/14 = 2.0$ in Try $14 \times 2\frac{1}{8}$-in. plate (35 × 6 cm)
Web plate thickness = $D\sqrt{f_b}/23{,}000 = (50)(\sqrt{20{,}000})/23{,}000 = 0.307$ in. (0.78 cm) (AASHTO 1.7.70)
Web plate $50 \times \frac{5}{16}$ in. $t = 0.3125$ (127 × 0.8 cm)
Limits are $D/170$ and $D/165 = 0.303$ in. Satisfactory.
(0.77 cm)

Composite girder design: Midspan section
Section for slab and girder loads (see Fig. 8.6):

Member	Assumed Size (in.) (cm)	Area (in.^{2})	y (in.)	Ay (in.^{2})	Ay^2 (in.^{4})
① Top flange	$1\frac{1}{8} \times 12$ (3 × 30)	13.50	52.69(y_1)	711	37,479
② Web	$\frac{5}{16} \times 50$ (0.8 × 127)	15.63	27.125(y_2)	424	11,500
③ Bottom flange	$2\frac{1}{8} \times 14$ (6 × 35)	29.75	1.0625(y_3)	32	34
		Σ = 58.88		1167	49,013

$y_b = \Sigma Ay/\Sigma A = 19.82$ in. $I_{\text{o-o web}} = td^3/12$ = 3,255
$y_t = 33.43$ in. $I_{\text{steel about base}}$ = 52,268
Less $y_b^2 \Sigma A$ = −23,129
$I_{\text{steel about neutral axis}}$ = 29,138 in.^{4} (1,212,000 cm^4)

Section for curb, railing, and utilities:

	Area	y	Ay	Ay^2
	Σ = 58.88		1167	49,013
④ Concrete section ($3n$) = $12d(d)/3n = 12(8^2)/30$	= 25.60	58.125(y_4)	1488	86,490
	Σ = 84.48		2655	135,503

$I_{\text{o-o web}}$ = 3,255
$I_{\text{slab}} = 12(8^4)/30(12)$ = 137
= 138,895

$y_b = \Sigma Ay/\Sigma A = 2655/84.48 = 31.43$ in. (79.8 cm)
$y_t = 53.25 - 31.43 = 21.82$ in. (56.2 cm)
Less $y_b^2 \Sigma A$ = −83,453
$I_{\text{composite about neutral axis}_{DL}}$ = 55,442 in.^{4} (2,307,000 cm^4)

Section for live load:

	Area	Ay	Ay^2
Previous totals	58.88	1167	49,013
⑤ Concrete section$_{LL}$ ($n = 10$) = $12(8^2)/10$ =	76.80	4464	259,470
	Σ = 135.68	5631	308,483

$I_{o-o\ web}$	=	3,255
$I_{slab} = 12(8^4)/10(12)$	=	410
$I_{about\ base}$	=	312,148

$y_b = \Sigma A/\Sigma A = 5631/135.68 = 41.50$ in.

$y_t = 53.25 - 41.50 = 11.75$ in.

Less $y_b^2 \Sigma A$	=	−233,675
$I_{composite\ about\ neutral\ axis_{LL}}$	=	78,473 in.^{4} (3,270,000 cm^4)

Moments at center of span:

Slab and girder = w $L^2/8 = (1.362)(82^2)/8$ = 1145 ft-kips (1550 mkN)
Curb and utility = $(0.18)(82^2)/8$ = 151 ft-kips (205 mkN)
Live loads = (1201)(1.242)(0.94) = 1402 ft-kips (1900 mkN)

Stresses at midspan:		Bottom Flange	Top Flange
Slab and girder:			
Bottom flange:	1145(12)(19.82)/29138 =	9.35 ksi (64.5 MPa)	
Top flange:	1145(12)(33.43)/29138 =		15.76 ksi (108.6 MPa)
Curb and utility:			
Bottom flange:	151(12)(31.43)/55442 =	1.03 (7.1)	
Top flange:	151(12)(21.82)/55442 =		0.71 (4.9)
Live loads:			
Bottom flange:	1402(12)(41.50)/78473 =	8.90 (61.4)	
Top flange:	1402(12)(11.75)/78473 =		2.52 (17.4)
	Total stress bottom flange =	19.28 ksi (133.0 MPa)	
	Total stress top flange =		18.99 ksi (130.9 MPa)

Meets 20 ksi working-stress limit.

Quarter-span section.
Section for slab-and-girder loads:

Member	Assumed Size	Area (in.^{2})	y (in.)	Ay (in.^{2})	Ay^2 (in.^{4})
① Top flange	$\frac{3}{4} \times 12$	9.00	51.88(y_1)	467	24,228
② Web	$\frac{5}{16} \times 50$	15.63	26.50(y_2)	414	10,971
③ Bottom flange	$1\frac{1}{2} \times 14$	21.00	0.75(y_3)	16	12
		= 45.63		897	35,211

$y_b = \Sigma Ay/\Sigma A = 897/45.63 = 19.66$ in.

$y_t = 52.25 - 19.66 = 32.59$ in.

$I_{o-o\ web}$	=	3,255
I_{base}	=	38,466
Less $y_b^2(\Sigma A) = 19.66^2(45.63)$	=	−17,637
$I_{steel\ about\ neutral\ axis}$	=	20,829 in.^{4} (867,000 cm^4)

Section for curb, railing, and utilities:

Member	Assumed Size	Area (in.^{2})	y (in.)	Ay (in.^{2})	Ay^2 (in.^{4})
		$\Sigma = 45.63$		897	35,211
④ Concrete section ($3n$) $= 12d(d)/3n = 12(8^2)/30 =$		25.60	57.5(y_4)	1472	84,640
		71.23		2369	119,851
			$I_{o-o\ web}$	=	3,255
			$I_{slab} = 12(8^4)/30(12)$	=	137
					123,243 (5,130,000 cm^4)

$y_b = \Sigma Ay/\Sigma A = 2369/71.23 = 33.26$ in.
$y_t = 52.25 - 33.26 = 18.99$ in.

Less $y_b^2 \Sigma A = 33.26^2(71.23) = -78,797$
$I_{\text{composite about neutral axis}_{DL}} = 44,446$ in.^{4} (1,850,000 cm^4)

Section for live load:

Member	Area (in.^{2})	y (in.)	Ay (in.^{2})	Ay^2 (in.^{4})
	$\Sigma = 45.63$		897	35.211
⑤ Concrete section$_{LL}$ ($n = 10$) $= 12(8^2)/10 =$	76.80	57.5(y_4)	4416	253,920
	122.43		5313	289,131
		$I_{o-o\ web}$	=	3,255
		$I_{slab} = 12(8^4)/10(12)$	=	410
				292,796 (1,219,000 cm^4)

$y_b = \Sigma Ay/\Sigma A = 5313/122.43 = 43.40$ in.
$y_t = 52.25 - 43.40 = 8.85$ in.

Less $y_b^2 \Sigma A = 43.4^2(122.43) = -230,604$
$I_{\text{composite about neutral axis}_{LL}} = 62,192$ in.^{4} (2,590,000 cm^4)

Stresses at quarter-span: (Moments 75% of midspan moments)

Slab and girder:
- Bottom flange: 859(12)(19.66)/20,829 = 9.73 ksi (67.1 MPa)
- Top flange: 859(12)(32.59)/20,829 = 16.13 ksi (111.2 MPa)

Curb and utility:
- Bottom flange: 113(12)(33.26)/44446 = 1.01 (7.0)
- Top flange: 113(12)(18.99)/44446 = 0.58 (4.0)

Live loads:
- Bottom flange: 1052(12)(43.40)/62192 = 8.81 (60.7)
- Top flange: 1052(12)(8.85)/62192 = 1.80 (12.4)

Total stress, bottom flange = 19.55 ksi (134.8 MPa)

Total stress, top flange = 18.51 ksi (127.6 MPa) Satisfactory.

Shear connectors.

Shear range:

Shear caused by L.L. at support = 63.8 k. At midspan = 27.8 k (123 kN).
Maximum shear$_{LL+I}$ at support = 63.8(1.242)(0.955) = 75.7 k (337 kN)
Maximum shear$_{LL+I}$ at midspan = 27.8(1.300)(0.955) = 34.5 k (154 kN)

Horizontal shear range (at support):

$S_{r(LL+I)} = V_r Q/I = 75.7(76.8)(14.1)/62{,}192 = 1.32$ k/in.

$Q_{LL} = Ay = A(y_t + 10 - [t + 4]) = 76.8(8.85 + 10 - [0.75 + 4])$
$= 76.8(14.10)$

$S_{r(\text{Railing})} = 7.38(25.6)(24.24)/44{,}446 = 0.10$ k/in.

$Q_{\text{Railing}} = 25.6(18.99 + 10 - [0.75 + 4]) = 25.6(24.24)$

Horizontal shear range (at midspan):

$S_{r(LL+I)} = 69.0(76.8)(16.63)/78473 = 1.2$ k/in.

$Q_{LL} = 76.8(11.75 + 10 - [1.125 + 4]) = 76.8(16.63)$

Shear connector spacing:

$$\text{Pitch of connectors} = \frac{\text{Range of horizontal shear of all connectors at girder cross section}}{\text{Horizontal range of shear per linear inch}}$$

$P = \Sigma Z_r/S_r$

Assume: Two $\frac{7}{8}$-in.-diam. studs per section
Fatigue capability of 500,000 cycles
H = height of connector stud = 6 in.

then

$H/d = 6/7/8 = 7$

which is greater than the required 4

Z_r = allowable design load range = $\propto$ diam^2
$= 10{,}600(\frac{7}{8})^2 = 8115$ lb $= 8.1\ k$

Spacing at supports = P = 8.1(2)/1.32 = 12.27 in. (Pairs of $\frac{7}{8}$-in.-diam. studs)
Use 12-in. center-to-center in end quarters. Total required = 84 studs.
Spacing at midspan = P = 8.1(2)/1.12 = 14.46 in. Use 14$\frac{1}{2}$ in. (1.21 ft) (36 cm)
Use 14$\frac{1}{2}$ in. center-to-center in midspan half. Total required = 41(2)/1.21 = 68 studs.
Total number of studs for beam, required by fatigue = 84 + 68 = 152

Check ultimate strength **(AASHTO 1.7.100)**

Check midspan section as the most severe location.

$$P_1 = A_s f_y$$

A_s = total area of steel section = 58.88 in.^{2}
f_y = steel yield strength = 36,000 psi

$$P_1 = 58.88(36.0) = 2120\ k\ (9430\ \text{kN})$$

$$P_2 = 0.85 f'_c bc$$

f'_c = 3250 psi (22.4 MPa)
b = effective flange width = 96 in. (240 cm)
c = depth of slab = 8 in. (20 cm)

$$P_2 = 0.85(3.25)(96)(8) = 2122\ k\ (9440\ \text{kN})$$

Minimum P to be used = 2120 k (9430 kN)
Number of connectors required = $N = P_{\min}/\phi S_u$

$\phi = 0.85$
$S_u = 930d^2 \sqrt{f'_c} = 930(0.875^2)(\sqrt{3250}) = 40{,}592 \text{ lb} = 40.6\ k\ (180 \text{ kN})$
$N = 2120/0.85(40.6) = 61$ studs for center half of span.

Since fatigue requires 68 studs in this area, fatigue governs. Use pairs of $\frac{7}{8}$-in.-diam. studs at $14\frac{1}{2}$ in. (36 cm) through the center of the span and pairs at 12-in. (30-cm) centers in the end quarters.

8.9 COMPOSITE BOX GIRDERS

8.9.1 Concrete Box Girders

Concrete box girders naturally work compositely by their very nature. The only care that must be taken to insure the composite action is that there be a proper shear bond and connection between the girders and the deck slab. Box girders are normally placed in at least two pours. It is common to place the bottom slab and the girders in one pour, let it set, and then remove the interior forms. Lightweight disposable forms are then placed to support the deck slab, and these forms are normally left in place. Several days may elapse between the time when the girders and the deck are placed. It is essential that the joint between the pours be clean and roughened sufficiently to provide good shear contact. It is a common practice to sand-

Fig. 8.12. Example of curved composite bridge. (*Courtesy California Department of Transportation.*)

blast or wire-brush this surface while the concrete is green in order to remove all laitance and provide clean aggregate to which the deck slab concrete can adhere. Chips, sawdust, and dirt should be blown out with compressed air just before the pour is started, and the concrete surface should be thoroughly wet. With these precautions, there should be no difficulty in assuring composite action.

8.9.2 Steel Box Girders

Considering composite action in steel box girders is merely an extension of the composite action of beams. The rules and restrictions provided for these steel box girders in the AASHTO Specifications have for some time been the only available guides to design. It should be noted that these specifications are restricted to bridges of "moderate span." Practically speaking, this is about a 400-ft (120-m) span. Some spectacular failures of long-span steel boxes around the world seem to indicate that these longer spans are pressing the limits of knowledge of the action of large plates and boxes. Designers working on steel box girders with spans over 400 ft (120 m) should consult the latest available information, such as the British Merrison Report and subsequent American adaptions of that report and its findings.

As an item of practical experience with steel box girders, a designer might be inclined to regard the composite section as very stable—having a three-sided steel box and the very stiff concrete deck on top, providing both stiffness and lateral support. However, during construction, the three-sided steel box is not a stiff torsion-resistant member, yet it is expected to carry the load of the wet concrete plus all the construction loads. Superelevation, curvature, and other spatial relationships tend to warp the box. This warping will cause both horizontal and vertical deflections and, under extreme conditions, might cause collapse. If unbalanced loads of any nature are anticipated upon the box girder before the concrete has attained its full strength (and it is hard to conceive of any situation where some unbalanced loads are not possible), the designer should provide a steel upper lateral system that is adequate to hold the steel box in shape until the concrete has set. This lateral system should be adequate to sustain all dead-load torsions as well as those created by construction loads.

For anyone designing a steel box girder of one or more cells, the AASHTO Specifications

Fig. 8.13. Kern River Bridge. *(Courtesy California Department of Transportation.)*

provide detailed design formulas for load distribution and the design of web and flange plates. Since these formulas are mainly concerned with the buckling of the plates and are only incidentally concerned with composite action, they will not be covered herein. The designer should review the AASHTO Specifications for this information.

One overall restriction is that the cantilever overhang of the deck slab, including curbs and parapets beyond the exterior girder, should be limited to 60% of the distance between the adjacent top steel flanges of adjacent box girders, but in no case greater than 6 ft (1.8 m).

8.10 HYBRID WELDED PLATE GIRDERS

The advent of welding changed the bridge girder picture materially to the point that the majority of steel girders used in bridges are now welded rather than being rolled beams. Some agencies are still using rolled beams with cover plates, but the rolled beams are not as economical as the welded beams which can be tailored to meet their stress needs. Welding has also made possible the hybrid steel beam, which combines a number of different stress grades of steel into one beam. The hybrid beam uses the grade of steel which matches the stress requirement, thus making possible large beams of constant physical section–large beams with nearly constant flange widths and flange thicknesses throughout their length.

As far as composite action is concerned, the interaction of the concrete deck with a hybrid girder is the same in theory as with an ordinary steel beam with the same grade of steel throughout. In fact, the action is so similar that the benefits of the high-strength steel are often lost because of the fatigue limitations. Thus, compositely speaking, the hybrid beams have little advantage over the lower strength beams.

BIBLIOGRAPHY

AASHTO (1973), *Standard Specifications for Highway Bridges*, American Association of Highway and Transportation Officials. Eleventh Edition.

AASHTO (1974), *Interim Specifications–Bridges*, American Association of Highway and Transportation Officials.

Aneja, I. and Roll, F. (1969), An Experimental and Analytical Investigation of a Horizontally Curved Box-Beam Highway Bridge Model, *Proceeding of Second International Symposium on Concrete Bridge Design*, ACI Publication SP-26, Detroit, Michigan.

California Department of Transportation (1961), *Manual of Bridge Design Practice*, California Department of Transportation Bridge Department. Third Edition.

Elliott, A. (1977), Bridges, *Structural Engineering Handbook*, Section 18, McGraw-Hill Publishing Co.

Heins, C. P. (1976), *Bending and Torsional Design In Structural Members*, Lexington Books, Boston, pp. 367.

Hsu, T. T. C. (1968a), Torsion of Structural Concrete–Plain Concrete Rectangular Sections, *Torsion of Structural Concrete*, ACI Publication SP-18, Detroit, Michigan.

Hsu, T. T. C. (1968b), Torsion of Structural Concrete–Behavior of Reinforced Concrete Rectangular Sections, *Torsion of Structural Concrete*, ACI Publication, SP-18, Detroit, Michigan.

Slutter, R. G. and Green, R. (1966), *Fatigue Strength of Shear Connectors*, Highway (no Transportation) Research Board, Washington, D.C.

Timoshenko, S. P. and Woinowsky-Krieger, S. (1959), *Theory of Plates and Shells*, McGraw-Hill Book Co., New York.

Viest, I. (1977), Composite Construction, *Structural Engineering Handbook*, Section 14, McGraw-Hill Publishing Co.

Vlassov, V. F. (1961), *Thin Walled Elastic Beams*, National Science Foundation Translation, Washington, D.C. Second Edition.

Wright, D. T. and Green, R. (1959), *Human Sensitivity to Vibration*, Queen's University, Kingston, Ontario. Report No. 7.

Wright, D. T. and Green, R. (1964), *Highway Bridge Vibrations, Part II, Ontario Test Program*, Queen's University, Kingston, Ontario, Canada. Report No. 5, May.

9

Composite behavior of walls and frames in multistory buildings

GAJANAN M. SABNIS, Ph.D., P.E.
Associate Professor of Civil Engineering
Howard University
Washington, D.C.

9.1 INTRODUCTION

The growth of multistory buildings in the last several decades is seen as part of the necessity for vertical expansion for businesses as well as for residences in major cities around the globe. On the other hand, the technological development of higher strength materials and new innovative ideas were equally responsible for such a growth. Eventually, in the late fifties and sixties, such vertical development in buildings became a symbol of prestige for many cities. A study undertaken by the author indicates that skylines of several cities around the world have building heights ranging from a low 20 stories to a mighty 120 stories, with the majority having a medium height of 40 to 50 stories (Sabnis 1976). Therefore, it appears to be sensible to further investigate the economics of the large number of tall structures around this medium height.

This chapter will attempt to bring out the main points of structural economy due to the widely discussed but little used interaction between the masonry walls (or nonstructural walls) and frames, which may be used by practicing engineers. Thus, the information presented here is intended to illustrate to designers some benefits of the composite frame and wall interaction, thereby permitting them to make their own judgments on the validity and use of such composite systems in design, since such information is not presented in standard practice.

The economics of the structural design of multistory buildings certainly is governed by the efficiency of the structural system used in transmitting the applied loads to foundation, particularly the horizontal loads due to the heavy winds and in some places due to earthquake forces. In order to design the structure for various applied forces, one could simply separate the load-carrying elements in the structure on an appropriate basis to allow columns and slabs to resist the vertical loads and (shear) walls to carry horizontal loads. An economical combination of columns, walls, and floors should constitute the best structural system. However, development of the conventional load-bearing wall structures and the frame structures with shear wall took place separately due to other influencing factors. Many frame systems were developed and successfully used in the 1960s; they were excellent from the point of view of designing "tall" buildings of over 100

Fig. 9.1. Sears Tower Building, Chicago, Illinois. *(Courtesy Harr, Hedrich-Blessing, Chicago, Illinois.)*

stories,* after the earlier development of structural masonry, due to its massiveness, was found to be out of phase with buildings taller than 15 to 20 stories. Despite this shortcoming, masonry walls were later used for structural purposes, after higher strength materials were developed and combined with floors of various types in buildings up to 25 stories. Excellent examples are the Sears Towers of Chicago (140 stories), which uses the modified *frame* ("tube-in-tube") *structure* (Fig. 9.1), and the Park Lake Towers in Denver with 20 stories using *structural masonry walls and reinforced concrete floors* (Fig. 9.2).

*Excellent reviews of such systems may be found in Fintel (1974) and the Conference on Planning and Design of Tall Buildings (1973) and have been briefly overviewed in Chapter 1.

Medium-height buildings with 30 to 40 stories can be designed and built successfully using the composite behavior of frame (either steel or concrete) and "infill"-type walls, which give the additional stiffness required for part of the structural action. In tall buildings, the concept of the modified frame (or tubular system) is based on the rigidity and the stiffness of the frame itself, which makes the size of frame members quite large. In medium-height buildings, on the other hand, the bare frames are fairly flexible and often have to depend on (structural) shear walls for transmitting horizontal loads to foundation. The infill walls can stiffen such flexible frames using suitable structural elements as masonry or reinforced concrete panels by interacting with them as a composite system. Interaction can take place among frame, floors, and walls, the three main components of the structure. A brief discussion of these components follows.

Frames by themselves are normally designed in low-rise buildings to resist both the vertical and horizontal loads. For medium-height buildings, however, a combination of frame and shear wall constitutes a very useful system. A study by Konig (1972) shown in Fig. 9.3 indicates the effectiveness of the sharing of loads between shear walls and frames.

The *wall*, in the present context, is considered either as a shear wall or as an infill wall. Although the effectiveness of the latter depends considerably on the connections between the infills and the surrounding frames, they nevertheless represent a large percentage of stiffness in a composite system between a frame and the wall, and hence carry out the function of a member resisting lateral load.

Floors at various levels serve as diaphragms necessary to transfer the horizontal forces to the lateral load-resisting elements and are further useful for the stability of columns. They may be used either as cast-in-place floors or other similar topping on the precast ribbed metal floor units.

The interaction among the three main struc-

Fig. 9.2. Twenty-story masonry building, Park Lake Towers, Denver, Colorado. (*Courtesy A. Yorkdale, Brick Institute of America, McLean, Virginia.*)

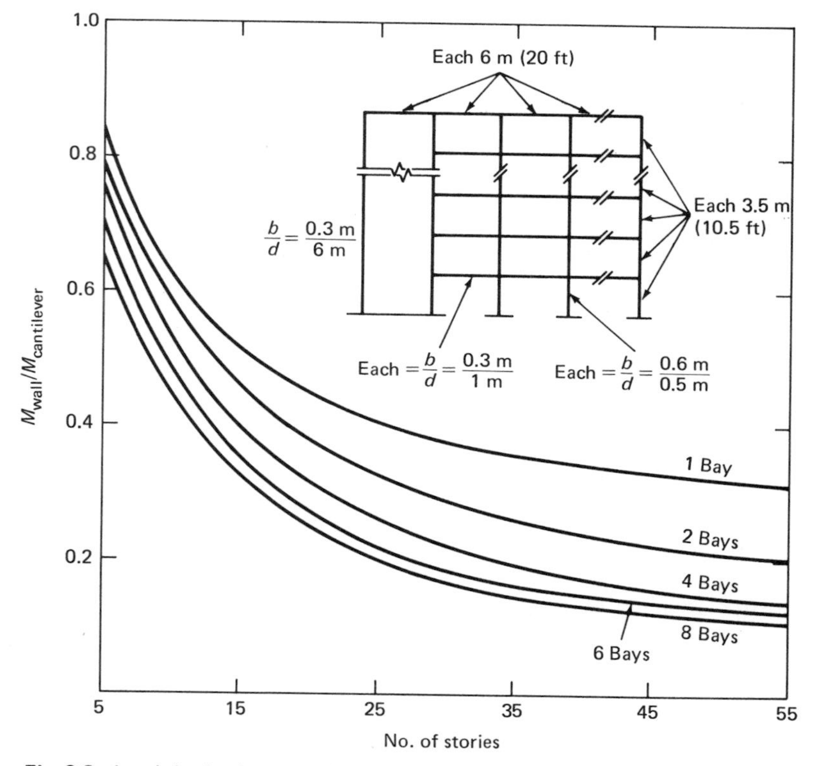

Fig. 9.3. Load-sharing between shear wall and frame members. (*After Konig 1973.*)

Fig. 9.4. John Hancock Building, Chicago. (*Courtesy Hedrich-Blessing, Chicago, Illinois.*)

tural components takes place due to the continuity as well as a large variation of their stiffnesses and implicit distribution of the externally applied forces in the structure. Amrhein (1976) points out several advantages of properly designed nonmassive masonry wall systems over other types of structures in areas where earthquakes frequently occur; these include their stiffness and ability to sustain such loadings with very little or no damage. The stiffness often reduces drift or sway of the building in the event of a horizontal load; the trend toward braced frames or frames with shear walls indicates the need to consider such interaction between the frame and the walls. Although bracing for a ductile frame may be provided in many different ways including externally as in the John Hancock Building in Chicago (Fig. 9.4), infill walls or nonstructural panels within the frames can also be used successfully. Interacting infilled frame systems thus have two characteristics: first, the frame is strengthened by

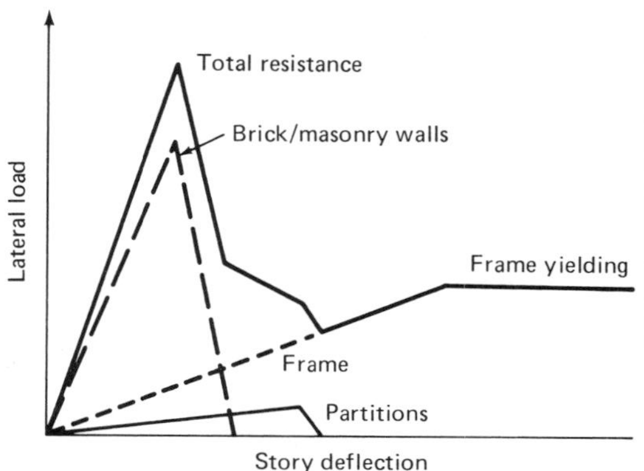

Fig. 9.5. Schematic plot of load vs. deflection in conventional building. (*Adapted from Fintel 1974.*)

stiff panels to form a composite shear wall; second, the shear wall thus formed by the infills is in turn strengthened by the frame itself. The advantages of the flexible frame and the rigid shear-wall system are combined to give stiffness to the composite system that is greater than that of either of the components. Not surprising, many buildings with properly designed infill wall-frame systems have survived the earthquake disasters and sustained only minor cracking in panels.

Infill walls contribute considerably to the resistance of the structure; thus, the proper consideration to it will make the structure economical. These walls will also contribute to the built-in factor of safety in a structure, particularly against the horizontal loads. A schematic presentation of the resistance expressed as lateral load vs. the story deflection is shown in Fig. 9.5. It is apparent that even though the

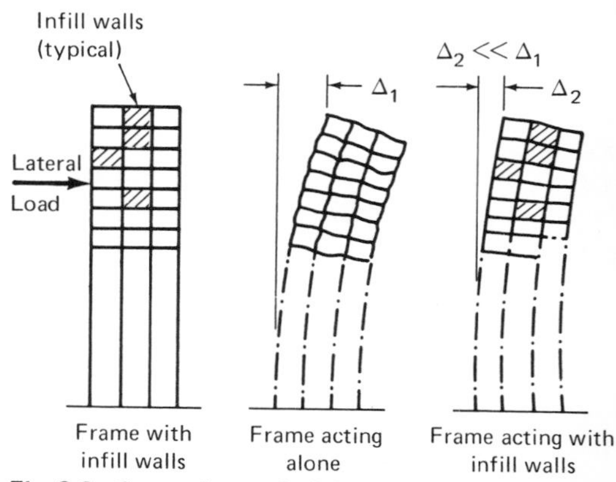

Fig. 9.6. Comparison of deflections of frames with and without infill walls.

frames may be designed to sustain horizontal loading, the major portion is resisted by the masonry or brick infill walls, thus stressing the importance of the interaction between the two. These walls also help improve serviceability of the structure as lateral deflections will be considerably reduced due to the additional shear rigidity. This aspect of interaction is schematically shown in Fig. 9.6.

9.2 DIFFERENT TYPES OF INTERACTING WALLS AND FRAMES

Walls may be constructed of masonry, brick, or concrete blocks and concrete panels, each of which can be either unreinforced or reinforced, the latter having more advantages. Steel (or metal) plate cladding or precast concrete cladding used as formwork for the exterior frames and mechanically connected precast concrete facade panels can also be used as satisfactory interacting elements. Frames of either structural steel or reinforced concrete may be used. The selection of a particular material in the medium range of heights of buildings will be governed by individual choice rather than economics. Detail description and design have been considered in Chapter 7.

9.2.1 Steel-Plate Cladding*

The concept of steel-plate cladding, evolved by Mies Vander Rohe, was a departure from the conventional masonry cladding in steel buildings, which was commonly used until the 1940s. This was mainly due to its flexibility for forms by architects for an expressive facade. The common form of such cladding is obtained by a flat or corrugated diaphragm of steel plate supported by a frame of light structural shapes used as mullions and extending over the entire height of the building. The exposed surface may be painted or of weathering steel. Such claddings are generally attached to steel frames by means of cast-in-place concrete, with shear connectors attached to the cladding to obtain a good bond and continuity to ensure full interaction between panel and the surrounding frame.

The cladding generally is $\frac{1}{4}$-in. (6-mm) thick common plate or corrugated light-gauge steel plates, welded to each other to form a continuous, weather-resistant, and leakproof cladding. The concrete filling behind these panels helps their stability and adds to the stiffness of the whole system, and in turn reduces the additional structural steel required for drift control in tall buildings. Scalzi and Arndt (1972) report comparisons of the frame-stiffening effect of such composite action and indicate that the cladding contributes from 20% to 50% of the lateral stiffness and stability to the structure depending on the geometry of the building and the composition of the frames. In the structural design of these frames, such a cladding can take up lateral forces, thus allowing the frame to carry primarily the gravity loads, with the lateral drifts reduced using the composite properties of the frames.

Structural contribution of such claddings is disallowed by the present building codes due to the lack of both experimental data and fire protection. As such claddings are used more commonly and the experimental data become available either through laboratory tests or by the instrumentation of actual structures and the development of suitable fire-protection techniques, these panels will have a very promising future. Additional research is thus desirable from the point of view of strength and overall economy.

Typical details of such cladding are shown in Chapter 7, Fig. 7.15.

9.2.2 Precast Concrete Formwork Cladding

Similar to the earlier type of cladding, precast concrete claddings are generally attached to the framework with bearing angles and steel strongbacks, in which form they do not add any significant stiffening to the frame. By using them as formwork for the exterior beams and columns, their composite action with frames may be achieved. The size of these panels depends on capability of handling and the possibility of cracking due to thermal stresses. Similar to the steel-plate cladding, this type also can be more

*Much of the description of claddings is based on personal correspondence with S. H. Iyengar (author of Chapter 7), to whom the author is greatly indebted.

effectively used for the stability of columns, rather than the overall strength. The labor associated with later installation of cladding is reduced if it is used as a formwork for the exterior frame, with the overall savings in building costs.

In terms of structural participation, the cladding formwork is not included for strength because resilient joints are usually provided at intervals to reduce cracking due to thermal movements. However, since these joints occur only once in each floor and once in a bay, the composite properties can be included in member stiffness calculations. They reduce drift or sway when used in conjunction with steel frames. Figure 9.7 shows a possible cross section of a steel frame with precast formwork. Without such joints, the composite section can

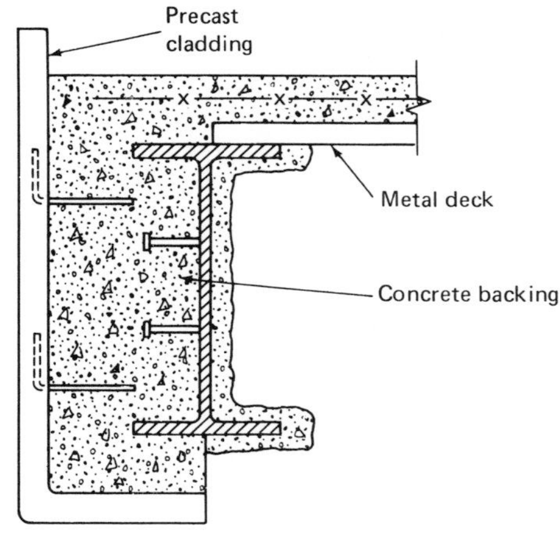

a. Spandrel section

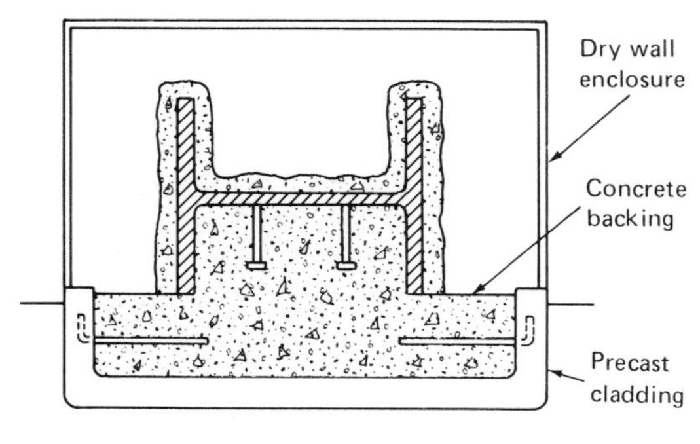

b. Column section

Fig. 9.7. Composite precast cladding. (*From Iyengar 1977.*)

also be used for strength. Problems arising from the crack potential of the cladding under different climatic conditions and those related to the creep effect between the cladding and poured-in-place concrete must be investigated to achieve the proper structural composite action between the claddings and frames.

Various examples of using precast concrete claddings may be cited here. Abdallah (1972) and Payton (1972) discuss projects of the MLC Tower in Sydney, Australia, and the Collins Place Project in Melbourne. In the former project, claddings measuring approximately 25 by 6 ft (8 by 2 m) were used; in the latter case, trough form T-shaped frame units for exterior columns were used with *in situ* concrete. Figure 9.8 shows an application in a composite tubular system in the CDC building in Houston, with a complete elevation of the building in Fig. 9.9. The exterior column consists of steel columns, which are integrated with a reinforced concrete column, and the precast concrete cladding panel provides the formwork for this concrete. The elevation shows a typical panel unit consisting of an entire window bay, covering columns, spandrel, and a window. Figure 9.8 also shows formwork for the side and bottom surfaces of the column beam and the spandrel beam, respectively. Reinforcement cages for columns and beams are erected first, followed by the cladding unit. The exterior steel columns are used as stays for the reinforcing cage and cladding unit. Backforms for columns and spandrel beams are then provided, after which the concrete for columns and beams is poured in one monolithic pour.

9.2.3 Mechanically Attached Precast Concrete Cladding

Reinforced concrete panels acting as bracing panels on the interior of buildings have been used as structural elements resisting wind loads, provided they are appropriately connected with the steel frame. These panels are different from the earlier precast panels; the former are attached to the steel frame at discrete points by means of steel-plate connections instead of through concrete. When solid exterior panel without windows can be used, there is less of a

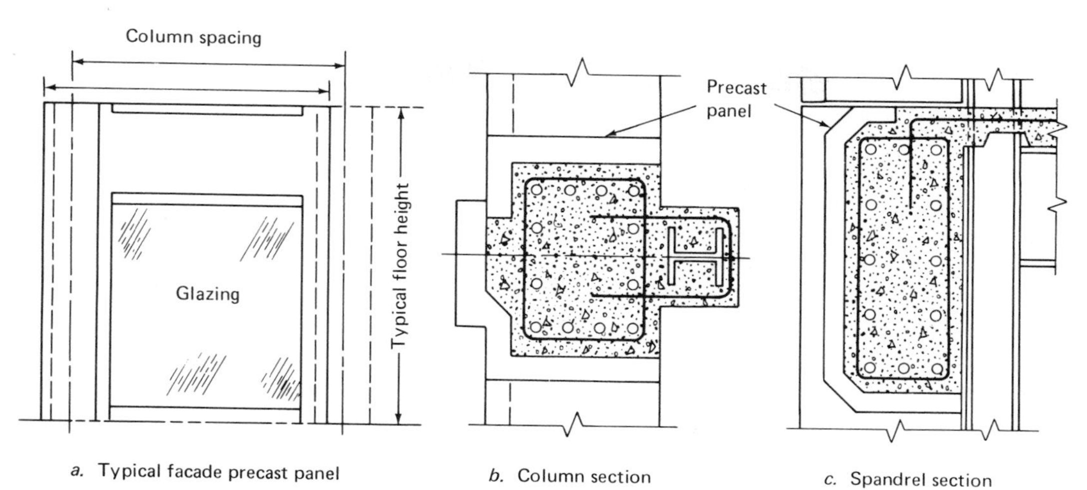

Fig. 9.8. Precast cladding formwork. (*From Iyengar 1977.*)

problem than when windows are needed. Thus, considerable potential exists for using this type of cladding in panelized form.

Weidlinger's (1973) studies indicate that such panels be reinforced for wind forces from either direction, and therefore two-way reinforcements are required in the panels. In a solid panel, this reinforcing system provides strength for direct tension and compression, while concrete primarily resists shear. In the case of panels with penetrations, the suitable reinforcement must be provided to take up the induced secondary stresses. Weidlinger (1973), using a finite-element analysis, points out the effectiveness of panels in their interaction with frames. He recommends a connection between the spandrel beam and panel to transmit only the horizontal forces parallel to the edge of the panel and not the vertical forces. With such connections, boundary strain is induced into the panel due to the column shortening and results in localized compressive stresses in the panel corners that can be predicted.

Fig. 9.9. Precast cladding formwork. (*From Iyengar 1977; photograph courtesy of Ezra Stoller [ESTO], Chicago, Illinois.*)

9.2.4 Masonry Infill Walls

Unlike the previous type, masonry walls are constructed in the same plane as the frame. In that respect, with a little extra detailing, they offer a large contribution to the frame design by virtue of their in-plane shear rigidity. ("Extra detailing" refers to the connections between the frames and walls to minimize their separation.) This detailing and suitable reinforcement in the wall constitute an effective composite frame from both the strength and serviceability point of view. The limitations on the use of both concrete block and brick masonry appear from the point of view of their strength, which may vary considerably with the construction quality.

Another disadvantage of these walls is that they take a longer time to build. Unfortunately, the same can be said about any concrete system; precasting or prefabricating the panels seems to be the solution for such a problem. Yorkdale (1972) discusses these techniques in his recent work. Accordingly, the recent research and development in many countries resulted in various procedures and methods of prefabricating masonry related to their use either as load-bearing panels or as curtain walls. Some of the methods used in prefabricating these walls are:

1. hand-laid systems on forms under plant conditions
2. machine fabrication with brick- and/or block-laying machines
3. casting operations—face up, face down, and vertical
4. thin masonry slabs on lightweight or normal-weight concrete panels
5. thin-skin panels of masonry unit faces on fiberglass-reinforced polyester resin.

It should not be overlooked that most of the preceding systems were possible due to the development of high-strength cements, grouts, and adhesives. Quickset cements have been developed and marketed throughout the world. Mortar additives such as latex and Saran have increased bond strengths. There are several systems of reinforcing, prestressing, and post-tensioning of masonry wall elements. All this information merely indicates the potential of such infill panels, since only a few have been used in the United States. Figures 9.10 through 9.12 show various aspects of unlimited shape possibilities and uses of infill masonry panels.

9.3 COMPOSITE ACTION BETWEEN WALLS AND FRAMES

Composite action between the infill walls and the frames depends basically on their acting integrally together, i.e., it depends on the area of contact between the two in order to enable the infill to distribute the stresses in the flexible frames. Thus, the composite action will be either complete or partial depending on proper bonding between the two. Research carried out so far has generally addressed itself to the interaction when the two components are not inte-

Fig. 9.10. Prefabricated brick panel being placed in the frame. *(Courtesy A. Yorkdale, Brick Institute of America, McLean, Virginia.)*

Fig. 9.11. Different shape brick panel. *(Courtesy A. Yorkdale, Brick Institute of America, McLean, Virginia.)*

Fig. 9.12. Prefabricated brick panel a a erection. *(Courtesy A. Yorkdale, Brick Institute of America, McLean, Virginia.)*

gral. In these conditions, the structure is considered as a truss, shown in Fig. 9.13, in which the diagonal of the truss is the intact portion of the wall still in contact with the surrounding frame. In recent work by Mallick and Garg (1971), shear connectors have been used to increase the effectiveness of these walls, thus improving further economy and efficiency of the structure.

The composite action further depends on the relative stiffness of the two interacting components. The stiffness response of the infill–and in turn that of the composite frame–is related again to the length of contact between the two; thus, stresses in the infill and strength of frame will also depend upon it.

Stafford-Smith and Carter (1969) show this length of contact to be a function of the relative stiffness of the frame members in flexure and the infill member in compression; the stiffer the frame relative to the infill, the longer the length of contact results in smaller lateral deflection and better serviceability of the structure.

The contribution of wall and its interaction with the frame was demonstrated by the actual test of a bare frame and an infilled frame by Ockleston (1955), as shown in Fig. 9.14. The reduction, by as much as 75% in deflections at various load levels, indicates clearly the effective interaction. Such marked reduction in deflections suggests that an adequate prediction of the horizontal stiffness can be achieved by considering only the direct forces in the frame and in the equivalent diagonal truss and ignoring the bending in the frame members. This allows a simplified determination of the horizontal stiffness at any floor level, using infills as equivalent struts at that particular load and computing

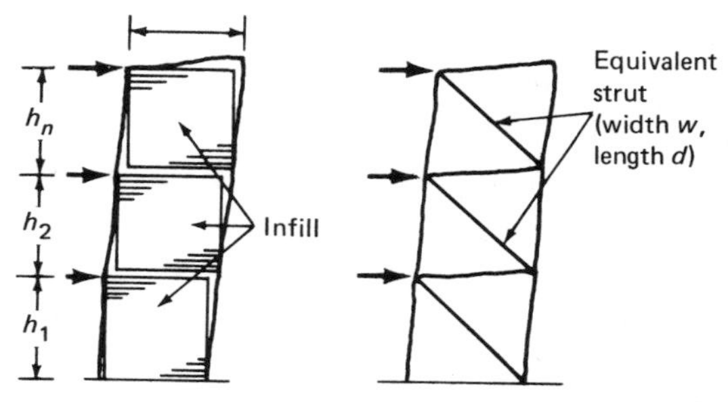

Fig. 9.13. Laterally loaded infilled frame and equivalent truss (frame). *(From Stafford-Smith and Carter 1969.)*

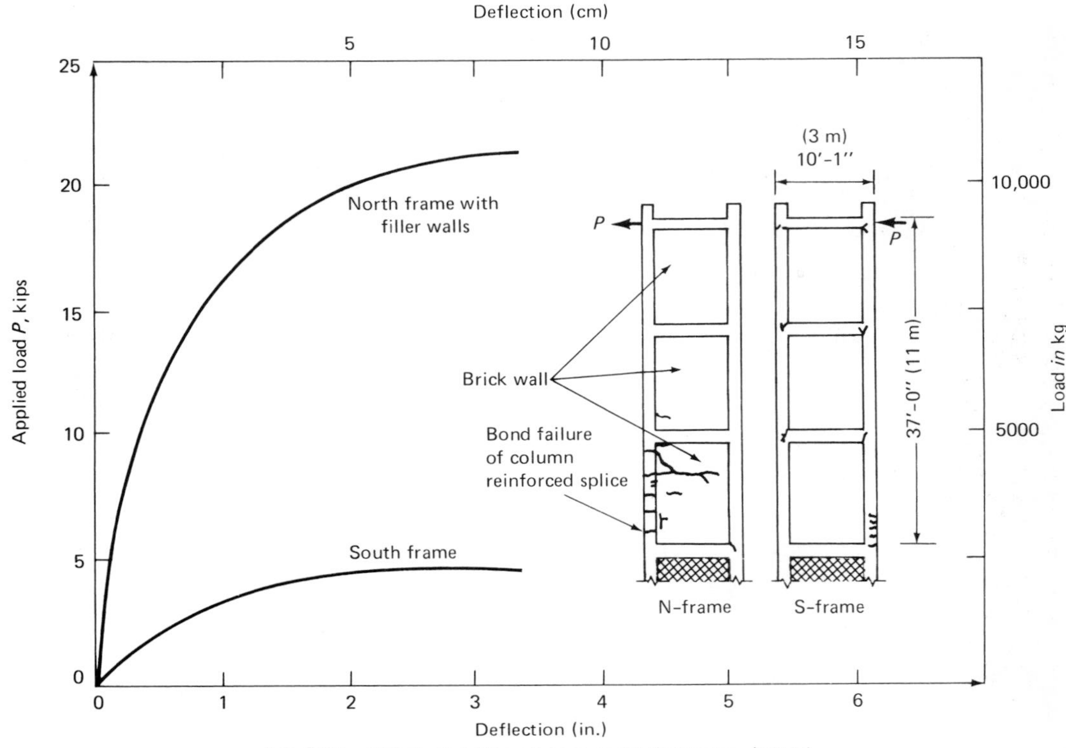

Fig. 9.14. Effect of infill wall in tests by Ockleston (1955).

$$\sum_{F} PU\ell/AE$$

where

P = force in struts due to unit load replacing the applied load

U = force in struts due to unit load at the point where displacement is required

A = cross-sectional area of members including diagonal strut (the larger the value of A, the better it will be to reduce the value of the above expression)

ℓ = length of the member

E = modulus of elasticity of frame members and infill wall (initial tangent modulus).

As the number of stories in a building increases, the axial force P assumes even greater significance over the bending in the frame, thus justifying the use of a simplified equivalent frame analysis or truss analogy. This can be seen from Fig. 9.13. The parameter related to various properties of materials used for both the infill and frame, and the length of contact between the two components, will be considered later in detail.

As mentioned previously, the infilled frames are extremely useful in obtaining stiff composite structure and in minimizing the drift or later sway due to heavy winds. In the case of earthquakes, the stiffness helps prevent large displacement and other nonstructural damage to the structure, thus requiring the analysis of interaction under dynamic loading. Many buildings with infill frames have withstood several earthquakes quite well, which has given rise to a recent interest in analyzing such structures for dynamic loading and in determining the safety and the possible future use of infills as structural components. Some of these studies will be discussed in the next section.

An important factor influencing the stiffness of infill, which affects interaction between infill and the wall, is whether there are any openings. If such openings are quite small, as may be the case for running electrical cables or ducts, then their effect may be negligible. On the other hand, for large openings such as windows and doors, the effect may be pronounced, and the interaction may be considerably reduced. Several related studies will be presented in the review of literature in the following section.

9.4 REVIEW OF LITERATURE

As discussed previously, the topic of infilled frames has gained importance in the last couple of decades following the development of tall buildings. This has resulted in a steady but significant amount of literature in this field, essentially to achieve economy and at the same time to investigate properties of the various components of a structure subjected to different loading conditions. In this section, the state-of-the-art is presented, with a large number of references pointed out at the end for additional information. Also, as indicated in the previous section, the present codes do not have any provisions for such interaction; therefore, information presented here should be useful to designers in their designs.

Although at present there are no code provisions, it's the author's hope that in the not-too-distant future such a move will be made.

Work by Thomas (1952) and Ockleston (1955) may be cited as early major contributions to the understanding of the interaction between wall and frame. Thomas pointed out the arbitrary considerations used prior to his tests for the strength of brickwork, and their variability due to lack of quality control; he conducted tests essentially to demonstrate experimentally the additional strength that infill walls add to the surrounding frames. The tests were only on single-bay and single-story frames made of concrete-encased steel infilled with various materials. The boundary conditions and loadings were simulated as in multistory frames. The results of Thomas's tests are shown in Fig. 9.15, which clearly demonstrates the benefits of infillings of brickwork and other building materials over the bare frame. The stiffness of the frame was increased considerably, even with a relatively weak infilling that might be used as partition. The maximum lat-

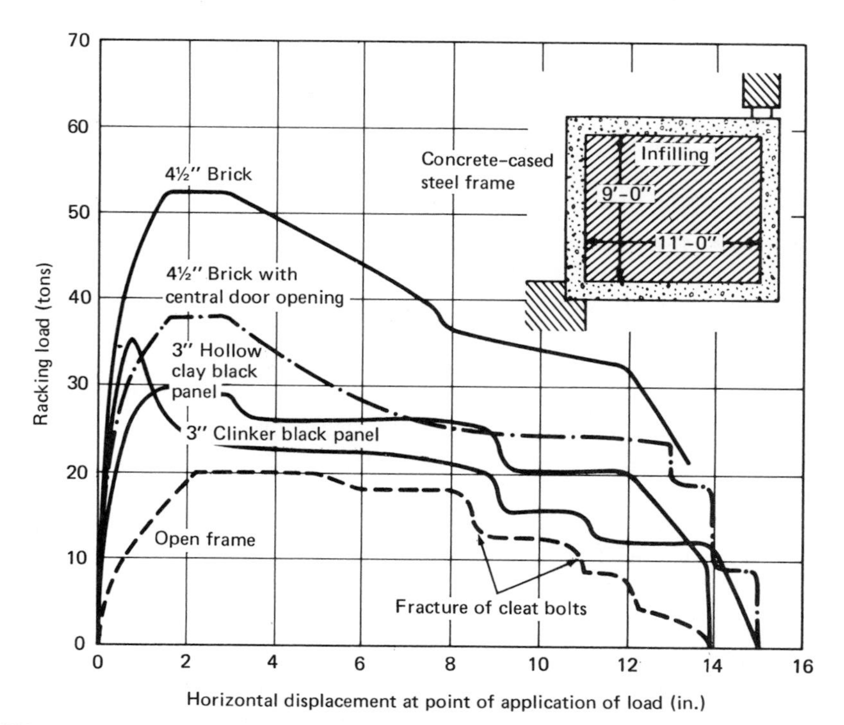

Fig. 9.15. Effect of infill panels on concrete cased steel frame subjected to racking load. (*From Thomas 1953.*)

eral (racking) load on the composite frame with a $4\frac{1}{2}$-in. (12-cm) brick infilling was over twice that which could be supported by the encased steel frame alone, and much of it was retained even when a door opening was located centrally in the panel. Thomas's tests were probably the first experimental evidence to indicate that allowance for the lateral strength due to walls and partitions might often eliminate the necessity for providing special considerations or bracing in structural frameworks for resisting lateral loads.

Ockleston (1951) investigated the effectiveness of infill walls in the behavior of three-story frames during the demolition of a hospital in Johannesburg, South Africa. The end frames made with reinforced concrete were separated from the rest of the structure, instrumented, and tested to failure; of these, one was with and one without infill brickwork. The infilled frame showed approximately 400% more strength as well as considerably more stiffness as indicated by the load-deflection curve of the bare and infilled frames.

In the United States, Benjamin and Williams (1957, 1958) carried out tests on frames infilled with brick and concrete shear walls, some with openings. They concluded that the stiffness and strength of infilled frames was independent of frame stiffness, provided that the frame was strong enough to produce infill failure. They used a lattice analogy approach in the analysis of their frames and developed relationships to calculate stiffness and strength of masonry-infilled frames, based on shear considerations and full integration between the frame and infill.

Sachanski (1960) reported tests on infilled model and prototype frames; he developed a method of analysis, based on the theory of elasticity. The interacting forces between frame and infill were replaced by a large number of redundant reactions. A solution for these reactions was obtained considering the compatibility of displacements of two components assuming a continuous bond.

In the Soviet Union, Polyakov (1956, 1960) conducted a number of tests on frames of up to three stories to investigate stresses in infills and developed approximate methods for their design. His design method included the maximum shear stress in the infill as related to the material shear strength. He also developed empirical formulas and design curves, taking into account the openings and variations in the infill dimensions.

Wood (1958) pointed out the difficulties in predicting the behavior of frames when the contribution of cladding (or other infills) in stiffening the tall building was neglected. He suggested that the design of frames be best accomplished by a "collapse design" method using an empirical interaction formula, which related the individual strengths of infill and frames in the composite action.

Holmes (1961, 1963) tested steel frames infilled with brick masonry and reinforced concrete walls and developed a semiempirical design method for laterally loaded infilled frames based on the *equivalent strut* concept. His tests indicated that the reinforced concrete infill wall was the most effective in increasing the strength of bare frame. It increased the strength by as much as 400% and the brick masonry by up to 100%. He also found that the presence of vertical load increased the strength by about 15% and that openings in walls, such as doors or windows, might reduce strength up to 40%, based on the composite behavior. In his method, Holmes considered only the compressive mode of failure in infill. The load carried by infill at failure was calculated by multiplying the compressive strength of material by the area of equivalent strut. Holmes approximated width of the equivalent strut as one-third the diagonal length of the infill, which resulted in the infill strength being independent of frame stiffness. The load carried by the frame was then calculated by assuming that the strut was shortened by an amount, which was its length multiplied by the strain at failure in the infill material.

Considerable work on interaction of frames and infill walls is reported by Stafford-Smith (1962, 1966, 1967), Stafford-Smith and Carter (1969), and Riddington (1974). In his initial work, Stafford-Smith (1962, 1966) tested a number of infilled frames subjected to diagonal loading and, like Holmes, used the diagonal-strut concept. He developed a design method using a finite difference method to analyze infills, loaded as described previously as well as under various assumed load distributions. He prepared design curves to calculate the effective width of strut, the compressive failure load and the diagonal failure load, as related to frame stiffness and infill aspect ratio. Later, Stafford-Smith and Carter (1969) included the effects on nonlinearity of material (i.e., the variation in modulus of elasticity of infill with increasing stress levels) in their analysis by finite-element method. They investigated various modes of failure and concluded that shear cracking is the predominant mode of failure. Carter also produced a number of useful design curves using the mode of failure.

Recently, Riddington (1974) reanalyzed the previous work and developed a simple design method in terms of limit-states design. He used modes of failure by the shear cracking, and the crushing and spalling in the corner of the infill and defined these as *serviceability* and *collapse limit* states. By comparing his method with several full-scale as well as scaled-model tests, he concluded that infilled frames had substantial strength even after separation (boundary cracking), provided that a good initial fit of the infills was achieved. Riddington also recommended that vertical loading, if included in the analysis of the composite frame, be done with extra care, since the behavior under vertical loading was more sensitive to initial fit and frame prestressing than to the lateral loads.

Later work at the University of Illinois was reported by Yorulmaz and Sozen (1968), Fedorkiw and Sozen (1968), and Fiorato, et al. (1970). The entire project was devoted to theoretical and experimental work to verify the load and energy-absorption capacity of laterally loaded reinforced concrete frames with and without infill walls. Yorulmaz and Sozen (1968) conducted tests on $\frac{1}{8}$-scale models of up to five-story frames. Load-deflection behavior as well as various failure modes were investigated, and predictions were made using conventional methods, which according to the authors established safe lower bounds for energy-absorbing capacity. Based on the experimental work, they also showed that the type of failure was dependent on the percentage of reinforcement present: low percentage caused a mechanism to occur in the compression column, while high percentage resulted in shear failure in tension column. The work of Fiorato et al. consisted of a series of tests on infilled frames subjected to both lateral and vertical loading. Various observed phenomena on resistance mechanism and the failure modes were discussed quantitatively and related to the geometric and material properties of the wall-frame composite. Fedorkiw and Sozen (1968) developed an analytical procedure and made a systematic evaluation of the energy capacity of an infilled frame by representing it as a discrete physical model. The model consisted of a finite number of elements connected with springs to form a lumped spring-mass system. Although the initial development of this procedure was aimed toward comparing the results of static tests, this investigation is useful for wind or earthquake loads.

Many tests in the early work were based on the infills with no openings and no special bonding between the walls and frames. Mallick and Garg (1971) investigated the effect of openings and of the shear connectors on the behavior of infill panels. They showed that openings at either end of a loaded diagonal are undesirable in regard to strength, since earthquake or wind loads may act from either side of the frame. Accordingly, if openings are necessary, they should be located in the middle third of the panel rather than near the end; this may also be helpful for transferring gravity loads above the openings by arch action and to prevent progressive collapse in such infilled frames. They used a finite-element approach for the analysis and considered a range of properties of infill and other geometrical parameters in their studies. Their work also indicated that shear-connector spacing used for a better connection between the infill and frame does not appreciably affect the capacity of the system.

Dawson (1972) investigated the influence of imperfectly fitting unbonded infills on frame behavior under lateral loads. It is interesting that even loosely fitting walls resulted in increased frame stiffness. Dawson recommended the inclusion of stiffeners close to corners of the frame to improve the interaction and in turn the stiffness of the composite system. Dawson also used a finite-element approach with the equivalent-strut concept. He found that the variation of the modulus of elasticity had no effect on the magnitude of the equivalent-strut area used in the analysis and that the influence of the Poisson ratio was negligible. He recommended that, for design purposes, the equivalent-strut areas be considered as a function of the wall geometry alone.

In addition to these studies, a number of other investigations have also been reported. These include work by Wood (1952, 1958), Karamanski (1967), Coull and Stafford-Smith (1967), Mallick and Severn (1968), Kazimi (1967), Tamura (1969), Liauw (1970), Franklin (1970), Mainstone (1971), Lefter (1974), and Kadir (1974). A number of symposia that have taken place in the last 10 years or so contain a wealth of information in their proceedings—e.g., Coull and Stafford-Smith (1967), Johnson (1969), Lee and Karasudhi (1974), Jessop and Ward (1976), Conferences on Planning and Design of Tall Buildings (1972), and Leige (1976) certainly indicate the extent of interest in the profession.

9.5 BEHAVIOR OF INFILL FRAMES

The behavior of infill frames in general is similar to the cantilever beam made of different materials but acting in a composite manner. Columns at the end of the frame form compression and tension flanges with walls, together with floors acting as stiffened shear webs. The actual behavior, however, will depend on various parameters, including the properties of the material and its dimensions (which in turn will affect the respective stiffnesses), the extent to which both the walls and frames are in contact, and the type of loading (either static or dynamic). Openings and reinforcement in walls and the use of shear connectors between the frames and walls will also affect the behavior. It can be represented by a load vs. lateral deflection plot, which becomes nonlinear with the cracking of wall(s). Until any cracking takes place, the system acts as elastic, and the behavior of such a frame can be calculated and predicted according to the elastic theory.

In general, a (lateral) load-deflection behavior may be presented as shown in Fig. 9.16. It consists of an initial linear portion *01*, followed then by a nonlinear portion *12*, up to and including failure. The nonlinear part of the curve will essentially depend on the various parameters mentioned earlier, and failure may be ductile or brittle depending on the properties of various members and loading distribution. Detailed experimental investigations by Fiorato et al. (1970) present the influence of a number of variables on the behavior of frames of one and five stories with and without openings. They include:

1. height or number of stories
2. width or number of bays
3. column loads
4. wall openings
5. amount, quality, and arrangement of reinforcement in frames (since all frames were reinforced concrete).

Much of the behavior discussed in this section is based on Fiorato's work, which leads to general conclusions about the behavior of infill frames with more than five stories and the proper method of analysis and design of the frames.

In the system of infilled frames, forces can be resolved into three major components: (vertical) axial loads, (overturning) moments, and (horizontal) shears, the magnitudes of each of these varying with the combination of loads. Axial loads and moments will be taken care of by columns, their contribution depending on size, location, and other characteristics, while

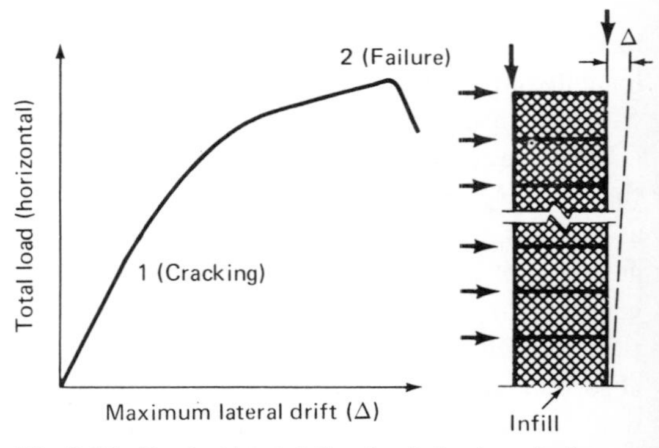

Fig. 9.16. Typical load-deflection behavior of a frame.

shear will act predominantly on the wall. Due to substantially larger stiffness, a considerable portion of the shear will be taken up by the walls. The (lateral) load-deflection behavior, therefore, will vary as shown in Fig. 9.17, which indicates:

1. calculated response for the frame alone (based on inelastic, nonlinear behavior)
2. calculated response for the composite infill frame (based on elastic, uncracked section)
3. experimentally observed response from frame alone
4. experimentally observed response of the composite infill frame.

It may be noted that calculated response (*1*) even for a frame alone is somewhat conservative (the actual capacity is larger) due to the difference between the assumed and the actual failure mechanism of the frame. This response (*1*) is also the lowest bound, as may be expected, due to the considerably greater flexibility of the bare frame compared to that of a composite frame. Theoretical curve (*2*) represents an upper bound (and coincides with the initial response observed experimentally), since no cracking or nonlinearity has been taken into account. In order to predict the observed behavior, the various stages of failures must be properly accounted. Departure of curve (*4*) from curves (*2*) and (*1*) in Fig. 9.17c will depend on the relative contributions of wall and frame to the carrying capacity of the total frame. The contributions, being a function of individual stiffnesses, can be calculated on a simplified basis. The stiffness of the frame will vary as a function of cracking in the case of a reinforced concrete frame and will progress with the lateral load increase. On the other hand, in the case of steel frame, it will depend on the development of mechanisms as loading progresses. As for the walls, the stiffness will be reduced with the development of cracking in walls and also with the continued separation of portions in contact with columns and beams. Stafford-Smith and Carter (1970) observed that this will cause an eventual truss action (Fig. 9.13). The diagonal of such truss near failure would have a width 10% to 15% of its length, the rest being cracked and damaged, and would not be considered useful. At service loads, however, this value could be higher, thus contributing to the strength of the infilled frame, especially when walls are reinforced. Proper accounting of this stiffening of frames by walls as diagonal braces should be made in the strength calculations of these infill frames.

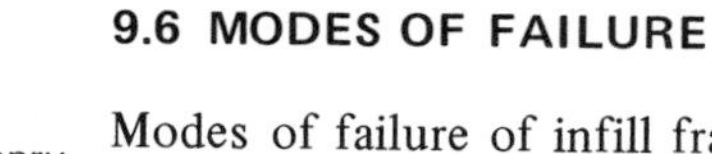

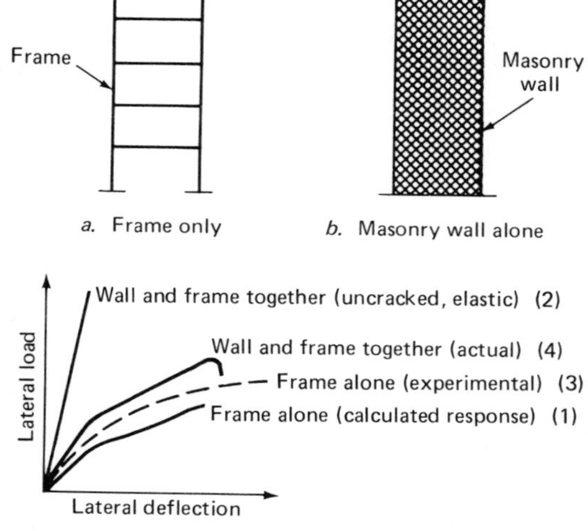

Fig. 9.17. Limits of calculated and actual behavior of infilled frames with and without infills. (*Based on Fiarato et al. 1970.*)

9.6 MODES OF FAILURE

Modes of failure of infill frames may be classified as flexural, shear, or axial failures, depending on the many variables discussed previously. General characteristics of flexural and axial modes are similar to those in reinforced concrete elements. The shear-failure mode generally differs from that in reinforced concrete beam due to the presence of walls. The modes are discussed first, followed by the presentation of some experimental observations.

9.6.1 Flexural Failure

Flexural failure takes place when the predominant action in the equivalent frame (i.e., cantilever action) is due to flexure. This happens when the height-to-width ratio of the building is approximately greater than 5 : 1. In such an event, shear stresses in the masonry wall are too low to cause any distress in the web as in the case of cantilever beams of like proportions. This failure is further characterized by failure in the tension column. If the initial cracking in the tension column does not stress its steel to yielding, then this crack may not be opened and is generally difficult to observe.

The failure may also be further evidenced by the crushing of the compression column or by extensive deformation due to tensile yielding, but will give sufficient warning prior to the failure. A typical mode of failure is shown in Fig. 9.18a.

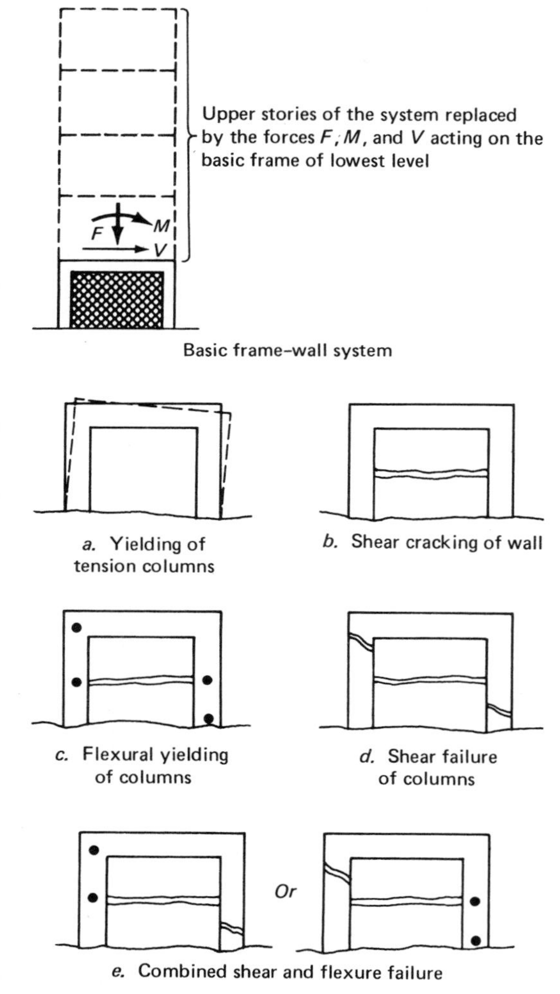

Fig. 9.18. Various modes of failure of infilled frame. (*Based on Fiarato et al. 1970.*)

9.6.2 Shear Failure

Shear failure is initiated by the formation of a shear crack in the wall when its capacity is exceeded prior to the columns, as in the previous case. The shear crack will cause separation of the wall into two or three parts, causing a knee bracing for the frame. (See Fig. 9.18b.) The failure, unlike in the case of unreinforced concrete beams, is not incipient but may go into a flexural mode of failure due to the bracing action. This action has been used by Fiorato et al. (1970) and Lefter (1974), who proposed a theory based on hinges in the columns. In such situations, however, the individual stiffness of the frame and walls will be quite important and should be accounted properly. Reinforcing of panels will also be effective in delaying the undesirable shear cracking of panels.

Shear-failure modes may involve failure of individual columns in flexure or in shear. A portion of both columns may also develop yield hinges leading to a plastic mechanism as shown in Fig. 9.18c; if, on the other hand, its shear strength exceeds the flexural strength, the column might fail in flexure, as shown in Fig. 9.18d. Certainly, combinations of both the above modes are also possible leading to a shear-flexural mode of failure in the columns, as in Fig. 9.18e.

9.6.3 Axial Failure

In the case of multistory infill frames, axial forces in columns are significant. Because of the large overturning moment on the lowest story in the structure, a considerable thrust is developed in the columns. With a relatively low percentage of steel, the column may result in the yielding of steel prior to shear cracking in the walls. Often, this failure—at the base of the column in the lowest story—may be transferred to the top of that story due to the bracing action of the wall in its full height. The shear capacity of tension column in such frames is exceeded before sufficient shear forces could be developed to cause further cracking of wall panel. The loss of shear transfer from tension column to wall panel must be emphasized in multistory frames due to the presence of column thrusts.

9.7 DIFFERENTIAL MOVEMENT BETWEEN FRAME AND WALL

Frames (either in structural steel or reinforced concrete) and infill walls (either in brickwork or masonry) are subjected to different elastic strains under loads, due to their different material properties. Particular concern is in reinforced concrete frames, which will have creep strain as well as elastic strain, causing additional shortening under the sustained loading. In addition to live, dead, or other applied loads, even seasonal temperature variation will cause differential thermal movement in the various components of a structure. The structure must therefore be constructed with proper accounting of the differential movement; otherwise, they may cause distress in some part(s) or even failure in the structure.

The Brick Institute of America has published technical notes to aid calculations of thermal movements outlining causes, effect, and remedial solutions in various types of structures (BIA 1963). This section is mostly based on this available information. Table 9.1 indicates the average values of thermal movements for various building materials. A proper allowance must be made for such movements in composite frames, as high stresses will develop from any restraints to these movements.

In a multistory building, vertical shortening of the structural frame frequently imposes very high stresses in the masonry infills or facings. Failures of walls are sometimes characterized by bowing, by horizontal cracks at angles, and by spalling of masonry units at window heads, shelf angles, and other points where stresses are concentrated. Excessive shortening may take

Table 9.1. Thermal movements in various building materials. (Based on BIA 1963.)

Material	Average Coefficient of Linear Thermal Expansion in Millionths (0.0000001) per °T		Thermal Movement, in./100 ft (30 m) for 100°F (40°C) Temperature Variation	
	/°F	/°C	in.	mm
Clay masonry				
Clay or shale brick	3.6	6.5	0.43 ($\frac{7}{16}$)[a]	10.9
Fire clay brick or tile	2.5	4.5	0.30 ($\frac{5}{16}$)	7.6
Clay or shale tile	3.3	6	0.40 ($\frac{3}{8}$)	10.2
Concrete masonry				
Dense aggregate	5.2	9.5	0.62 ($\frac{5}{8}$)	15.7
Cinder aggregate	3.1	5.6	0.37 ($\frac{3}{8}$)	9.4
Expanded shale aggregate	4.3	7.8	0.52 ($\frac{1}{2}$)	13.2
Expanded slag aggregate	4.6	8.4	0.55 ($\frac{9}{16}$)	14
Pumice or cinder aggregate	4.1	7.5	0.49 ($\frac{1}{2}$)	12.4
Concrete				
Gravel aggregate	6.0	10.9	0.72 ($\frac{3}{4}$)	18.3
Lightweight aggregate	4.5	8.2	0.54 ($\frac{9}{16}$)	13.7
Metal				
Aluminum	12.8	23.3	1.54 (1-$\frac{9}{16}$)	39.1
Steel	6.7	12.2	0.80 ($\frac{13}{16}$)	20.3
Structural plaster				
Gypsum	7.6	13.8	0.91 ($\frac{15}{16}$)	23.1
Perlite	5.2	9.5	0.62 ($\frac{5}{8}$)	15.7
Vermiculite	5.9	10.7	0.71 ($\frac{11}{16}$)	18

[a]Approximate values in sixteenths of an inch.

place both in structural steel and concrete frames, the latter being more prone due to the nature of material properties, such as shrinkage and creep.

When infill walls are used in frame structures, care must be taken to anchor the walls to frames in a manner that will allow each to move relative to the other. The frames will undergo greater deformations than walls due to their load-carrying action, while their reactions to moisture and to the thermal movements will also differ. As shown by experimental evidence, the beneficial interaction between the wall and frame will take place despite the above differential movement.

In multistory construction, it is generally not necessary to anchor walls to columns; anchorage to spandrel beams is usually sufficient. The following solutions are recommended to account for movement in major structural parts to prevent any distress:

1. *Floor-wall joint*
Sufficient gap must be provided at the top of wall, below the upper beam level, to allow the wall to breathe and the frame to shorten under loads. Such shortening in the frame should be calculated both for short-term and sustained load, and joint should be provided at every few stories using softer material, which will give way under the load and not let it be transferred to the wall below. Since the infill walls are not load-bearing and since they have large in-plane rigidity, this joint will enable wall and frame to act together to resist any lateral load but not any vertical load. Periodic inspection (every few years) should be carried out to insure that the joint is working properly.

2. *Partial Reinforcement*
As mentioned elsewhere, it is advisable to increase the strength of plain masonry to accommodate or distribute high stresses, by providing reinforcement. This will also aid in minimizing cracking and will prevent any falling of masonry units to cause nonstructural and undesirable occupancy damage. This will be particularly important if the structure is in an earthquake region where high stresses may exist for a short duration.

9.8 BEHAVIOR OF INFILL FRAMES UNDER DYNAMIC LOADS

There are several situations in which the walls (or the entire structure) may be subjected to dynamic loads. In order to predict the behavior of the structure, several tests have been conducted to study it under cyclic or dynamically applied loads. In seismic regions, walls (or, in general, a structure) must be able to withstand cyclic loading due to in plane shear. Due to the sophistication required in the equipment and the relatively high cost, only a limited number of tests have been reported. The variables generally include number of cycles, stress level, and type of loads, and the energy-absorption capacity of a structure under such high-intensity but low-cycle loading.

It has been found that the inclusion of infill walls has a major influence on the dynamic response of a frame. This is primarily due to the multifold increase in stiffness of the composite structure, which in turn causes reductions in strains in the frame and changes in the frequency composition of the response. This, in part, is due to the better damping characteristics of masonry walls. The higher the value, the better will be the energy-absorbing capacity of the structure; such a structure should provide a better supporting system to the lateral forces. The composite structure between two structural components, frames and walls, not only offers a better structure, but it also has a higher elastic working range capable of performing much more work before it is damaged. The reason for such performance is the higher factor of safety built in the walls compared to that in the flexible frames. Research work to demonstrate these characteristics was carried out by Mallick and Severn (1968), Ward (1969), Dawson (1972), Williams (1971), Meli (1973), Mayes and Clough (1975), Priestly and Bridgeman (1974), and Scrivener (1976).

Mallick and Severn (1968) did considerable work in infill frames both analytically as well as experimentally, part of which is discussed in other sections. In their work, they present experimental determination of the damping characteristics of the vibrating infill frames. They express damping as a coefficient of equivalent viscous damping, which is obtained, first, by half-cyclic load tests on infill frames both with and without shear connectors, and second, by forced vibration tests on the same range of frames. For a height-to-span ratio greater than 3, their work indicates that the infill frames behave like a cantilever rather than a shear structure when subjected to axial forces in the columns.

Ward (1969) presented an analysis to obtain four modes and frequencies of vibration of a 15-story reinforced concrete building with flat slabs and shear walls and correlates with some measurements on the actual structure. Measured frequencies of the building were found to be within the calculated frequency bounds for a mathematical model used. He further demonstrated that the frequencies of the lower modes of vibration may be decreased further due to the foundation rocking.

Dawson (1972) subjected four-story model infill frames to earthquake-type loading with a predetermined time history on a shake table. His observations on the time history of strain indicate that the infill walls caused a significant reduction in the amplitudes (Fig. 9.19d) to show the pressure of high-frequency response more clearly. The figure also indicates that the response of the structure close to the fundamental modes was spread over a wide bandwidth *when the walls were present*, which indicates the increased damping resulting from the addition of walls. The response of the system was very similar even with *gaps* between

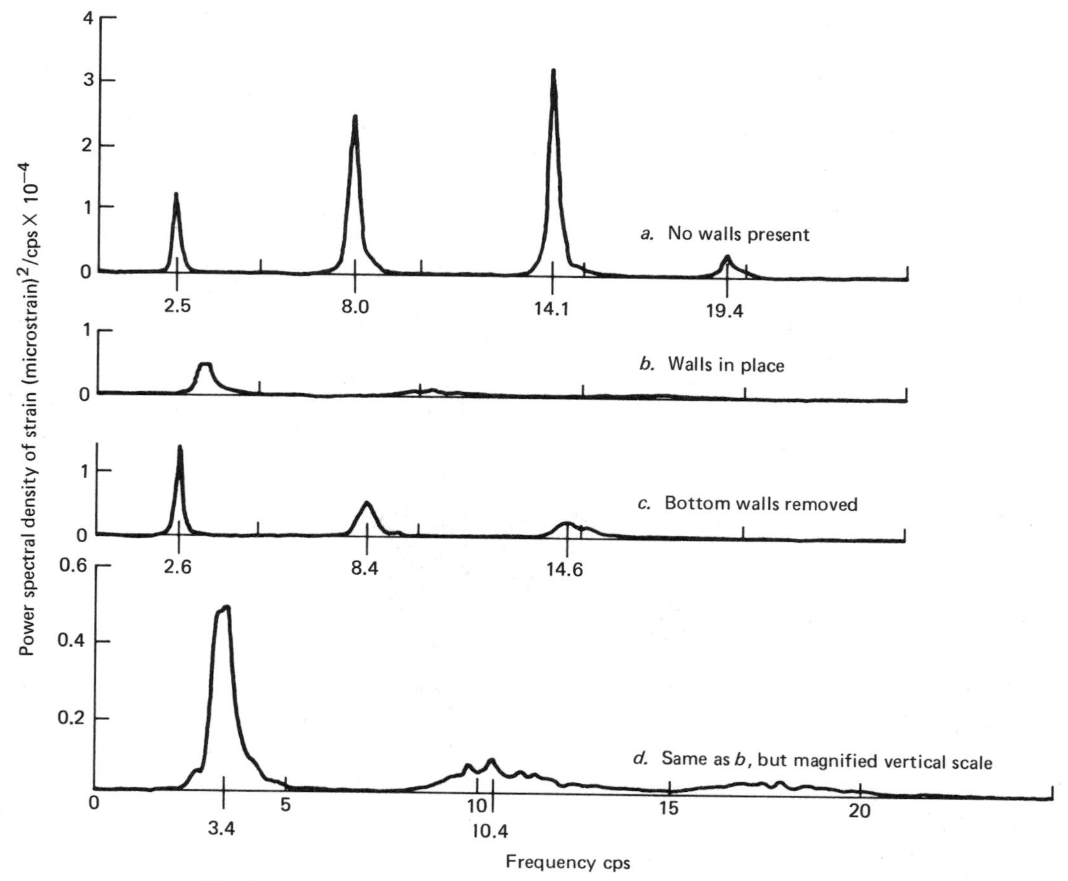

Fig. 9.19. Power spectral density of strain in frames with walls present.

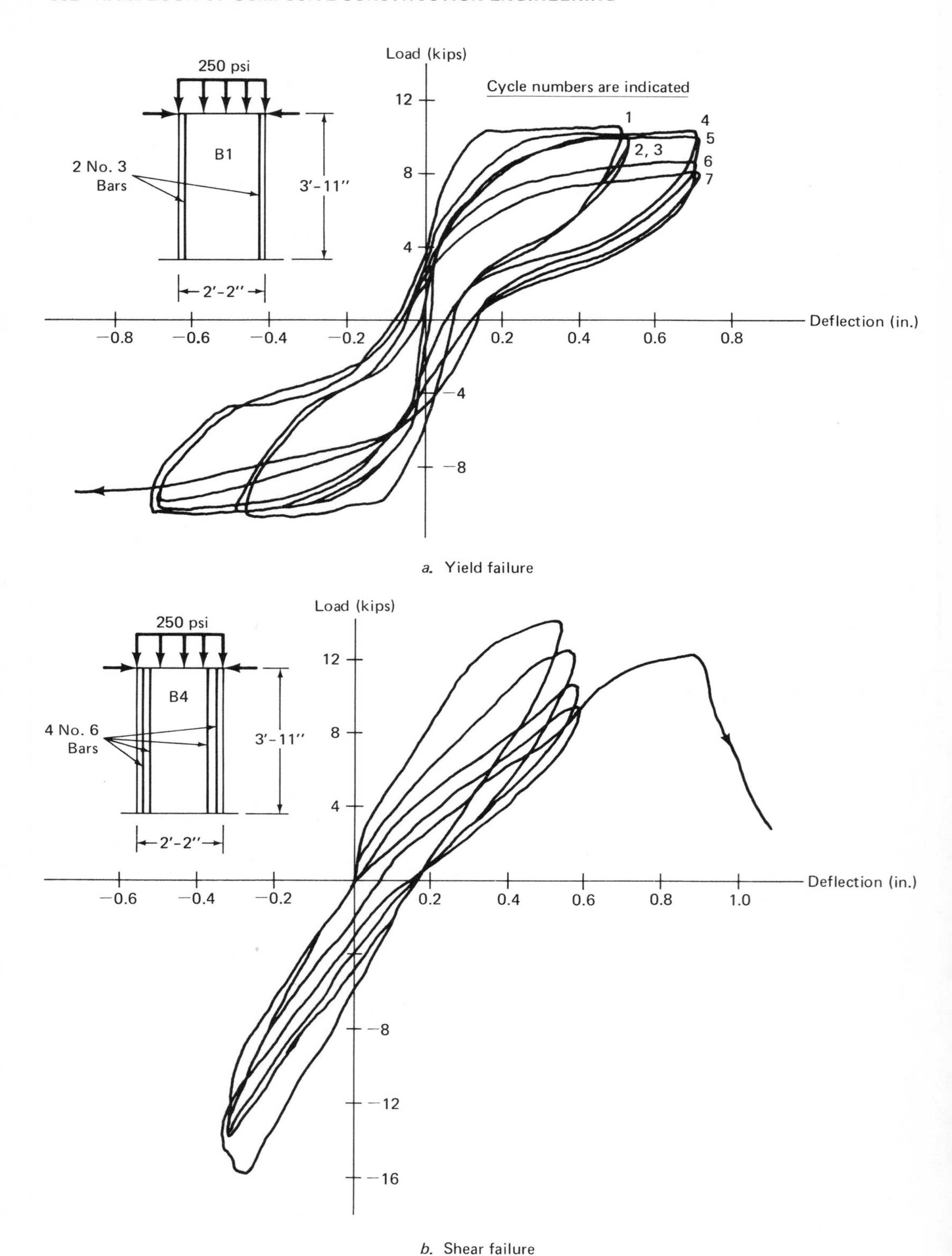

a. Yield failure

b. Shear failure

Fig. 9.20. Load-deflection response for flexural and yield failures. (*From Williams 1971.*)

wall and beam, which often has to be done in such structures to account for long-term effects on frames and walls.

Williams (1971) performed a series of cyclic load tests on cantilever walls made of varying materials and aspect ratios. Typical load-deflective curves representative of flexural or yield failure and shear failure are shown in Fig. 9.20a and b, respectively. William's discussion related to stiffness degradation, effect on bearing load, wall geometry, and reinforcing and dynamic loading. His conclusions were similar to those discussed in other sections and by Mallick and Severn (1968).

Meli (1973) summarized much of the research in Mexico and of his own work on dynamic testing of both reinforced and unreinforced walls. He concluded that the failure of walls may be divided into bending and diagonal shear cracking; however, in the latter case, there was a considerable deterioration subsequent to the diagonal cracking. He also indicated that this behavior could be improved by using higher ratios of vertical and horizontal closely spaced reinforcement or, more effectively, by confining grout and vertical reinforcement in the corners by means of closely spaced ties. In short, according to Meli, for adequate behavior under cyclic dynamic loads, the layout, aspect ratio, and reinforcement of walls should be chosen in such a manner as to force the system to fail in bending. Mayes and Clough (1975) also found similar results from their tests on unreinforced and reinforced block walls and piers. Priestly and Bridgeman (1974) conducted a thorough investigation on the effect of horizontal and vertical reinforcement on the shear strength of cantilever piers due to a confining plate in the mortar joints at the compression toes, which was found to considerably improve the inelastic behavior of the piers. They derived expressions for effectiveness of horizontal and vertical bars (as shown in Fig. 9.21) as follows:

$$\left.\begin{aligned} &P_{max} \text{ (due to horizontal bars)} \\ &\qquad = 0.785 n d^2 f_y \\ &P_{max} \text{ (due to vertical bars)} \\ &\qquad = 0.240 n d^2 f_y \end{aligned}\right\} \quad (9.1)$$

where

n = number of bars across the crack
d = diameter of bar
f_y = yield strength of steel used.

The preceding equations show that the horizontal bars are three times more effective. This may be due to the reduced flexural resistance of vertical bars as dowels and to the direct tension in the horizontal bars, which add protection for strain hardening at large crack widths.

Scrivener (1976) found that walls that showed some degradation in static tests (i.e., that failed in brittle shearlike manner) behaved similarly under dynamic loading. But, interestingly, walls with ductile flexural behavior under static loads performed much less satisfactorily in dynamic tests. Figure 9.22 indicates the pronounced load degradation obtained with increasing cycles of dynamic load, whereas a similar wall (Fig. 9.22) maintained a constant load.

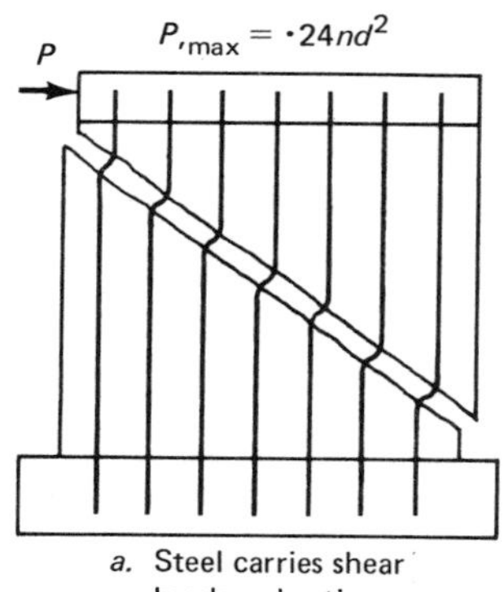

a. Steel carries shear by dowel action

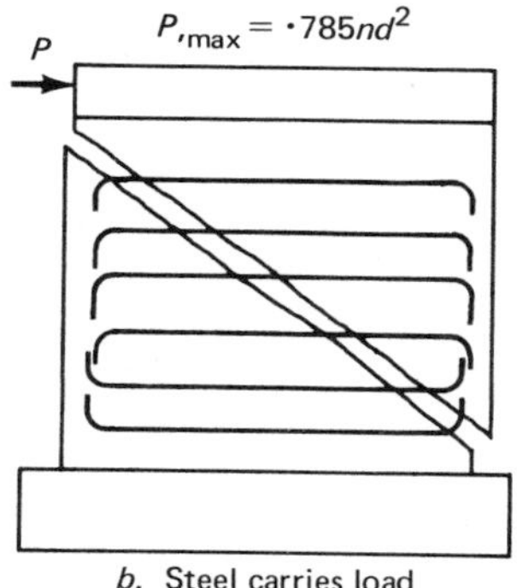

b. Steel carries load by tension

Fig. 9.21. Relative effectiveness of vertical and horizontal shear steel. (*From Meli 1972.*)

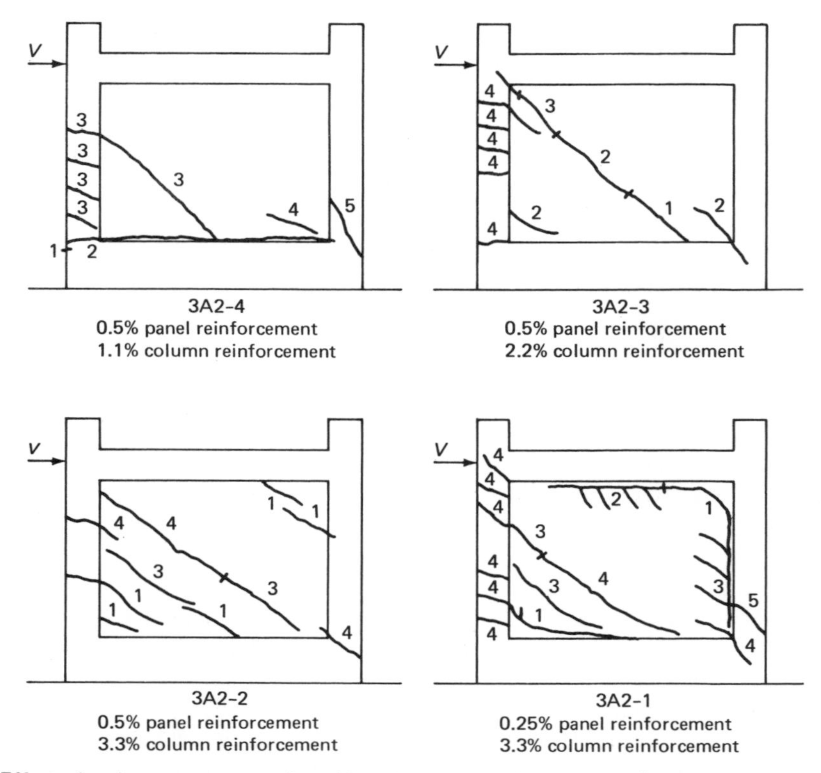

Fig. 9.22. Effect of various percentages of steel in column on cracking of walls (Benjamin and Williams 1957).

Scrivener reasoned that, in the toe areas, pieces of masonry and mortar that are cracked and crushed in the static tests drop out of the wall due to the wall motion. Accordingly, the vertical edge reinforcing would lose lateral support of masonry and cause it to buckle in the following half-cycle when in compression. Its load-carrying capacity thus is due to a reduced steel couple.

9.9 REINFORCED VS. UNREINFORCED INFILL PANELS

It becomes apparent from the section on the behavior of infill frames that it is desirable to reinforce the infill masonry. If panel is unreinforced, the first crack will rapidly spread toward the tension column and perhaps to the foundation, depending on the location of panel and load combination acting on it. A further increase in shear load, which initiates this crack, will produce a crack along the tension column to the beam, thus causing partial separation between the two, unless shear connectors are provided. Thus, both the reinforcement in the panel and the shear connectors will help the composite frame carry additional load. The type of failure of the entire structure may vary with the relative stiffnesses of wall and frame, but, in general, will be similar to that of the unreinforced parallel structure.

Reinforcement in the infill walls does a number of things. To start with, it certainly provides increased strength. It prevents any local but incipient failure that might occur in the nonreinforced masonry walls and would also delay the spreading of such initial crack. Action will be very much like the shear reinforcement in the reinforced concrete cantilever beam, and failure through the wall will be controlled by the amount and distribution of this reinforcement. Furthermore, it performs other important functions. It provides the means of tying together the masonry units and other structural elements. Hence, in an earthquake or other dynamic loading, such a structure may act as a unit without the danger of individual masonry units falling out or separating from each other. Reinforcing can also increase the ductility of a structure, if the structural mode of carrying load allows steel to yield.

It should be pointed out that the interest in non-load-bearing masonry panels was substantially increased due both to the increasing use of frames with such panels and the earthquake disasters in the last decade. In the United States, extensive damage of masonry structures during the San Fernando, California earthquake in 1971 caused investigations of the remaining buildings in the state and modifications to various applicable codes.

Scrivener (1976) and Mayes and Clough (1975) summarize much of the work done in tests of both unreinforced and reinforced masonry and their behavior through failure under both static and dynamic loads. One of the early tests was conducted by Benjamin and Williams (1957). They investigated the influence of the amount of reinforcement (0% to 1.5%) and its distribution in the panels on the behavior of the reinforced concrete panels; they also varied column reinforcement to study its influence. Their work, presented in Figs. 9.22 and 9.23, indicates that the behavior of all walls is essentially the same before cracking. Reinforcing in the panels becomes effective only after cracking begins. As the percentage increases, the number of cracks before ultimate load increases with decrease in the individual width of each crack. Distribution and pattern of reinforcing was found to be most efficient in a simple rectangular grid rather than a diagonal one. An interesting observation was that heavy concentrations of steel along diagonals produced large shrinkage cracks along the panel diagonals and reduced the wall rigidity. In the case of masonry panels, the mortar used should be strong enough to prevent cracking along the joints.

Considerable work has been done by Scrivener (1968, 1970, 1976) on reinforced and unreinforced masonry panels under static, dynamic, monotonic, and cyclic loads. Results from Scrivener's tests on these walls under static monotonic loads are presented in Fig. 9.24, which clearly demonstrates the many advantages mentioned earlier. On the distribution of panel reinforcements, he arrived at conclusions simi-

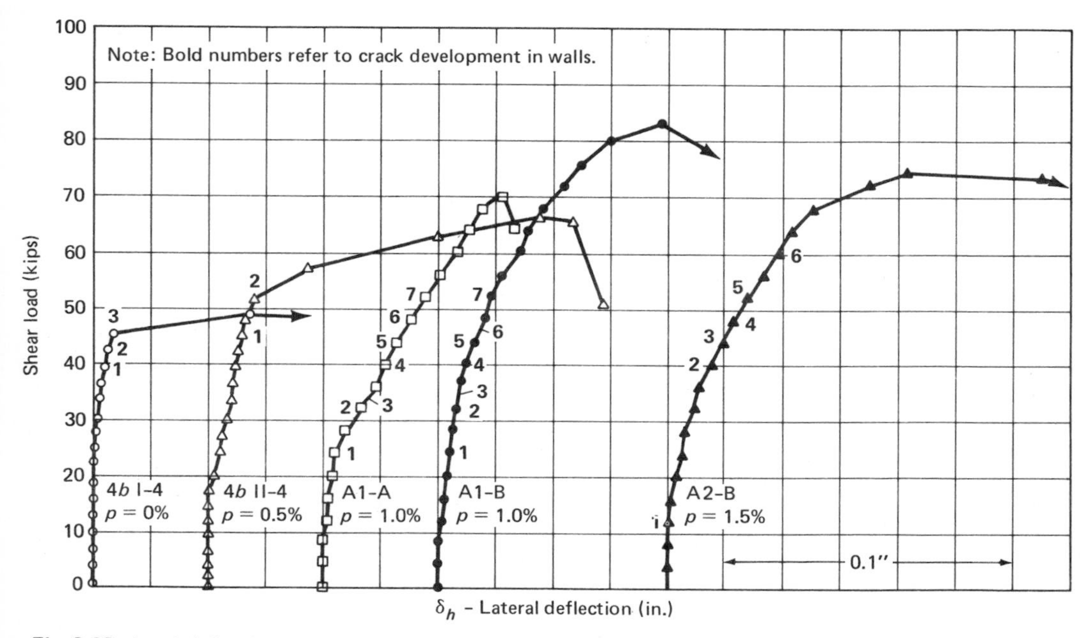

Fig. 9.23. Load-deflection curves for walls with varying steel ratios, p, in percent (Benjamin and Williams 1957).

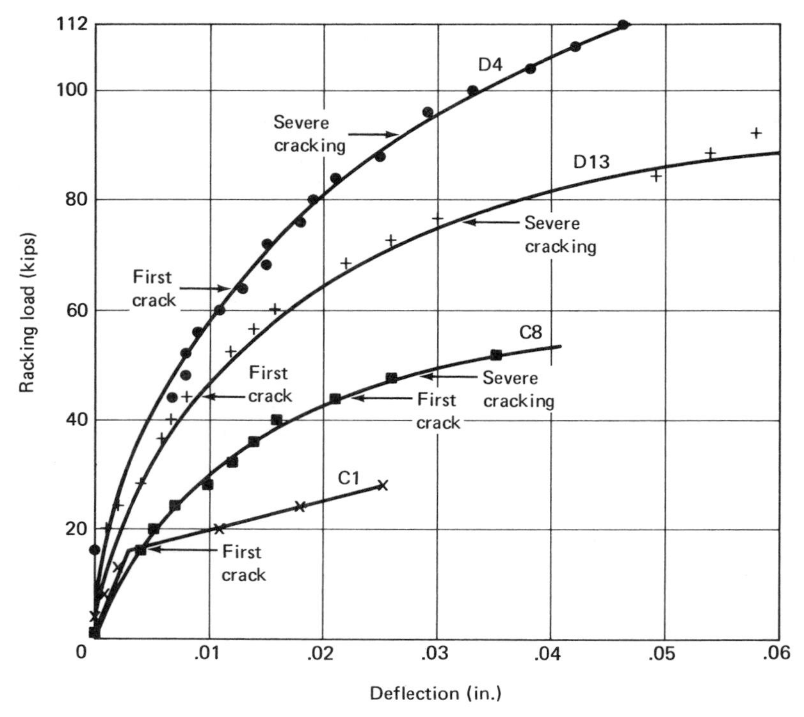

Fig. 9.24. Typical load-deflection curves of infills (Scrivener 1968).

lar to those of Benjamin and Williams, and of Franklin. The percentage of reinfocement higher than 0.3% was found to be ineffective in obtaining higher ultimate strength. Scrivener considered various parameters including wall aspect ratio, the bearing load, and the amount and disposition of the reinforcing steel in cyclic load tests. A complete cycle of 100% load was applied with considerable yielding of steel. The tests produced a range of results. Ductile flexural behavior similar to long cantilevers was obtained with walls of high aspect ratio, low bearing load, and low quantity of vertical edge reinforcing. Results of such walls shown in Fig. 9.25 show long yield plateaus and relatively large hysteresis loops indicating good energy dissipation desirable for the seismic-resistant structures. On the other hand, reducing aspect

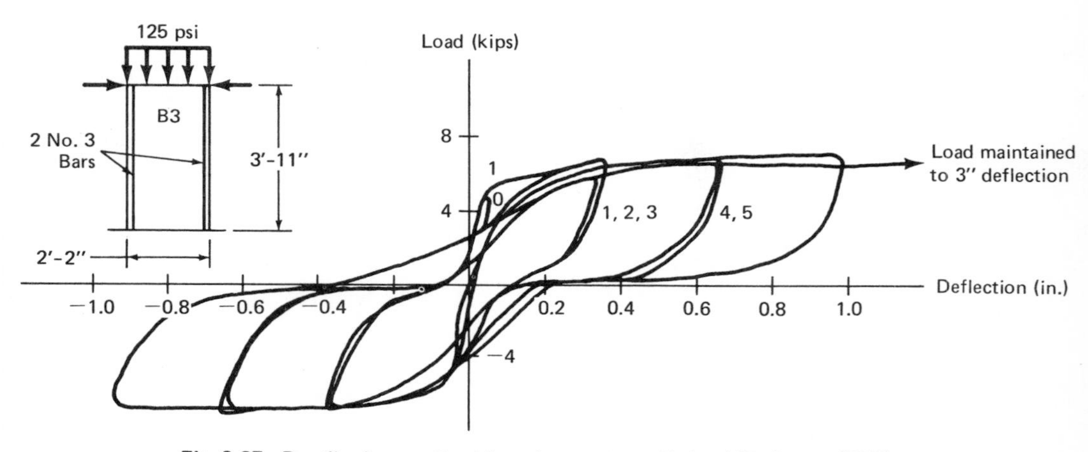

Fig. 9.25. Ductile shear wall subjected to static cyclic load (Scrivener 1976).

ratio, increasing bearing load, and increasing the vertical edge steel all tended to produce brittle shearlike failure characterized by diagonal cracking. If the reinforcement percentage was very small and away from vertical edges, the failure was sudden and caused considerable damage to the wall. Results in Fig. 9.26 indicate that although initial behavior is flexural, the subsequent loading produced smaller and smaller cycles and narrower hysteresis loops with relatively small energy dissipation; combined with load degradation, this does not represent a desirable aseismic behavior. When panels were put together with the surrounding frames and tested similarly to the walls described earlier, they showed a considerable strength increase (of over 400%) and higher initial stiffness.

It should be apparent that the maximum benefit of infill panels is derived only if they are reinforced (even nominally). The tests also indicated that face shells of these walls did not break even after a number of cycles (of very heavy loads to represent catastrophic earthquakes). This is significant because often the major cause of death in earthquakes is flying debris and shattered pieces. Therefore, it is important that all units be tied together with light reinforcement to reduce this tendency. Various codes recommend reinforcements in masonry walls to achieve some of these advantages. These recommendations may be referred to, in the publications by the National Concrete Masonry Association (NCMA 1973), the Veterans Administration (1973), and the other codes. It is recommended that all masonry walls be reinforced (see Scrivener 1976) as follows:

> Both horizontal and vertical reinforcing will be provided with a minimum total of 0.1% of gross area of section. Maximum vertical and horizontal reinforcing spacing to be 24 to 32 inches, respectively.

9.10 EFFECT OF OPENINGS ON THE INTERACTION

Although it may be possible to have an infill wall without opening, such advantage discussed earlier cannot be expected all the time. Some walls are bound to have openings: if small for running cables or ducts, the effect may also be negligible; however, if large, such as for windows and doors, the effect may be pro-

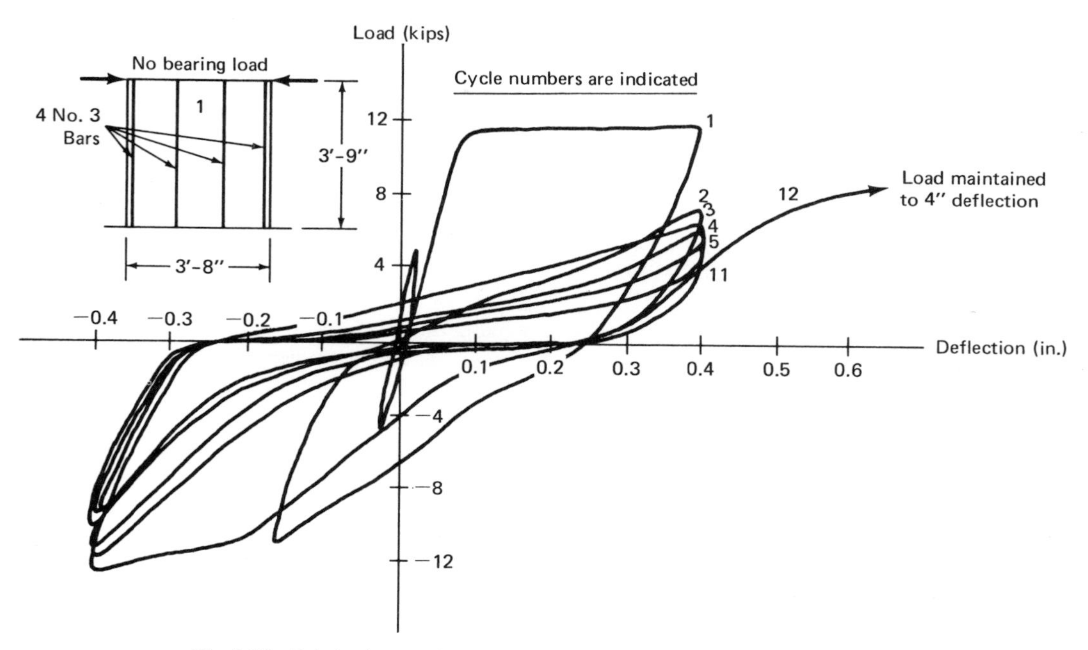

Fig. 9.26. Brittle shear wall subjected to static cyclic load (Scrivener 1976).

nounced, and interaction between the wall and frame may be considerably reduced.

The work Benjamin and Williams (1958) on reinforced concrete shear walls can also be related to masonry walls with some modification. As mentioned previously, the behavior is affected by a full or partial contact between walls and frame, which is the case with concrete wall. Benjamin and Williams considered the effects of large openings in walls, such as doors and windows. It was demonstrated that the behavior of the composite system can be predicted by taking into account all deformations resulting from bending, shear, and axial loading. It was shown that the total wall strength is equal to the sum of the strengths of individual piers assumed as separate walls. This result was later used by Lefter (1974) in his work to predict the behavior of masonry walls. Another important conclusion reached was that a proper corner anchorage between the walls and frame would increase the strength and rigidity of a wall containing an opening after cracking had begun.

Coull and Stafford-Smith (1967) tested infilled steel frames with central openings and infill not connected with the frames. The additional reinforcement around the openings was one of the variables in their tests. They showed that there was a considerable decrease in strength and stiffness of the system as much as 70% without reinforcement and to 45% by using reinforcements around these openings. The failure was generally due to crushing of the loaded corner, and a considerable cracking prior to failure was observed.

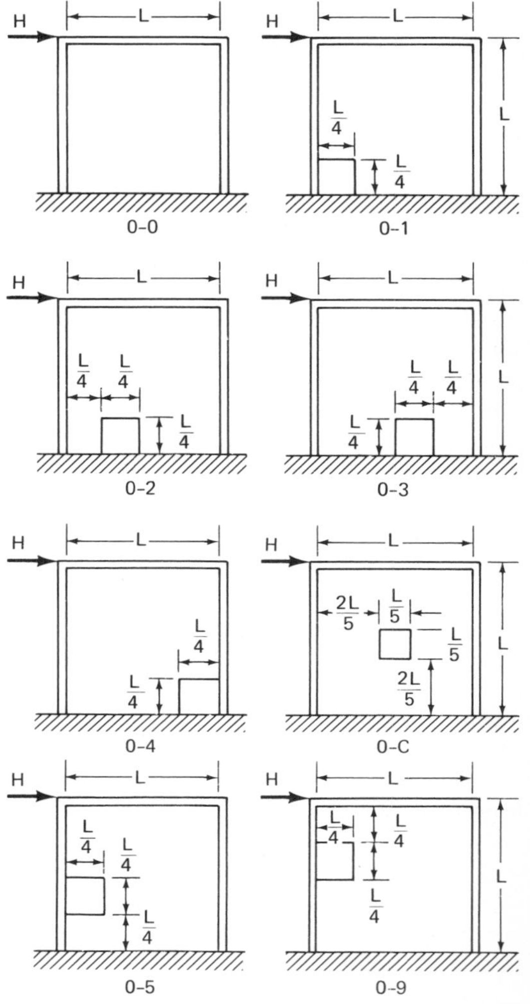

Fig. 9.27. Different locations of openings in the infills; results shown in Table 9.2. *(From Mallick and Garg 1971.)*

More research on infilled frames with openings was presented by Mallick and Garg (1971). They investigated a wide range of variables affecting the frame behavior; the major variables were size, shape, and location of openings and also the shear connectors between the wall and frame. Figure 9.27 taken from their work indicates the extent of openings considered. Based on experimental work and using the finite element method for analysis, Mallick and Garg recommended proper locations of such openings. Their investigation also demonstrated that the shear connectors should be provided, which, in addition to giving more stiffness, forces the failure initiated by the diagonal crack and followed by crushing of loading corner. This kind of failure gave sufficient warning before the composite frame reached its ultimate strength. It is also worth noting that spacing of shear connectors did not appreciably affect the capacity; however, additional work is necessary to determine the optimum spacing of shear connectors. This work points out the diagonal strut action dealt with by the

Table 9.2. Comparison of theoretical and experimental behavior of infilled frames (from Mallick and Garg 1971).

Infill: 24 × 24 × 0.75 in. (60 × 60 × 2 cm)
Frame: 0.75 × 0.75 in. (2 × 2 cm)

		Frame *without* Connectors				Frame *with* Connectors			
		Experimental		Theoretical		Experimental		Theoretical	
Opening Type and Position	Openings (in.)	(kg/cm × 10^3)	(kip/in.)	(kg/cm × 10^3)	(kip/in.)	(kg/cm × 10^3)	(kip/in.)	(kg/cm × 10^3)	(kip/in.)
0–0	No opening	20	231	18.2	210	33.4	385	39.2	452
0–1	6 × 6	23	265	17.8	205	30.8	355	34.5	398
0–2	6 × 6	17.85	206	17.7	204	46.5	536	39.13	451
0–3	6 × 6	15.4	178	17.35	200	36.4	420	34.5	398
0–4	6 × 6	3.12	36	7.85	90	12.2	141	35	404
0–C	4.8 × 4.8	16	184	15.2	175	30.8	355	35.4	408
0–5	6 × 6	23.2	267	17.6	203	33	381	38.9	449
0–9	6 × 6	16.1	186	17.4	201	37.8	436	36.8	424
0–5.9	12 × 6	-	-	15.3	176	-	-	35.9	414

earlier methods of calculations and cautions that the strength may be reduced as much as 75% if door or window openings are near the corners. This reduction in strength is much smaller when the openings are located near the middle of the frame span. This is also desirable for transferring the gravity loads above such openings by *arch action* into the columns rather than to the lower story beams; this also will prevent progressive collapse in such infill frames. Test results of frames are shown in Fig. 9.27 and in Table 9.2, which indicate experimental evidence of the lowering of frame capacity due to openings and also the desirable locations of such openings.

According to the tests by Fiorato et al. (1970), infilled frames with openings show more tendency for flexural- or hinging-type failures compared to those without, due to their greater flexibility resulting from openings. Like the earlier tests, they also showed that the size, shape, and location of openings had a definite influence on the location of hinges developed at failure. Another significant finding was that the prediction equation developed for solid infill frame underestimated the capacity of these frames. One reason can be that these openings would force the crack pattern to a more suitable failure pattern, and the cracked portion of wall could still carry more load due to friction and mechanical interlock between the frame and wall below the openings.

9.11 METHODS OF ANALYSIS OF INFILL FRAMES

Several methods have been developed based on the equivalent-frame concept, the diagonal-strut action of the infill, and some based on elastic analysis using the finite element analysis approach with high-speed computers. The use of any particular method will be up to the individual, depending on the nature of the structure and the available resources. Equivalent-frame concept is simpler to use and less time-consuming, but it is approximate and also not quite suitable for frames with openings in walls, when considerable discrepancies are obtained. On the other hand, due to the development of computers and considerable progress in the use of the finite-element method, this method seems more appropriate. It allows more variations in the dimensions and properties of the infill including openings and can be used to reduce data in a tabular form for use by design engineers. The recently developed TABS program at the University of California at Berkeley by Wilson and Dovey (1972) can be conveniently used for analysis. This program is available with a user's manual and is not too difficult.

A brief outline of these methods follows.

9.11.1 Equivalent-Frame (Truss) Method

The equivalent frame presented by Stafford-Smith and Carter (1969) is made up of truss elements, as shown in Fig. 9.13 in which diagonal-strut properties should be determined. As the horizontal load is increased, the width of this strut will be reduced due to partial separation between the wall and frame. It has been shown that the diagonal stiffness and strength of an infill panel depends on the length of contact with the frame, along with its own properties. This length of contact, α, is given approximately by:

$$\alpha/h = \pi/2\lambda h \qquad (9.2)$$

in which,

$$\lambda = \sqrt[4]{\frac{E_I t \sin 2\theta}{4EIh'}} \qquad (9.3)$$

where

E = approximate value for modulus of elasticity for frame members and infill (in the case of infill, this will be the initial tangent modulus)

α = length of contact between infill and frame member

θ = slope of infill diagonal to the horizontal

h = story height

λh = nondimensional parameter expressing the relative stiffness of frame to that of infill

E_I = modulus of elasticity of infill

t = thickness of the infill

h' = height of infill

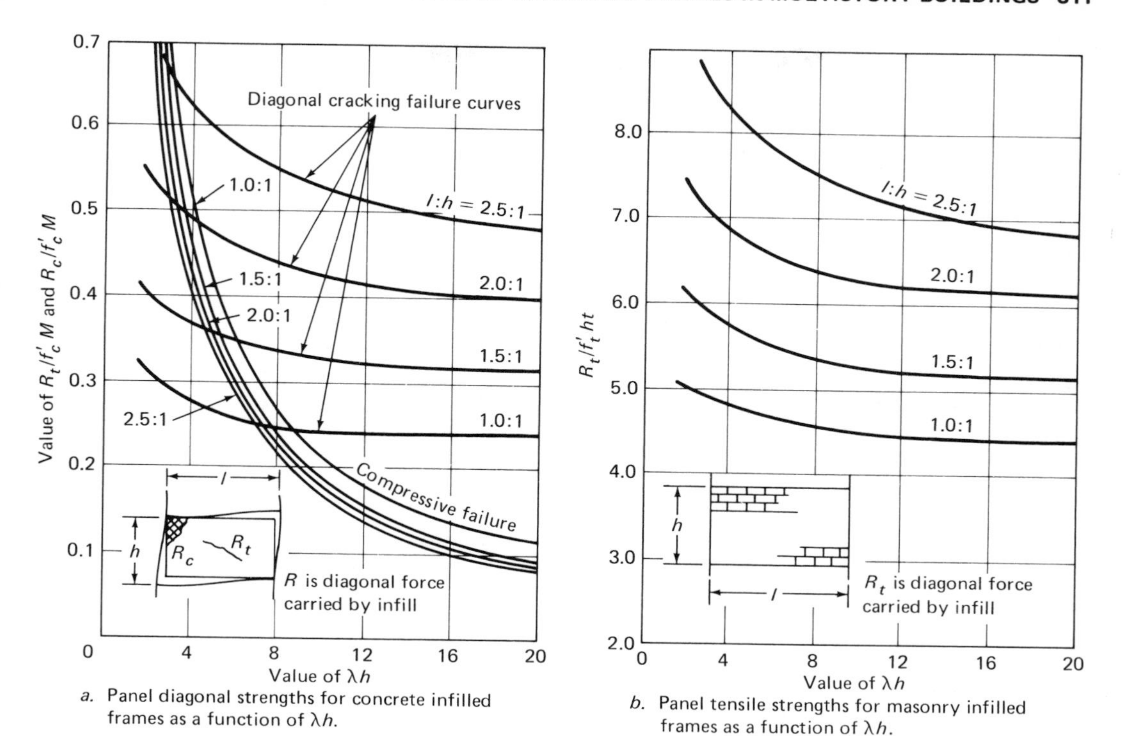

a. Panel diagonal strengths for concrete infilled frames as a function of λh.

b. Panel tensile strengths for masonry infilled frames as a function of λh.

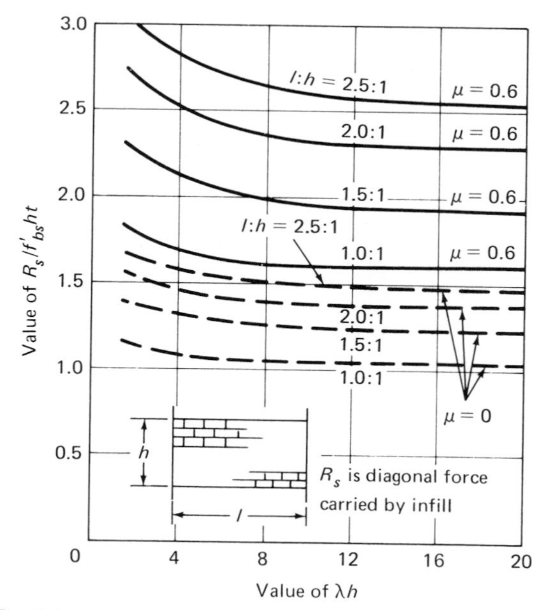

c. Panel shear strengths for masonry infilled frames as a function of λh.

Fig. 9.28. Theoretical evaluation of various properties of infill frames. (*From Stafford-Smith and Carter 1969.*)

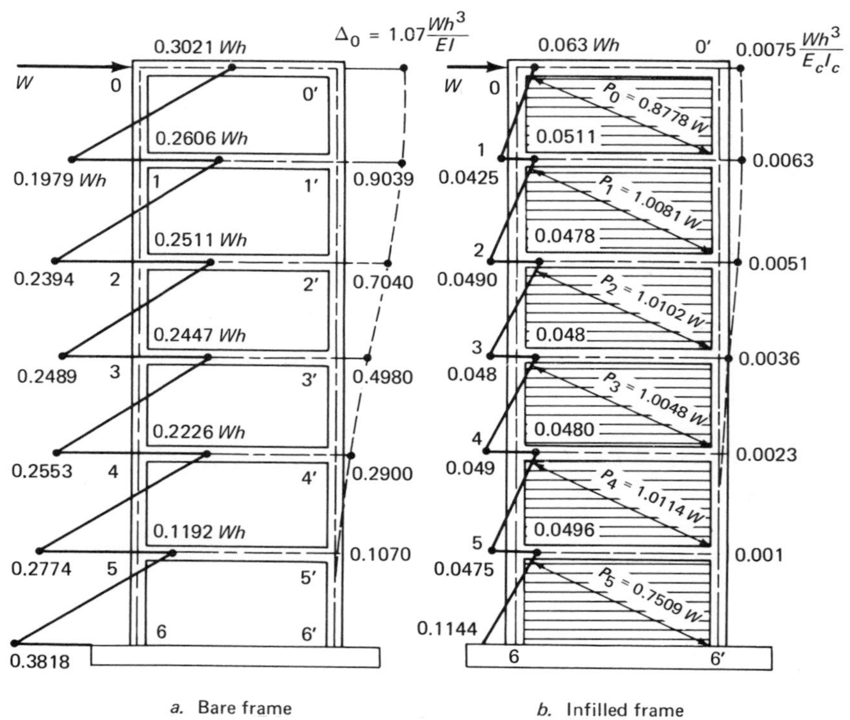

a. Bare frame *b.* Infilled frame

Fig. 9.29. Computed values of moments, forces in infills, and deflections in model frames. (*From Smolira 1974.*)

I = moment of inertia of column.

In Eq. 9.2, λh represents essentially a nondimensional parameter expressing the relative stiffness of frame to that of the infill. Equations 9.2 and 9.3 are expressed in algebraic or graphical form as a function of other parameters by Stafford-Smith and Carter and reproduced in Fig. 9.28. In Fig. 9.28a, diagonal struts for concrete infill are shown as a function of λh. It shows relationships between different length-to-height ratios and the corresponding governing failures, compressive and diagonal cracking of the strut. In Fig. 9.28b, a similar plot is shown for the masonry infill. An additional parameter, α, representing the coefficient of friction between the masonry and mortar, is introduced to account for failure occurring at joint by its shear strength. Mathematically derived equations for these are also shown in these figures. In the case of masonry infill, its tensile strength, which will also cause diagonal-strut failure, should be considered as a mode of failure; it is shown in Fig. 9.29c. These three figures can be used to obtain approximate strength and design of a composite frame.

9.11.2 Force-Displacement (Truss) Method

Smolira (1973) developed his method similar to the last one, but on a different basis. Basically, it is a truss solved by the force-displacement method, by setting up a matrix formulation, in which statically indeterminate values are forces (i.e., moments in the frame and direct forces in the infill) and linear displacements of joints. It is assumed that the frame and infill are in contact at diagonally opposite corners and that the frame is subjected to bending moments (as well as forces), whereas the infill wall develops a resultant diagonal force. In Smolira's method, the structure is assumed to deflect to its final position defined by lateral displacement; this causes discontinuity at the joints, which are closed by reapplying all moments and panel forces of such magnitude as necessary to simultaneously satisfy compatibility of deformations and equilibrium of forces at all joints. The unknown forces are four quantities: two bending moments (*ma* and *mb*) at two corners, the direct force p, and deflection A, per story. The final stiffness matrix size is four times the number of stories, which is solved on a computer to get all four quantities at each story. An example from Smolira (1973) is shown in Fig. 9.29 for a six-story model reinforced concrete frame with brick infill. It demonstrates clearly the advantages of infill for reducing the frame moments and the lateral displacements. The experimental values showed a discrepancy of 18% to 33% for moments and deflections, respectively, which was attributed to the several inaccuracies and material parameters of model frame.

9.11.3 Finite-Element Approach

Mallick and Garg (1971) have used a finite-element approach to analyze the infill frames with and without shear connectors. In this approach, the concept of discretization of a continuum by means of small finite elements is used following the basic theory of elasticity. Each element represents the same part in the continuous system and has properties of the original structure. The composite nature, the properties of the two components–namely, infill and the frame and openings, if any–can also be considered in this analysis. The main disadvantage is that a large number of elements are required to get a desirable accuracy. The stiffness of each element is developed by accounting for the displacement continuity between the adjacent elements (particularly between the infill and frame) in a suitable manner. All such stiffnesses are finally combined to form a stiffness matrix for the entire system.* A typical element and stiffness matrix of an in-plane rectangular element is shown in Fig. 9.30. Using previous expressions developed by Stafford-Smith and Carter (1969) for the length of contact between the wall and frame, interaction can be carried out to determine the stresses and displacements in this method. When shear connectors are present, the method is used with an assumption that the surrounding frame transfers loads between the infills and the contact surfaces through these shear connectors.

9.11.4 Computer Analysis (TABS)

A three-dimensional analysis of building systems program (TABS) was developed at the University of California at Berkeley by Wilson and Dovey (1972), and has been used to analyze building frames with different parameters of shear walls and configurations. It can take into account both static as well as earthquake (or

*Details of the finite element method may be found in any standard textbook on this subject.

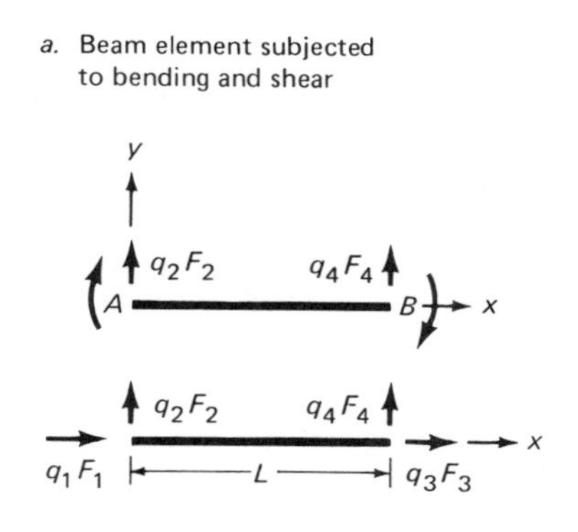

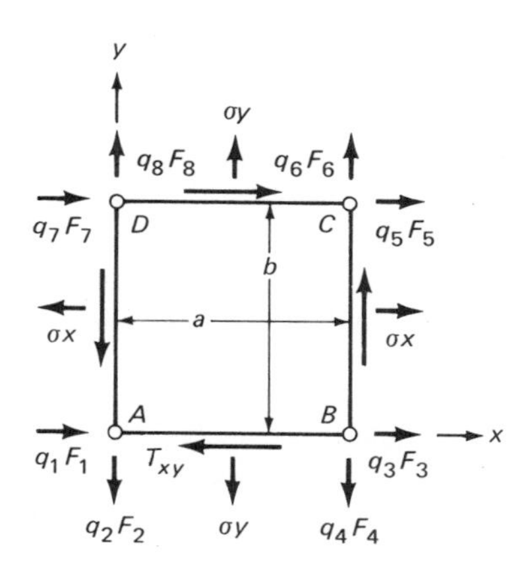

Fig. 9.30. Details of various elements and forces used in the finite element method. (*From Mallick and Garg 1971.*)

dynamic) loading. In this program, the building is idealized by a system of independent frame and shear-wall (panel) elements interconnected by floor diaphragms, which are rigid in their own plane. It can evaluate the response of the frame under independent or combined static and earthquake loading (input as a time-dependent ground acceleration or as an acceleration spectrum response), and, along with three-dimensional mode shapes and frequencies, it gives an output in terms of story shears, moments, and lateral deflections. Frames and shear walls are considered as substructures in the basic formulation; therefore, input data preparation for many structures is minimized and significantly reduces the computation effort.

The TABS as well as many other available programs may be used to analyze the composite frames to fully achieve advantages of the interaction of walls and frames and computational ease to achieve the overall economy. The structures, already designed as structural steel and reinforced concrete frames using conventional shear walls, can be made composite by replacing reinforced-concrete shear walls by masonry walls. The additions of such infill walls can be constructed in practice without sacrificing safety or reliability.

9.12 CONCLUDING REMARKS

From the discussion of various topics related to the composite interaction between frames and infill walls, it becomes apparent that such shear-wall-frame action can be handled systematically for medium-height buildings, and the strength of the composite frame under the action of horizontal loads may be accounted properly. The variation of stiffness of frames and walls should be considered to determine the optimum strength for combinations of stiffness to design the structure accordingly. The finite-element approach mentioned earlier should prove very useful for further analytical studies including the effect of openings, shear connectors, and other parameters. Recent failures due to progressive collapse have drawn considerable attention from engineers. These studies should be tied to the interaction to design against such failures when masonry infills are used. Combination of load-bearing walls as infill walls with the recent techniques of tubular frame system should be studies to elevate the uses of infill walls to heights not attempted earlier. Such combination will be particularly economical in countries where frame materials are more expensive and the labor less expensive. This technique can be successfully used to design and construct structures, wherein they may be made taller due to the composite action of walls with frames. In summary, it is hoped that the composite action between the infill walls and the frame will be applied by engineers to design resulting structures more economically.

REFERENCES

Abdallah, N. (1972), *The MLC Tower*, Planning and Design of Tall Buildings Conference, Sydney, Australia, August 14–17.

Amrhein, J. (1976), A Tower with Its Top in the Heavens, *Proceedings of the First Canadian Masonry Symposium*, Calgary, Alberta, p. 251.

Benjamin, J. R. and H. A. Williams (1957), The Behavior of One-Story Reinforced Concrete Shear Walls, *Journal of the Structural Division*, ASCE, Proc. Paper 1254, Vol. 83, No. ST3, May, pp. 1254.1–1254.39.

Benjamin, J. R. and H. A. Williams, "Behavior of One-Story Reinforced Concrete Shear Walls Containing Openings," Journal of the American Concrete Institute, Proceedings, Vol. 30, No. 5, November, 1958, pp. 605–618.

Brick Institute of America BIA (1963), *Differential Movement*, Technical Note No. 18, published by Brick Institute of America, McLean, VA (3 parts).

Conference on Planning and Design of Tall Buildings (1972), *Proceedings of Conference on Planning and Design of Tall Buildings*, Sydney, Australia, August 14–17.

Conference on Planning and Design of Tall Buildings (1972), Lehigh University, *Proceedings of Conference on Planning and Design of Tall Buildings*, published by ASCE, New York (5 volumes).

Coull, A. and B. Stafford-Smith, eds. (1967), *Tall Buildings*, Pergamon Press, Oxford.

Dawson, R. V. (1972), An Analytical and Experimental Study of the Lateral Load Response of Framed Structures Before and After the Inclusion of Infill Walls, Ph.D. Thesis, The University of Calgary, December.

Fedorkiw, J. P. and M. A. Sozen (1968), *A Lumped-Parameter Model to Simulate the Response of Reinforced Concrete Frames with Filler Walls*, University of Illinois, Civil Engineering Studies, Structural Research Series No. 338, Urbana, June.

Fintel, M. (1974), Multistory Structures (Chapter 10) in *Handbook of Concrete Engineering* (M. Fintel, ed.), Van Nostrand Reinhold, New York, pp. 287–344.

Fiorato, A. E., M. A. Sozen, and W. L. Gamble (1970), *Behavior of Five-Story Reinforced Concrete Frames with Filler Walls*, Interim Report to the Department of Defense Office of the Secretary of the Army, Office of Civil Defense, Urbana, January.

Franklin, H. A. (1970), *Nonlinear Analysis of Reinforced Concrete Frames and Panels*, University of California, Department of Civil Engineering, SESM Report No. 70-5, Berkeley, March.

Holmes, M. (1961), Steel Frames with Brickwork and Concrete Infilling, *Proceedings of the Institution of Civil Engineers*, Vol. 19, August, pp. 473–478.

Holmes, M. (1963), Combined Loading on Infilled Frames, *Proceedings of the Institution of Civil Engineers*, Vol. 25, May, pp. 31–38.

Jessop, E. L. and Ward, M. A., eds. (1976), *Proceedings of the First Canadian Masonry Symposium*, held at Calgary, Alberta, Canada, June, p. 408.

Johnson, F. B., ed. (1969), *Designing, Engineering and Constructing with Masonry Products*, Gulf Publishing Co., Houston, Texas.

Kadir, M. R. A. (1974), The Structural Behavior of Masonry In-Fill Panels in Frame Structures, Ph.D. Thesis, Edinburgh University.

Karamanski, T. (1967), Calculating In-filled Frames by the Method of Finite Elements, *Symposium Proceedings on Tall Buildings* (Coull, A. and Stafford-Smith, B. S., eds.) Pergamon Press.

Kazimi, S. M. A. (1967), Photo Elastic Study of One Story Shear Walls Containing Openings, *Indian Concrete Journal*, December, p. 463.

König, G. (1972), Cast-in-Place Reinforced Concrete Systems, *Proceedings of the Conference on Planning and Design of Tall Buildings*, Lehigh University, Vol. Ia, p. 510.

Lee, S. L. and Karasudhi, P., eds. (1974), *Proceedings of the Regional Conference on Tall Buildings*, Sponsored by the Asian Institute of Technology, Bangkok, Thailand.

Lefter, J. (1974), Reinforcing Existing Buildings to Resist Earthquake Forces, M.S. Thesis. University of Maryland, College Park, Maryland.

Liauw, T. C. (1970), Elastic Behavior of Infilled Frames, *Proceedings of the Institution of Civil Engineers*, Vol. 46, July, pp. 343–349.

Mainstone, R. J. (1971), On the Stiffness and Strengths of Infilled Frames, *Proceedings of the Institution of Civil Engineers*, London, England, Supplement Paper No. 73605, pp. 57–90.

Mallick, D. V. and Garg, R. P. (1971), Effect of Openings on the Lateral Stiffness of Infilled Frames, *Proceedings of the Institution of Civil Engineers*, London, England, Vol. 49, June, pp. 193–209.

Mallick, D. V. and Severn, R. T. (1968), Dynamic Characteristics of In-filled Frames, *Proceedings of Institution of Civil Engineers*, London, England, Vol. 39, pp. 261–287.

Mayes, R. L. and Clough, R. W. (1975), *A Literature Survey-Compressive, Tensile, Bond and Shear Strength of Masonry*, EERC Report No. 75-15, University of California, pp. 199.

Meli, R. (1973), Behavior of Masonry Walls under Lateral Loads, *Proceedings of Fifth World Conference on Earthquake Engineering*, Rome, Italy.

NCMA (1973), *Specification for the Design and Construction of Concrete Masonry*, National Concrete Masonry Association, McLean, VA.

Ockleston, A. J. (1955), Load Tests on a Three Story Reinforced Concrete Building in Johannesburg, *The Structural Engineer*, Vol. 33, No. 10, October, 1951, pp. 304-322.

Payton, J. J. (1972), Collins Place Project—Melbourne, *Planning and Design of Tall Buildings Conference*, Sydney, Australia, August 14–17.

Polyakov, S. V. (1956), *Masonry in Framed Buildings* (English Translation by G. L. Cains, 1963).

Polyakov, S. V. (1967), On the Interaction Between Masonry Filler Walls and Enclosing Frame When Loaded in the Plane of the Wall, *Transactions in Earthquake Engineering*, Earthquake Engineering Research Institute, San Francisco, California, pp. 36–42.

Priestly, M. J. N. and Bridgman, D. O. (1974), Seismic Resistance of Brick Masonry Walls, *Bulletin New Zealand Society for Earthquake Engineering*, Vol. 7, No. 4, pp. 167–187.

Proceedings of the Conference on Planning and Design of Tall Buildings (1973), ASCE, New York (5 volumes).

Riddington, J. R. (1974), The Composite Behavior of Walls Interacting with Flexural Members, Ph.D. Thesis, University of Southampton, England.

Sabnis, G. M. (1976), Interaction between Masonry Walls and Frames in Multistory Structures, *Proceedings of the First Canadian Symposium on Masonry Structures*, Calgary, Alberta, p. 324-341.

Sachanski, S. (1960), Analysis of the Earthquake Resistance of Frame Buildings Taking into Consideration the Carrying Capacity of the Filling Masonry, *Proceedings of the Second World Conference on Earthquake Engineering* Vol. 3, Tokyo, pp. 2127–2141.

Scalzi, J. B. and Arndt, A. P. (1972), Plate Wall Cladding, State of the Art Report 3B, TC17, *Proceedings of the Conference on Planning and Design of Tall Buildings*, Lehigh University, Vol. Ia, p. 567.

Scrivener, J. C. (1968), Static Racking Tests on Masonry Walls, *Designing, Engineering and Construction with Masonry Products*, F. B. Johnson (ed.), Gulf Publishing Co., Houston, Texas.

Scrivener, J. C. (1969), Static Racking Tests on Masonry Walls, *Designing Engineering, and Construction with Masonry Products*, F. B. Johnson (ed.) Gulf Publishing Company, Houston, Texas, pp. 185–199.

Scrivener, J. C. (1976), Reinforced Masonry in a Seismic Area-Research and Construction Development in New Zealand, *Proceedings of the First Canadian Symposium*, Calgary, Alberta, pp. 371–382.

Smolira, M. (1973), Analysis of Infilled Shear Walls, *Proceedings of Institution of Civil Engineers*, London, December, pp. 895–911.

Stafford-Smith, B. S. (1962), Lateral Stiffness of Infilled Frames, *J. Structural Division*, ASCE, Vol. 88, No. ST6, December, 1962, pp. 183–199.

Stafford-Smith, B. S. (1966), Behavior of Square Infilled Frames, *J. Structural Division*, ASCE, Vol. 91, No. ST1, February, 1966, pp. 381–403.

Stafford-Smith, B. S. (1967), Methods for Predicting the Lateral Stiffness and Strength of Multi-Story Infilled Frames, *Building Science*, Vol. 2, Pergamon Press, November, pp. 247–257.

Stafford-Smith, B. S. and Carter, C. (1969), Structural Behavior of Masonry Infilled Frame Subjected to Raking Loads, *Proceedings of the International Conference on the Masonry Strut System*, University of Texas, Austin, pp. 226–233.

Stafford-Smith, B. S. and Carter, C. (1969), A Method of Analysis for Infilled Frames, *Institution of Civil Engineers (London), Proceedings*, Vol. 47.

Stafford-Smith, B. S. and Carter, C. (1970), The Diagonal Tensile Strength of Brickwork, *The Structural Engineer*, No. 4, Vol. 48, June.

Tamura, R. et al., (1969), A Vibration Test of a Large Model Steel Frame with Precast Concrete Panel until Failure, *Proceedings of Fourth World Conference on Earthquake Engineering*, Chile, pp. 15–30.

Thomas, F. C. (1952), The Strength of Brickwork, *The Structural Engineer*, Vol. 33, No. 2, February, pp. 35–46.

Veterans Administration (1973) Specifications, Earthquake Resistant Design Requirements for VA Hospital Facilities, V.A. Administration, Washington, D. C., 36 pages.

Ward, H. S. (1969), Dynamic Characteristics of a Multi-Story Concrete Building, *Proceedings of Institution of Civil Engineers*, London, p. 553–560.

Weidlinger, P. (1973), Shear Field Panel Bracing, *J. Structural Division*, ASCE, Vol. 99.

Wilson, E. L. and Dovey, H. H. (1972), *Three Dimensional Analysis of Building Systems* (TABS), EERC Report No. 72-8, University of California, Berkeley.

Williams, D. (1971), Seismic Behavior of Reinforced Masonry Shear Walls, Ph.D. Thesis, University of Canterbury, New Zealand.

Wood, R. H. (1952), *Studies in Composite Construction*, National Building Studies Research paper 13, HMSO, London.

Wood, R. H. (1958), The Stability of Tall Buildings, *Proceedings of the Institution of Civil Engineers*, Vol. 11, September, pp. 69–102.

Yorkdale, A. (1972), Masonry Building Systems, *Proceedings of the Conference on Planning and Design of Tall Buildings*, Lehigh University, Vol. Ia, p. 567.

Yorulmaz, M. and M. A. Sozen (1968), *Behavior of Single-Story Reinforced Concrete Frames with Filler Walls*, University of Illinois, Civil Engineering Studies, Structural Research Series No. 337, Urbana.

10

Composite construction in wood and timber

JAMES R. GOODMAN, Ph.D., P.E.
Professor of Civil Engineering
Colorado State University
Fort Collins, Colorado

RICHARD M. GUTKOWSKI, Ph.D., P.E.
Associate Professor of Civil Engineering
Colorado State University
Fort Collins, Colorado

10.1 INTRODUCTION

In engineering applications, layered systems of various materials are used to fabricate beams, floor and roof structures, plates, and shells. Interconnection of the layers in such systems is accomplished in a variety of ways. Consequently, the structural behavior is greatly dependent on the exact nature of the interconnection. If the layers are fastened together with rigid adhesives as in the case of most plastic-based composites, welded assemblies, and laminated wood systems, an assumption of rigid interconnection between layers is common and reasonable. In such instances, interlayer movement, caused by connector deformations, is insignificant, and the composite behavior can be assessed by well-established procedures. In some widely used wood and timber systems, however, the interconnection is achieved mechanically by employing nails or elastomeric adhesives. The assumption of rigid interconnection is incorrect for such systems, and its use leads to improper assessment of the structural performance.

A general theory and design procedure for wood composite systems with nonrigid interconnection resulting in incomplete composite action will be presented. Detailed examples for various structural wood systems will be given. Included in these systems will be examples of layered-beam and column structures. Use of composite analysis procedures and the resulting assessment of the performance of typical wood systems, such as wood joist floor structures, will be presented. Recent developments in combining wood and other materials in such systems as wood-concrete or wood-steel composite structures will be discussed. A novel bridge structure consisting of glued-laminated (glulam) girders acting in composite behavior with glulam decking will be described. Aspects of code requirements for use of composite wood and timber systems will be discussed, and references to current research in composite behavior of wood systems will be presented.

10.2 STATE-OF-THE-ART

The degree of composite action in wood systems depends on the type and number of connectors used to attach the component parts and on the characteristics of the wood itself. Unlike other composite systems, in which a fully rigid interconnection is generally assumed, "interlayer slip" results in incomplete composite action. The occurrence of interlayer slip can significantly affect the overall behavior of wood structures in both the service-load and overload ranges. Until recently, only limited consideration had been given to the effects of interlayer movements occurring in such systems. New developments have greatly advanced the state-of-the-art, and appropriate methods of analysis are available.

General theories for rigidly connected, layered beams, such as welded steel or glulam timber beams, are well known and can be found in basic mechanics of material texts such as Popov (1968). These types of systems do not constitute composite construction in the usual sense. Detailed descriptions of glulam wood systems can be found in the *Timber Construction Manual* (American Institute of Timber Construction [AITC] 1974) and in *Wood Structures: A Design Guide and Commentary* (ASCE 1975). General theories for rigidly connected layered plates and shells, including anisotropic effects, as developed by Ambartsumyan (1964), Dong (1962), and others, are well established. Layered beams and plates fastened by rigid connectors at discrete points also have been analyzed by Clark (1954, 1959). These solutions provide useful upper bounds for the behavior of the more general systems that have interlayer slip. Clearly, the lower bound solution in such problems corresponds to a system of separate layers on top of one another with no friction between them. These bounds are far apart, as will be shown for specific problems in later sections of this chapter.

The number of early studies of systems with interlayer slip is rather small. Granholm (1949) investigates composite beams and columns made of timber, and Pleshkov (1952) also considers composite wooden structures and investigates the interlayer slip. At approximately the same time, in the United States, Newmark et al. (1951) and Siess et al. (1952) analyzed the slip action between steel and concrete in composite beams. An extensive review of past literature can be found in Goodman (1967).

More recently, a number of theoretical and experimental studies have been conducted to develop rational procedures for the analysis and design of wood systems including the effects of interlayer slip and other variables. Goodman (1967) and Goodman and Popov (1968) develop the basic theory for layered beam, plate, and shell systems and discuss experimental studies used to verify the proposed mathematical model. A method based on sandwich theory for beams was developed by Norris et al. (1952) and extended by Kuenzi and Wilkinson (1971). Other studies have been performed by Amana and Booth (1967), Polensek (1969, 1972), and Sliker (1972). Both Polensek and Sliker are concerned with wood-joist floor system analysis under certain assumptions. Work on wood columns has been done by Rassam (1969), Rassam and Goodman (1970, 1971, 1972), and Van Dyer (1976). Analysis of dynamic behavior of layered beam systems with interlayer slip was developed by Henghold (1972) and further reported by Henghold and Goodman (1972).

A recent large-scale study of wood-joist structural systems (Vanderbilt et al. 1974) has resulted in numerous publications concerning analysis and design procedures. A series of theses by Ko (1973), Patterson (1973), Kuo (1974), DeBonis (1974), Liu (1974), Dawson (1974), Tremblay (1974), McLain (1975), Fezio (1976), and Sazinski (1978) comprise a detailed description of the analytical and experimental verification of the composite behavior in wood-beam and joist structural systems with nailed or glued connections. Most recently, a finite-element analysis procedure has been developed for such systems by Thompson et al. (1975, 1976).

The preceding review is, of necessity, brief, and additional references may be found within the papers cited. Further references for specific problems in wood composite systems will be cited as each type of system is discussed.

10.3 CONCEPTS OF INTERLAYER SLIP BEHAVIOR AND INCOMPLETE COMPOSITE ACTION

In considering the behavior of composite systems in wood and timber structures, the presence of interlayer slip must be recognized. To understand this phenomenon, consider the T-beam shown in Fig. 10.1a. Considering the

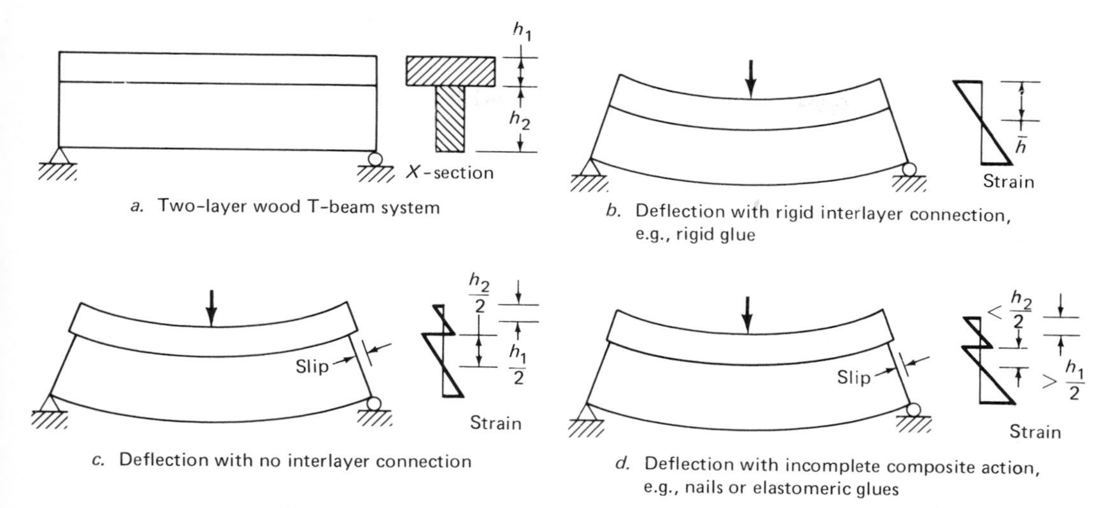

Fig. 10.1. Composite action as affected by interlayer connection.

beam to be made of homogeneous materials, the behavior of the system is a function of the degree of rigidity of the interlayer connection between the stem and flange of the beam. If a rigid glue or some other rigid connection system is used, the behavior of the T-beam is well known as the strains will vary as the distance from the centroid of the cross section as shown in Fig. 10.1b. If no connection is made between the flange and the stem as shown in Fig. 10.1c, each layer (neglecting friction, which is very small) simply slides over the other, and, assuming the layers remain in contact with each other, the strains in each layer will vary about their own midheight. The conditions shown in Fig. 10.1b and c represent the upper and lower limits of behavior of composite systems, i.e., complete interaction and no interaction, respectively. For many, if not most, wood-beam systems with mechanical connections or fastened with elastomeric glues, an intermediate behavior exists as shown in Fig. 10.1d. The interlayer connection offers a partial restraint to the slippage that would occur if no interlayer attachment was provided. The resulting behavior is generally described as "incomplete composite action." The need to recognize this behavior is clear when one notes that the deflection of a beam made of three identical layers with no interconnection is nine times that of the same beam with rigid interconnection. The potential for improving the performance of layered systems through effective interlayer connections and the recognition of composite behavior is, therefore, very important.

10.3.1 Wood-Connector Interaction

For wood systems connected with mechanical fasteners such as nails, bolts, and screws, the interaction of the connector and the wood is the fundamental factor controlling the rigidity of the interlayer connection. The behavior of mechanical fasteners in wood is complicated by the multiplicity of events that take place when the fastener bears against the surrounding wood. The bearing stress between the wood and the connector causes deformation in the wood material as well as shear and bending in the connector. Clearance, intentional or unintentional, between the connector shank and the wood permits the occurrence of a certain degree of virtually unrestrained interlayer slip. The relative significance of the action of each component of the joint depends on numerous factors. A list of the major variables influencing the lateral resistance of a joint as given by McLain (1975) is shown in Table 10.1.

Several researchers (e.g., Kuenzi 1955; Wilkinson 1971, 1972a, 1972b, 1974a, 1974b; Foschi 1974; Patterson 1973; McLain 1975) have conducted studies of the interaction be-

Table 10.1. Selected important variables influencing the lateral resistance of a mechanically connected wood (after McLain 1975).

Material	Physical and chemical properties Mechanical properties (including creep) Species Specific gravity Defects Moisture content (including changes)	May be different for all members connected
Connector	Size Shape Mechanical properties Surface condition	
Joint configuration	Type of connection (single, double, multiple shear) Pressure between contacting surfaces Finish of contacting surfaces Member thickness Connector penetration Prebored holes Angle of connector to surface	
Loading	Direction (perpendicular, parallel, or inclined to grain) Duration of load (rate, length of sustained load) Time between fastening and testing Type of loading (static, dynamic, random) Tension, compression, bending	

tween the wood and the connector. This work has resulted in a variety of theoretical approaches. One common analysis procedure has been to treat the behavior of the connectors in wood as a beam on an elastic foundation. Empirical relationships, including a number of the important variables listed in Table 10.1, have been developed to predict the load-slip behavior of mechanically connected wood joints. In particular, equations developed by Wilkinson (1971, 1972a, 1974a) and McLain (1975) offer the designer a means of evaluating the performance of various wood-connector systems. Examples of these prediction methods are presented in a subsequent section.

10.3.2 Force-Deformation (Load-Slip) Diagram for Wood Connector Systems

The analysis of wood systems with incomplete composite action requires the evaluation of the load-slip relationships for the various interlayer connection methods. This evaluation has generally been accomplished through the use of double- or single-shear tests as shown in Fig.

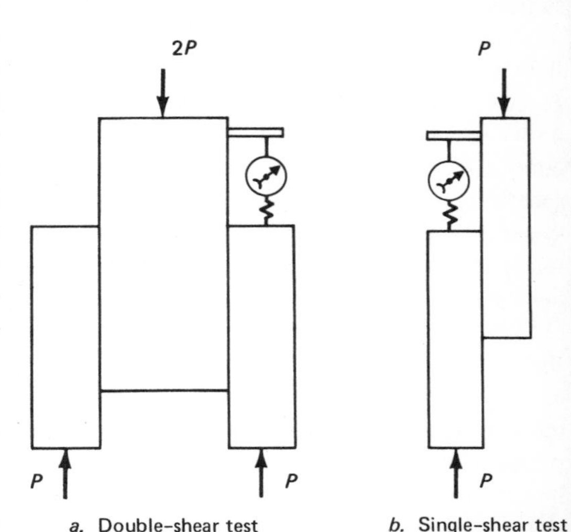

Fig. 10.2. Tests for evaluating load-slip behavior of wood connection systems.

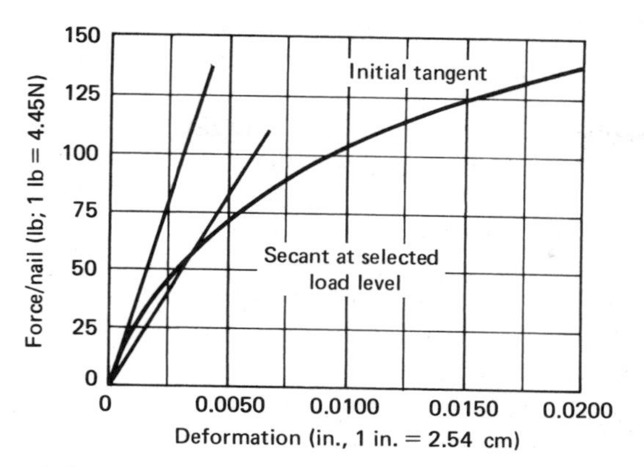

Fig. 10.3. Load-slip test result for 6d nails in Douglas fir (parallel to grain loading).

10.2. A special device for nailed connections that provides excellent alignment and experimental control is described by McLain (1975).

A typical result of a load-slip test is shown in Fig. 10.3. Some nonlinearity in the load-slip behavior, as is evident in this example, is generally exhibited. Various approximation methods have been used, particularly at service-load levels, to model this nonlinearity. For static loading, a secant slope at an appropriate connector-load level is usually recommended. For buckling and possibly dynamic problems, the initial tangent should generally be used. In either case, the slope (i.e., force divided by deformation) is a measure of the stiffness of the connector-wood system. This parameter is commonly referred to as the "slip modulus." The value of the slip modulus is the key variable in the analysis of incomplete composite action in wood systems. As with all wood-associated parameters, a natural variation in the value of the slip modulus may be expected. Quantification of this variation is important in fixing the limits of performance associated with a wood structural system. Additional information on values of the slip modulus and some expected variations are presented subsequently.

10.4 CONNECTIONS FOR COMPOSITE WOOD SYSTEMS

The type, number, and size of connectors used in composite wood systems are the primary influences on the degree of composite action that results when individual components are fastened together. While a number of recent developments in connectors have been made, as noted in *Wood Structures* (ASCE 1975), they are primarily associated with concerns such as withdrawal resistance, which have little bearing on composite behavior. The concern in composite wood systems, from the structural viewpoint, is the lateral (shear) properties of the wood-to-wood connection. Thus, discussion here will be limited to connections commonly used to develop lateral load strength and stiffness.

Three primary classifications of connection types are currently used in structural systems. The most common connectors for light-frame construction are nails or spikes. For larger structural systems, bolts, lag bolts, or timber connectors are used. Increasingly, elastomeric adhesives are being employed, particularly in housing-floor construction. References with general information concerning connectors are well summarized in *Wood Structures* (ASCE 1975). In addition, certain codes such as the *National Design Specification for Stress-Grade Lumber and Its Fastenings* (*NDS*) (NFPA 1973) and trade publications such as *Timber Construction Manual* (*TCM*) (AITC 1974) provide additional details as to connection strength and other design requirements. Basic information on slip modulus (lateral stiffness) for each connector type is more difficult to obtain. A more detailed description of this aspect of designing for composite action in wood systems will be discussed next.

10.4.1 Mechanical Connectors

10.4.1.1 Nails and Spikes

The nailed or spiked joint has always been, and undoubtedly will continue to be, predominant in connecting wood structural systems, particularly in light-frame construction. Two main concerns must be investigated when nailed connections are used to produce composite action in wood systems: (1) ultimate or design capacity of the laterally loaded connectors themselves, and (2) lateral stiffness (slip modulus) behavior of the resulting joint. Both of these design aspects will be discussed.

The *Wood Handbook* (USDA 1940, 1955) states that the allowable load for a bright, common-wire nail driven into side grain is expressed by

$$P = Kd^{3/2} \tag{10.1}$$

where

K = specific gravity-dependent constant
d = nail diameter (in.).

The ultimate load is about $6P$ for softwoods; in hardwoods, it rises to about $11P$. Furthermore, these references indicate that the proportional limit load for softwoods and hardwoods is about $1.6P$.

The *Wood Handbook* (USDA 1974) retains Eq. 10.1, but with different values of K, and redefines P as the lateral load in pounds per nail at a joint slip of 0.015 in. (0.0381 cm). This slip is taken as occurring approximately at the proportional limit load. The corresponding ultimate lateral load is $3.5P$ and $7.0P$ for softwoods and hardwoods, respectively.

Most major code authorities in the United States utilize the *NDS* as the basis for allowable design loads for nailed joints. In reality, the *NDS* has employed most of the recommendations of the 1955 *Wood Handbook* as a foundation and then has modified the data as time and need required.

The *NDS* breaks down the major construction species into four groups, generally by specific gravity ranges, and then assigns an allowable nail load for each group. Table 10.2 cites the allowable nail loads that may be found in the NDS. These loads are valid only for the following conditions:

1. a single, common-wire nail is driven in the side grain of a two-member joint
2. both members are of approximately equal density, are seasoned, and remain dry in service
3. duration of loading is normal (10 yr)
4. the depth of penetration into the member holding the point is not less than the length specified in Column 2 of Table 10.2
5. relative thickness of joint components is in accordance with $T = 50\phi^2$ where T = thickness of side member (in.) and ϕ = nail diameter (in.).

The values in Table 10.2 are subject to adjustments for conditions other than normal duration of load and two-member joint configurations, and may be increased 25% if the side member is a metal side-plate. (These corrections and

Table 10.2. Allowable lateral nail loads from *National Design Specification* (NFPA 1973).

	Load (lb/nail)										
	Size of Common Nail										Required Penetration
Group	6d	8d	10d	12d	16d	20d	30d	40d	50d	60d	(ϕ = nail diameter)
I	78	97	116	116	132	171	191	218	249	276	10 ϕ
II	63	78	97	94	107	139	154	176	202	223	11 ϕ
III	51	64	77	77	88	113	126	144	165	182	13 ϕ
IV	41	51	62	62	70	91	101	116	132	146	14 ϕ

Note: 1 lb = 4.45 N.

others concerning the service conditions may be found in *NDS* and most building codes.)

The quantification of slip-modulus values for wood joints connected with nails or spikes has been the subject of considerable research. No single unified approach has been accepted by code authorities as yet. A thorough review of recent research in this area is given by McLain (1975). In particular, theoretical methods, based on the concept of a beam on an elastic foundation, developed by Kuenzi (1955) and later supplemented by Wilkinson (1971, 1972a, 1972b, 1974a, 1974b), are described and offer one possible means of predicting slip modulus for nailed or spiked connections. Alternatively, techniques that offer promise for empirical evaluation of slip modulus and the load-slip relationship, especially as developed by McLain (1975), are presented in the same reference. Each of these methods will be discussed.

Wilkinson (1971) gave

$$\frac{P}{\delta} = 0.1667E^{1/4}(\alpha E_{\ell})^{3/4}d^{7/4} \qquad (10.2)$$

as a prediction equation for the slip modulus of nailed joints, where

P = nail load (1 lb)
δ = slip (in.)
αE_{ℓ} = elastic bearing constant
E = modulus of elasticity of the nail (psi)
d = diameter of the nail (in.).

This equation is assumed to be valid to a stated "proportional limit" (approximately 0.011-in. [0.0279-cm] slip) and is for round nails loaded parallel to the grain. The recommended value of αE_{ℓ}, obtained by regression analysis, is

$$\alpha E_{\ell} = 3{,}200{,}000 \cdot (SG) \qquad (10.3)$$

where SG = specific gravity of wood members.

In 1972b, Wilkinson extended this analysis to additional conditions. Table 10.3 summarizes the results for smooth shank nails based on tests of three species of wood at 12% moisture content. This work was extended to joints constructed of members with dissimilar properties in a later paper (Wilkinson 1974). A comparison of this method with an empirical approach by McLain (1975) will be given later.

As stated previously, many empirical methods

Table 10.3. Relationship between elastic bearing constant and specific gravity (SG) for several joint variables with smooth shank nails (after Wilkinson 1972).

Loading to Grain	Lead Hole	Relationship
Parallel	Yes	$\alpha E_{\ell} = 3{,}200{,}000 \cdot (SG)$
Parallel	No	$\alpha E_{\ell} = 2{,}144{,}000 \cdot (SG)$
Perpendicular	Yes	$\alpha E_{\ell} = 1{,}280{,}000 \cdot (SG)$
Perpendicular	No	$\alpha E_{\ell} = 1{,}280{,}000 \cdot (SG)$

have also been used to evaluate the load-slip relationship for nailed or spiked joints. After considerable research involving many possible choices of equations to represent the slip phenomenon, McLain (1975) chose, as statistically preferable to the numerous other forms, the expression,

$$P = A \log_{10}(1 + B\delta) \qquad (10.4)$$

where

P = load (lb)
δ = slip (in.)
A = constant predicted from material properties
B = constant obtained from load-slip curve or predicted data.

It was also found that the effect of previous cyclic loading was not significant to the load-slip relationship for loads beyond the limit of the cycling. Thus, this form has the advantage of wide applicability. An additional advantage lies in the fact that the curve is valid up to 0.1-in. (0.245-cm) slip, and thus a complete range of behavior is given by the single expression. The slope of the curve at any point, given by

$$\frac{dP}{d\delta} = \frac{AB}{2.303(1 + B\delta)} \qquad (10.5)$$

may be used to obtain a tangent slip modulus, or, given P and δ values, a secant modulus, $\frac{P}{\delta}$, at any load level may be calculated. McLain (1975) suggested the following equations for parameter A:

$$A = 248.6 - 42.1/SGS + 41.36(SGS^2 \times SGM) \qquad (10.6)$$

where

SGS = specific gravity of side member
SGM = specific gravity of main member

or

$$A = 85.3 + 2.52/ES^2 + 7.34 \times 10^{-19}(EM^2 \times ES) \quad (10.7)$$

where

ES = modulus of elasticity of side member (psi)
EM = modulus of elasticity of main member (psi)

Parameter B is obtained by substituting known or predicted values of P vs. δ into Eq. 10.4 and solving.

EXAMPLE 10.1

Assume that it is desired to know the load-slip relationship of a joint consisting of a Douglas-fir side member fastened to a red oak main member with an 8d nail. Both members are oriented parallel to the grain, and the specific gravity of red oak is 0.619 and that of Douglas fir is 0.459. Substituting these values into Eq. 10.6 gives

$$A = 248.6 - 42.1/0.459 + 41.36(0.459^2 \times 0.619)$$
$$= 156.6$$

If the load at 0.015-in. (0.0381-cm) slip is known from either published data or experimental evidence, then B can be determined by solution of Eq. 10.4. In this case, referring to Fig. 10.4, $P = 144$ lb (640.8N) at 0.015 in. (0.0381 cm) slip. The resulting function is then

$$P = 156.6 \log_{10}(1 + 490.1\delta)$$

which is shown graphically in Fig. 10.4 as the dotted line. The solid line indicates the fitted curve based upon an experimental joint tested by McLain (1975). The dashed line in Fig. 10.4 indicates the theoretical slope of the joint based upon Wilkinson's (1972a) elastic analysis. This slope is, of course, only valid to a slip of 0.015 in. (0.0381 cm) beyond which it overpredicts experimental values. The slope appears to be quite adequate for the description of the curve below 0.015 in. (0.0381 cm), considering joint

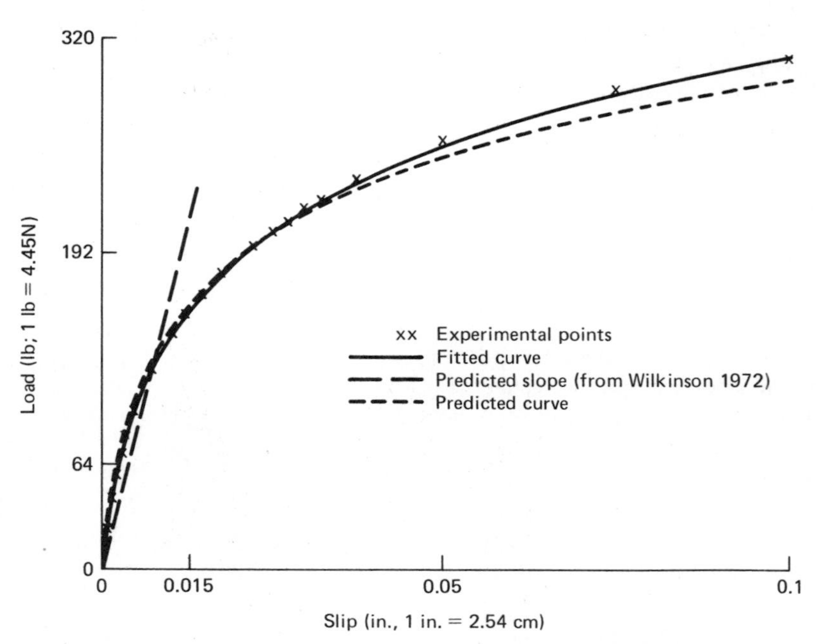

Fig. 10.4. Example of load-slip curve prediction. (*After McLain 1975.*)

variability. However, Eq. 10.4 provides the user a clear relationship beyond this limit.

10.4.1.2 Bolts and Lag Bolts

The behavior of bolted connections in wood systems has received much attention, particularly in regard to strength under loading. However, only recently has research on laterally loaded joints included measurement of slip-modulus values. A good summary of bolted-connection design information is given in Gurfinkel (1972). In particular, some load-slip data are given, indicating a slip modulus of 80,000 lb/in. (14,000 KN/m) for a $\frac{1}{2}$-in. ϕ (1.27-cm) bolt in parallel-to-the-grain loading with metal side plates. A very extensive study of connection behavior as related to layered or spaced column design was conducted by Van Dyer (1976). His study of connectors included nails, spikes, bolts to $\frac{1}{2}$-in. ϕ (1.27 cm), and split rings of $2\frac{1}{2}$-in. ϕ (6.35 cm). Tests in a double-shear configuration were conducted to obtain slip-modulus values for each type of connector. Since buckling phenomena were being considered, the values of slip modulus were given as the initial tangent modulus to the load-slip curves. In the case of buckling, this approach is correct; however, for use in beam-bending analysis, a secant modulus at the value of connector load is more nearly correct. Van Dyer concluded that the method presented originally by Kuenzi (1955) gives good predictions of the initial tangent modulus. He presented several charts, which simplify the calculation of slip modulus for a wide range of nailed, spiked, or bolted connections. The curves are limited to the case of nominal 1- and 2-in. (2.54- and 5.08-cm) wood thickness (equal thickness on each side of the joint) and are given for ranges of modulus of elasticity, E, of wood between 1.0×10^6 psi and 3.0×10^6 psi (6900 MPa and 20,700 MPa). Connector stiffness is taken as 25×10^6 psi (172,000 MPa). Figures 10.5 through 10.8, taken from Van Dyer (1976), are reproduced here.

To determine the initial tangent slip modulus using these figures, the modulus of elasticity, E, for the wood members, member thickness, and connector diameter are required. The value of E is determined as the average for members in the joint considered. For timber joints having members of unequal thicknesses, the least member thickness determines the appropriate curve to be used. Using the least member thickness to determine connector modulus does not ap-

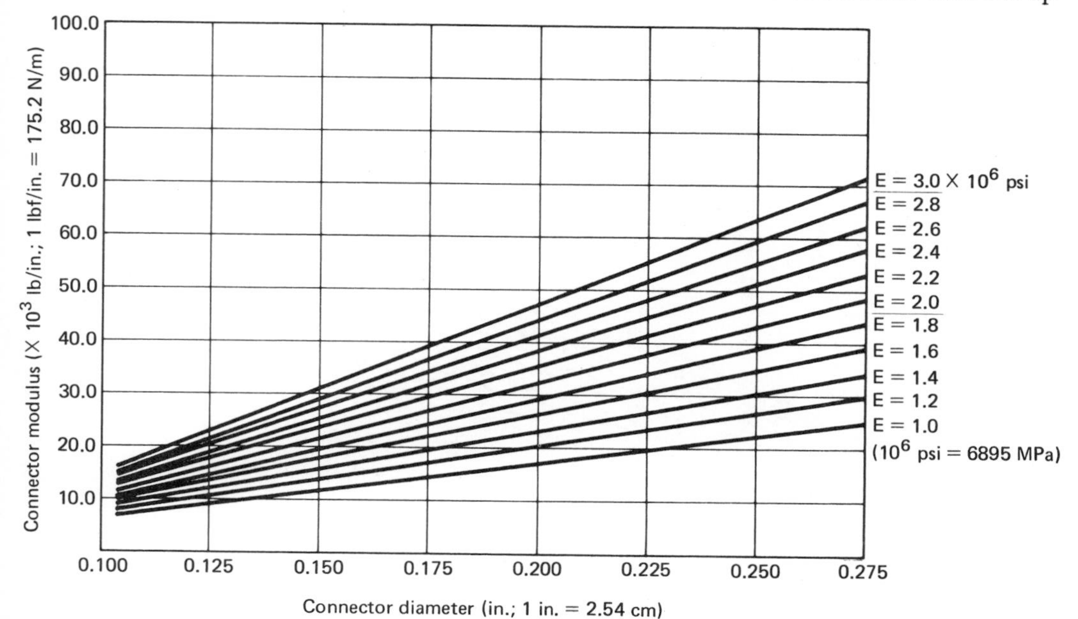

Fig. 10.5. Connector modulus vs. connector diameter for nominal 1-in.-thick wood—connector diameter 0.100 to 0.275 in. (*After Van Dyer 1976.*)

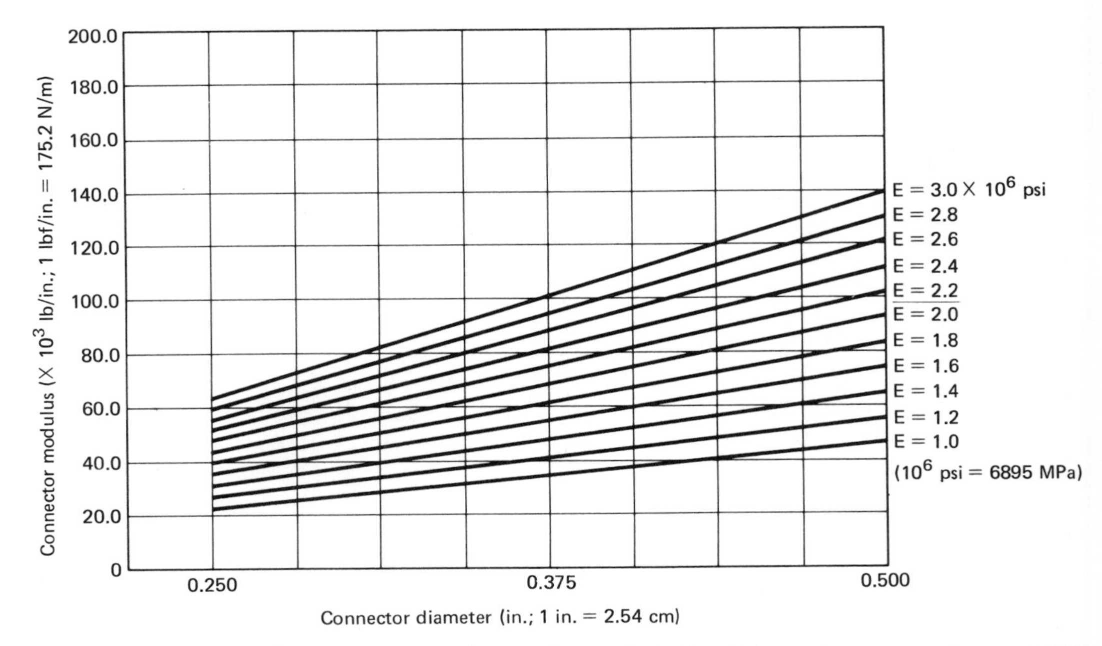

Fig. 10.6. Connector modulus vs. connector diameter for nominal 1-in.-thick wood—connector diameter 0.250 to 0.500 in. (*After Van Dyer 1976.*)

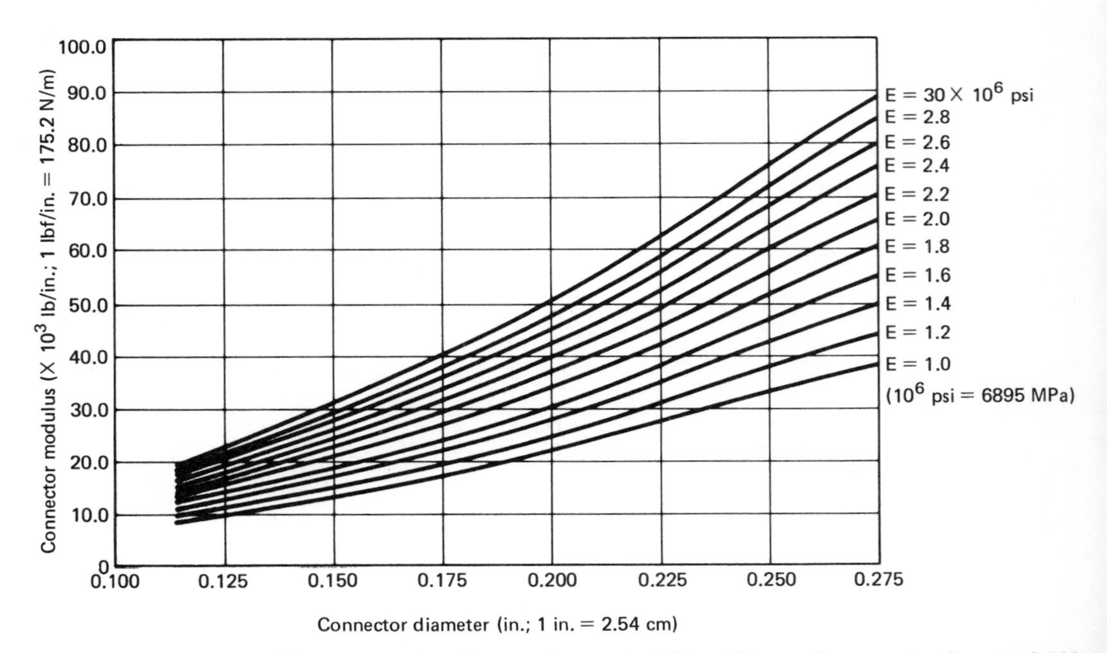

Fig. 10.7. Connector modulus vs. connector diameter for nominal 2-in.-thick wood—connector diameter 0.100 to 0.275 in. (*After Van Dyer 1976.*)

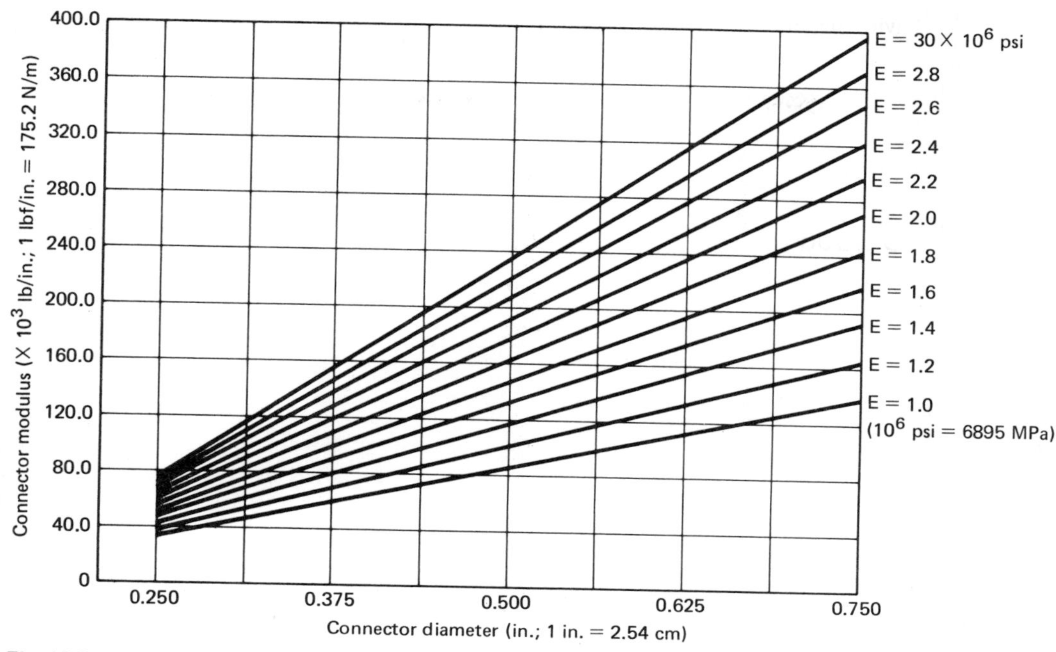

Fig. 10.8. Connector modulus vs. connector diameter for nominal 2-in.-thick wood—connector diameter 0.250 to 0.500 in. (*After Van Dyer 1976.*)

pear to significantly affect the result. These curves provide a means of estimating an upper limit on connector slip modulus for a wide range of conditions. Actual testing of individual joints is recommended for conditions not included in the curves.

10.4.2 Structural Adhesives

The horizontal shear stress at the interface of layers in composite systems with adhesive interlayer connection is distributed over the entire bonded area as opposed to the localized transfer in mechanically fastened systems. When strong, rigid adhesives are used, such as in glulam construction, complete interaction of composite behavior may be assumed. In certain other plant-manufactured products, such as plywood stress-skin panels or assemblies in modular housing, rigid glues may be used with controlled conditions of pressure and temperature for curing the adhesive. However, under field assembly involving conditions of wet materials, varying temperatures, incomplete assembly pressures, and so on, no currently used rigid adhesives have been approved. Thus, composite designs for field assembly are currently based on the use of one of several types of elastomeric adhesives that have been approved by the Federal Housing Administration and others. A review of the development of "construction adhesives" was presented by Gillespie and River (1972). Early work in this area was done by Krueger (1962, 1965) and Krueger and Blomquist (1964). Recent work has been done by Hoyle (1970, 1971, 1973). Tests of glued floors have been done by the American Plywood Association (see Rose 1969), by the National Association of Home Builders (1970), and by Vanderbilt et al. (1974). Reports on approved systems for use in floor construction are given by the American Plywood Association (1970, 1971). Thus, the use of elastomeric "construction adhesives" is increasing and rapidly becoming a standard method for wood-joist floor construction.

Unfortunately, engineering data required for the application of composite design with elastomeric adhesives have not kept pace with construction developments. Limited data are available on slip modulus, the key parameter required for proper design of such systems. Since most systems are nail-glued (i.e., nails are used to apply the pressure to the glued assembly during curing), the effective slip modulus must reflect the combined action of the glue and nails. In addition, some elastomerics exhibit considerable creep behavior depending to some degree on glue-line thickness. Research to obtain the necessary engineering data for composite behavior of elastomeric adhesives is currently progressing, and results should enable a more rational design procedure.

10.5 LAYERED WOOD BEAMS

The basic concepts of layered composite systems of wood connected by mechanical methods or nonrigid adhesives have been discussed in previous sections. As noted, the presence of interlayer slip in such systems has a critically important effect on their behavior under loading. Typical load-deflection curves, which indicate the effect of interlayer connection on the performance of layered beams, are shown in Fig. 10.9. As indicated, as interlayer connection is increased, the performance of the layered beam is improved toward that of a rigidly connected beam. Analysis and design methods to account for this effect will be presented in this section.

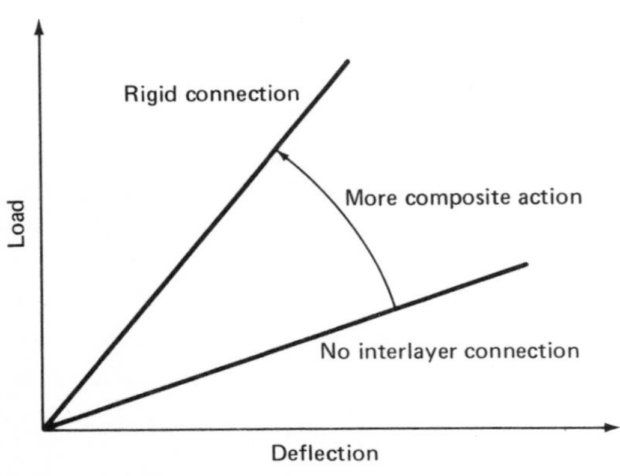

Fig. 10.9. Effect of interlayer connection on layered beam behavior.

10.5.1 Mathematical Theory of Layered Systems with Interlayer Slip

The basic theory of layered beam systems is best illustrated by considering a two-layer composite beam system. A typical system of this type is illustrated in Fig. 10.10. In this figure, the following is used:

$\bar{h}$ = distance to the centroid of the fully composite system
r_i = distance from the system centroid to the centroid of ith layer
h_i = the depth of the ith layer.

The differential element shown is composed of two layers of differing material properties, which must share in resisting the forces and moments caused by the transverse loads. In the general case, the deformations that result must accommodate any interlayer slip in addition to the usual flexural and axial strains.

Neglecting higher order terms, the equilibrium equations for the element are

$$\Sigma F_y = 0: \frac{dv}{dx} = -w \tag{10.8}$$

$$\Sigma F_x = 0: \frac{dF_1}{dx} = -\frac{dF_2}{dx} \tag{10.9}$$

$$\Sigma M = 0: V - \frac{dM_i}{dx} - \frac{dM_2}{dx} + \frac{dF_1}{dx} r_1 - \frac{dF_2}{dx} r_2 = 0 \tag{10.10}$$

The expression for the tensile strain at the interface for each layer may be written as

$$\text{Upper layer:} \quad \epsilon_1 = \frac{M_1}{E_1 I_1} \frac{h_1}{2} + \frac{F_1}{E_1 A_1} \tag{10.11}$$

$$\text{Lower layer:} \quad \epsilon_2 = -\frac{M_2}{E_2 I_2} \frac{h_2}{2} + \frac{F_2}{E_2 A_2} \tag{10.12}$$

where

I_i = moment of inertia of the ith layer about its own centroid
A_i = cross-sectional area of the ith layer
E_i = modulus of elasticity of the ith layer.

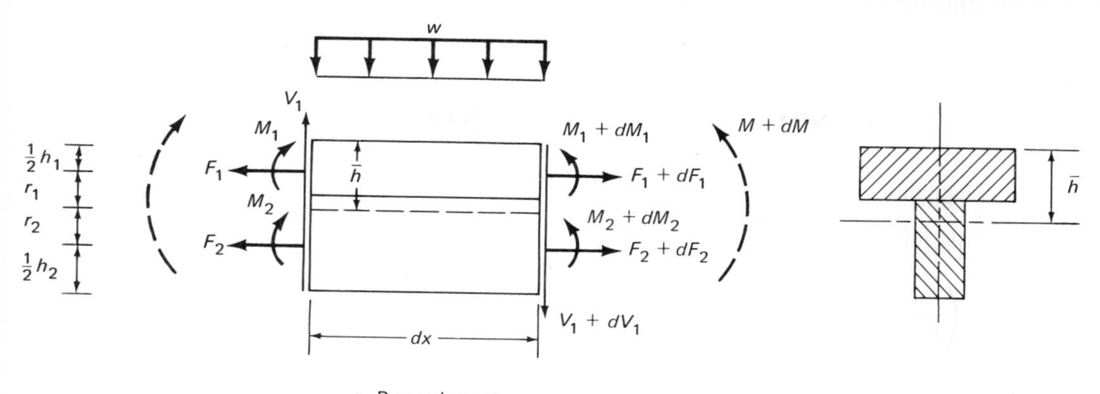

a. Beam element

b. Cross-section

Fig. 10.10. Two-layer beam system.

Compatibility requires that these strains be consistent with the interlayer slip, which implies

$$\int_0^x \epsilon_2 dx - \int_0^x \epsilon_1 dx = \Delta_s \qquad (10.13)$$

where Δ_s is the interlayer slip.

It is common to assume that the slip of a connector is proportional to the shear force, Q, which it transmits. The constant of proportionality is the slip modulus, k. Consequently,

$$\Delta_s = \frac{Q}{k} \qquad (10.14)$$

If the connectors are assumed to have equal moduli and are equally spaced at a distance s along the beam, it is convenient in some applications to use a "smeared" slip modulus, k^*, where

$$k^* = \frac{k}{s} \qquad (10.15)$$

Similarly, it is convenient to replace the connector forces with an equivalent shear flow between layers

$$q_{12} = \frac{nQ}{s} \qquad (10.16)$$

where

q_{12} = the shear flow
n = the number of connectors per row.

Equations 10.14 and 10.16 are combined to give

$$\Delta_s = \frac{s}{kn} q_{12} \qquad (10.17)$$

Horizontal equilibrium of the upper layer (see Fig. 10.11) requires

$$q_{12} = -\frac{dF_1}{dx} \qquad (10.18)$$

Substitution of Eq. 10.18 into Eq. 10.17 and recognition, from Eq. 10.9, that $F_1 = -F_2 = F$, yields

$$\Delta_s = -\frac{s}{kn}\frac{dF}{dx} \qquad (10.19)$$

Equations 10.11 and 10.12 and 10.19 are inserted into Eq. 10.13, and the result is differentiated to give

$$\sum_{i=1}^{2} \left[-\frac{M_i}{E_i I_i}\frac{h_i}{2} + (-1)^i \frac{F_i}{E_i A_i} \right] = -\frac{s}{kn}\frac{d^2F}{dx^2} \qquad (10.20)$$

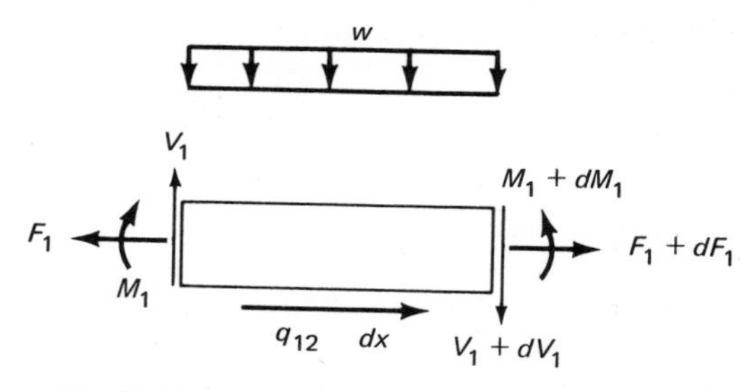

Fig. 10.11. Free body diagram of the upper layer.

If no vertical separation of layers is assumed, the curvature of each layer must be identical and

$$\frac{M_i}{E_i I_i} = -\frac{d^2y}{dx^2} \qquad (10.21)$$

Again recognizing the equality of F_1 and $-F_2$, Eq. 10.21 permits expressing Eq. 10.20 as

$$\frac{s}{kn}\frac{d^2F}{dx^2} - \left(\frac{1}{E_1A_1} + \frac{1}{E_2A_2}\right)F = -\left(\frac{h_1 + h_2}{2}\right)\frac{d^2y}{dx^2} \qquad (10.22)$$

To render the system analytically determinate, one of the two variables, F or y, must be eliminated from Eq. 10.22. Observing from Fig. 10.10 that

$$M = M_1 + M_2 - F_1r_1 + F_2r_2 \qquad (10.23)$$

it can be stated that

$$M_1 + M_2 = M + \left(\frac{h_1 + h_2}{2}\right)F \qquad (10.24)$$

which, in view of Eq. 10.21, becomes

$$-\frac{d^2y}{dx^2}\sum_{i=1}^{2} E_iI_i = M + \left(\frac{h_1 + h_2}{2}\right)F \qquad (10.25)$$

Equation 10.25 can be employed, by direct substitution, to eliminate either variable from Eq. 10.22.

An appropriate method of solution can be accomplished by substitution of Eq. 10.25 into Eq. 10.22 and making a change of variable $z = x/L$, normalizing the beam length L, which gives

$$\frac{1}{L^2}\frac{s}{kn}\frac{d^2F}{dz^2} = \left(\frac{1}{E_1A_1} + \frac{1}{E_2A_2} + \frac{C_{12}^2}{\Sigma E_iI_i}\right)F - \frac{C_{12}M(z)}{\Sigma E_iI_i} \qquad (10.26)$$

where $C_{12} = \dfrac{h_1 + h_2}{2}$.

Recognizing that the transformed section concept can be applied to convert to a common material modulus of elasticity, E,

$$\frac{d^2F}{dz^2} = \frac{knL^2}{Es}\left(\frac{1}{A_1^*} + \frac{1}{A_2^*}\right)\left[1 + \frac{C_{12}^2}{\Sigma I_i^*\left(\frac{1}{A_1^*} + \frac{1}{A_2^*}\right)}\right]F - \frac{knL^2}{Es}\frac{C_{12}}{\Sigma I_i^*}M(z) \qquad (10.27)$$

where

A_i^* = transformed area of layer i
I_i^* = moment of inertia of the transformed area.

Generalization may be obtained by introducing the concept of a transfer constant (T.C.) defined such that

$$\text{T.C.} = \frac{I}{\Sigma I_i^*} = 1 + \frac{C_{12}^2}{\Sigma I_i^*\left(\frac{1}{A_1^*} + \frac{1}{A_2^*}\right)} \qquad (10.28)$$

where I = moment of inertia of an equivalent rigidly connected beam section.

Thus, Eq. 10.27 becomes

$$\frac{d^2F}{dz^2} - C_1F = -C_2M(z) \qquad (10.29)$$

where

$$C_1 = \frac{knL^2}{Es}\left(\frac{1}{A_1^*} + \frac{1}{A_2^*}\right)(\text{T.C.})$$

$$C_2 = \frac{kn}{Es}\frac{C_{12}}{\Sigma I_i^*}$$

Therefore,

$$F = \frac{C_2}{C_1}M(z) + \frac{1}{C_1}\frac{d^2F}{dz^2} \qquad (10.30)$$

Now, Eq. 10.25 may be rewritten, by using Eq. 10.30, as

$$\frac{d^2y}{dz^2} = L^2\left(-\frac{1}{E\Sigma I_i^*} + \frac{C_2}{C_1}\frac{C_{12}}{E\Sigma I_i^*}\right)M(z) + \frac{1}{C_1}\frac{C_{12}}{E\Sigma I_i^*}\frac{d^2F}{dz^2} \qquad (10.31)$$

After algebraic simplification, this equation may be reduced to

$$\frac{d^2y}{dz^2} = -\frac{M(z)}{EI} + \frac{1}{C_1}\frac{C_{12}}{E\Sigma I^*}\frac{d^2F}{dz^2} \tag{10.32}$$

Direct integration of Eq. 10.32 results in

$$y = y_s + \frac{1}{C_1}\frac{C_{12}}{E\Sigma I_i^*}F(z) \tag{10.33}$$

where

y_s = deflection of an equivalent solid beam (in.)
$F(z)$ = solution of Eq. 10.29 with appropriate boundary conditions.

In this form, it is clear by examining the coefficient of $F(z)$, i.e.,

$$\frac{1}{C_1} = \frac{E_s}{knL^2}\frac{1}{\left(\dfrac{1}{A_1^*} + \dfrac{1}{A_2^*}\right)(\text{T.C.})}$$

that when $k \to \infty$ for a rigid connector, $y \to y_s$, and the solution limits properly at the upper bound. The lower bound may be obtained by setting F equal to zero in the governing equations resulting in

$$\frac{\Sigma EI_i^*}{L^2}\frac{d^2y}{dz^2} = -M(z) \tag{10.34}$$

which will give the proper solution for a layered system with no interconnection between the layers.

EXAMPLE 10.2

Consider applying the developed equations to the cross section shown in Fig. 10.12; namely, two equal size layers having identical material properties. For the condition of no interlayer slip (i.e., a solid rectangular cross section), the developed relationships should reduce to conventional forms.

By differentiating Eq. 10.23 and comparing with Eq. 10.10, it is apparent that for the layered system,

$$V = \frac{dM}{dx} \tag{10.35}$$

remains a valid condition.

With no interlayer slip, Eq. 10.13 reduces to

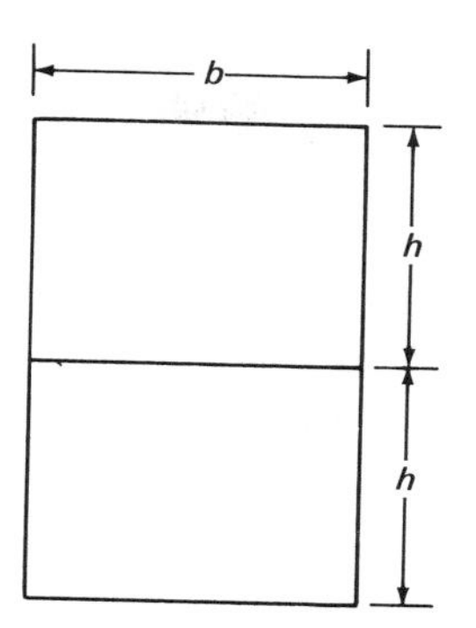

Fig. 10.12. Rectangular cross-section.

the condition that the strains ϵ_1 and ϵ_2, expressed by Eqs. 10.11 and 10.12, must be identical, giving

$$\left(\frac{M_1 + M_2}{I_L}\right)\frac{h}{2} = \left(\frac{-F_1 + F_2}{A_L}\right) \tag{10.36}$$

where

$I_L = I_1 = I_2$ = moment of inertia of one layer about its own centroid
$A_L = A_1 = A_2$ = cross-sectional area of one layer.

Equation 10.36 may be substituted into the third equilibrium condition, Eq. 10.10, giving

$$V - \frac{d}{dx}(-F_1 + F_2)\frac{2I_L}{hA_L} + \frac{dF_1}{dx}r_1 - \frac{dF_2}{dx}r_2 = 0$$

Recognizing that $r_1 = r_2 = h/2$ and $F_1 = -F_2 = F$, result is

$$V + \frac{dF}{dx}\left(\frac{4I_L}{hA_L}\right) + \frac{dF}{dx}h = 0 \tag{10.37}$$

Solving for the shear flow (at the interface)

$$\frac{dF}{dx} = -\left(\frac{hA_L}{4I_L + h^2A_L^2}\right)V \tag{10.38}$$

or, because $A_L = bh$ and $I_L = bh^3/12$,

$$-\frac{dF}{dx} = +\frac{3}{4h}V \tag{10.39}$$

It is straightforward to demonstrate that the right-hand side is identical with the usual shear-flow computation

$$q_{12} = -\frac{dF}{dx} = \frac{VQ}{I} \tag{10.40}$$

The usual governing equation for the beam deflection may also be obtained. Using the relationship of Eq. 10.36, Eq. 10.37 may be rewritten as

$$V - \left(1 + \frac{h^2 A_L}{4 I_L}\right) \frac{d}{dx} (M_1 + M_2) = 0 \tag{10.41}$$

or

$$V = 4 \frac{d}{dx} (M_1 + M_2) \tag{10.42}$$

With the further substitution of Eq. 10.21 and differentiation of both sides, the result is

$$\frac{dV}{dx} = 8 E_L I_L \frac{d^4 y}{dx^4} \tag{10.43}$$

Introducing Eq. 10.8 and noting that $I = 8 I_L$, where I is the moment of inertia of the entire solid cross section, gives the final form

$$EI \frac{d^4 y}{dx^4} = -w \tag{10.44}$$

Finally, letting $k \to \infty$ in Eq. 10.22 and substituting into Eq. 10.25, the result for the beam in Fig. 10.12 is

$$-\left[\left(\sum_{i=1}^{2} E_i I_i\right) + \left(h^2 \frac{E_L A_L}{2}\right)\right] \frac{d^2 y}{dx^2} = M \tag{10.45}$$

or

$$-2E[I_L + A_L (H/2)^2] \frac{d^2 y}{dx^2} = M \tag{10.46}$$

Equation 10.46 is easily recognized as being equivalent to the general expression

$$EI \frac{d^2 y}{dx^2} = -M \tag{10.47}$$

10.5.2 Finite-Element Analysis of Layered Beams

10.5.2.1 Structural Equations

The governing equations (10.30 and 10.31) constitute a classical analysis of layered beams including the effects of interlayer movement. Unfortunately, the solutions are strictly correct only if the material properties of the individual layers are constant along the entire span. For wood, this is not true, and the designer must be content with solutions based upon average values for the necessary properties. For situations where either variation in properties along the length of the beam is important to model or where one or more layers is discontinuous (gapped), a more versatile analytical approach is necessary. The finite-difference method or finite-element method offers the capability of accounting for such effects. A finite-element approach formulated by Thompson et al. (1975), which has proven successful for a number of systems (Tremblay 1974, Vanderbilt et al. 1974), will be described.

The potential energy, π_p, of a structural system is expressed by

$$\pi_p = U + W + L \tag{10.48}$$

where

U = strain energy
W = work of the applied loads
L = energy losses.

For the layered beam, Eq. 10.48 takes the form

$$\pi_p = \sum_{i=1}^{n_L} \int_0^{\ell} \left[\frac{1}{2} E_i I_i \left(\frac{d^2 y}{dx^2}\right)^2\right] dx + \sum_{i=1}^{n_L} \int_0^{\ell} \left[\frac{1}{2} E_i A_i \left(\frac{du_i}{dx}\right)^2\right] dx + \sum_{i=1}^{n_L - 1} \int_0^{\ell} \frac{1}{2} \left(\frac{k_i n_i}{s_i}\right) \left[(u_{i+1} - u_i) - \frac{1}{2}(h_{i+1} + h_i) \frac{dy}{dx}\right]^2 dx - \int_0^{\ell} w(y)\, dx \tag{10.49}$$

where

i = "for layer i"
n_L = number of layers
E_i = modulus of elasticity of layer i
I_i = moment of inertia of layer i
A_i = cross-sectional area of layer i
k_i = slip modulus of connector between layers i and $i + 1$

n_i = number of rows of connectors between layers i and $i + 1$
s_i = spacing of connectors between layers i and $i + 1$
h_i = depth of layer i
w = beam loading
x = length along the beam
y = vertical displacement of the beam
ℓ = beam length
u_i = axial displacement of layer i.

The first and second terms in Eq. 10.49 are the flexural and axial strain energies, respectively, and the last term is the work due to w. Energy losses due to interlayer slip are accounted for in the third term, which was developed by Thompson et al. (1975). In formulating the first and fourth terms, it is necessary to assume that the individual layers have identical curvature.

The principle of minimum potential energy requires that the first variation of π_p be equal to zero for the equilibrium position of the structure. Consequently,

$$\delta\pi_p = 0 \qquad (10.50)$$

or

$$\sum_{i=1}^{n_L}\left[\int_0^{\ell} E_i I_i \left(\frac{d^2y}{dx^2}\right)\delta\left(\frac{d^2y}{dx^2}\right)dx + \int_0^{\ell} E_i A_i \left(\frac{du_i}{dx}\right)\delta\left(\frac{du_i}{dx}\right)dx\right]$$
$$+ \sum_{i=1}^{n_L-1}\left\{\int_0^{\ell}\left(\frac{k_i n_i}{s_i}\right)\left[(u_{i+1} - u_i) - \frac{1}{2}(h_{i+1} + h_i)\frac{dy}{dx}\right]\delta(u_{i+1} - u_i)\,dx\right.$$
$$\left. - \int_0^{\ell}\left(\frac{k_i n_i}{s_i}\right)\left[(u_{i+1} - u_i) - \frac{1}{2}(h_{i+1} + h_i)\frac{dx}{dy}\right]\frac{1}{2}(h_{i+1} + h_i)\,\delta\left(\frac{dy}{dx}\right)dx\right\}$$
$$- \int_0^{\ell} w\delta y\, dx = 0 \qquad (10.51)$$

The finite-element approach consists of dividing the layered beam into a series of longitudinal elements interconnected at node points as shown in Fig. 10.13. The Rayleigh-Ritz procedure (i.e., using approximation functions for the various displacements needed to evaluate Eq. 10.51) is applied to each element. As described by Thompson et al. (1975), suitable functions in matrix form are

$$y_j = [N_y]\,\{Y\}_j \qquad (10.52)$$

$$u_{ij} = [N_u]\,\{U_i\}_j \qquad (10.53)$$

in which

$[N_y]$ = shape functions for a cubic approximation to y for element j
$[N_u]$ = shape functions for a linear approximation to u for element j
$\{Y\}_j$ = unknown vertical deflections and slopes at the ends of element j
$\{U_i\}_j$ = unknown axial deformations at the ends of layer i of element j.

Substitution of Eqs. 10.52 and 10.53 into Eq. 10.51 permits expressing the variation of potential energy attributed to element j in the form

$$\delta\pi_{pj} = \{\delta\Delta\}_j^T [k_e]_j \{\Delta\}_j - \{\delta\Delta\}_j^T \{f\}_j \qquad (10.54)$$

in which

$\{\Delta\}_j$ = column matrix containing the unknown displacement quantities
$\{k_e\}_j$ = stiffness matrix for element j,
$\{f\}_j$ = an array for nodal forces for element j as produced by the applied loads.

The system variation of potential energy is obtained by summing, figuratively, the contributions of all elements to obtain

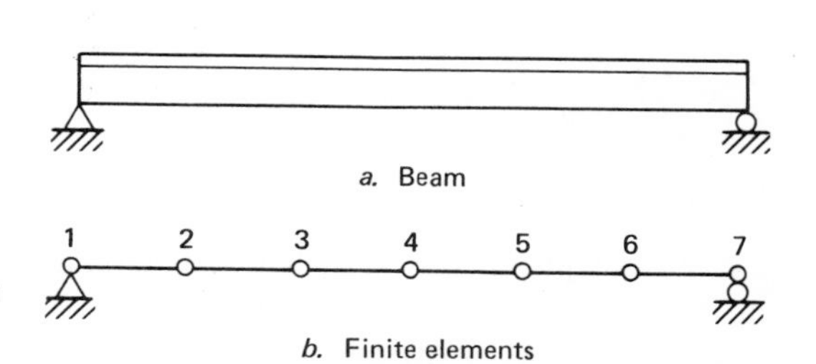

Fig. 10.13. Finite element model.

$$\delta\pi_p = \{\delta\Delta\}^T [K] \{\Delta\} - \{\delta\Delta\}^T \{F\} \tag{10.55}$$

The condition that Eq. 10.55 equates to zero for all disturbances $\delta\Delta$ requires that

$$[K]\{\Delta\} - \{F\} = 0 \tag{10.56}$$

where

$[K]$ is the structure stiffness matrix
$[F]$ is the entire set of nodal forces
$[\Delta]$ is the entire set of unknown displacement quantities

which are the equilibrium conditions one expects to obtain by minimizing the potential energy. Application of boundary conditions and solving for the displacement quantities follows conventional procedures. The summation referred to in obtaining Eq. 10.55 is actually accomplished by building $[K]$ by the usual "direct element method" of feeding the element stiffness matrices into their proper locations.

10.5.2.2 Gaps

Two techniques for modeling gaps in the layers of the beam have been employed by Thompson et al. (1975). Either a "special" element or a "soft" element can be employed. The special element consists of releasing the continuity of axial force at the gapped location. This is accomplished by making the axial force of each of the two elements adjacent to the gap independent of each other. A soft element is an element of finite, but small, length with a low modulus of elasticity, placed at the gap location. Even at low modulus values, instability is never approached provided at least one layer is continuous at the gap.

10.5.2.3 Application

Finite-element modeling of the behavior of the layered beam makes possible the consideration of all major factors that affect its response. Because many wood systems contain layered components as key elements, development of this type of model has great merit. Extensive research studies have been made possible for a number of these systems. Subsequent sections describe these advances and their impact on future designs.

10.5.3 Design Method for Static Loading of Two-Layer Beams

The theoretical developments of Section 10.5.1 enable the construction of generalized design charts for two-layer beams. These charts greatly simplify the calculations needed to evaluate the influence of interlayer slip on the behavior of the system. Solutions of the governing differential equations are not necessary if the design chart is used, and only the physical parameters of the system are required to predict the behavior of the layered system.

The final form of the solution of the governing equation for the general two-layer beam system has been stated in Eq. 10.33 as

$$y = y_s + \frac{1}{C_1} \frac{C_{12}}{E\Sigma I_i^*} F(z)$$

where

y_s = deflection of an equivalent rigidly connected beam
$F(z)$ = solution of Eq. 10.29 with appropriate boundary conditions.

As an example, a design chart is constructed for the case of a central concentrated load applied to a simply supported beam. For this case, the solutions for $F(z)$ and y have been developed by Goodman (1967) and may be stated as

$$y = y_s\{1 + [(\text{T.C.}) - 1]\} \frac{3}{\nu^2} (1 - \frac{1}{\nu} \tan h\nu \tag{10.57}$$

where

$$\nu = \frac{\sqrt{C_1}}{2} = \frac{K(\text{T.C.})}{2}$$

$$K = \frac{knL^2}{Es}\left(\frac{1}{A_1^*} + \frac{1}{A_2^*}\right)$$

$$y_s = \frac{PL^3}{48EI}.$$

Thus, a design chart may be constructed in terms of the basic geometric, material, and slip-modulus parameters of the system. This chart

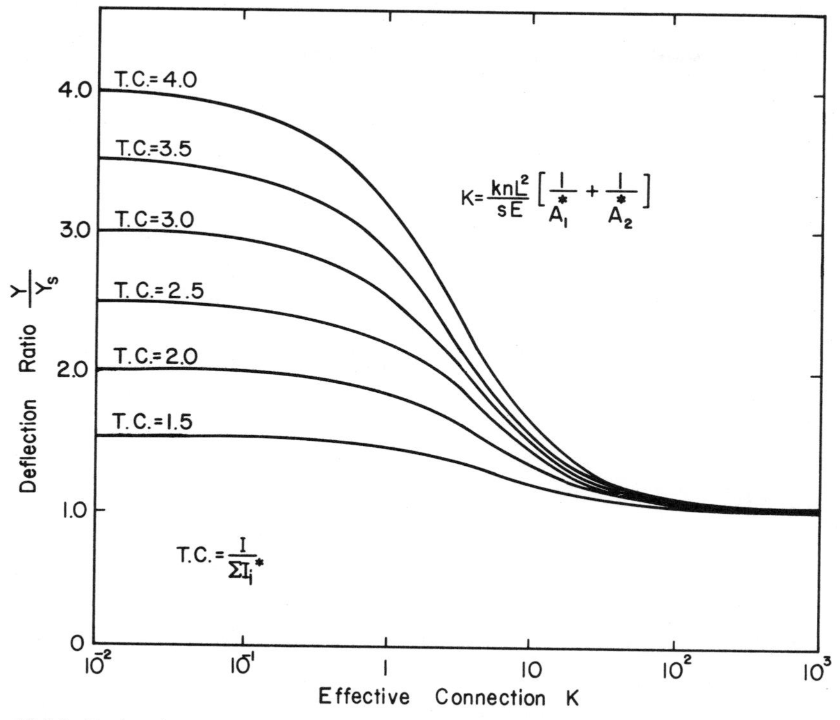

Fig. 10.14. Design chart for deflection for a centrally loaded, simply supported, two-layer beam.

is shown in Fig. 10.14 for various values of the transfer constant (T.C.), and in terms of the deflection ratio y/y_s and the parameter K. This parameter includes the slip data, $\frac{kn}{s}$, obtained from connector tests as described previously. The chart also illustrates that as K becomes large (i.e., the connectors become more rigid or more closely spaced), the ratio y/y_s limits properly to unity. On the other hand, if k is small, the unconnected case is obtained.

EXAMPLE 10.3

Assume a two-layer system with given data as follows:

Layer 1: width = 12 in. (30.48 cm), thickness (h_1) = 2 in. (5.08 cm), modulus of elasticity (E_1) = 1×10^6 psi (6900 NPa)

Layer 2: width = 2 in. (5.08 cm), thickness or depth (h_2) = 12 in. (30.48 cm), modulus of elasticity (E_2) = 2×10^6 psi (13800 MPa)

Connector Properties: slip modulus (k) = 25,000 lb/in. (4380 kN/m), spacing (s) = 8 in. (20.32 cm), number per row (n) = 1

Beam Span: 10 ft (3.05 m)

Parameters (dimensionless)—

Effective Connection Parameter

$$K = \frac{knL^2}{sE}\left(\frac{1}{A_1^*} + \frac{1}{A_2^*}\right)$$

$$= \frac{25{,}000(1)(120)^2}{8(1 \times 10^6)}\left(\frac{1}{12(2)} + \frac{1}{4(12)}\right)$$

$$K = 2.813$$

where the layer 1 modulus of elasticity (E_1) is used to calculate the transformed section properties.

Centroid of Transformed Section ($\bar{h}$) = 5.67 in. (14.4 cm) (from top).

$$\text{Transfer Constant (T.C.)} = \frac{I}{\Sigma I_i^*} = 2.343$$

where

$$\Sigma I_i^* = \tfrac{1}{12}(12)(2)^3 + \tfrac{1}{12}(4)(12)^3$$

$$= 584 \text{ in.}^4 \ (24{,}300 \text{ cm}^4)$$

$$I = 584 + 24(4.66)^2 + 48(2.34)^2$$

$$= 1368 \text{ in.}^4 \ (56{,}940 \text{ cm}^4).$$

For the beam data given, assume the design requirements specify a deflection of L/360 for a centrally loaded simple beam. Thus,

$$\text{L}/360 = \frac{120}{360} = 0.333 \text{ in. } (0.846 \text{ cm})$$

Enter Fig. 10.14 with $K = 2.813$ and T.C. = 2.343; interpolate to read $y/y_s = 1.85$. The equivalent solid-beam deflection y_s is given by

$$y_s = \frac{PL^3}{48EI} = 0.333/1.85$$

$$= 0.18 \text{ in. } (0.457 \text{ cm}) \quad \text{(allowable)}$$

Therefore, the allowable load

$$P = \frac{48(1 \times 10^6)(1368)(0.18)}{(120)^3}$$

or

$$P = 6840 \text{ lb } (30{,}440 \text{ N})$$

Stresses may be checked by referring to the governing equations of the system.

Additional design charts for other loading conditions may be constructed by solution of the basic governing equations for the case required.

10.5.4 Dynamic Problems for Two-Layer Composite Wood Beams

The development of the governing equations and a design chart for the fundamental vibration frequencies of a two-layer composite beam system with interlayer slip is considered in this section. The basic assumptions for the analysis are the same as those cited for the static problem. Of special importance for the dynamic analysis is the additional assumption of neglected rotary inertia and friction. A review of damping and a modification of the governing equations for the approximate solution of the damped case are extensively covered by Henghold (1972). Only the design and analysis problems associated with the determination of fundamental frequencies will be treated here.

The major modification of the basic theory—as derived in Section 10.5.1—required to treat dynamic problems is to use Newton's second law in place of equilibrium equations. The resulting equations are

$$\frac{\partial F_1}{\partial x} + \frac{\partial F_2}{\partial x} = 0 \qquad (10.58)$$

$$\frac{\partial V}{\partial x} = \sum_{i=1}^{2} \rho_i A_i \frac{\partial^2 y_i}{\partial t^2} \qquad (10.59)$$

where

ρ_i = mass per unit volume of the ith layer
A_i = area of the ith layer

$$V = \sum_{i=1}^{2} \left(\frac{\partial M_i}{\partial x} + r_i \frac{\partial F_i}{\partial x} \right) \qquad (10.60)$$

Maintaining the assumption that each layer has the same curvature, the characteristic equation for y may be found by conventional procedures of dynamic analysis. The outcome is

$$\frac{d^6 y}{dz^6} - K \left[1 + \frac{(r_2 - r_1)^2}{E\Sigma \left(\frac{1}{A_1^*} + \frac{1}{A_2^*} \right)} \frac{d^4 y}{dz^4} \right] - N\omega^2 \frac{d^2 y}{dz^2} + KN\omega^2 y = 0 \qquad (10.61)$$

where

$$K = \frac{knL^2}{Es} \left(\frac{1}{A_1^*} + \frac{1}{A_2^*} \right)$$

and

$$N = \frac{1}{E} \sum_{i=1}^{2} \frac{(\rho_i A_i L^4)}{I_i^*}$$

This equation may be generalized by introducing the concept of a transfer matrix (T.C.) defined previously in Eq. 10.28 resulting in

$$\frac{d^6 y}{dz^6} - (\text{T.C.})K \frac{d^4 y}{dz^4} - N\omega^2 \frac{d^2 y}{dz^2} + KN\omega^2 y = 0 \qquad (10.62)$$

The solution of this equation proceeds in the usual manner, and details can be found in Henghold (1972). Final results for the n natural frequencies of the system for a simply supported beam are

$$\omega^2 = \frac{(n\Pi)^4}{N} \frac{(n\Pi)^2 + (\text{T.C.})K}{(n\Pi)^2 + K} \qquad (10.63)$$

If K is very large, such as in the case where the layers are rigidly connected, Eq. 10.60 reduces to

$$\omega^2 = \frac{(\text{T.C.})(n\Pi)^4}{N}$$

which upon substitution of the equations for (T.C.) and N results in

$$\omega_s = (n\Pi)^2 \sqrt{\frac{EI}{pAL^4}} \qquad (10.64)$$

where

I = moment of inertia of the equivalent solid beam

pA = mass per unit volume times the total cross-sectional area for the equivalent solid beam.

This equation is recognized as the frequency equation for a homogeneous solid beam on simple supports, and thus the equations developed limit properly for the case of rigid connections.

Using a ratio ω/ω_s, where ω_s is the fundamental frequency of an equivalent solid beam given by Eq. 10.64, a design chart may be constructed as shown in Fig. 10.15. The frequencies of a layered system may be determined easily by entering the chart with appropriate values of K and the transfer constant (T.C.).

EXAMPLE 10.4

Consider the beam as described in Section 10.4.1 where $K = 2.813$ and T.C. = 2.343. To determine the effect of mechanically connecting the beam layers as given in this example, enter Fig. 10.15 with the parameters K and (T.C.), and read $\omega^2/\omega_s^2 = 0.54$. Thus, the mechanically connected system has an $\omega = 0.735\,\omega_s$. By adjusting the parameters of the system, any desired natural frequency can be obtained with the range shown on the chart.

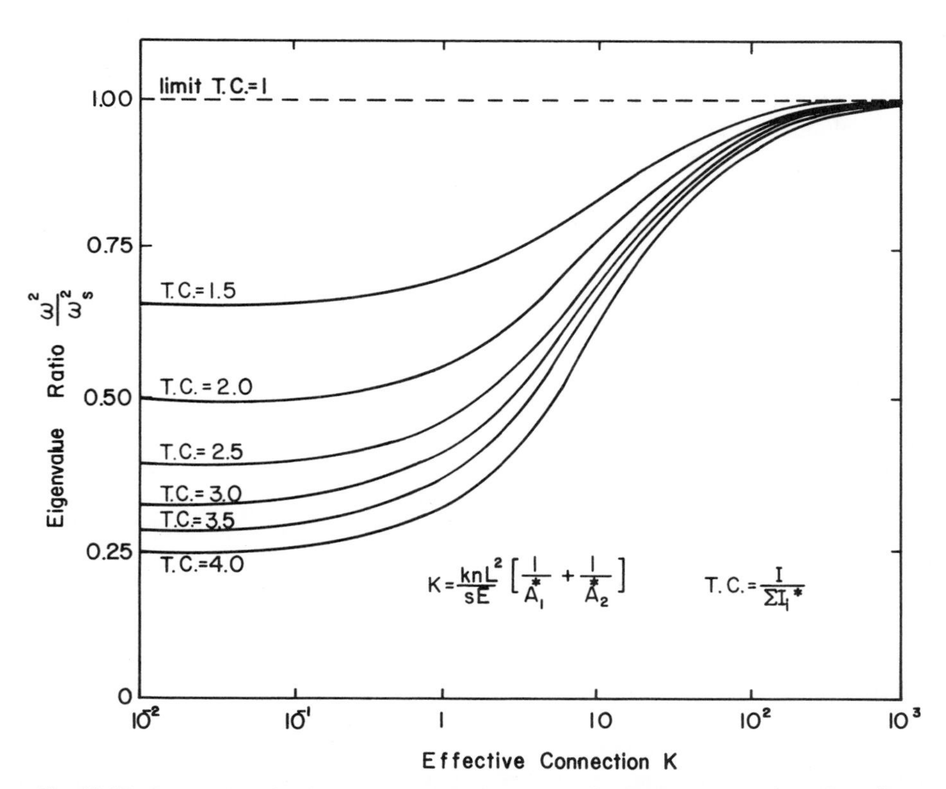

Fig. 10.15. Design chart for first-mode natural frequency of a simply supported two-layer beam.

The dynamic behavior of composite wood systems can thus be determined by rational analysis. Experimental verification of this analysis has been made by Henghold (1972). Implications of the reduced natural frequency of such systems as compared with rigidly connected systems can be important in dynamic applications such as bridges and other structures subjected to vibration or dynamic loading. "Tuning" to desired natural frequency is possible in the design of such systems, and thus control under dynamic loading situations is feasible. Systems with interlayer slip also have greatly increased damping over rigidly connected systems. Thus, their use would seem to be favorable under conditions of critical dynamic loading.

10.6 COMPOSITE WOOD COLUMNS

To use wood economically, wood columns are often constructed of layers in various configurations generally assembled by means of nailing or bolting. The buckling capacity of such columns are affected to a large extent by the type and spacing of the connectors used in the system. The behavior of mechanically connected wood columns varies between that of a fully rigidly connected system, as an upper limit, and a set of individual members, as a lower limit. A rational design procedure for such columns requires the use of concepts not normally considered in column buckling theory for other building materials.

Several studies have been directed at solving the problem of column buckling for layered and built-up wood columns. Early studies were by Pleshkov (1952), Granholm (1949), and Niskanen (1961). More recent investigations were by Rassam (1969), Rassam and Goodman (1970, 1972), and most recently a very extensive study by Van Dyer (1976). In his study, Van Dyer included inelastic as well as elastic buckling phenomena. Furthermore, developments made by Pleshkov (1952), Niskanen (1961), and Malhotra (1969) were incorporated to develop a rational design procedure for a wide variety of column types. Approximately 400 columns including a wide range of slenderness ratios and types were tested as experimental verification of this design procedure. This work and the earlier studies of column theory provide a rational basis for design procedures suggested in subsequent sections.

10.6.1 Basic Theory and Design Method for Two-Layer Composite Column

The two-layer case will be discussed as a simple case to illustrate the use of the theoretical procedures for layered systems with interlayer slip as applied to the analysis of long columns. Generalization to other cases and to inelastic analysis will follow.

The basic assumptions for the development of the theory for buckling load for layered columns with interlayer slip are the same as those for the static beam problem. In addition, the columns are assumed to be long—i.e., the theory provides an equivalent Euler load for the layered system. Thus, upper and lower bounds for a two-equal layer, pin-ended column are, respectively,

$$P_{cr} = \frac{n^2\pi^2 EI}{L^2} \tag{10.65}$$

and

$$P_{cr} = \frac{2n^2\pi^2 EI}{L^2} \tag{10.66}$$

where

P_{cr} = critical buckling load for the nth mode
I = moment of inertia of equivalent solid column
I_L = moment of inertia of each layer of two-equal layer columns.

The effect of interlayer slip will produce critical buckling loads somewhere between those given by Eq. 10.65 (rigid connection between layers) and by Eq. 10.66 (the unconnected case).

Development of the theory for the two-unequal-layer case including the effects of slip follows methods similar to those presented in the derivations for the static and dynamic beam cases. The primary difference is in the equilibrium equation needed for the column problem. The governing characteristic equation for y is

$$\frac{d^4y}{dz^4} - K(\text{T.C.})\frac{d^2y}{dz^2} + \lambda^2 \frac{d^2y}{dx^2} - \lambda^2 Ky = 0 \tag{10.67}$$

where

$$K = \frac{knL^2}{sE}\left(\frac{1}{A_1^* + A_2^*}\right)$$

and

$$\lambda^2 = \frac{PL^2}{E\Sigma I_i^*} \text{ (eigenvalues)}$$

The details solution of this equation can be found in Rassam (1969). Final results for the first-mode critical buckling load for a simply supported column are

$$P_{cr} = \frac{\Pi^2 E\Sigma I_i^*}{L^2} \frac{[\Pi^2 + K(\text{T.C.})]}{K + \Pi^2} \tag{10.68}$$

Therefore, the buckling load may be calculated for any two-layer system with mechanical connection, given the input data of column geometry and properties and the slip modulus for the connectors.

Using a ratio of P_{cr}/P_{cr_s}, where P_{cr_s} is the first-mode buckling load for an equivalent, long, rigidly connected column, a design chart may be constructed as shown in Fig. 10.16. The critical loads may be determined easily by entering the chart with appropriate volume of K and the transfer constant (T.C.). It is interesting to note that the curves in Figs. 10.16 and 10.15 are identical in shape due to the analogy between buckling loads and fundamental dynamics frequencies.

EXAMPLE 10.5

Consider the following two-layer column:

Layer 1: width = 12 in. (30.48 cm), thickness, (h_1) = 2 in. (5.08 cm), modulus of elasticity $(E_1) = 1 \times 10^6$ psi (6900 MPa)

Layer 2: width = 2 in. (5.08 cm), thickness or depth (h_2) = 12 in. (30.48 cm), modulus of elasticity $(E_2) = 2 \times 10^6$ psi (13,800 MPa)

Connector Properties: slip modulus (k) = 25,000 lb/in. (4380 kN/m), spacing (s) =

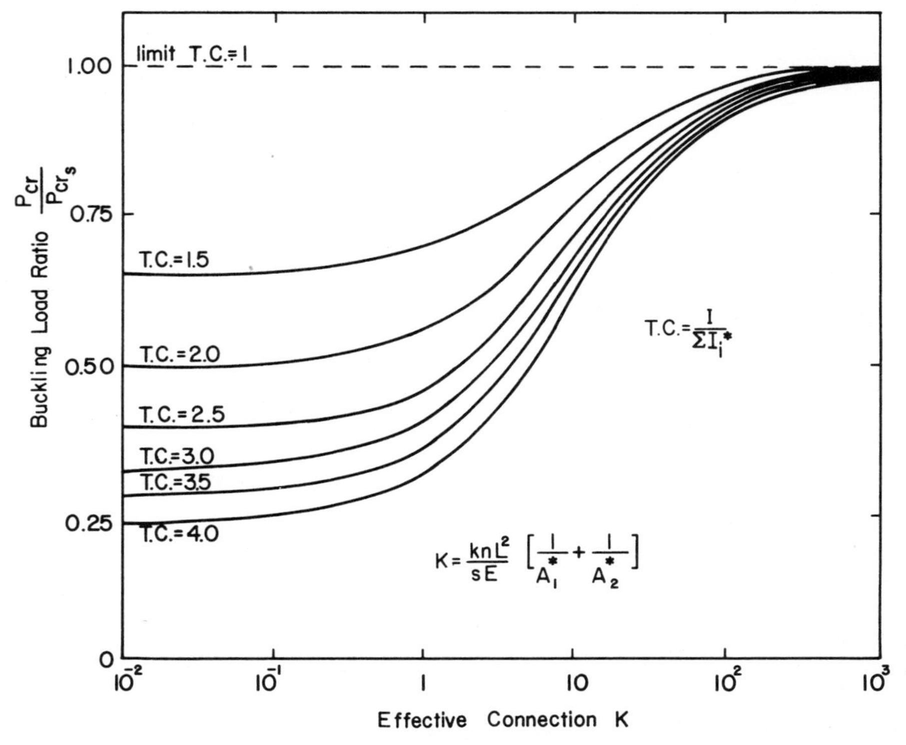

Fig. 10.16. Design chart for first-mode buckling load of a simply supported two-layer column.

8 in. 20.32 cm), number per row (n) = 1
Column Length = 10 ft. (3.05 m)
Thus,

$$K = \frac{knL^2}{sE}\left(\frac{1}{A_1^*} + \frac{1}{A_2^*}\right)$$

$$= \frac{25{,}000(1)(120^2)}{8(1 \times 10^6)}\left[\frac{1}{12(2)} + \frac{1}{4(12)}\right]$$

$$K = 2.813$$

where

Layer 1 modulus of elasticity was used to calculate the transformed section properties
Centroid of Transformed Section ($\bar{h}$) = 5.67 in. (14.4 cm) from top

$$\text{Transfer Constant (T.C.)} = \frac{I}{\Sigma I_i} = 2.343$$

$$\Sigma I_i^* = \tfrac{1}{12}(12)(12)^3 + \tfrac{1}{12}(4)(12)^3$$

$$= 584 \text{ in.}^4 \ (24{,}300 \text{ cm}^4)$$

$$I = 584 + 24(4.66)^2 + 48(2.34)^2$$

$$= 1368 \text{ in.}^4 \ (56{,}940 \text{ cm}^4)$$

Entering Fig. 10.16 with $K = 2.813$ and T.C. = 2.343 given $P_{cr}/P_{cr_s} = 0.54$.
Noting that P_{cr_s} is given by

$$P_{cr_s} = \frac{\pi^2 EI}{L^2}$$

$$= \frac{\pi^2(1 \times 10^6)(1368)}{(120)^2}$$

$$P_{cr_s} = 7813 \text{ lb } (34{,}800 \text{ N})$$

then

$$P_{cr} = 0.54(7813)$$

$$P_{cr} = 4220 \text{ lb } (18{,}780 \text{ N})$$

By adjusting parameters in the design of the system—for example, decreasing the spacing of the connectors—a greater buckling resistance may be obtained.

A similar procedure for long columns of three layers has been by Rassam (1969). A parameter, ψ, is developed such that the buckling load of the column may be represented as

$$P_{cr} = \psi P_u \tag{10.69}$$

where

P_u = buckling load of the unconnected column.

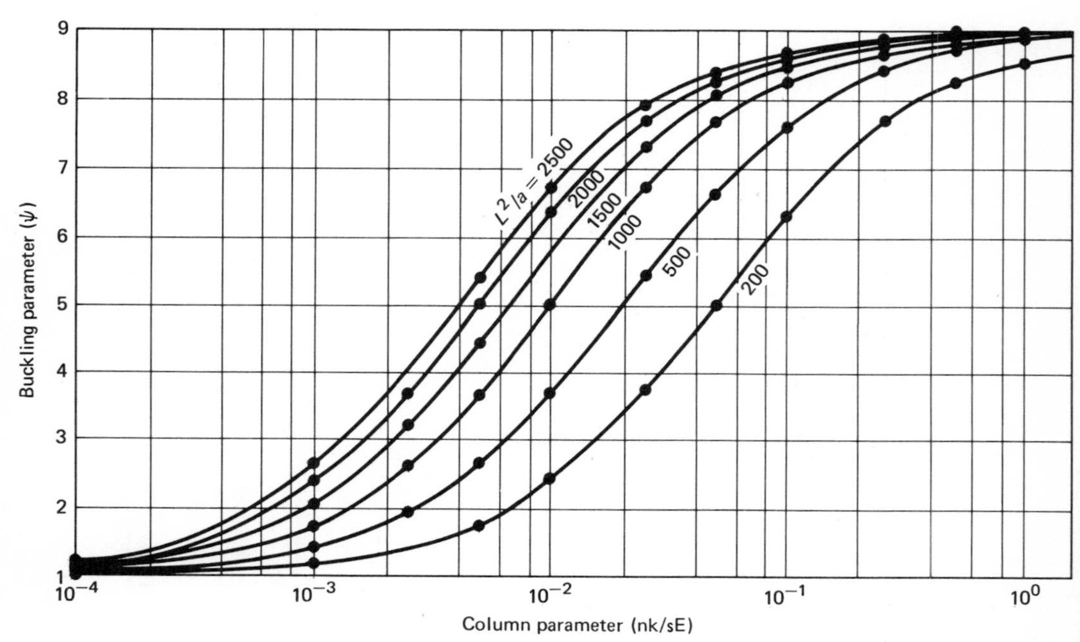

Fig. 10.17. Design chart for three-layered column for various values of L^2/a. *(After Rassam and Goodman 1972.)*

For three equal layers, ψ ranges between 1 for unconnected layers and 9 for rigidly connected layers. For the special, although common, case of three identical layers, the buckling parameter, ψ, may be simplified to

$$\psi = \frac{\pi^2 + 9\dfrac{L^2}{a_1}\dfrac{nk}{sE}}{\pi^2 + \dfrac{L^2}{a_1}\dfrac{nk}{sE}} \tag{10.70}$$

where

a_1 = area of one layer
n = number of connectors per row
k = slip modulus of an individual connector
s = spacing of connectors.

A design chart for ψ can be developed for this case and is shown in Fig. 10.17.

EXAMPLE 10.6

The buckling load for a column consisting of three identical layers with the following given data will be calculated:

h = 1 in. (2.54 cm)
b = 4 in. (10.16 cm)
$E = 1 \times 10^6$ psi (6900 MPa)
s = 6 in. (15.24 cm) between rows
n = 3 connectors/row
$k = 2 \times 10^4$ lb/in. (3500 kN/m) per connector
L = 70 in. (1.78 m)

Solution

$$\frac{nk}{sE} = \frac{3 \times 2 \times 10^4}{6 \times 10^6} = 10^{-2}$$

$$\frac{L^2}{a_1} = \frac{70 \times 70}{4} = 1220$$

$$\frac{nkL^2}{sEa_1} = 10^{-2} \times 1220 = 12.2$$

$$\Sigma I_i^* = 3 \times \frac{bh^3}{12} = \frac{3 \times 4 \times 1^3}{12}$$

$$= 1.0 \text{ in.}^4 \text{ } (41.62 \text{ cm}^4)$$

Using Fig. 10.17 with $nk/sE = 10^{-2}$ and $L^2/a_1 = 1220$, ψ is found to be 5.4 and, since

$$P_u = E\frac{\pi^2}{L^2}\Sigma I_i^* = 10^6 \times 1 \times \frac{\pi^2}{70 \times 70}$$

$$= 2014 \text{ lb (8960 n)},$$

then

$$P_{cr} = P_u\Psi = 2014 \times 5.4$$

$$= 10900 \text{ lb (48400 n)}$$

The resulting critical stress, σ_{cr}, is obtained by dividing by the total cross-sectional area, and thus

$$\sigma_{cr} = \frac{P_{cr}}{A}\frac{10,900}{3 \times 4 \times 1} = 906 \text{ psi (6.2 MPa)}$$

which should be checked to ascertain that it is below the elastic limit for the wood used in the column. It should be clearly noted that the value of P_{cr} as given here does not include any factor of safety that must be considered to determine the allowable design load from this theoretical buckling load.

10.6.2 Concepts for Inelastic Design Procedures for Composite Wood Columns

The fundamental concepts regarding the buckling performance of layered columns, including the effects of interlayer slip at connections, were introduced for the two- and three-layer systems in the previous section. While that introduction serves to describe basic theoretical considerations for such problems, the described procedures are limited because only long columns that buckle in the elastic stress range were considered. A more complete approach, based on previous work by Malhotra (1969) and Malhotra and Kwan (1971), was taken by Van Dyer (1976). Van Dyer treated a wider range of column types and included a study of inelastic buckling with extensive experimental verification. Results of Van Dyer's research and his suggested, albeit approximate, design procedures are presented to indicate the current state-of-the-art. It is likely that some version of this procedure will be adopted by code authorities in the future.

The general formula for inelastic buckling of composite wood columns is postulated by Van Dyer (1976) as

$$F_{cr} = \frac{\pi^2 E_t}{\lambda^2} B \qquad (10.71)$$

where

$F_{cr} = \frac{P_{cr}}{A}$ critical column buckling stress
A = area of entire cross section
$\lambda = L/r$ = slenderness ratio of the column
$r = \sqrt{I/A}$ = radius of gyration
I = moment of inertia of entire column cross section
E_t = tangent modulus of elasticity at some stress level on the stress-strain curve of the column material
B = parameter involving connector properties, column geometry, etc.

The key variables, E_t and B, must be determined for the specific situation (column type, connector case, and set of material properties) under consideration. Various approaches exist for doing this, but only one is described herein.

As an example of Van Dyer's approach, consider a layered column of which the cross section is shown in Fig. 10.18. Following a theroetical procedure developed by Pleshkov (1952), an approximate value for the buckling parameter, B, is derived as

$$B = \frac{1 + \left[\dfrac{\alpha \pi^2 E A_r a}{2(m-1)k_1 L^2}\right]}{1 + \left[\dfrac{\pi^2 E A_r a}{2(m-1)k_1 L^2}\right]} \qquad (10.72)$$

where

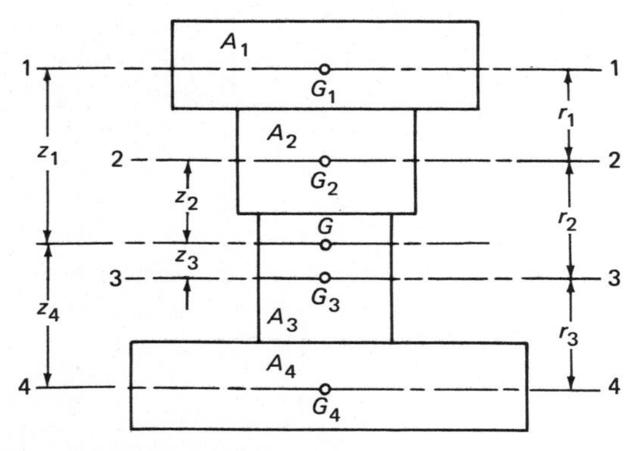

Fig. 10.18. Layered column cross section. (*After Van Dyer 1976.*)

$$\alpha = \frac{\Sigma I_i^*}{I} = \frac{1}{\text{T.C.}}$$

E = modulus of elasticity
m = number of layers
a = connector spacing

$$A_r = \frac{A_1 Z_1}{r_1} + \frac{A_1 Z_1 + A_2 Z_2}{r_2} + \cdots + \frac{A_1 Z_1 + A_2 Z_2 + \cdots + A_m Zm}{r_m}$$

$k_1 = Ga$ = shear flow
G = average connector modulus per layer.

The tangent modulus, E_t, must be determined at the level of stress to which the column is subjected. Its evaluation requires knowledge of the inelastic behavior of the wood used in making the column. Several methods, such as using stress-strain curves developed by O'Halloran (1973), could be employed. However, Van Dyer modified an expression developed by Malhotra (1969) to obtain the value of E_t to be used in assessing the inelastic behavior of the wood. This method assumes the compression stress-strain curve for wood is given by

$$\epsilon = \frac{1}{E}\left[cF - (1-c)F_u \log_e \left(1 - \frac{F}{F_u}\right)\right] \qquad (10.73)$$

where

ϵ = strain
F = stress at any level
F_u = ultimate compressive stress
c = constant dependent on the shape of the curve.

From this assumption and after considerable development, a value of the ratio F_{cr}/F_u can be obtained. It should be noted that the constant c in Eq. 10.73 depends on species and ideally should be determined experimentally. Eastern white spruce was used in the research and the constant c was obtained as 0.9. Realizing that $E_t = \frac{dF}{d\epsilon}$, the parameter E_t in Eq. 10.71 may be replaced with equivalent terms involving stress. The resulting expression enables

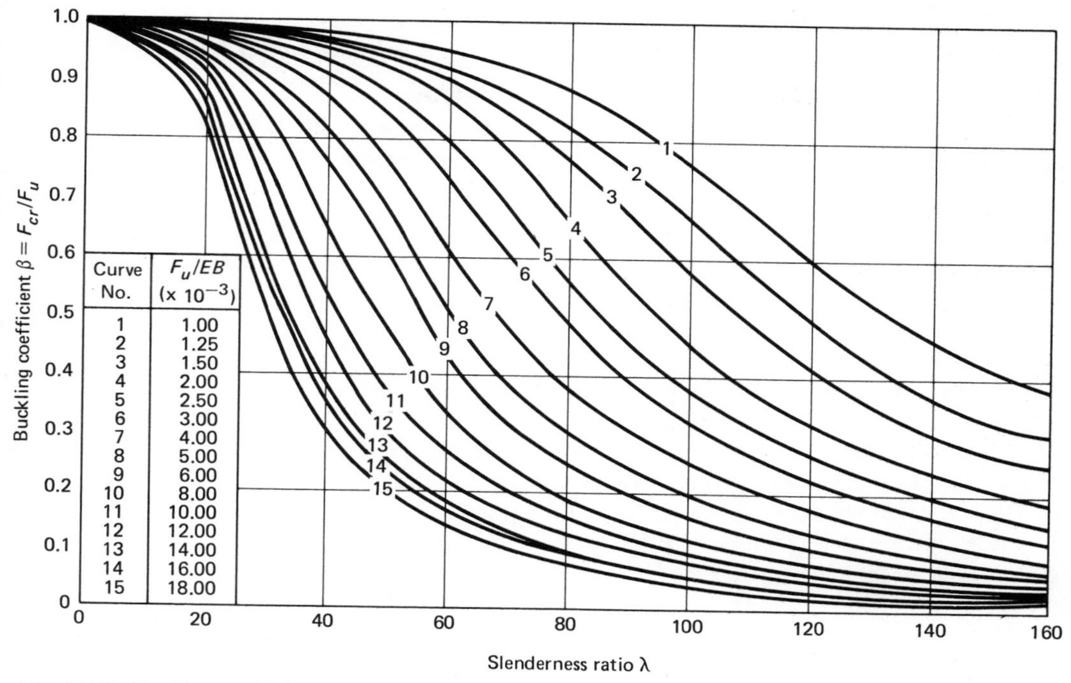

Fig. 10.19. Buckling coefficient vs. slenderness ratio curves for various F_u/EB values. (*After Van Dyer 1976.*)

the construction of a "design" chart, such as shown in Fig. 10.19. To use this chart, one calculates the value of $\frac{F_u}{E(B)}$ and the slenderness ratio, λ, for the column (using, for example, Eq. 10.72 to evaluate B for a layered column system), and then enters Fig. 10.19 to read $\beta = F_{cr}/F_u$ from the appropriate curve. Thus, the critical stress, F_{cr}, and the margin of safety of the chosen column can be determined. Additional examples for various types of spaced columns (see Fig. 10.20), were also included in Van Dyer's research. It is likely that the column design highlighted herein will become a standardized procedure in the future. Presently, the method at least offers the designer methods to evaluate composite wood columns using an experimentally verified, rational approach.

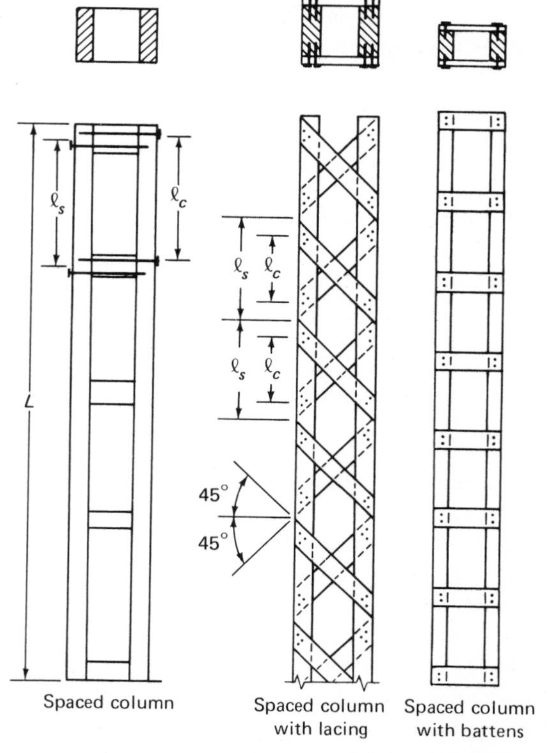

Fig. 10.20. Spaced columns. (*After Van Dyer 1976.*)

10.7 COMPOSITE WOOD SYSTEMS IN LIGHT-FRAME CONSTRUCTION

Layered wood systems are widely used in housing and other wood light-frame construction.

Wood-joist and sheathing construction, commonly used in housing for floors, walls, and roof systems, is an important example. Although wood-joist and sheathing systems are simple in concept and construction, they are extremely complex to analyze. Not only are they usually multilayered, with each layer made of orthotropic materials, but also, in nailed and elastomerically glued systems, the degree of composite action is a function of the presence of interlayer slip. In addition, gaps at joints in sheathing layers produce discontinuities that have a significant effect on the composite action of the system. These factors, coupled with the variability of the materials in each component, require the use of sophisticated analysis techniques to predict system performance.

A brief review of literature related to composite behavior in wood systems has been given previously in Section 10.1. Analytical concepts for layered wood beams have been described in Section 10.5. Since layered beams represent the key component in composite wood systems, their analysis is fundamental to predicting the behavior of such systems. References germane to each type of wood composite system will be cited as required.

10.7.1 Wood-Joist Floor Systems

10.7.1.1 Introduction

The current method of design generally employed for wood-joist structural systems is based on the grossly simplified assumption of independent action of each component in the structural system. The inadequacies of this method have long been recognized as evidenced by remarks made by Whittemore in a National Bureau of Standards publication (1948).

> Houses have never been designed like engineering structures. Since prehistoric times, safe house constructions have been found by the tedious and wasteful method of trial and error. If the modern research that has proved so successful in the solution of other problems had been applied to houses, not only would homes be more satisfactory as dwellings, but, much more important, the cost would be much less. This would be an outstanding contribution to the problems of providing acceptable houses for the low-income groups in this country.

As recently as 1970, the situation was not greatly improved, as evidenced by the following quotation by Johnson and Angleton (1970) concerning the wood-joist floor system.

> The current piece-by-piece, simplifying assumption-by-simplifying assumption, design procedure imposed on frame floor systems penalizes wood. On the basis of laboratory and field research conducted by the National Association of Home Builders (NAHB) Research Foundation (1970) and others, we know that there is an additive performance effect of the additional materials and fastenings typically used in frame floor systems. Currently, there are no recognized methods or even adequate data to provide criteria for this improved performance for wood frame floor systems.

A series of research programs that began in the early 1970s—by Polensek (1969, 1972), Sliker (1972), Onysko (1970), and a recent large-scale investigation by Vanderbilt et al. (1974)—have all been aimed at developing a rational analysis-and-design procedure for wood-joist floor systems. Extensive theoretical studies along with a full-scale verification test program as reported by Kuo (1974), Liu (1974), Vanderbilt et al. (1974), and Goodman et al. (1973) have led to a verified mathematical model that can be used to predict the behavior of wood-joist floor systems. This model recognizes the interaction of the many material properties and two-way action. A review of the procedures used for the verified model will be given, and key results of its use will be demonstrated in subsequent sections.

10.7.1.2 Development of a Mathematical Model for Wood-Joist Floor Systems

The basic scheme for the analysis of wood-joist floors is the idealization of the complex system as a set of crossing beams. A schematic for this method is shown in Fig. 10.21. Each of the T-beam strips consists of a joist plus a sheathing flange (Fig. 10.21b), usually taken to be equal

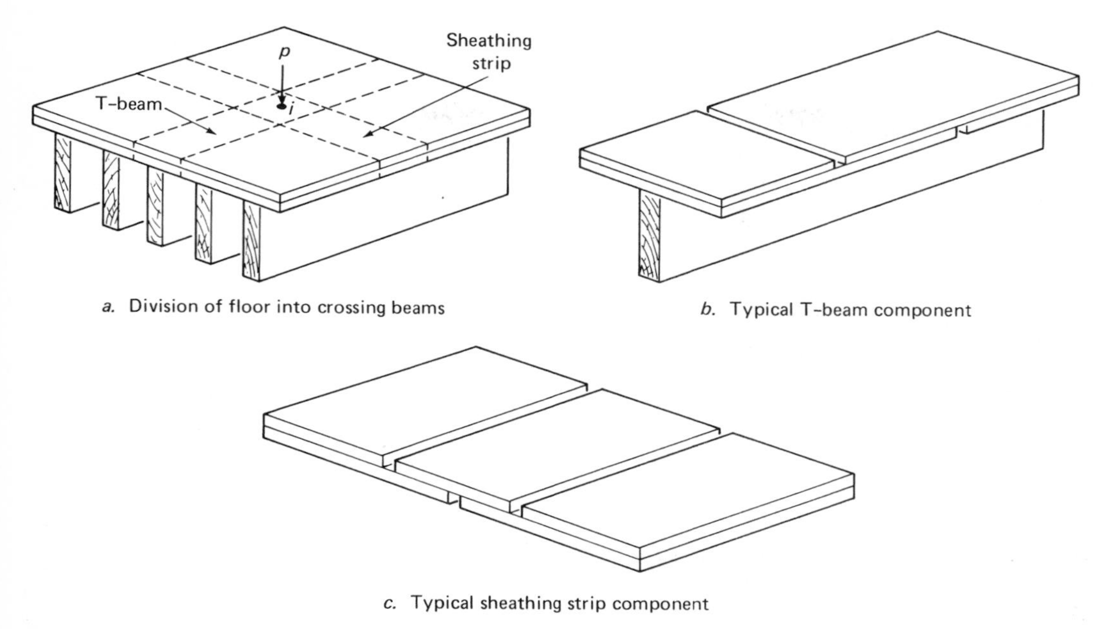

a. Division of floor into crossing beams

b. Typical T-beam component

c. Typical sheathing strip component

Fig. 10.21. Crossing beam model of wood-joist floor systems: *a*, division of floor into crossing beams; *b*, typical T-beam component; *c*, typical sheathing-strip component.

in width to the joist spacing, although any effective width may be used. Each of the sheathing strips for a two-layer system consists of a continuous beam of sheathing spanning in the direction transverse to the joists and of any chosen width (Fig. 10.21c).

Assuming the floor to be represented by a system of crossing beams ignores the contribution of the torsional stiffnesses of the sheathing and T-beams. As the G/E ratio for plywood is small, neglect of its torsional stiffness is believed to represent a small error. Likewise, the torsional stiffness of a T-beam section is small compared to its bending stiffness. Deflections computed using the crossing-beam model may be expected to be slightly greater than if torsional stiffness were included.

The mathematical model based on the concept of crossing beams can be developed by the use of either the flexibility (consistent deformation) or stiffness methods. Details of the flexibility approach are given by Goodman et al. (1973) and by Vanderbilt et al. (1974). The use of the finite-element analysis method for the T-beam and sheathing components as developed by Thompson et al. (1975) has led to a stiffness solution, which reflects the effects of gaps, if any, in the sheathing layers. Details of the finite-element model used in this latter analysis procedure were discussed in Section 10.5.3. A computer program (FEAFLO) has been developed for the analysis. A listing and example of its use are given in a paper by Thompson et al. (1976).

10.7.1.3 Verification Studies for Mathematical Model of Wood-Joist Floor Systems

The results of the full-scale floor tests conducted for 22 cases using 16 basic floor specimens were used to verify the mathematical model as described by Vanderbilt et al. (1974). Deflections computed using the finite-element form of the crossing-beam mathematical model were compared with measured values for working load levels. A concentrated load ranging from 600 to 1000 lb maximum acting over the midspan of the center joist was applied to each floor. Measured deflections at selected positions along the centerline of each floor specimen were plotted against the computed results. Percentage errors of the predicted versus the measured

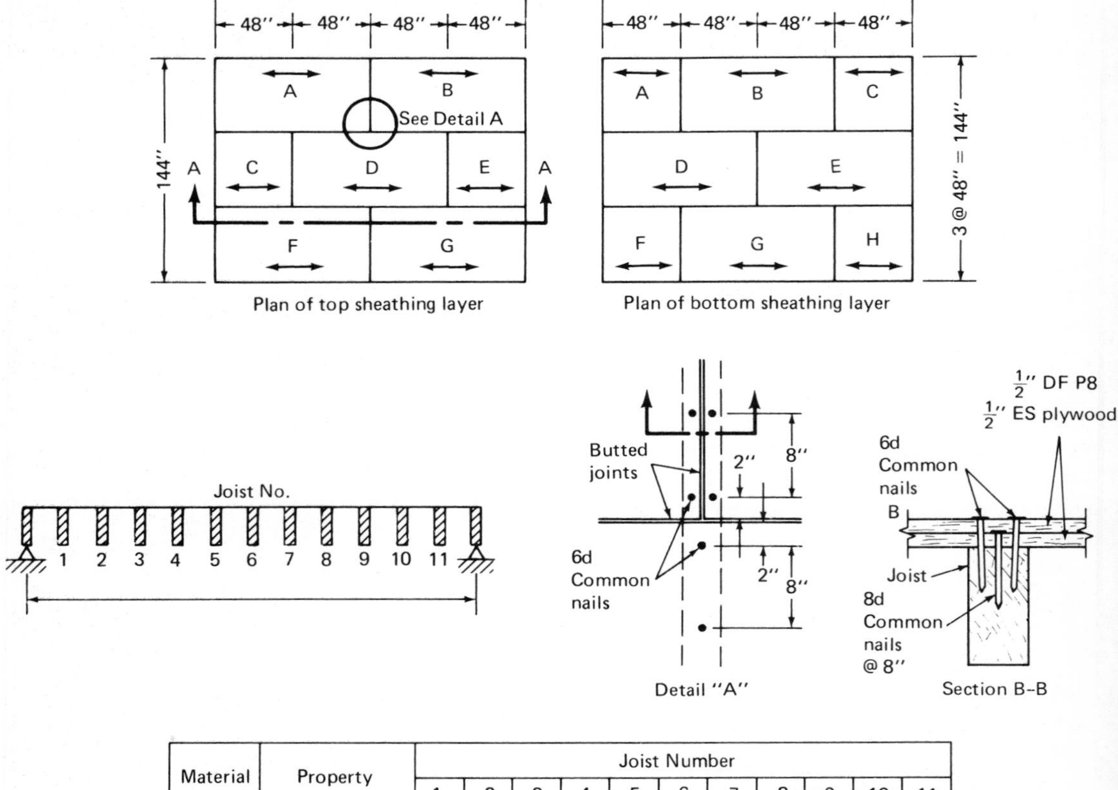

Material	Property	Joist Number										
		1	2	3	4	5	6	7	8	9	10	11
2 X 8 ES joist	Actual size, in (width X depth)	1.51 X 7.20	1.50 X 7.20	1.54 X 7.38	1.47 X 7.24	1.50 X 7.23	1.48 X 7.26	1.51 X 7.19	1.54 X 7.15	1.50 X 7.23	1.51 X 7.19	1.50 X 7.27
	Average MOE ∥ Grain, 10^6 psi	1.16	1.12	1.13	1.14	1.06	1.67	1.24	1.21	1.18	1.23	1.27

Joist Data

Layer	Material	Thickness in.	Gap	Connector	Average MOE, 10^6 psi															
					A		B		C		D		E		F		G		H	
					∥	⊥	∥	⊥	∥	⊥	∥	⊥	∥	⊥	∥	⊥	∥	⊥	∥	⊥
Bottom	$\frac{1}{2}$ ES plywood	$\frac{3}{8}$	$\frac{1}{16}''$ T&G	8d nails @ 8″	1.36	0.21	1.36	0.20	1.36	0.21	1.41	0.23	1.40	0.23	1.45	0.23	1.43	0.24	1.45	0.23
Top	$\frac{1}{2}$ DF P8	$\frac{1}{2}$	Butted no gap	6d nails @ 8″	0.42	0.31	0.53	0.41	0.45	0.35	0.46	0.33	0.45	0.35	0.42	0.33	0.51	0.40	—	—

Sheathing Data

Fig. 10.22. Layout and data of a typical floor test specimen (1 in. = 2.54 cm, 10^6 psi = 6895 MPa). (*After Vanderbilt et al. 1974.*)

deflection and additional results are available in Liu (1974).

The theoretical analysis of the behavior of each floor requires certain input data to the mathematical model. These data include floor span, floor length, numbers of T-beams and sheathing strips, dimensions and MOE (modulus of elasticity) values of all components, slip-modulus values, data on gaps in sheathing layers, and load data. While the joist span and floor length were held constant at 144 and 192 in., respectively, other physical and geometric properties varied widely. Material properties for each component of the floor were determined prior to the construction of each specimen. Flexible gap evaluations were used for those gaps that were tightly butted or glued, while open gaps were assumed where the sheathing panels were not butted or glued. An example of the detailed arrangement of a typical floor is given in Fig. 10.22. Slip moduli values were developed by experiment for each case. Adjustments were made for nail spacing and attachment of the third layer (top layer) to the floor. Two examples of typical comparisons between measured and predicted results for "working-load levels" of concentrated load at the center of the floor are given in Figs. 10.23 and 10.24. In general, good results are obtained using the predictions of the mathematical model. The average algebraic error for the center deflections under concentrated load was about +3%, indicating that the model over-predicts the deflections slightly as expected.

Additional verification studies for uniform loading are given by Sazinski (1978).

10.7.1.4 Composite and Two-Way Action in Wood-Joist Floor Systems

To demonstrate the degree of composite and two-way action in wood-joist floor systems, a series of parameter studies of important system

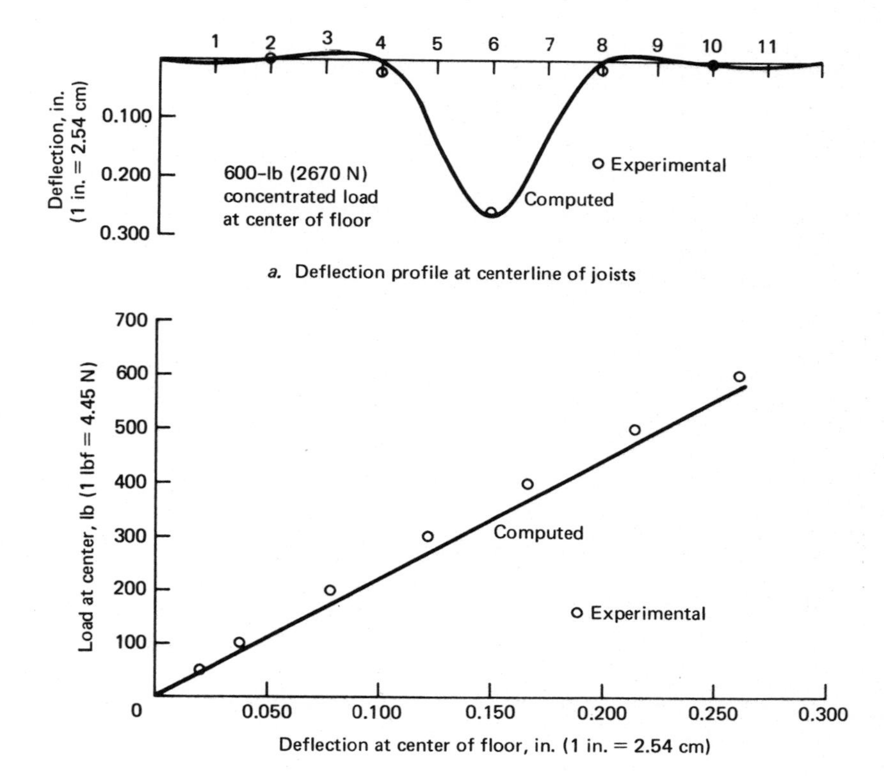

Fig. 10.23. Computed vs. measured results of floor specimen with one layer attached. (*After Vanderbilt et al. 1974.*)

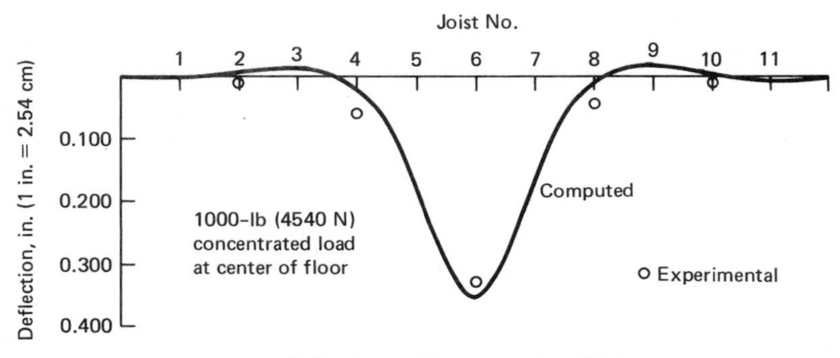

a. Deflection profile at centerline of joists

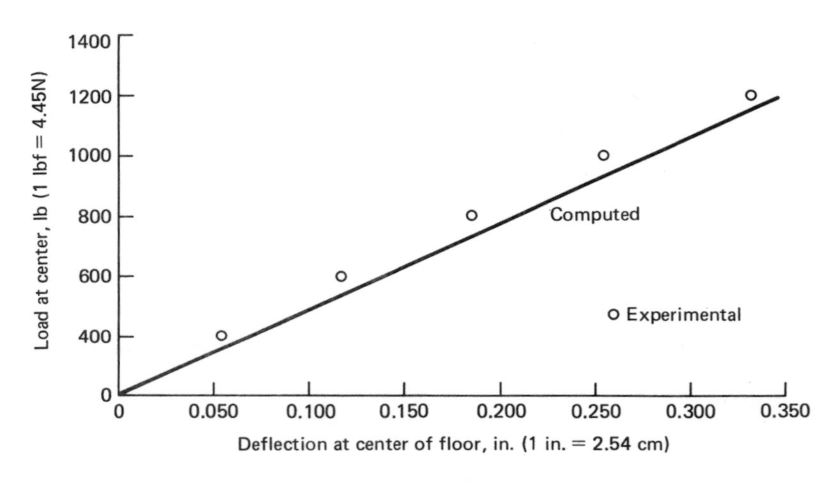

b. Load–deflection behavior

Fig. 10.24. Computed vs. measured results of floor specimen with two layers attached (nails of top layer driven into joists). (*After Vanderbilt et al. 1974.*)

variables is presented. These studies enable an evaluation of the critical aspects of the behavior of the system and point the way to improved design procedures.

The dimensions and fixed parameters for the study are shown in Fig. 10.25. Joist and plywood MOEs are specified on the figures for each example. Fixed data include 2-by-8-in. (5.08-by-20.32-cm) nominal joist sizes, nail spacing of 8 in. (20.32 cm) with one nail per row (8d nails), and floor dimensions of 12-ft (3.66-m) span and 16-ft (4.88-m) width. The face grain of the sheathing was assumed to be perpendicular to the joist span for all cases. The floor contains 11 T-beam and 11 sheathing-strip components. Simple supports are assumed at all edges of the floor. The values for the fixed data were arbitrarily chosen, and, although not necessarily representative of actual data, they serve to illustrate the effects of key variables.

To demonstrate the full effects of composite and two-way action, the deflections of an example floor with varying joist MOE were computed using the mathematical model for the case of uniform load. The results are shown in Fig. 10.26. This example was computed for four conditions: (1) joists only carrying loads (no sheathing), (2) distribution of loads by two-way action but neglecting composite or T-beam action (two-way, no T-beam), (3) composite action due to 8-in. (20.32-cm) nail spacing and $k = 30{,}000$ lb/in. (5260 kN/m) with no two-way action, and (4) complete composite and

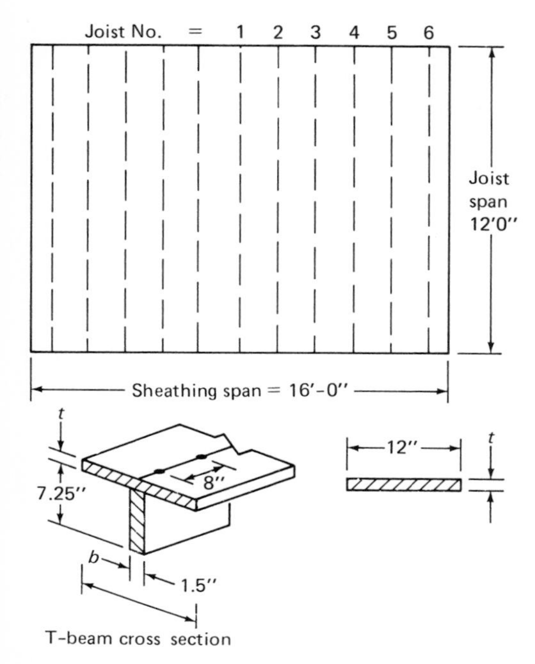

Fig. 10.25. Floor studied in parameter studies.

two-way action as predicted by the complete mathematical model. In addition, the MOEs of the joists were varied as shown to simulate a possible actual case. The results of this analysis clearly demonstrate the importance of the composite and two-way behavior of wood-joist floor systems. As shown, the differences in deflection as predicted by the various assumptions range up to greater than 100% when deflections of joists acting alone are compared with those predicted by the complete mathematical model. Since design of wood-joist floors is currently based on neglecting the contributions of the composite and two-way action, except for repetitive member increases, the deflections predicted can be greatly overestimated in many cases. Variations of joist stiffness from values assumed in design also contribute inaccuracies, of course. In addition, the smoothing out of deflection behavior by two-way action is demonstrated to be significant as evidence by comparison of the various cases. This example shows the need for considering the complete behavior of the system including full recognition of both composite and two-way action.

10.7.1.5 Parameter Studies of Composite Action

By using the capabilities of the complete mathematical model, any aspect of the static behavior of the system may be studied by varying the basic parameters. These studies enable the easy evaluation of design or material choices. Focusing on composite action, the fundamental variable is that of the interlayer slip modulus, k. The "composite action curves" shown in Figs. 10.27 and 10.28 show the effect that varying k has on the behavior of the complete floor. For either uniform or concentrated load, it is seen that for low values of k, less than $k = 1000$ lb/in. (175.2 kN/m), the sheathing does not act as an integral flange, and the joist alone must carry almost all the load. While no composite action is present for these low values of k, two-way action is present. For k values of 10^6 lb/in. (175,200 kN/m) and above, nearly full composite action is present. Values of k between 10^3 and 10^6 lb/in. (175.2 and 175,200 kN/m) represent the region of incomplete composite action. All practical fasteners used in wood construction except the most rigid glues lie in this region. Thus, either neglecting slip or assuming complete composite action will result in significant errors in predicting the performance of the system. It should be pointed out that a change in nail spacing from 8 to 4 in. (20.32 to 10.15 cm) has the same effect as doubling k. In the practical range of $k = 30{,}000$ to 60,000 lb/in. (5256 to 10,512 kN/m) (typical for eight penny nails), this difference in k for the uniform load case amounts to about a 10% to 15% difference in deflection performance.

Other parameter studies and results are presented in Goodman et al. (1973) and in Liu (1974). Probabilistic analysis of the effects of variability of materials on wood-joist floor-system performance are given by Dawson (1974), Dawson and Goodman (1976), and Fezio (1976).

10.7.1.6 Implications for Improved Design

The developments in analysis methods described in the previous sections now make possible new national design procedures for wood-joist floor

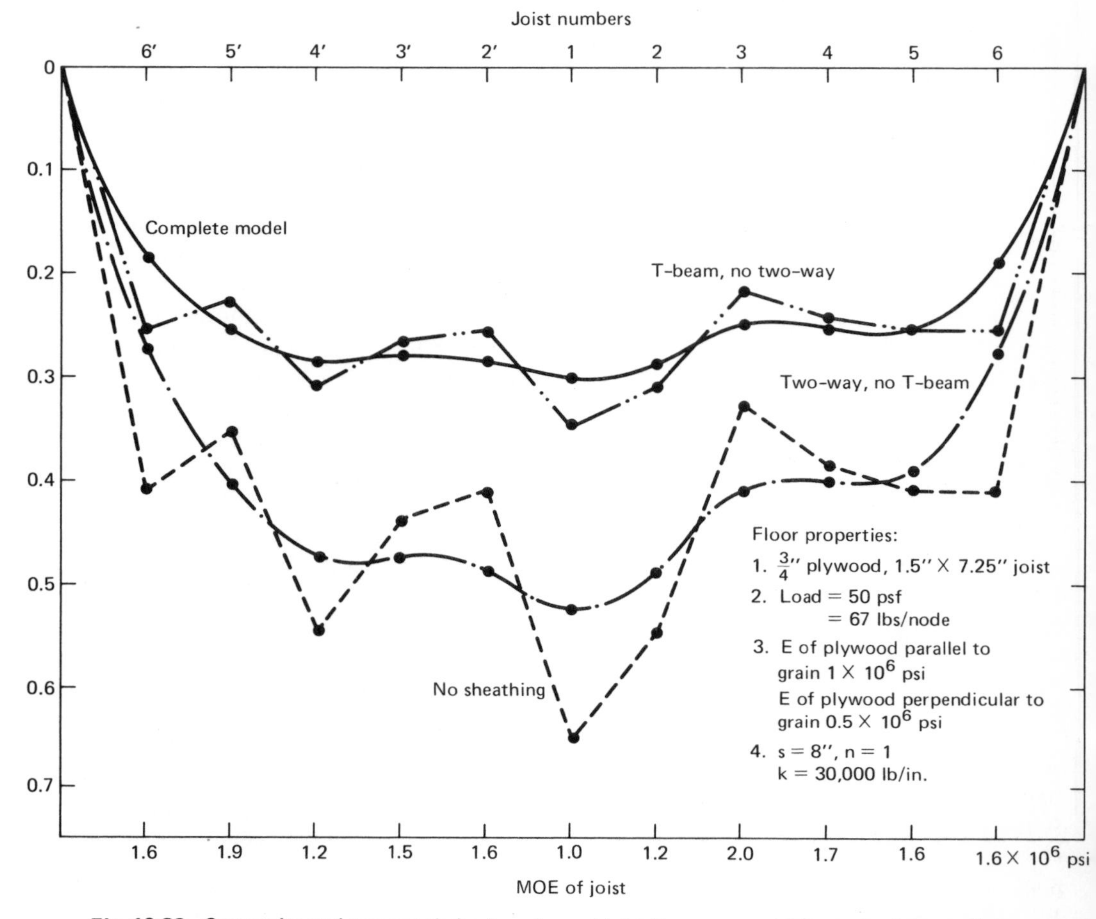

Fig. 10.26. Composite and two-way behavior of wood-joist floor system with varying joist moduli.

systems. The existence of a verified mathematical model as quantified by the computer program, FEAFLO (Thompson et al. 1976) allows the application of accepted engineering practices to light-frame wood floor systems such as used in housing. Research and application of this rational procedure to design problems was accomplished by Sazinski (1978). It is anticipated that some form of limit states or probability-based design procedure will emerge. Implications for conserving materials and insuring reliable performance offer much promise for the future use of wood in light-frame systems.

10.7.2 Wood-Stud Wall Systems

Wood-stud wall systems with various covering materials are standard components in light-frame and housing construction. Current design methods for such systems are based on an oversimplified model, which assumes a set of isolated, identical beam-columns. As in other wood systems, the current design procedure does not account for existing partial interaction between studs and wall sheathing and neglects any effects of load sharing that may exist. According to Polensek (1976), the current method overestimates the maximum stresses and deflections, and thus produces uneconomical designs.

An extensive study of wood-stud wall systems has been done by Polensek (1975, 1976) and others at Oregon State University. The study has resulted in the development of a procedure that accounts for partial composite action, load sharing, and variable material and joint properties. Stresses at service loads and ultimate load

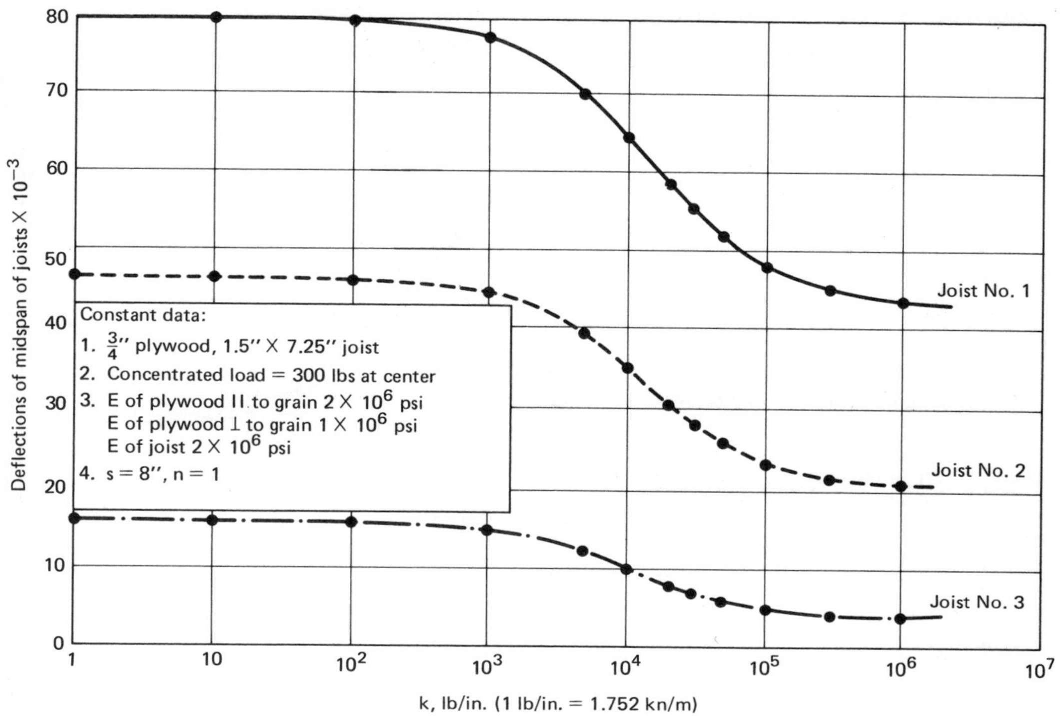

Fig. 10.27. Deflections vs. *k*, concentrated load.

capacity of wood-stud walls are also treated. Methods similar to those described for wood-joist floor systems, involving finite-element analysis and including effects of semirigid inter-layer connection, were used to predict the behavior of the wood-stud wall system. A summary of this work is presented subsequently. Extensive details may be found in Polensek and Atherton (1973) and Polensek (1975, 1976).

10.7.2.1 Finite-Element Model of Wood-Stud Wall

The theoretical approach to the solution of problems involving the service and ultimate load capacity of wood-stud walls is based on a semi-iterative finite-element model. Polensek (1976) describes the various features of the model. The basic components of the model consist of I-beam and plate elements as shown in Fig. 10.29. These components are superimposed in the analysis to properly represent the complete composite and load-sharing behavior of the system. Total wall stiffness is a combination of the bending-axial stiffness of studs and wall coverings in the x-direction and the bending stiffness of the wall coverings in the y-direction. To reduce the number of structural equations, axial effects are treated by an iterative process instead of by inclusion in the stiffness matrix. Composite action is included in the I-beam-column elements.

Mathematical details of the development of the finite-element analysis may be found in Polensek (1976), including work by Amana and Booth (1967a, 1967b), which is extended and used as the basis for finite-element formulation for the plate elements. Load-deflection results from tests of the stud material and other available sources were used to assess their inelastic behavior. Other material properties, for wall coverings, etc. were obtained experimentally for use as input to the mathematical model.

10.7.2.2 Experimental Verification Studies

A series of tests of I-beams and walls were conducted to verify the mathematical model. The

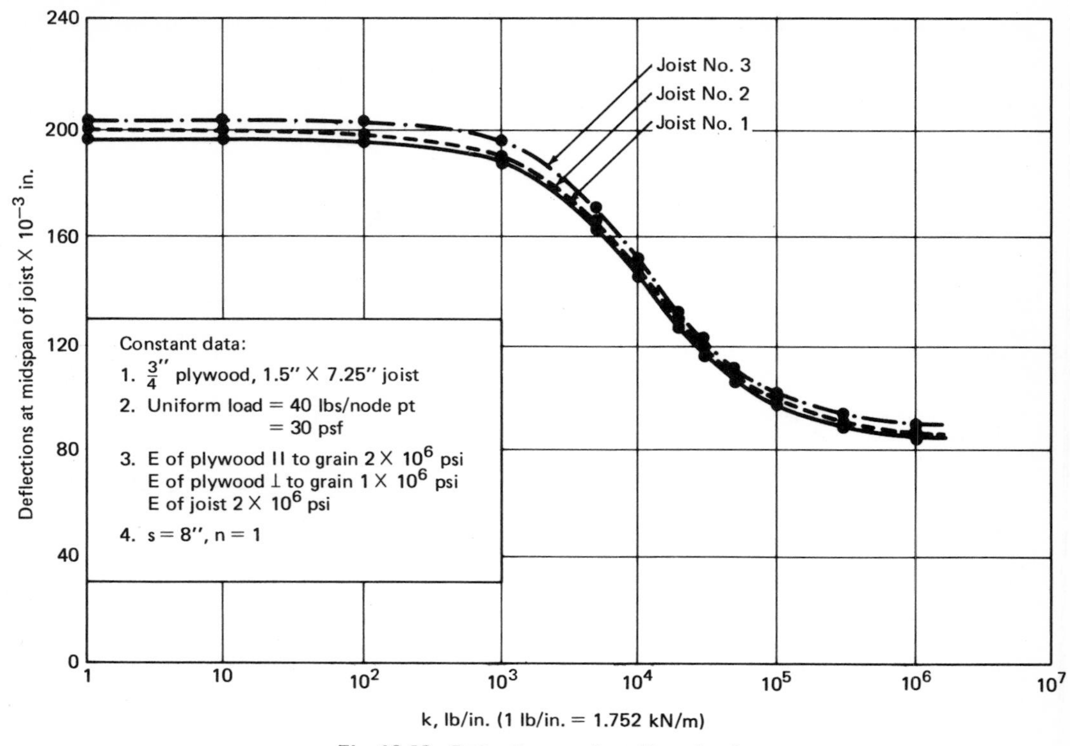

Fig. 10.28. Deflections vs. k, uniform load.

I-beam tests were used to verify the finite-element method for this component and generally showed good agreement with predicted deflections and slip displacements.

Three 8-ft-high (2.4-m) and 12-ft-long (3.7 m) walls were tested. The general wall construction conformed to current standards. Interior tension wall covering was a $\frac{1}{2}$-in. (1.27-cm) gypsum board with 8-ft (2.4-m) edges positioned horizontally. The gypsum was attached to studs with 4d drywall nails, spaced 8 in. (20.32 cm) along the stud. Joints between the sheets were taped and spackled, which provided for full continuity in the tensile wall covering. Exterior (compression) wall covering was $\frac{3}{8}$-in. (0.95-cm) sheathing plywood attached to studs with galvanized 6d nails, spaced 12 in. (30.480 cm) at intermediate locations and 8 in. (20.32 cm) on edges. The testing apparatus enabled simultaneous application of concentrated vertical loads and uniform lateral load (through use of an air bag). The lateral load was increased at levels of 1 psf (48 N/m^2) per min. until the wall failed. Midspan deflections and slip displacements were recorded.

Figure 10.30 shows the test setup and summarizes the stud properties for one wall of the test series. Other properties are given in Tables 10.5 and 10.6.

Typical results of wall testing and analysis, shown in Figs. 10.31 and 10.32, display good agreement between theoretical and experimental values. Theoretical analysis predicted that stud 3 of wall 1 would be the first to fail at 90 psf (4300 N/m^2), as was actually observed in the test. The experimental load of 115 psf (5500 N/m^2) at the time the next stud (stud 6) failed could not be accurately predicted theoretically, and was attributed to insufficient knowledge of the load-deflection relation of stud 3. Furthermore, stud 3 in the actual wall probably did not fail completely at 90 psf (4,300 N/m^2) and was still carrying some load afterward. The MOE of stud 3 was not known for the higher loads, and no attempt was made to predict the second stud failure.

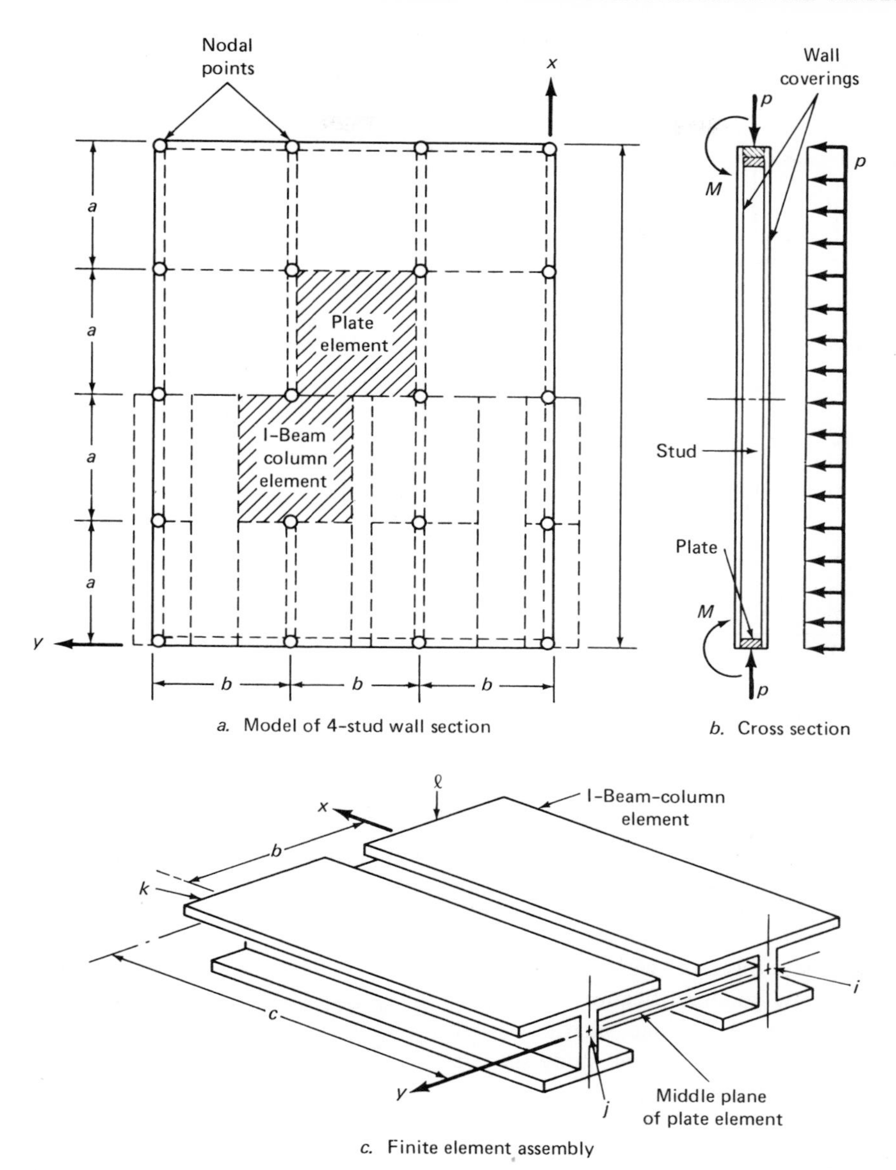

Fig. 10.29. Finite element model of wood-stud wall. (*Personal communication, Polensek 1976.*)

Minor deviations between theoretical and experimental deflections shown in Fig. 10.31 are considered acceptable. They probably were caused by local variations in material and joint properties. Deflections of stud 6 deviate the most, probably because of errors associated with its stiffness. This stud failed during the wall test, and its load-deflection trace had to be approximated for the theoretical analysis.

Deflections at service loads are compared in Fig. 10.32. The agreement between the experimental and theoretical deflections is good considering that the shear-slip behavior is more complex at small loads than at high loads. In particular, frictional effects, present at small loads in stud-covering interlayers, are not known exactly. In addition, the modulus of plywood stud joints employed was approximate.

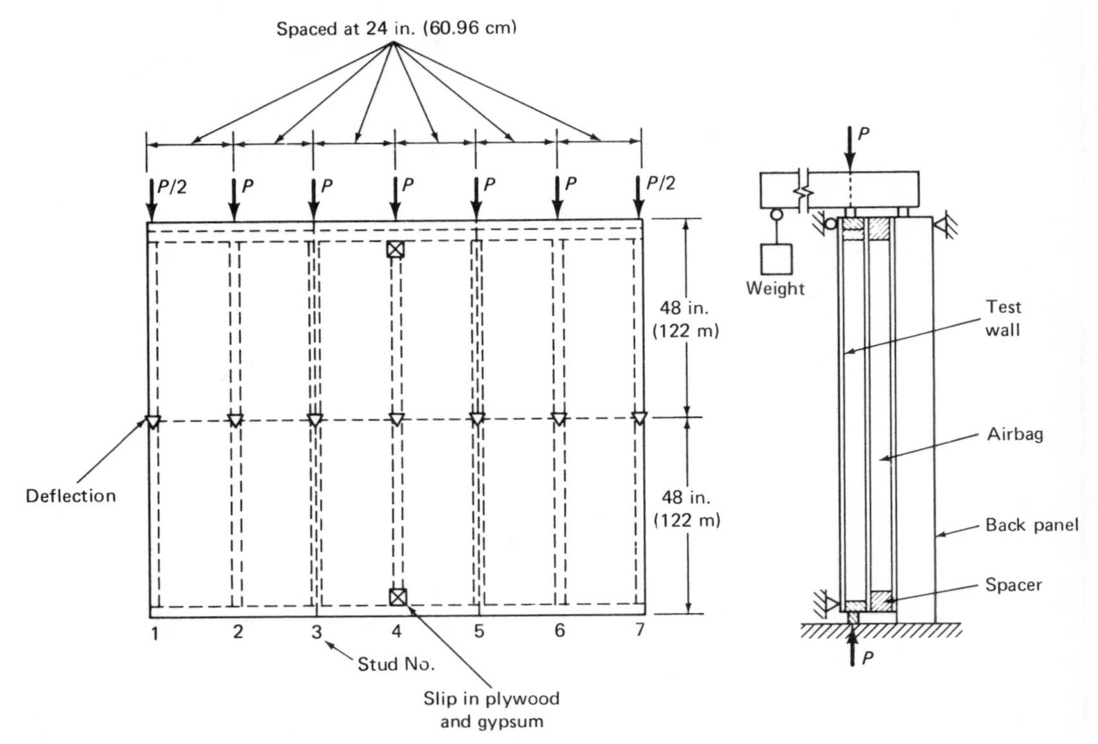

Fig. 10.30. Testing arrangement for Wall 1. *(After Polensek 1976.)*

10.7.2.3 Summary

A finite-element method for modeling wood-stud walls developed by Polensek (1976) has been described, together with experimental results that show good verification of the predicted service and ultimate load behavior. Other results, including a parameter study and a probabilistic simulation technique, have been presented in Polensek (1975). Future developments in this area of research are expected to enable designers to properly account for partial composite action, load sharing, and material variability and nonlinearily, in the design of wood-stud wall systems.

10.8 STRUCTURAL COMPONENTS OF WOOD COMBINED WITH OTHER MATERIALS

A number of systems that have been used where wood is combined with other materials to form structural components will be discussed. Some

Table 10.5. Properties of wall coverings.

Property	Gypsum Board		Plywood	
	Bending	Tension	Bending	Compression
E_x, in kips/in.^{2}	230.	93.	1730.	1090
E_y, in kips/in.^{2}	360.	163.	290.	
G_{xy}, in kips/in.^{2}	159.		150.	
ν_{xy}	0.37		0.025	
ν_{xy}	0.23		0.146	
Rupture σ_x in kips/in.^{2}	0.301	0.096		

Note: 1 ksi = 6.9 MN/m^2.

Table 10.6. Connector Moduli, in kips/in./nail between studs and wall coverings.

Connection	Moduli for Interval of Uniform Load, p			
	0 psf–3 psf	3 psf–8 psf	8 psf–20 psf	20 psf–ultimate
Stud-gypsum	10^4	70	7.8	1.2
Stud-plywood	10^4	100	16.7	3.0

Note: 1 kip/in. = 0.175 MN/m; 1 psf = 47.9 N/m^2.

of these systems are widely used, while others have mainly been subjects of laboratory research to investigate their structural and economic feasibility. A thorough review of wood used in combination with other materials is given in *Wood Structures* (ASCE 1975). References to specific systems and components will be given as each is presented.

10.8.1 Reinforced Wood Components

Reinforced wood components have been the subject of research studies dating back to the late 1800s. Although many systems have been tried, no commercially acceptable system has emerged to date. Recently, extensive research by Krueger (1967, 1973), Krueger and Sandberg (1974), and Krueger and Eddy (1974) has provided criteria for the design of reinforced wood systems. Krueger states, in *Wood Structures* (1975):

In general, it was found that the material should be highly elastic, desirably ductile at ultimate, it should have a highly predictable fracture stress at ultimate, it should be resistant to environmental effects and fire resistant. Just as important as the material properties are factors such as cost, handling characteristics, adaptability to standard laminating production techniques, material preparation requirements, bonding characteristics, and most important, the flexibility with which the properties can be varied, particularly the effective area of the reinforcement.

The theory of reinforced wood components is generally a straightforward application of the concepts of transformed section assuming that adequate shear resistance is developed to avoid interlayer slip between the wood and the reinforcement. In general, due to wood's relative weakness in tension, reinforcement of the component in areas of tension stress is the

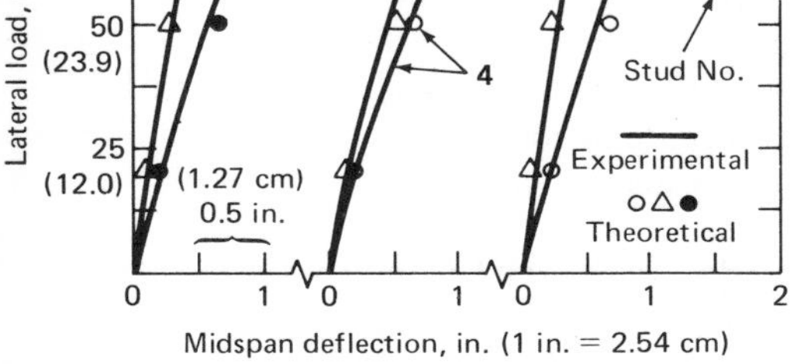

Fig. 10.31. Wall 1, theoretical and experimental deflections. (*After Polensek 1976.*)

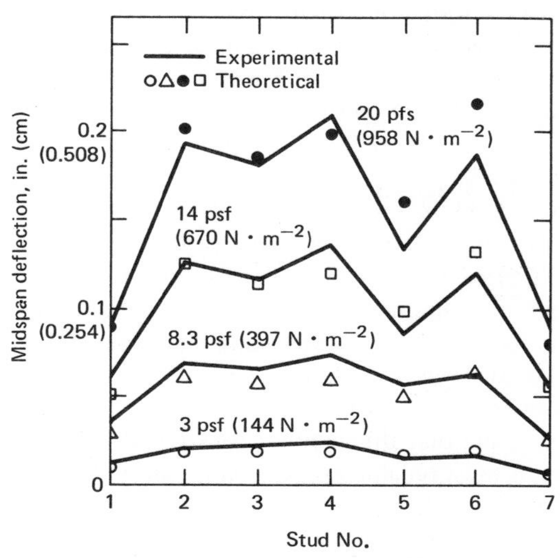

Fig. 10.32. Wall 1, theoretical and experimental deflections at service loads. (*After Polensek 1976.*)

most important consideration. However, additional stiffness, as predicted by the transformed section concept, can be achieved by reinforcing both the top and bottom of the beam to accomplish a composite section using a high modulus of elasticity material. Krueger and Sandberg (1974) developed an ultimate-strength criteria for reinforced wood components, similar to those used in reinforced concrete, which were verified by laboratory tests.

Studies have been made on various types of reinforcement for wood components. Granholm (1954) and Lantos (1964, 1970) have reported on the use of steel for reinforcement. Research on fiberglass reinforcement has been reported by Biblis (1965). Wire reinforcement embedded in polyester or epoxy resin has been studied by Bystrom (1968), Krueger and Sandberg (1974), and Schollet (1970). This latter system appears to offer the greatest potential for further development. Use of high-strength, wire-reinforced plastic to reinforce glulam beams appears to be promising, particularly when recognition is made of the current problem of a reduced amount and quality of tension laminating stock.

An additional area of recent study has been that of flitch beams. Flitch beams are made by vertically laminating wood members with layers of metal. Work by Kumar et al. (1972) has provided some criteria for the use and construction of this system. While improved shear capacity results, the tension stress in the extreme fiber of the wood is not reduced as it is in horizontally reinforced systems.

10.8.2 Prestressed Wood Components

The concept of prestressing, widely applied in concrete systems, has been studied for wood-beam components. In particular, research by Bohannan (1962, 1964) and Peterson (1965, 1968) have established criteria for applying prestressing concepts to wood members. Recognizing that the compressive strength of wood is its most reliable property and that it generally exceeds the tensile strength, the process of prestressing to take advantage of this characteristic offers promise.

Methods of post-tensioning appear to be the most practical means of achieving prestressing. A patented procedure has also been used involving use of spring steel bonded to the timber member.

The performance of prestressed timber beams as reported by Bohannan (1962, 1964) and Peterson (1965, 1968) generally show strength increases of nonprestressed beams in the range of 30%. Allowable stresses can be increased substantially over currently used values. Tests under long-term and cyclic loading indicate good performance of such components.

The future use of prestressed wood members depends on further developments, particularly the use of E-rated wood materials, which allow the designer to take full advantage of the concept. Posttensioning may also be used as a means of field repair of damaged or overstressed beams. It is expected that further use of the concept of prestressing will be made in the future.

10.8.3 Wood-Concrete Components

Mechanically connected wood-concrete composite systems have been used commonly in bridge construction for at least 40 years, and standard plans are available for their design and construction (Bureau of Public Roads 1969, Canadian Institute of Timber Construction 1970). Further discussion of these systems is given in Section 10.9.3b.

Other types of wood-concrete systems involving use of epoxy-bonded composite connection between timber girders and the concrete deck, have been researched by Pincus (1969, 1970). Design criteria must include recognition that the allowable stresses, in tension for the wood girders, in compression for the concrete deck, and in shear at the interface, must not be exceeded. Transformed section concepts (see Pincus in *Wood Structures* (1975)) are used in the stress analysis. Continued research is required to assess the effects of long-term loading, fatigue, temperature effects, etc., to allow these new systems to be utilized.

10.8.4 Open-Web Composite Wood-Steel Systems

Factory-manufactured systems involving open-web truss systems with top and/or bottom chords of wood and steel web members are popular and economical components for roof and floor systems. Their use has greatly increased in recent years, particularly where longer span components are required. The record span wood building has recently been constructed at the University of Idaho (see *Civil Engineering*, June 1976), with an arch span of 400 ft (122 m). This structure uses an open-web wood-steel composite truss system,* which was assembled and erected on the site. Recognition of this structure as the Outstanding Structural Engineering Achievement of 1976 has been given by the American Society of Civil Engineers.

The concept of the open web, wood-steel composite system is to take advantage of the nailability and stiffness of the wooden chord members combined with the high tensile strength and simple connection of the steel web members. Design of the system is generally done on a "custom" basis by the various suppliers. Standard truss analysis methods are employed, and the usual wood and steel specifications are applied. Design concepts are reviewed by Nelson in *Wood Structures* (1974). Special recommendations are made for deflection limitations of L/360 for spans up to 15 ft, $\frac{1}{2}$ in. (1.27 cm), for spans 15 to 20 ft (4.58 to 6.1 m), and L/480 for spans beyond 20 ft (6.1 m). Longer span structures require particular attention to camber requirements to avoid possible sagging under load. Adequate erection precautions for lateral bracing during erection are particularly important for open-web truss systems.

10.8.5 Plywood-Lumber Components

A variety of plywood-lumber composite systems are available, either as specialty items from various suppliers, or as "custom" designs by individual structural engineers. The design of such components for the case of factory-built units using rigid adhesives follows the usual transformed section concepts. Special consideration must be given to the properties of the plywood in analyzing the system since the layered, oriented nature of this material must be recognized. Since a variety of component types and systems exist, a general table (Table 10.7) is presented to aid the designer in understanding the various features of each type. References for each component are also given to guide the designer to more specific design data. The latest edition of the *Plywood Design Specification* should be consulted for detailed recommendations as well. A recent paper by Bower (1974) and a bibliography on wood and plywood diaphragms by Carney (1975) give the designer considerable additional information on lateral load analysis of plywood-lumber components. Other specific literature is available from manufacturers of each type of system and designers should consult suppliers for special applications of these systems.

*TRUS JOIST Corporation, Boise, Idaho.

10.9 MODERN TIMBER BRIDGES

One of humanity's earliest structures was probably the timber bridge. Since the time a log was first used to cross a waterway, timber bridges have been built in many forms and types. In the United States, their use was at its peak near the end of the nineteenth century when covered bridges exhibited widespread popularity. About 1880, with the advent of steel, interest in timber bridges began to lag behind other systems. Coupled with this event, concern for the longevity of timber structures greatly contributed to their declining usage.

Despite the concerns of designers, many nineteenth-century covered bridges have remained in service until today. The paramount reason for this remarkable service life, has been the overhead protection of the superstructure from the elements that normally lead to decay. With regard to this aspect, design and construction of timber bridges have progressed significantly in the modern era (see Williamson 1975). Today, the general use of pressure-treated glulam members has lead to a revival of interest in the use of timber in highway bridge construction (AITC 1973). This modern construction material offers versatility and durability, and has greatly influenced bridge design.

10.9.1 Glulam Bridge Systems

Today, glulam bridges are rapidly increasing in popularity due to low material cost, ease of

Table 10.7. Plywood-lumber components.

Component	Description	Primary Use	References
Flat plywood stressed-skin panels	Flat panels with lumber stringers covered with plywood skins.	Floor and roof panels panelized roof systems, folded plates. (Single skin up to 30-ft (9.15 m) spans.)	*Plywood Design Specification–Supplement No. 3–Design of Plywood Stressed-Skin Panels*, U813, American Plywood Association, Tacoma, Washington, 1972. *Fabrication of Plywood Stressed-Skin Panels–SS-8*, V340, American Plywood Association, Tacoma, Washington, 1974.
Curved plywood stressed-skin panels	Curved panels with lumber stringers covered with plywood skins.	Roof panels for spans up to about 20 ft (6.1 m).	*Plywood Design Specification–Supplement No. 1–Design of Plywood Curved Panels*, S811, American Plywood Association, Tacoma, Washington, 1968. *Fabrication of Plywood Curved Panels–CP-8*, V306, American Plywood Association, Tacoma, Washington, 1974.
Sandwich panels	Flat or curved panels with plywood skins or honey-comb-type core.	Roof, wall, and thin shell systems. Spans vary.	*Plywood Design Specification–Supplement No. 4–Design of Plywood Sandwich Panels*, U814, American Plywood Association, Tacoma, Washington, 1970. *Fabrication of Plywood Sandwich Panels–SP-61*, V309, American Plywood Association, Tacoma, Washington, 1974.
Glued or nailed plywood beams	Laminated or lumber flanges with plywood web I-beams or box-beams.	Beams, garage-door headers, light-frame construction systems, etc. Spans vary.	*Plywood Design Specification–Supplement No. 2–Design of Plywood Beams*, S812, American Plywood Association, Tacoma, Washington, 1972. *Fabrication of Plywood Beams*, V335, American Plywood Association, Tacoma, Washington. *Nailed Plywood Beams–Design Example*, American Plywood Association, Tacoma, Washington.
Prefabricated wood joists	Lumber or laminated plywood flanges with plywood webs.	Beams, garage-door headers, light-frame construction systems, etc. Spans vary.	Manufacturer's literature

erection, and natural beauty. The most common type is the longitudinal stringer or straight girder bridge (Fig. 10.33), which can provide economic spans of 20 to 80 ft (6.1 to 24.4 m). Longer spans of greater than 100 ft (30.5 m) are possible in certain situations. While the stringer bridge is most often employed, other configurations have been used effectively. Bow-string truss bridges are probably most economical up to 100-ft (30.5-m) spans. They are comprised of a curved compression chord and lower tension chord, which are glulam members and either glulam or solid sawn webs. Parallel chord trusses are also employed and, when used as a through span, permit practical designs in the 100-to-250-ft (30.5-to-76.8-m) range. The deck arch bridge is used to achieve efficient, clear spans of up to 300 ft (91.5 m). Glulam arches are used as the main element with glulam bents to transfer load from the glulam girder-deck system to the arches. The Keystone Wye interchange on U.S. Highway 16 near Mount Rushmore, South Dakota, is a recent example. Other popular glulam bridge systems are the king-post truss and the two-hinged flat arch, which are used in spans of 100 to 150 ft (30.5 to 45.8 m).

The Canadian Forest Service has been active in developing more efficient bridge designs. One bridge in Vancouver, British Columbia, uses a system of king posts and tension cables. The stringers are designed to carry only dead-weight, and the cables, tensioned over the posts, accommodate live load. This bridge spans 125 ft (38.1 m), but queen-post bridges, using two posts, have been built to span 180 ft (54.9 m).

The fallacy that timber is restricted to minor structures is probably based on the fact that trees are limited in size. Although this limits the size of solid-sawn members, the advent of glued-laminated timbers has presented designers with new alternatives. The Canadians have also designed and constructed clear-span bridges over 150 ft (45.8 m) long using glued-laminated girders approximately 7 ft (2.14 m) deep. The girders resemble typical I-beams with flange and web sections. In particular, over 300 glulam bridges have been built in British Columbia, (see Scarisbrick 1976), of which 200 are on forest roads. A number of the others are serving railways, as reported in *Railway Track and Structures* (1974). For most of the logging bridges, the typical bridge cross section contains large glulam I-beams as the main element (see Fig. 10.34). Typical I-beams have depths from 66 to 86 in. (168 to 218 cm), web widths from 9 to 12 in. (22.9 to 39.48 cm), flanges 4 to 8 in. (10.16 to 20.32 cm) wider than the webs, and flange depths a little less than the flange width. Clear spans range up to 150 ft (45.8 m). Cantilevered-beam cable-suspended systems and an array of unusual trussed girder arrangements are described by Scarisbrick (1976).

Possibly, the record span for timber construction is the Luzanne-Barrage suspension bridge in Switzerland. It spans more than 540 ft (165 m). The top and bottom chords are of glulam construction.

Fig. 10.33. Example of a glulam bridge.

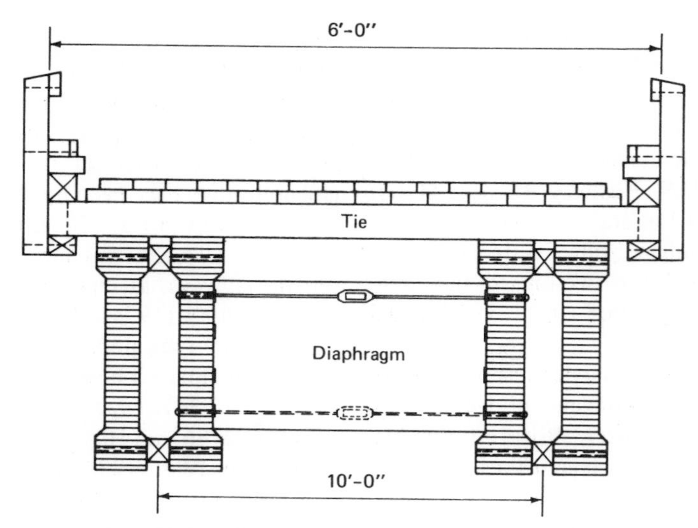

Fig. 10.34. Typical bridge cross section (1 ft = 0.305 m). (*After Scarisbrick 1976.*)

10.9.2 The Glulam Stringer Bridge

Although various timber bridge configurations are being built, relatively short timber stringer bridges have been used most frequently. The modern version of this bridge is a system that contains glulam girders and deck panels as the major components. The timber industry has developed the necessary design criteria and construction concepts for this system. This development was spurred by recommendations, made at the 1971 Systems Bridge Conference cosponsored by the Virginia Highway Research Board, to reduce timber bridge construction time by use of a systems approach. Details of this novel system are presented in subsequent sections.

The superstructure of the typical glulam bridge is basically composed of glulam girders and decking, their interconnection devices, usually an asphalt wearing and protective surface, and, depending upon the span, interior and end diaphragms. Figure 10.35 illustrates a general floor plan. The individual deck panels are usually manufactured 4 ft (1.22 m) wide and placed transverse to the stringer system. The panel length is the same as the width of the bridge. This sytem contrasts with the past, when bridge deck usually consisted of a nailed-laminated assembly. Nominal 2-in. dimension lumber was fastened by through nailing of the laminations and toe-nailing the laminations to the stringers below.

To fabricate a glulam deck panel, 2-in. dimension lumber is glued together to form a vertically laminated flat panel. The depth of the panel depends upon the loading and the girder spacing. When connected together, the panel system behaves as an orthotropic plate. The deck panels are anchored to the stringers either by lag bolts, as shown in Fig. 10.36, or by alternate positive anchoring methods such as bridge spikes. Panel-to-panel shear and moment continuity is provided by steel dowels placed along the longitudinal face of the deck (see Fig. 10.37). Research studies by McCutcheon and Tuomi (1973, 1974) have resulted in a design procedure for the glulam deck panels. Prior to this, numerous interpanel connecting devices were examined in an investigation by Bohannan (1972). More complete construction and design details for the glulam bridge system are available from the American Institute of Timber Construction (1975). The AITC presents complete tables of typical design configurations for the dowel-connected, glulam, stringer-deck bridge.

Practice is not limited to the stringer-bridge system described by AITC. Stringer bridges, which employ plank decks or nailed-laminated

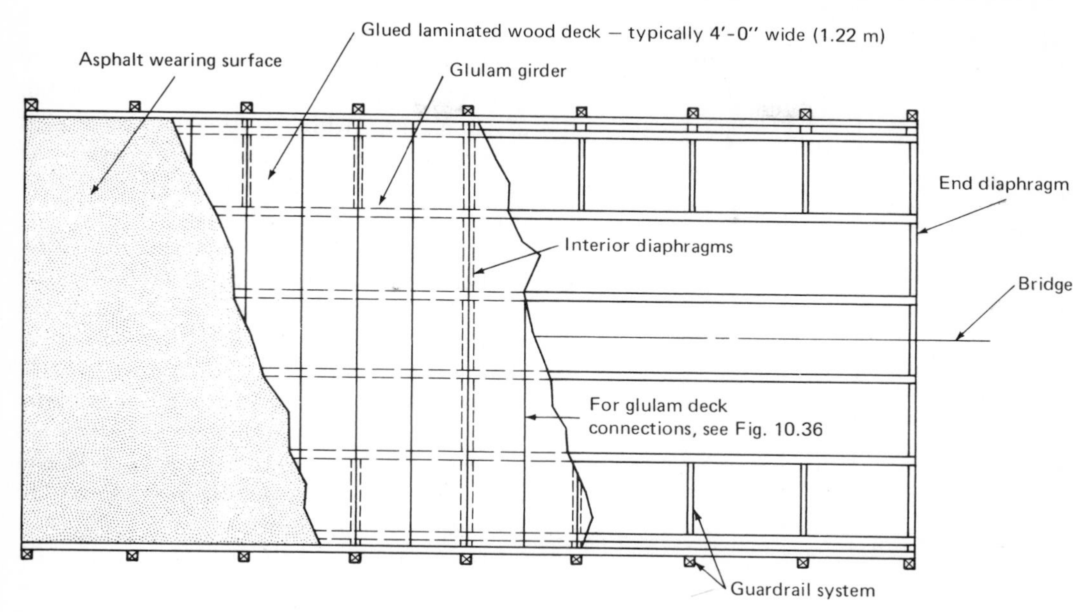

Fig. 10.35. Plan with deck partially removed. *(After AITC 1975.)*

decks, are also in use. One particular manufacturer* has developed a panelized bridge system that uses glulam deck panels in a special way. This system is illustrated in Fig. 10.38. Its noteworthy features are the absence of panel-to-panel connectors and the use of a cast aluminum deck bracket (Fig. 10.39), which engages a longitudinal slot in the girders. Special vertical bolts are used to mate the ends of the deck panel with the deck bracket.

10.9.3 Composite Behavior in Timber Bridges

10.9.3.1 Introduction

Timber bridge technology has been greatly advanced by the use of glulam timber for large structure members, development of the glulam panel deck, and treatment with modern pressure-impregnated preservatives. However, the concept of composite action, despite its common and accepted use for bridge systems constructed of other building materials, has not been extensively developed for timber bridges and is virtually ignored in "all-wood" systems.

The failure to account for component interaction in all-wood systems has been due largely to several factors: (1) interest in timber bridges declined at the turn of the century, and, simultaneously, research and development lagged behind studies of other materials, (2) until the advent of large deck panels, little was to be gained by its consideration, (3) theories for treating partial composite action did not exist, and (4) designers expressed concern for long-term effects.

10.9.3.2 Wood-Concrete Systems

Composite action in timber systems is no longer being ignored by researchers and designers. Composite timber-concrete elements have been used effectively in bridge construction. Component interaction in all-wood bridge structures is under evaluation by researchers and will be discussed subsequently.

Composite timber-concrete construction in use today commonly consists of one of two types:

1. T-beams: timber stringer stems attached to concrete slab flanges
2. Slab-decks: nailed-laminated plank deck over which concrete is cast monolithically.

T-beams are generally employed for simple span bridges, whereas slab-deck is more common in

*Weyerhaeuser Company, Tacoma, Washington.

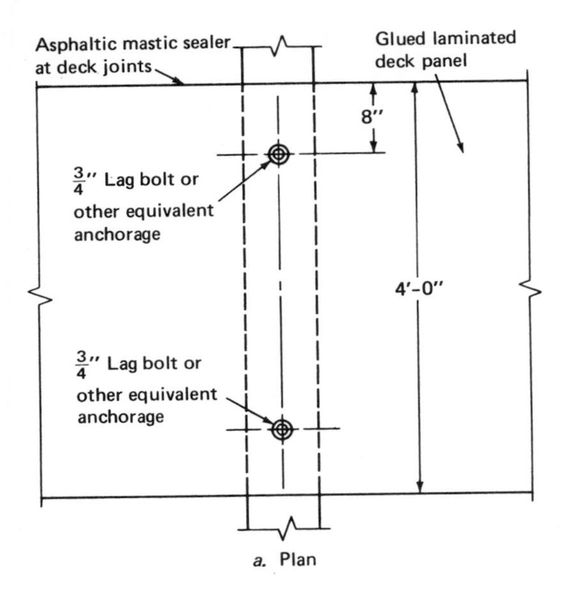

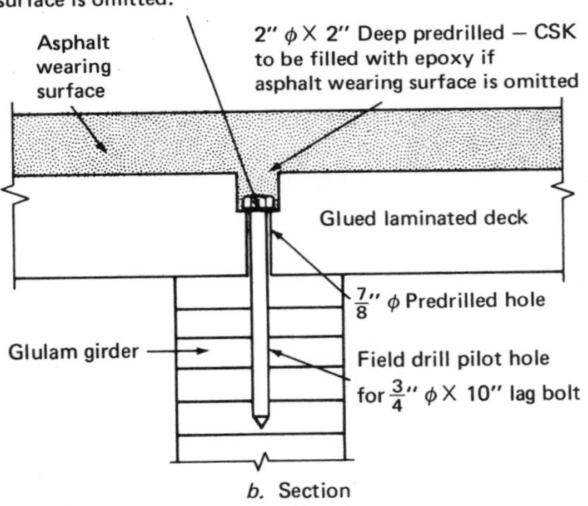

Fig. 10.36. Anchorage details for deck. *(After AITC 1975.)*

continuous span bridges. Other methods of achieving composite action exist and generally involve the use of metal shear developers. As an example, shear developers installed in the Keystone Wye Bridge are shown in Fig. 10.39.

In T-beams, a series of notches, $\frac{1}{2}$ to $\frac{3}{4}$ in. (1.27 to 1.91 cm) deep, cut into, together with a series of nails and spikes driven into, the top of the stringers provide resistance to horizontal shear and vertical separation. Shear resistance in slab-deck is usually achieved either by patented triangular steel-plate "shear developers" driven into precut slots in the grooves or by $\frac{1}{2}$-in. (1.27-cm) deep notches cut into the top of all planks. With the former, separation is prevented by vertical nails or spikes; with the latter, each raised lamination has grooves milled over the full length of each face.

Slab-deck usually consists of 2-in. dimension lumber (with the wide face upward) arranged with alternate planks raised or changed in size to form longitudinal channels. A concrete section is then cast in place atop the wood deck. Figure 10.40 illustrates one type of composite timber-concrete slab deck. Grooves in the sides of the higher laminations resist uplift, and horizontal shear is resisted by grooves cut into the projecting laminations.

Effects of differing thermal-expansion coefficients must be considered in the T-beam system. In slab-deck, temperature change is not a problem provided that expansion joints are made in the concrete slab. Except for these considerations, the basic analysis for both systems is that of the transformed area method commonly used when interlayer slip is ignored. Early work on such systems was performed by Baldock and McCullough (1941). Details of the current design procedure for both are available in the *Timber Construction Manual* (AITC 1974). Standard designs are given by the Bureau of Public Roads (1969) and the Canadian Institute of Timber Construction (1970). Design procedures have also been detailed by the American Wood Preservers Institute (1969). As an example problem, a slab-deck system will be analyzed with accompanying comments where necessary. The procedure is based upon the recommendations given by AITC as adopted from the *Standard Specifications for Highway Bridges* (AASHTO 1973). Computations are in units consistent with that code. More complete examples are available in that source.

EXAMPLE 10.7

Analyze a simple span concrete slab-deck with a 12-ft span for an H-20 truck loading. The deck is comprised of alternating nominal 2-by-6-in. and 2-by-4-in. nailed-laminated lumber. A con-

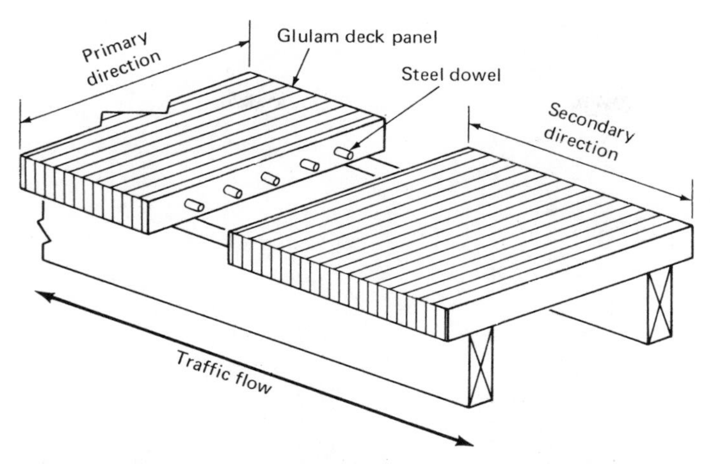

Fig. 10.37. Glued-laminated deck panel with steel dowel connectors. (*After McCutcheon and Tuomi 1973.*)

crete slab provides 5-in. of cover above the 2-by-6-in. pieces. Shear developers are spaced at 12 in. and have a shear capacity of 1800 lb, $E_c/E_w = 1.0$ and $E_s/E_w = 18.75$ [AASHTO-1.3.5C]

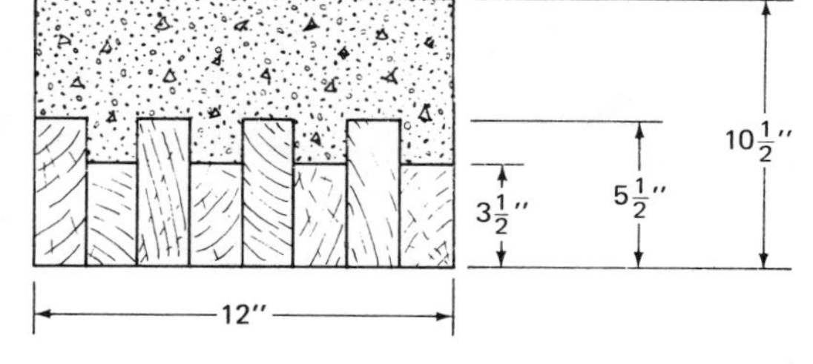

Allowable Stresses:

Wood: $F_b = 1600$ psi Concrete: $f_c' = 2500$ psi

$F_v = 95$ psi

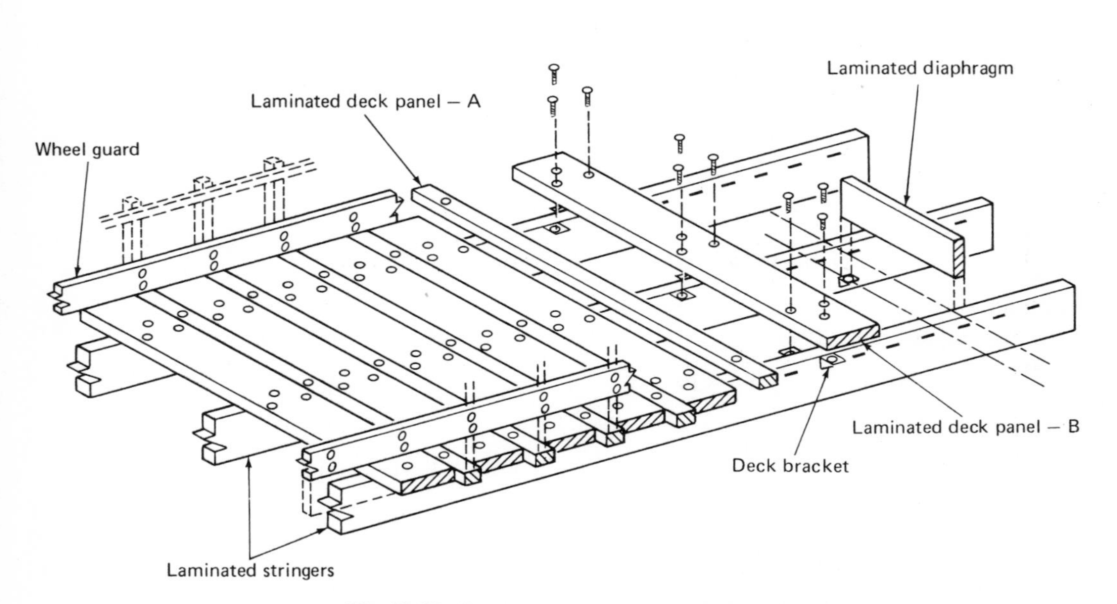

Fig. 10.38. Panelized timber stringer bridge.

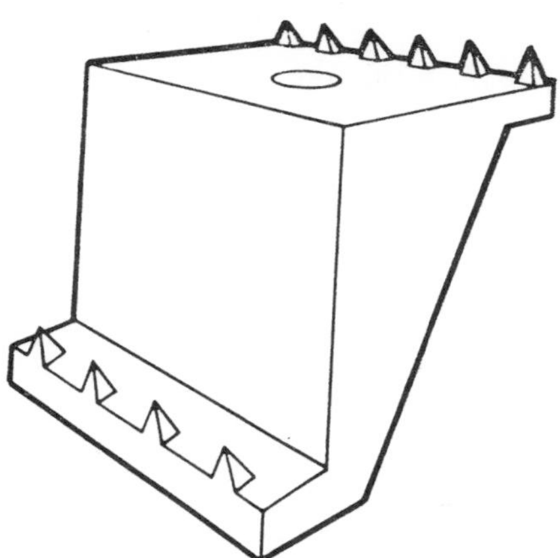

Fig. 10.39. Deck bracket for panelized stringer bridge.

1. Compute dead load:

$$\text{Wood } \tfrac{4.5}{12}\,(50 \text{ psf}) = 19 \text{ psf}$$

$$\text{Concrete } \tfrac{6}{12}\,(150 \text{ psf}) = \underline{75 \text{ psf}}$$

$$94 \text{ psf}$$

2. Bending moment (on a 1-ft-wide section):

$$M_D = (94)\,\frac{(12)^2(12)}{8} = +20{,}300 \text{ in-lb}$$

3. Section properties (wood alone)

$$\bar{y} = \frac{(5.5)(5.5)/2 + (3.5)(3.5)/2}{(5.5) + (3.5)}$$

$$= 2.36 \text{ in.}$$

$$I = 4\left[\tfrac{1}{12}\,(1.5)(5.5)^3 + (5.5)(1.5)(0.39)^3 + \tfrac{1}{12}\,(1.5)(3.5)^3 + (3.5)(1.5)(0.61)^3\right]$$

$$= 117.4 \text{ in.}^4$$

Fig. 10.40.

4. Dead-load stress

$$f_{b_{TOP}} = \frac{Mc_t}{I} = \frac{20{,}300(3.14)}{117.4} = 543 \text{ psi}$$

$$< 1500 \text{ psi}$$

$$f_{b_{BOT}} = \frac{Mc_b}{I} = \frac{20{,}300(2.36)}{117.4} = 408 \text{ psi}$$

$$< 1500 \text{ psi}$$

5. Compute bending moments after curing. The rear wheel load of an H-20 truck is 16,000 lb per wheel [AASHTO-1.25B] and is to be distributed over 5 ft [AASHTO-1.3.5A].

$$P = \frac{16{,}000}{5} = 32{,}000 \text{ lb per 1-ft section}$$

$$M = \frac{PL}{4} = \frac{3200(12)(12)}{4}$$

$$= 115{,}200 \text{ in.-lb}^4$$

6. Section properties (transformed section)

$E_c/E_w = 1$, thus tranformed section is same as actual section

$$y = 5.25 \text{ in.}$$

$$I = \frac{12(10.5)^3}{12} = 1158 \text{ in.}^4$$

7. Determine total stress. Include impact allowance of 30% (assumed) when computing concrete stress [AASHTO-1.3.5B]

$$f_{b_{TOP}} = \frac{115{,}200(5.25)}{1158} = 522 \text{ psi}$$

$$f_{b_{TOTAL}} = 408 + 522$$

$$= 930 \text{ psi} < 1500 \text{ psi}$$

$$f_{b_{TOP}} = 1.3(522) = 679 \text{ psi}$$

$$< 0.45(3000) = 1125 \text{ psi}$$

8. Determine vertical shear
Load placement (*Timber Construction Manual*, AITC 1974)

$$3d = 3(10.5) = 31.5 \text{ in.} \leftarrow \text{controls}$$

$$\ell/4 = 12(12)/4 = 36 \text{ in.}$$

Wheel load distributed over 4-ft width (AASHTO 1.3.5A)

$$P = \frac{16{,}000}{4} = 4000 \text{ lb per 1-ft section}$$

$$V = 1.3P\left(\frac{\ell - 3d}{\ell}\right)$$

$$= (1.3)(4000)\left(\frac{12 - 2.63}{12}\right)$$

$$= 4060 \text{ lb (includes impact)}$$

9. Check horizontal shear stress

$$f_v = \frac{3}{2}\frac{V}{A} = \frac{3(4060)}{2(12)(10.5)} = 48 \text{ psi} < 95 \text{ psi}$$

10. Determine shear developer spacing

n = No. of channels formed by lamination

$$= \frac{12}{2(1.5)}$$

$$= 4 \text{ per ft of width}$$

$$\text{Shear per ft of channel} = \frac{f_v}{n} = \frac{48(12)(12)}{4}$$

$$= 1728 \text{ lb}$$

$$\therefore \text{spacing at load point} = \frac{1800}{1728}(12)$$

$$= 12.5 \text{ in.}$$

∴spacing of 12 in. is probably acceptable to most designers.

10.9.3.3 Research Studies

Present-day design codes do not reflect the presence of composite action in the overall glulam bridge structure. Stringers and decking (whether the deck itself is composite or not) are treated as individual, self-supporting components. This status reflects the fact that analytical methods for predicting the effects of incomplete composite action are recent developments, and application to real systems is in the embryo stage.

As a layered system, the glulam bridge is a structure that fits the mathematical models developed in earlier sections. Glulam deck panels are discontinuous and connected to the supporting stringers by lag bolts, which cannot completely prevent interlayer movement. In particular, research work done by Gutkowski and Goodman (1975, 1976) and Pault (1977) demonstrates the applicability of the finite element method in predicting the effects of partical interaction on glulam bridge performance. These studies adapted Thompson's (1975) finite element approach for floor-system analysis to the bridge problem. Theoretical and experimental studies of large-size, double T-beams (representative of real bridge-system components) have shown that recognition and quantitative evaluation of the composite action in glulam bridges are feasible. Theoretically, composite design could result in up to as much as 50% reduction in deflection as opposed to current design procedures that neglect this component interaction.

From both the conceptual and performance point of view, it is vital to recognize and include the effect of interlayer slip and gaps in the analysis as they are the key parameters that influence the behavior of layered systems. Qualitative assessment of these parametric influences is best made through the use of "composite action curves." A typical example of such a curve for large members is presented in Fig. 10.41. This figure illustrates the influence of the degree of composite action developed by the flange-to-beam connection on the deflection behavior of a particular T-beam. It is evident that as the degree of composite action, as measured by the slip modulus of the connectors, k, increases, the deflection of the T-beam decreases. Using Fig. 10.41 as an example of the expected behavior of a typical bridge T-beam, the predicted deflection, for a slip modulus of 42,000 lb/in., is shown to be about 76% of the 1.14 in. deflection exhibited if the composite action gained by the connection to the deck is ignored.

In addition, it must be noted that the general shape of the curve and its plotted position are a function of flange-gap condition. As the gap condition changes from a periodically discontinuous flange (open gaps), to a partially continuous flange (e.g., dowel connectors at various locations), to a completely continuous flange (no gaps), the T-beam stiffness for the full range of slip moduli increases, and consequently lower deflections are realized.

The primary goal in the work of Gutkowski and Goodman (1977) and Gutkowski et al. (1978) and Pault (1977) was to establish composite action curves as verification of their test program and to extrapolate the observations to the understanding of composite behavior in full-size bridge structures. Curves illustrated in Fig. 10.42 are for one of the Southern pine T-beam specimens studied in that work. The curves are derived for the loading condition used in the structural testing–i.e., a pair of concentrated loads, one 2 ft to each side of midspan. The deflection at midspan is plotted versus slip modulus values for several conditions. The abscissa is put in nondimensional form by normalizing the displacement values relative to the theoretical noncomposite deflection (Δ_N). Each of the curves is representative of a particular interpanel gap condition, and each demonstrates the influence of slip modulus on the composite action. The lower curve represents

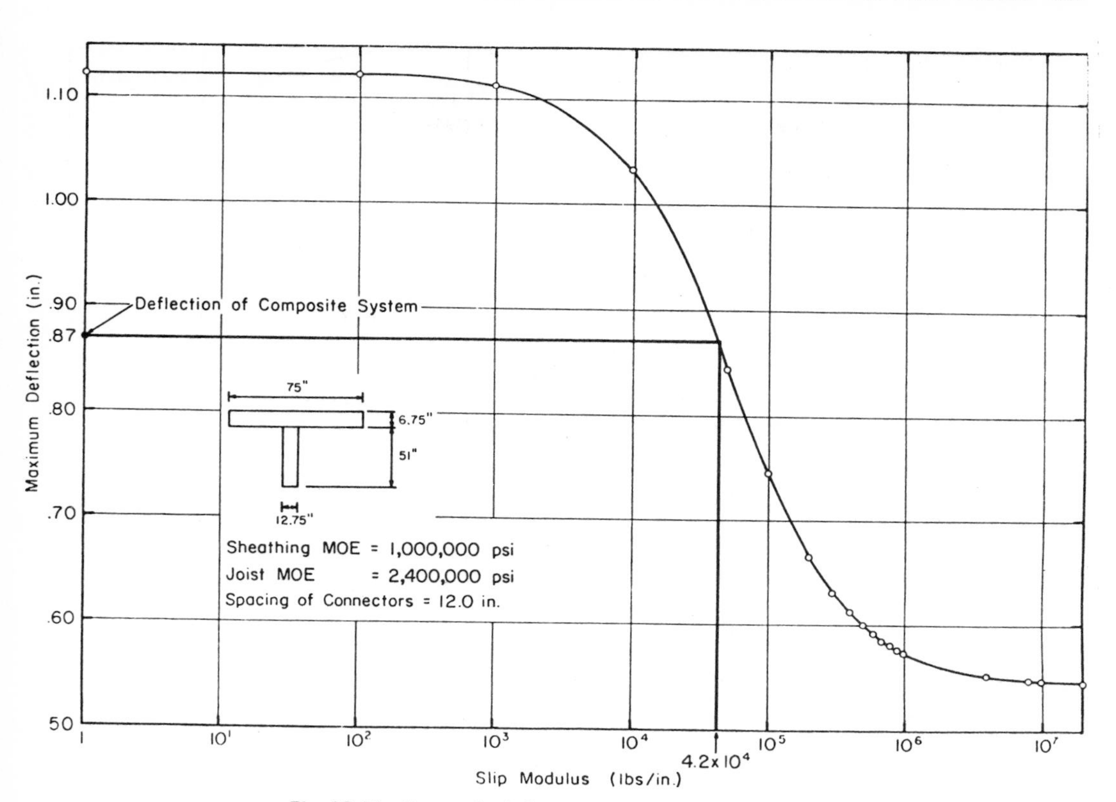

Fig. 10.41. Composite behavior of typical bridge T-beam.

a no-gap deck system. The upper curve represents an open-gap deck system in which the individual deck panels have no interfacial transfer of moment or shear, i.e., open space between panels. The intermediate curve constitutes a gap condition that is between the two extremes.

As expected, for any of the three curves, as the slip modulus increases, the stiffness of the system increases. At low k values, essentially no composite action is realized regardless of the gap condition. At an infinite k value, all three converge to the rigid system (fully composite) deflection value. The vertical distance between these two extreme levels constitutes the maximum possible percentage reduction in deflection. This magnitude of reduction can be achieved only under ideal conditions—namely, no gaps and a monolithic cross section. For the particular structure represented in Fig. 10.42, a 22% reduction is possible at the ideal limit.

Due to the presence of mechanical fasteners, actual test specimens do not achieve the total reduction in deflection that is ideally possible—i.e., incomplete interlayer connection and gap conditions reduced the efficiency of the system. The composite action curves in Fig. 10.42 graphically display this behavior. The vertical distance above the *NO GAPS* composite action curve constitutes the percentage reduction in deflection available for given k values. This represents the loss of stiffness attributable to the use of a specified number and size of lag bolts for the deck-to-stringer connection. In addition, the presence of gaps in the deck further affects the composite action. As E/L, a measure of gap stiffness, decreases, the composite action curve assumes a higher position. Thus, in effect, for a given k value, the observed composite action decreases as E/L is lowered (the nature of the gaps is worsened), reaching a limit at the *GAPS* composite action curve. The vertical distance above the *GAPS* composite action curve indicates that, for large k values, some composite behavior is guaranteed.

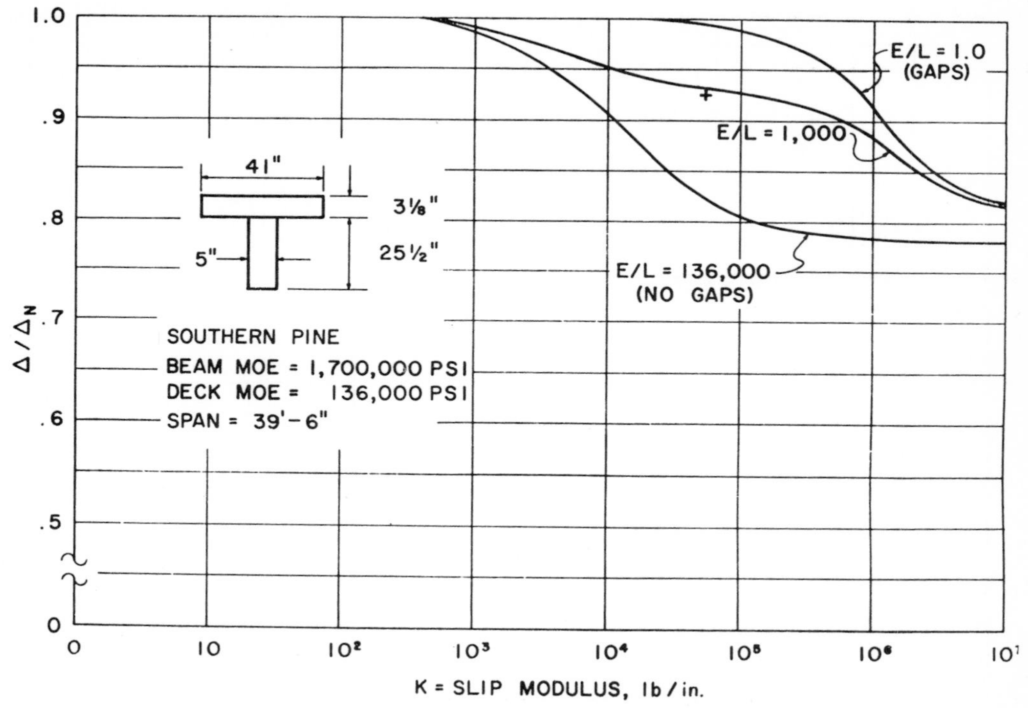

Fig. 10.42. Composite action curve-Specimen SP-25.5.

The test program and mathematical studies conducted by Gutkowski and Goodman (1977) and Gutkowski et al. (1978) and Pault (1977) have resulted in a physical and theoretical understanding of the glulam bridge system. Developed composite action curves give great insight into the nature of the composite-behavior layered T-beam specimens. Curves of the type shown in Fig. 10.42 were generated for a test-specimen cross section that had reduced scale dimensions. However, by extrapolating the dimensions to full-size cross sections and regenerating the composite action curves, valuable comparisons were made in their work.

Figure 10.43 illustrates composite action curves for a cross section whose dimensions are realistic for a 39′-6″ span. The dashed curves are the extreme-case composite action curves for this larger cross section. The most significant observation to be made is that at the ideal limit ($k = \infty$), a reduction in deflection of the order of 48% is possible. Clearly, this suggests that composite action is present to a significant degree in full-size glulam bridges.

One major deficiency in the glulam bridge system is the orientation of the glulam deck panels. Normally, the need to safely span from stringer to stringer necessitates having the significantly lower transverse MOE of the deck effective in the direction of the bridge span. Primarily, the effectiveness of the layered cross section is greatly reduced because the transformed area of the effective flange is quite small. Figure 10.43 emphatically demonstrates this point. The dashed curves in Fig. 10.43 are based upon a cross section in which the deck MOE is 140,000 psi (the transverse MOE for the test panels). The solid curves are generated for the same cross section except the deck MOE has been changed to 1,900,000 psi (the longitudinal MOE for the test panels). For the solid curves, the reduction in deflection at $k = \infty$ is in the order of 75%. The importance of this observation is that innovations in construction that would better orient the deck panels and take advantage of the high longitudinal MOE value would result in greatly improved composite behavior.

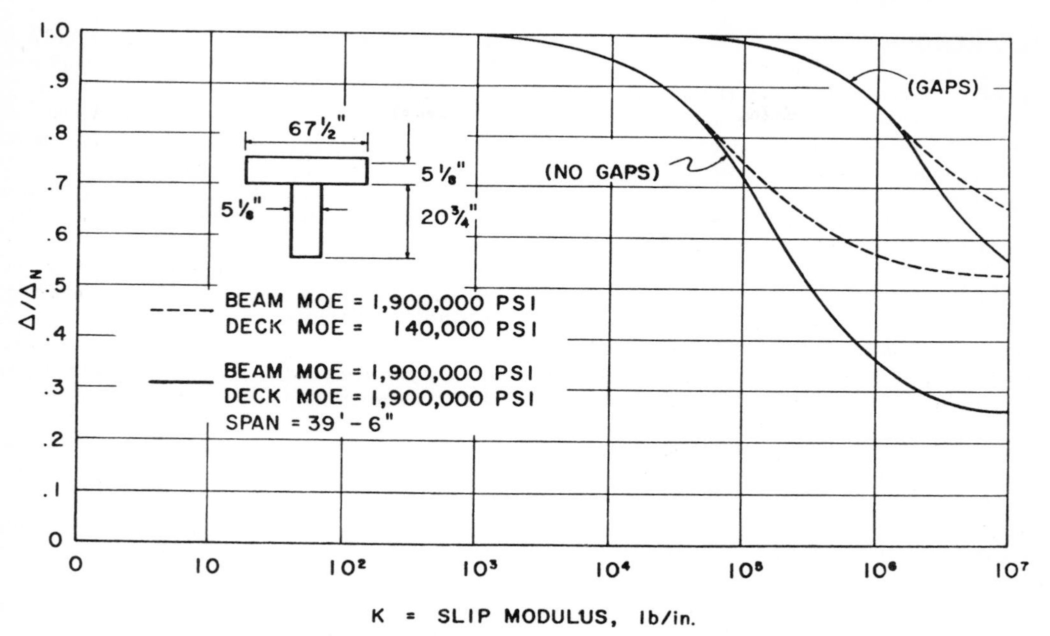

Fig. 10.43. Composite action curve for extrapolation studies.

Composite action curves demonstrate the difference in behavior between the condition of continuous deck (no gaps) and completely severed deck (complete gaps). For actual bridges, the composite action curve lies somewhere between the two extreme given curves. Its exact location depends on the particular dowel connection that exists for the system. The effect of the behavior of the dowel-connection properties on glulam bridge behavior is also being investigated by Gutkowski and Goodman. The slip modulus to be used to evaluate a particular bridge configuration depends upon the type of vertical connection between the deck panels and stringers. General laboratory methods to evaluate this parameter for large specimens are being developed. In particular, the lag bolt type, number, and location recommended by AITC (1975) is being studied in detail. It is expected that these current investigations will ultimately lead to a rational procedure for the consideration of partial composite action in glulam stringer bridges.

10.10 CLOSING REMARKS

The preceding sections clearly demonstrate that the majority of structural systems made of wood involve several interacting components. By means of various types of mechanical connectors and/or adhesives, individual pieces are assembled to form larger members or an entire structure. Methods of analysis exist, which accurately predict the degree of composite action that takes place between the components and, ultimately, the overall performance of the structural system. Design procedures are available for many of the applications discussed earlier. Despite this, the current state-of-the-art is such that only limited recognition of composite behavior in wood structures is evident in design codes. Furthermore, for some systems, the analysis procedures themselves are still in the development stage. However, it is likely that most of the procedures discussed in this chapter will gain acceptance and eventually be adopted by code authorities. The writers have striven to present an extensive and complete introduction to the composite behavior of wood structures. The presented material, together with the numerous references cited, are intended to give the practicing engineer a wide range of general information and a sound basis for more extensive individual studies of particular composite wood systems.

REFERENCES

Amana, E. J. and L. G. Booth (1967a), Theoretical and Experimental Studies of Nailed and Glued Plywood Stressed-Skin Components: Part I, Theoretical Study, *J. Institute of Wood Science*, Vol. 4, No. 1, pp. 43–69.

Amana, E. J. and L. G. Booth (1967b), Part II, Experimental Study, *J. Institute of Wood Science*, Vol. 4, No. 2, pp. 19–34.

Ambartsumyan, S. A. (1964), *Theory of Anisotropic Shell*, Technical Translation TTF-118, National Aeronautics and Space Administration, pp. 369.

American Association of State Highway Officials (1973), *Standard Specification for Highway Bridges*, Eleventh edition, Washington, D.C.

American Institute of Timber Construction (1973), *Modern Timber Highway Bridges*, A State-of-the-Art Report, Englewood, Colo.

American Institute of Timber Construction, *Timber Construction Manual* (1974), John Wiley & Sons, New York.

American Institute of Timber Construction (1975), *Glulam Bridge Systems Plans and Details*, Englewood, Colo.

American Plywood Association (1970), *APA Glued Floor System*.

American Plywood Association (1971), *Adhesives for Field-Gluing Plywood to Wood Framing*, Specifications AFG-01.

American Wood Preservers Institute (1969), Pressure-treated Wood Bridges Win Civil Engineering Achievement Award, *Wood Preservers Journal*.

Baldock and McCullough (1941), *Loading Tests on a New Composite-Type Short Span Highway Bridge Combining Concrete and Timber in Flexure*, Oregon State Highway Technical Bulletin No. 1., Salem, Ore.

Biblis, E. J. (1965), Analysis of Wood Fiber-Glass Composite Beams Within and Beyond the Elastic Region, *Forest Products Journal*.

Bohannan, Billy (1962), Prestressing Wood Members, *Forest Products Journal*, Vol. 12, No. 2, pp. 596–602.

Bohannan, Billy (1964), *Prestressed Laminated Wood Beams*, U.S. Forest Service Research Paper FPL 8, USDA, Forest Service, Forest Products Laboratory, Madison, Wis.

Bohannan, Billy (1972), FPL Timber Bridge Deck Research, *J. Structural Division*, ASCE, Vol. 98, No. ST3.

Bower, W. H. (1974), Lateral Analysis of Plywood Diaphragms, *J. Structural Division*, ASCE, Vol. 100, No. ST4, pp. 759–772.

Bureau of Public Roads, Department of Transportation (1969), *Standard Plans for Highway Bridges, Timber Bridges*, Vol. III, Washington, D.C.

Bystrom, R. C. (1968), *Ultimate Strength Behavior of Doubly-Reinforced Timber Beams*. Master of Science Thesis, Michigan Technological University, Houghton, Mich.

Canadian Institute of Timber Construction (1970), *Modern Timber Bridges*, Standard plans, third edition, Ottawa, Canada.

Carney, J. M. (1975), Bibliography on Wood and Plywood Diaphragms, *J. Structural Division*, ASCE, Vol. 101, No. ST11, pp. 2423–2436.

Civil Engineering (June 1976), University of Idaho Barrel-Arch Roof Wins Outstanding Structural Engineering Award, ASCE, pp. 48–51.

Clark, L. G. (1954), Deflection of Laminated Beams, *Transactions*, American Society of Civil Engineers, Vol. 199, pp. 721–736.

Clark, L. G. (1959), *General Small Deflection Theory of Elastic Bending and Buckling of Laminated Plates*, 4th Midwestern Conference on Solid Mechanics, University of Texas, pp. 127–151.

Dawson, P. R. (1974), *Variability Simulations of Joist Floor Systems*, Master of Science Thesis, Colorado State University, Fort Collins, Colo.

Dawson, P. R. and Goodman, J. R. (1976), Variability Simulations of Wood Joist Floor Systems, *Wood Science*, Vol. 8, No. 4, pp. 242–251.

DeBonis, A. L. (1974), Combined Loading Effects on Nailed Wood Joints, Master of Science Thesis, Colorado State University, Ft. Collins, Colo.

Dong, S. B. (1962), *Mechanics of Anisotropic Media*, Dissertation presented to the University of California at Berkeley, Calif., in partial fulfillment of the requirements for the degree of Doctor of Philosophy.

Fezio, R. V. (1976), *Material Variability and Wood Joist Floor Response*, Master of Science Thesis, Colorado State University, Ft. Collins, Colo.

Foschi, R. O. (1974), The Load-Slip Characteristics of Nails, *Wood Science*, Vol. 7, No. 1, pp. 69–77.

Gillespie, R. H. and River, B. H. (1972), Elastomeric Adhesives in Building Construction, Building Research Oct./Dec., pp. 11–23.

Goodman, J. R. (1967), *Layered Wood Systems with Interlayer Slip*, Dissertation presented to the University of California, at Berkeley, Calif., in partial fulfillment of the requirement for the degree of Doctor of Philosophy.

Goodman, J. R. and Gutkowski, R. M. (1975), *Research in Glulam Bridge Systems*, presented at Annual Meeting of the Forest Products Research Society, Toronto, Canada.

Goodman, J. R. and Popov, E. P. (1968), Layered Beam Systems with Interlayer Slip, *J. Structural Division*, ASCE, Vol. 94, No. ST11, pp. 2535–2547.

Goodman, J. R., Vanderbilt, M. D., and Criswell, M. E. (1973), Composite and Two-Way Action in Wood Joist Floor Systems, Presented at the Forest Products Research Society Annual Meeting. Publication in *Wood Science*, Vol. 7, No. 1, pp. 25–33 (July 1974).

Granholm, H. (1949), Om Sammansatta Balkar Och Pelare Med Sarskild Hansyn Till Spikade Trakonstruktioner (on Composite Beams and Columns With Particular Regard to Nailed Timber Structures), *Transac-*

tions, Chalmers Tekniska Hogskolas Handlingar, Gottenburg, Sweden, No. 88.

Granholm, H. (1954), Reinforced Timber, *Transactions of Chalmers University of Technology*, Gothenburg, Sweden.

Gutkowski, R. M. and Goodman, J. R. (1975), *Research Progress Report on Composite Behavior in Glulam Bridge Systems*, presented at the Annual Meeting of the American Society of Civil Engineers, Philadelphia, Pa.

Gutkowski, R. M., Pault, J. D., and Goodman, J. R. (1978), *Tests and Analysis For Composite Action in Glulam Bridges*, presented at the Annual Meeting of the Transportation Research Board, Washington D. C. (to be published in *Proceedings*).

Gutkowski, R. M. and Goodman, J. R. (1977), *Effect of Mechanical Connectors on Composite Action in Glulam Bridges*, presented at the Annual Meeting of the Forest Products Research Society, Denver, Colorado.

Gurfinkel, G. (1972), *Wood Engineering*, Southern Forest Products Association, New Orleans, La.

Henghold, W. M. (1972), *Layered Beam Vibrations Including Slip*, Dissertation presented to Colorado State University, Ft. Collins, Colo., in partial fulfillment of the requirements for the degree of Doctor of Philosophy.

Henghold, W. M. and Goodman, J. R. (1972), *Static and Dynamic Behavior of Layered Beams Including Slip*, presented and published, Sixth St. Louis Symposium on Composite Materials in Engineering Design.

Hoyle, R. J. (1970), *Design Method for Structures Bonded with Elastomeric Adhesives. Part 1–Shear Modulus of Adhesives*, Unpublished report, Washington State University College of Eng. Res. Div.

Hoyle, R. J. (1971), *Deflections of Twenty Experimental Wood Beams Using Method of Kuenzi and Wilkinson*, FPS-152, Wood Technology Section, College of Engineering, Washington State University.

Hoyle, R. J. (1973), *Behavior of Wood I-Beams Bonded with Elastomeric Adhesive*, Bulletin 328, Washington State University.

Johnson, R. J. and Angleton, H. D. (1970), Static and Dynamic Performance of a Minimum Wood Joist Floor Construction. *Proceedings of the Symposium on Research to Improve Design of Light-Frame Structures*, U.S. Forest Products Laboratory, Madison, Wis., Vol. 25–56.

Ko, M. F. (1973), *Layered Beam Systems with Interlayer Slip*, Master of Science Thesis, Colorado State University, Fort Collins, Colo.

Krueger, G. P. (1962), A Method of Determining the Modulus of Rigidity of an Adhesive in a Timber Joint, *Materials Research and Standards*, Vol. 2, No. 6, pp. 479–484.

Krueger, G. P. and Blomquist, R. F. (1964), *Performance of a Rigid and a Flexible Adhesive in Lumber Joints Subjected to Moisture Content Changes*, U.S. Forest Service Research Note FPL-076, USDA Forest Service, Forest Products Laboratory, Madison, Wis.

Krueger, G. P. (1965), *Experimental Techniques for Determining Mechanical Behavior of Flexible Structural Adhesives*, USDA Forest Service Res. Paper, FPL 21.

Krueger, G. P. (1967), *Ultimate Strength Design of Reinforced Timber*, State-of-the-Art, Department of Civil Engineering, Michigan Technological University.

Krueger, G. P. (1973), Ultimate Strength Design of Reinforced Timber. State-of-the-Art 1973, *Wood Science*.

Krueger, G. P. and Sandberg, L. B. (1974), Ultimate Strength Design of Reinforced Timber-Evaluation of Design Parameters, *Wood Science*, Vol. 6, No. 4.

Krueger, G. P. and Eddy, F. M. (1974), Ultimate Strength Design of Reinforced Timber-Moment Rotation Characteristics, *Wood Science*, Vol. 6, No. 4.

Kuenzi, E. W. (1955), *Theoretical Design of a Nailed or Bolted Joint Under Lateral Load*, USDA Forest Service Res. Paper, FPL No. D1951.

Kuenzi, E. W. and Wilkinson, T. L. (1971), *Composite Beams–Effect of Adhesives or Fastener Rigidity*, USDA Forest Service, Forest Prod. Lab., Research Paper FPL 152, 22 pp.

Kumar, V. K., Stern, E. G., and Szabo, T. (1972), *Built-Up Composite Beams*, V.P.I. Research Division Wood Research and Construction Lab. Report No. 110.

Kuo, M. L. (1974), *Verification of a Mathematical Model for Layered T-Beams*, Master of Science Thesis, Colorado State University, Ft. Collins, Colo.

Lantos, G. (1964), Test Results on Mild Steel Reinforced Glulam Timber Beams, TRADA (England), *Civil Engr. & Pub. Wks. Rev.*

Lantos, G. (1970), The Flexural Behavior of Steel Reinforced Laminated Timber Beams, *Wood Science*, Vol. 2, No. 3.

Liu, J. S. (1974), *Verification of a Mathematical Model for Wood Joist Floor Systems*, dissertation presented to Colorado State University, Ft. Collins, Colo., in partial fulfillment of the requirements for the degree of Doctor of Philosophy.

Malhotra, S. K. (1969), *Buckling Strength of Solid Timber Columns*, Ph.D. Thesis, Nova Scotia Technical College, Halifax, Nova Scotia.

Malhotra, S. K. and Swan, S. P. S. (1971), *Efficiency of Nailed Laminated Timber Columns*, presented at the Annual Meeting of the Engineering Institute of Canada, Quebec City.

McLain, T. E. (1975), *Curvilinear Load-Slip Relations in Laterally-Loaded Nailed Joints*, dissertation presented to Colorado State University, Ft. Collins, Colo., in partial fulfillment of the requirements for the degree of Doctor of Philosophy.

McCutcheon, W. J. and Tuomi, R. L. (1973), *Procedure for Design of Glued-Laminated Orthotropic*

Bridge Decks, USDA Forest Service Research Paper FPL 210, Forest Products Laboratory, Madison, Wis.

McCutcheon, W. J. and Tuomi, R. L. (1974), *Simplified Design Procedure for Glued-Laminated Bridge Decks*, USDA Forest Service Research Paper FPL 233, Forest Products Laboratory, Madison, Wis.

National Association of Home Builders (1970), *Performance of Glued Single-Layer Plywood-to-Wood Joist Floor Systems*, NAHB Research Foundation Report.

National Bureau of Standards, U.S. Dept. of Commerce (1948), *Strength of Houses, Application of Engineering Principles to Structural Design.*

National Forest Products Assn. (NFPA), (1973), *National Design Specification for Stress-Grade Lumber and its Fastenings*, Washington, D.C.

Newmark, N. M., Siess, C. P., and Viest, I. M. (1951), Tests and Analysis of Composite Beams with Incomplete Interaction, *Proceedings, Society for Experimental Stress Analysis*, Vol. 9, No. 1.

Niskanen, E. (1961), *Investigation of the Buckling of Compressed Columns Assembled by Nailing*, The State Institute for Technical Research, Finland.

Norris, C. B., Ericksen, W. S., and Kommers, W. J. (1952), *Flexural Rigidity of a Rectangular Strip of Sandwich Construction–Comparison Between Mathematical Analysis and Results of Tests*, USDA Forest Service, Forest Prod. Lab. Rep. 1505A.

O'Halloran, M. R. (1973), *Curvilinear Stress-Strain Relationship for Wood in Compression*, dissertation presented to Colorado State University, Ft. Collins, Colo., in partial fulfillment of the requirements for the degree of Doctor of Philosophy.

Onysko, D. M. (1970), *Performance of Wood Joist Floor Systems–A Literature Review*, Forest Prod. Lab., Ottawa, Canada, Information Report OP-X-24, 84 pp.

Patterson, D. W. (1973), *Nailed Wood Joints under Lateral Loads*, Master of Science Thesis, Colorado State University, Ft. Collins, Colo.

Pault, J. D. (1977), *Composite Action in Glulam Timber Bridge Systems*, Master of Science Thesis, Colorado State University, Ft. Collins, Colo.

Peterson, John (1965), Wood Beams Prestressed With Bonded Tension Elements, *Proceedings of American Society of Civil Engineers* 91(ST1) 1965.

Peterson, John (1968), The Time Dependent Behavior of Wood Beams Prestressed with Bonded Tension Elements, Unpublished report, Oregon State University, Corvallis, Oregon.

Pincus, G. (1969), Bonded Wood-Concrete T-beams, *J. Structural Division*, ASCE, Vol. 95, No. ST 10, pp. 2265–2279.

Pincus, G. (1970), Behavior of Wood-Concrete Composite Beams, *J. Structural Division*, ASCE, Vol. 96, No. ST 10, pp. 2009–2019.

Pleshkov, P. F. (1952), *Teoriia Rascheta Dereviannykh Sostavnnykh Sterzhnei* (*Theoretical Studies of Composite Wood Structures*), Moscow, 192 pp.

Polensek, A. (1969), *Frequency Analysis of Timber Joist Floors*, Master of Science thesis, Oregon State Univ., Corvallis, Ore., 42 pp.

Polensek, A. (1972), *Static and Dynamic Analysis of Wood-Joist Floors by the Finite Element Method*, dissertation presented to Oregon State University, Corvallis, Ore., in partial fulfillment of the requirements for the degree of Doctor of Philosophy.

Polensek, A. and Atherton, G. H. (1973), Compression-Bending Tests of Walls with Utility Grade Englemann Spruce Studs, Limited Distribution Report, Forest Research Laboratory, Oregon State University, Corvallis, Ore.

Polensek, A. (1975), *Rational Design Procedure for Wood-Stud Walls Under Bending and Compression Loads*, Limited Distribution Report, Forest Research Laboratory, Oregon State University, Corvallis, Ore.

Polensek, A. (1976), Finite Element Analysis of Wood-Stud Walls, *J. Structural Division*, ASCE, Vol. 102, No. ST7, pp. 1317–1335.

Popov, E. P. (1968), *Introduction to Mechanics of Solids*, Prentice-Hall Publishing Co., Englewood Cliffs, N.J.

Railway Track and Structures (1974), Glued-Laminated Bridges Gain in Canada.

Rassam, H. Y. (1969), *Layered Columns with Interlayer Slip*, dissertation presented to Colorado State University, Ft. Collins, Colo., in partial fulfillment of the requirements for the degree of Doctor of Philosophy.

Rassam, H. Y. and Goodman, J. R. (1970), Buckling Behavior of Layered Wood Columns, *Wood Science*, Vol. 2, No. 4, pp. 238–246.

Rassam, H. Y. and Goodman, J. R. (1971), Design of Layered Wood Columns with Interlayer Slip, *Wood Science*, Vol. 3, No. 3, pp. 149–155.

Rassam, H. Y. and Goodman, J. R. (1972), Spaced Columns with Nonrigid Connections, *Wood Science*, Vol. 4, No. 3, pp. 142–153.

Rose, John D. (1969), *Field Glued Plywood Floor Tests*, Laboratory Report 118 American Plywood Association, Tacoma, Wash.

Sazinski, R. J. (1978), *Behavior and Design of Wood Joist Floors*, dissertation presented to Colorado State University, Ft. Collins, Colo., in partial fulfillment of the requirement for the degree of Doctor of Philosophy.

Scarisbrick, R. G. (1976), Laminated Timber Logging Bridges in British Columbia, *J. Structural Division*, ASCE, Vol. 102, No. ST1, pp. 19–34.

Schollett, A. W. (1970), *Finite Element Analysis of a Reinforced Timber Beam Stressed to Flexural Failure*, Master of Science Thesis, Michigan Technological University, Houghton, Mich.

Seiss, C. P. and Viest, I. H., and Newmark, N. M. (1952), *Small Scale Tests of Shear Connectors of Composite T-Beams*, Bulletin 396, University of Illinois Expt. Sta. 49(45), 135 pp.

Sliker, A. (1972), Deflections Simulated by a Computer Program, *Forest Prod. J.*, Vol. 22, No. 9, pp. 71–73.

Thompson, E. G., Goodman, J. R., and Vanderbilt, M. D. (1975), Finite Element Analysis of Layered Wood Systems, *J. Structural Division*, ASCE, Vol. 101, No. ST12, pp. 2659–2672.

Thompson, E. G., Vanderbilt, M. D., and Goodman, J. R. (1976), FEAFLO: *A Program for the Analysis of Layered Wood Systems*, Presented at the Second National Symposium on Computerized Structural Analysis and Design, George Washington University. Accepted for publication in Computers and Structures.

Tremblay, G. A. (1974), *Nonlinear Analysis of Layered T-Beams with Interlayer Slip*, Master of Science Thesis, Colorado State University, Ft. Collins, Colo.

Vanderbilt, M. D. et al. (1974), *A Rational Analysis and Design Procedure for Wood Joist Floors*, Final Report on Grant No. GK-30853, Colorado State University, Ft. Collins, Colo. (available through National Technical Information Service).

Van Dyer, D. B. (1976), *Strength of Built Up Timber Columns*, Ph.D. Thesis, Nova Scotia Technical College, Halifax, Nova Scotia.

Wilkinson, T. L. (1971), Theoretical Lateral Resistance of Nailed Joints, *J. Structural Division*, ASCE, Vol. 101, No. ST5, pp. 1381–1398.

Wilkinson, T. L. (1972a), Analysis of Nailed Joints with Dissimilar Members. *J. Structural Division*, ASCE, Vol, 98, No. ST9, pp. 2005–2013, Proceedings papers #9189.

Wilkinson, T. L. (1972b), *Effect of Deformed Shanks, Prebored Lead Holes and Grain Orientation on the Elastic Bearing Constant for Laterally Loaded Nail Joints*, U.S.D.A. Forest Service Res. Paper, FPL 192.

Wilkinson, T. L. (1974a), *Elastic Bearing Constants for Sheathing Materials*, U.S.D.A. Forest Service Res. Paper, FPL 224.

Wilkinson, T. L. (1974b), *Elastic Bearing Constant of Wood: Effects of Moisture Content Conditions*, U.S.D.A. Forest Service Res. Paper, FPL 235.

Williamson, T. G. (1975), Bibliography on Timber Highway Bridge Design, *J. Structural Division*, ASCE, Vol. 101, No. ST1, pp. 1–10.

Wood Handbook (1940), USDA unnumbered handbook, U.S. Government Printing Office, 326 p.

Wood Handbook (1955, 1974), USDA Agricultural Handbook #72, U.S. Government Printing Office.

Wood Structures: A Design Guide and Commentary (1975), American Society of Civil Engineers (ASCE), New York, N.Y., 416 pp.

Index